# Everything Flows

# Everything Flows
*Towards a Processual Philosophy of Biology*

EDITED BY
Daniel J. Nicholson
and John Dupré

# OXFORD
## UNIVERSITY PRESS

Great Clarendon Street, Oxford, OX2 6DP,
United Kingdom

Oxford University Press is a department of the University of Oxford.
It furthers the University's objective of excellence in research, scholarship,
and education by publishing worldwide. Oxford is a registered trade mark of
Oxford University Press in the UK and in certain other countries

© the several contributors 2018

The moral rights of the authors have been asserted

First Edition published in 2018

Impression: 12

Some rights reserved. No part of this publication may be reproduced, stored in
a retrieval system, or transmitted, in any form or by any means, for commercial purposes,
without the prior permission in writing of Oxford University Press, or as expressly permitted
by law, by licence or under terms agreed with the appropriate reprographics
rights organization.

This is an open access publication, available online and distributed under the terms of a
Creative Commons Attribution – Non Commercial – No Derivatives 4.0
International licence (CC BY-NC-ND 4.0), a copy of which is available at
http://creativecommons.org/licenses/by-nc-nd/4.0/.

Enquiries concerning reproduction outside the scope of this licence
should be sent to the Rights Department, Oxford University Press, at the address above

Published in the United States of America by Oxford University Press
198 Madison Avenue, New York, NY 10016, United States of America

British Library Cataloguing in Publication Data

Data available

Library of Congress Control Number: 2017958461

ISBN 978-0-19-877963-6

Printed and bound by
CPI Group (UK) Ltd, Croydon, CR0 4YY

Links to third party websites are provided by Oxford in good faith and
for information only. Oxford disclaims any responsibility for the materials
contained in any third party website referenced in this work.

# Contents

| | |
|---|---|
| *Acknowledgements* | vii |
| *Contributors* | ix |
| *Foreword* | xi |

## Part I. Introduction

1. A Manifesto for a Processual Philosophy of Biology  
   *John Dupré and Daniel J. Nicholson* — 3

## Part II. Metaphysics

2. Processes and Precipitates  
   *Peter Simons* — 49

3. Dispositionalism: A Dynamic Theory of Causation  
   *Rani Lill Anjum and Stephen Mumford* — 61

4. Biological Processes: Criteria of Identity and Persistence  
   *James DiFrisco* — 76

5. Genidentity and Biological Processes  
   *Thomas Pradeu* — 96

6. Ontological Tools for the Process Turn in Biology: Some Basic Notions of General Process Theory  
   *Johanna Seibt* — 113

## Part III. Organisms

7. Reconceptualizing the Organism: From Complex Machine to Flowing Stream  
   *Daniel J. Nicholson* — 139

8. Objectcy and Agency: Towards a Methodological Vitalism  
   *Denis M. Walsh* — 167

9. Symbiosis, Transient Biological Individuality, and Evolutionary Processes  
   *Frédéric Bouchard* — 186

10. From Organizations of Processes to Organisms and Other Biological Individuals  
    *Argyris Arnellos* — 199

## Part IV. Development and Evolution

11. Developmental Systems Theory as a Process Theory — 225
    *Paul Griffiths and Karola Stotz*

12. Waddington's Processual Epigenetics and the Debate over Cryptic Variability — 246
    *Flavia Fabris*

13. Capturing Processes: The Interplay of Modelling Strategies and Conceptual Understanding in Developmental Biology — 264
    *Laura Nuño de la Rosa*

14. Intersecting Processes Are Necessary Explanantia for Evolutionary Biology, but Challenge Retrodiction — 283
    *Eric Bapteste and Gemma Anderson*

## Part V. Implications and Applications

15. A Process Ontology for Macromolecular Biology — 303
    *Stephan Guttinger*

16. A Processual Perspective on Cancer — 321
    *Marta Bertolaso and John Dupré*

17. Measuring the World: Olfaction as a Process Model of Perception — 337
    *Ann-Sophie Barwich*

18. Persons as Biological Processes: A Bio-Processual Way Out of the Personal Identity Dilemma — 357
    *Anne Sophie Meincke*

Index — 379

# Acknowledgements

The work leading to this volume is a central outcome of a grant from the European Research Council under the European Union's Seventh Framework Programme (FP7/2007–2013)/ERC Grant Agreement 324186, on which JD was the Principal Investigator and DJN was a Research Fellow. We are very grateful to the ERC for its support. More specifically, the majority of the papers herein originated at the workshop "Process Philosophy of Biology", the first major event funded by the grant, held in Exeter in November 2014. We would like to thank all the participants in that event, which marked the first step on the road to this publication.

We must also thank Stephan Guttinger and Anne Sophie Meincke, the other two Research Fellows on the project, who have been essential contributors to each stage of the development of the project, both intellectually and practically. Additionally, anyone who has run a large research project will know how important it is to have a capable administrator, and we have been very fortunate to have Chee Wong in that capacity. We are most grateful for her tireless contributions to the project's management.

Finally, we have had the privilege of inviting a very substantial number of colleagues from around the world to workshops, colloquia, and other kinds of collaborative research visits. We will not try to name them, if only because we would surely leave someone out, but we are grateful to them all. More even than is obvious from the many authors who have contributed to the volume, this is an output that has been influenced by an extended international and interdisciplinary academic community.

# Contributors

GEMMA ANDERSON, Centre for the Study of Life Sciences (Egenis), University of Exeter, Exeter, UK

RANI LILL ANJUM, Centre for Applied Philosophy of Science, School of Economics and Business, Norwegian University of Life Sciences, Ås, Norway

ARGYRIS ARNELLOS, IAS-Research Centre for Life, Mind, and Society, Department of Logic and Philosophy of Science, University of the Basque Country, San Sebastián, Spain

ERIC BAPTESTE, Institute of Biology Paris-Seine, Sorbonne University, Paris, France

ANN-SOPHIE BARWICH, Center for Science and Society, Departments of the Biological Sciences and Philosophy, Columbia University in the City of New York, New York, USA

MARTA BERTOLASO, Institute of Philosophy of Scientific and Technological Practice, Campus Bio-Medico University of Rome, Rome, Italy

FRÉDÉRIC BOUCHARD, Department of Philosophy, University of Montreal, Montreal, Canada

JAMES DIFRISCO, Konrad Lorenz Institute for Evolution and Cognition Research, Klosterneuburg, Austria

JOHN DUPRÉ, Centre for the Study of Life Sciences (Egenis), University of Exeter, Exeter, UK

FLAVIA FABRIS, Centre for the Study of Life Sciences (Egenis), University of Exeter, Exeter, UK

PAUL GRIFFITHS, Department of Philosophy and Charles Perkins Centre, University of Sydney, Sydney, Australia

STEPHAN GUTTINGER, Centre for the Study of Life Sciences (Egenis), University of Exeter, Exeter, UK

ANNE SOPHIE MEINCKE, Centre for the Study of Life Sciences (Egenis), University of Exeter, Exeter, UK

STEPHEN MUMFORD, Department of Philosophy, University of Durham, Durham, UK and Centre for Applied Philosophy of Science, School of Economics and Business, Norwegian University of Life Sciences, Ås, Norway

DANIEL J. NICHOLSON, Centre for the Study of Life Sciences (Egenis), University of Exeter, Exeter, UK

LAURA NUÑO DE LA ROSA, IAS-Research Centre for Life, Mind, and Society, Department of Logic and Philosophy of Science, University of the Basque Country, San Sebastián, Spain

THOMAS PRADEU, Immunology Unit, CNRS and University of Bordeaux, Bordeaux, France

JOHANNA SEIBT, Department of Philosophy and the History of Ideas, University of Aarhus, Aarhus, Denmark

PETER SIMONS, Department of Philosophy, Trinity College Dublin, Dublin, Ireland

KAROLA STOTZ, Department of Philosophy, Macquarie University, Sydney, Australia

DENIS M. WALSH, Department of Philosophy, Institute for the History and Philosophy of Science and Technology, Department of Ecology and Evolutionary Biology, University of Toronto, Toronto, Canada

# Foreword

> [T]here is really no 'thing' in the world.
> —David Bohm (1999: 12)

> [O]ur mind has an irresistible tendency to consider the idea it most frequently uses to be the clearest.
> —Henri Bergson (1946: 214)

There is a notable lack of substance, not in the writing you will find in this book, I assure you, but out there in the domain of the living. Let's face it: there is no thing in biology (or, as Bohm would have it, in the world). Things are abstractions from an ever-changing reality. Reality consists of a hierarchy of intertwined processes. If life is change, then the activities driving this change are what we must explain. Yet we lack concepts and experimental approaches for the study of the dynamic aspects of living systems. This severely limits the range of questions we ask, most of the time even without our realizing. The problem is so obvious it is rarely ever talked about. There are very few explicitly processual theories in biology today. As a practising biologist, I've always found this utterly baffling and disappointing. We remain strangely fixated on explanation in terms of static unchanging entities.

The prime example of this substance fixation in biology is our love affair with genes, those particulate agents of heredity and development. It is all too easy for biologists to slip into deterministic and preformationist language, where genes represent some sort of enduring essence of an ephemeral living body. As a result, the mysterious source of gene agency remains unexamined and unexplained. Another example is our insistence that proper 'mechanistic' explanations of living organisms must be formulated at the level of component molecules, which we take to be unchanging at the timescales relevant to the processes we study. James Ladyman and Don Ross (2007), in their book *Every Thing Must Go*, call this the metaphysics of 'microbangings', small entities causing their effects by bumping into each other. Ladyman and Ross point out that this view is outdated and inconsistent with the dynamic view of the world given to us by modern physics.

Our fixation on static things leads to fallacious patterns of reasoning, within biology and elsewhere. The French process philosopher Henri Bergson alluded to this in the quote above, while Alfred North Whitehead (1925: 52) put it more explicitly by calling it 'the fallacy of misplaced concreteness'. This consists in the unwarranted reification of objects, which become fundamental and replace the underlying dynamic reality in our thinking. This fallacy is deeply engrained in our cognitive habits. From a very early stage of development, we learn to distinguish objects, to isolate them from their context. Cognitive linguists George Lakoff and Mark Johnson (1980: 30–2) have suggested that this reflects a tacit commitment to a doctrine of 'containment': we treat the world as a container of objects that change properties or location and interact with one another. Each object is in turn a container with smaller objects inside, and so on. This doctrine is fundamental to our thinking; it forms the basis of set theory and relational logic. It is very deeply rooted in our human nature: all western languages share it, even ancient ones. To identify an object as a container, we must establish its boundaries as precisely as possible. Where and when does it begin? Where and when does it end? We instinctively crave for clear and rigorous answers to such questions.

However, modern science suggests that reality is simply not like that. The world is full of fuzzy boundaries. Seemingly unchanging entities keep on emerging and decaying if we consider them over a long enough time span. Moreover, it is impossible to say precisely when they truly become what they are and when they cease to be themselves. Or where they begin and where they end. This problem of identification and individuation is beautifully illustrated by the ancient Greek thought experiment about the ship of Theseus. According to the legend, the ship was preserved by the Athenians for centuries upon Theseus' return from his journeys. In the process, each plank of the hull was replaced when it started to rot, until none of the original planks was left. Just as in our own bodies, the substance that makes up the ship is constantly replaced. Does this mean that the ship changes over time, or does it remain the same? As this conundrum illustrates, we need criteria for recognizing, individuating, and classifying processes. We need more accurate and adequate thinking tools that let go of the abstraction of the object. In short, we need to transcend the limitations of substance-based thinking. This is what the book you have in your hands sets out to do.

This is not armchair philosophy, nor is it an exercise in speculative system building. This book outlines a processual research agenda for theoretical biology with direct and wide-ranging implications for practicing biologists. It connects to specific areas of inquiry, such as cancer genetics, evolutionary theory, developmental biology, and the neuroscience of olfaction. It is written in a language that makes it accessible not just to philosophers but also to experimentalists. And, perhaps most importantly, it challenges many of the substance-based assumptions that hamper progress in specific domains. These fundamental assumptions about the world shape the research questions we pursue and the explanations we accept as satisfactory.

Unfortunately, modern scientific curricula have long forgotten to teach students about these hidden aspects of science. Even worse, the format of scientific meetings and papers is designed deliberately to sweep these philosophical foundations under the rug. They have become invisible, barred from the conscious attention of many researchers. In ignorance of their own metaphysical assumptions, scientists are falling back on naïve, often neopositivist preconceptions that severely constrain their thinking and keep their minds closed to the possibility of unconsidered alternatives. This is a terrible shame. If this book succeeds in doing only one thing, I hope that it will be to ignite a lively and public discussion among researchers in the life sciences about our underlying philosophical worldviews and their limitations.

My own scientific trajectory has been inspired and shaped, in an absolutely crucial way, by such philosophical considerations. As a child, I was very strongly committed, both emotionally and intellectually, to a view of static preservation. I am writing these words while on vacation in my hometown of Tschiertschen, a small mountain village in the Swiss Alps. I can assure you that there is a strong and deeply ingrained resistance to change in rural Swiss society. Like many of my country people, I also wished to preserve the beauty of the mountain environment I grew up in and the wealth and orderliness of its society. It seemed perfect to me as it was. Thinking this through, however, I became aware of the suffocating dread of such a vision. This was a very visceral realization. Everything that is beautiful and exciting about the mountains I love has its basis in the dynamical processes that shape them: eon-long upheaval and erosion, the wild torrents so much appreciated by the Romantics, the unpredictability of the weather, and a tradition of tough high-altitude life, flexibly adapting to ever-changing and harsh environmental conditions. To me, static preservation, a freezing of the current state, no matter how precious, kills all that is beautiful, all that is exciting. The illusion of stability is just that: an illusion, and a perilous one at that. This realization was itself a slow and gradual process, not a sudden epiphany. And it has guided my journey of exploration ever since.

It guided me during my undergraduate training as a geneticist, which occurred in a staunchly reductionist molecular biology research environment. I suffered from the strongly antiphilosophical attitude around me, but was not able to express my dissatisfaction explicitly and convincingly. I wish I had known more about process thinking back then, to give my doubts and qualms focus and rigor. Who knows if this would have changed anything, as most of my colleagues didn't even feel that there was a problem. Worse still, they thought that molecular biologists didn't need any philosophy at all, since they were dealing with hard empirical facts! It didn't help to point out that this is itself a philosophical statement. In fact, nowadays scientists often use the term 'philosophical' in a derogatory manner, to describe questions that may or may not be interesting, but are definitely not answerable given our current state of research. Science, it is believed, will increasingly replace philosophy by making such questions answerable. This attitude has always bothered me. It creates a kind of intellectual monoculture that focuses only on the lowest-hanging fruit: the motto of science as the art of the feasible, taken to an unhealthy extreme.

Everybody around me was obsessed with the same question: how to decode the logic of gene expression during development by studying the regulatory sequences on the DNA that are thought to implement this logic. I felt that my colleagues ascribed an almost magical agency to those sequences. The central idea was (and to a large extent still is) that there is some sort of 'code' that can be read out of the DNA and that will result in a particular pattern in the embryo at some stage of development. Everybody was looking for the genetic program formed by this code: preformationist thinking par excellence! And yet very few people seemed to believe that their underlying assumptions were problematic and warranted philosophical scrutiny. When I looked for postgraduate advisors, I deliberately sought out (and was lucky to find) a number of exceptions to this widespread rule of wilful, self-imposed philosophical ignorance.

The most eclectic of these was Brian Goodwin, an unorthodox and open-minded thinker if there ever was one. Brian brought me into contact with process thought in the form of Husserl's and Merleau-Ponty's classic phenomenology, as well as with his own theory of biological structuralism (e.g. Webster and Goodwin 1996). On the one hand, I found these views tremendously fascinating and inspiring, fundamentally altering and refocusing my thinking about ways of becoming in embryology. But on the other, I felt that these approaches were a bit vague and detached from current experimental practices. Luckily, around the same time I learned the mathematical and conceptual tools of dynamical systems theory from Brian, Nick Monk, and my doctoral supervisor, John Reinitz. These tools could be combined in a powerful way with quantitative experimental work to study the processes of pattern formation. During this stage of my career my interests decidedly moved away from the molecular details and the substance-based approach of molecular developmental genetics.

This ended up causing a string of problems that I didn't anticipate at the time but which are obvious to explain with the benefit of hindsight. Many of my applications for postdoctoral fellowships, and then for grants that might fund my newly established independent research group, were rejected. Papers came back from journals too, often unread or with strange, uninformative, and even hostile reviews. It wasn't only that the editors and referees thought that my research was flawed. They didn't find it interesting at all, and mostly didn't even make an effort to understand the question. It took me a while to realize that the problem I had wasn't scientific but philosophical! Sadly, scientific reviewers are often so stuck in the habits and traditions of their field that they can't think of research being worthwhile if it does not neatly fit into one of their familiar categories.

This is when process thinking itself became a central and fixed part of my research agenda. Publishing our philosophical arguments has allowed me not only to detect weaknesses and find a better grounding for my own thinking, but also to better explain why I do what I do to my colleagues. And slowly I'm beginning to see an effect. Over the last decades I've been happy to observe interest shifting towards dynamical systems modelling in developmental biology. Reviewers who state that 'nothing can ever be learned from a model' still exist, but have become exceedingly rare these days. In fact they appear to be a species on the verge of extinction. An increasing number of my colleagues have overcome the scepticism they initially exhibited and now tolerate, or even actively support, the processual research agenda a small minority of us have been pursuing for years.

This recent trend is tremendously encouraging. Quite clearly, the time is ripe for more process thinking, not only in developmental biology but across the life sciences. This is why I am so excited about the collection of essays in this book. It is an important and timely endeavour. I hope it will inspire young biologists in particular to open their minds, to widen their intellectual horizons, and to adopt new philosophical perspectives. I also hope it encourages them to ask radically new questions, build new conceptual frameworks and theories, and develop new experimental approaches that directly address the fundamentally processual nature of living systems.

Enjoy the read! I certainly did.

**Johannes Jaeger**
Associate Researcher
Complexity Science Hub Vienna
Klosterneuburg, Austria, 16 August 2017

# References

Bergson, H. (1946). *The Creative Mind.* New York: Philosophical Library.
Bohm, D. and Biederman, C. (1999). *Bohm-Biederman Correspondence*, vol. 1: *Creativity and Science.* London: Routledge.

Ladyman, J. and Ross, D. (2007). *Every Thing Must Go: Metaphysics Naturalized.* Oxford: Oxford University Press.

Lakoff, G. and Johnson, M. (1980). *Metaphors We Live By.* Chicago, IL: University of Chicago Press.

Webster, G. and Goodwin, B. (1996). *Form and Transformation: Generative and Relational Principles in Biology.* Cambridge: Cambridge University Press.

Whitehead, A. N. (1925). *Science and the Modern World.* New York: Macmillan.

# PART I
# Introduction

# 1

# A Manifesto for a Processual Philosophy of Biology

*John Dupré and Daniel J. Nicholson*

## 1. Introduction

This book is a venture in the metaphysics of science, the exploration of the most basic features of the world implied or presupposed by science. One of its main aims is to demonstrate the fundamental importance of such an investigation. Getting this very general picture right makes a real difference to whether we do the science well and understand properly what it tells us. The particular metaphysical thesis that motivates this book is that the world—at least insofar as living beings are concerned—is made up not of substantial particles or things, as philosophers have overwhelmingly supposed, but of processes. It is dynamic through and through. This thesis, we believe, has profound consequences.

More specifically, we propose that the living world is a hierarchy of processes, stabilized and actively maintained at different timescales. We can think of this hierarchy in broadly mereological terms: molecules, cells, organs, organisms, populations, and so on. Although the members of this hierarchy are usually thought of as things, we contend that they are more appropriately understood as processes. A question that arises for any process, as we shall discuss in more detail below, is what enables it to persist. The processes in this hierarchy not only compose one another but also provide conditions for the persistence of other members, both larger and smaller. So, if we take for example a liver, we see that it provides enabling conditions for the persistence of the organism of which it is a part, but also for the hepatocytes that compose it. Outside a very specialized laboratory, a hepatocyte can persist only in a liver. And reciprocally, in order to persist, a liver requires both an organism in which it resides, and hepatocytes of which it is composed. A key point is that these reciprocal dependencies are not merely structural, but are also grounded in activity. A hepatocyte sustains a liver, and a liver sustains an organism, by doing things. This ultimately underlies our insistence on seeing such seemingly substantial entities as cells, organs, and organisms as processes.

These processes—which have so often been taken for things, or substances—themselves engage in more familiar-sounding processes such as metabolism, development, and evolution; processes that, again, often provide the explanations for the persistence of more thing-like, or continuant processes. Do we not need things as the subjects of such non-controversially processual occurrences as metabolism, development, and evolution? Should we not be dualists, endorsing a world of both things and the processes they undergo? There are many responses to this line of thought, but the minimal condition for a position to count as a form of process ontology is that processes must be, in some sense, more fundamental than things. What this means, in very general terms, is that the existence of things is conditional on the existence of processes. Our own preferred view, which we shall elaborate upon later, is that things should be seen as abstractions from more or less stable processes. Peter Simons, in his chapter, suggests that things are 'precipitates' of processes.

How might one argue for or against a thesis of this kind? One traditional philosophical answer appeals to pure argument or reason. Perhaps, as Kant thought, there are deep truths about the world that must be assumed if it is to be possible for us to gain empirical knowledge of it—presuppositions, therefore, of the possibility of science. Or perhaps the way our language works points to facts that are deeper and more universal than the local discoveries of science, facts therefore foundational to any specific empirical claims.

Without wanting to deny the value of such philosophical strategies, our conception of the metaphysics of science includes a conviction that such a project must proceed in dialogue with what science actually tells us about the world. In common with most of the contributors to this volume, we advocate a naturalistic metaphysics. That is to say, we think that the examination of scientific findings points us towards pictures of the world at a more abstract level than is the immediate target of scientific research. We believe that, as such pictures become established, they can in turn throw new light even on quite specific issues within the science on which they are grounded. To use a term that has become unfashionable in some philosophical quarters, the relation between science and metaphysics is dialectical.

We are commonly asked whether a processual philosophy of biology should really be an ontological project rather than, perhaps more modestly, an epistemological one. Is it not enough to claim that the idea of process provides a more effective conceptual instrument for approaching nature? Several authors in this collection advocate something like this position. But, in light of our naturalistic approach to metaphysics, we do not see a great difference between this position and the more metaphysical formulation that we prefer. Given this naturalism, metaphysics is generally to be established through empirical means, and is ultimately therefore answerable to epistemology. Scientific and metaphysical conclusions do not differ in kind, or in the sorts of arguments that can be given for them, but in their degree of generality and abstraction. If it turns out that process is indeed the right concept to make sense of nature, then this is as good a reason as we can expect for taking nature to be ontologically composed of process.

Although metaphysical conclusions are more abstract and general than what are normally taken to be scientific findings, they are not indefinitely so. The essays in this volume offer many reasons for thinking that the living world is a world of processes rather than a world of things, although we do not take this to demonstrate conclusively that the world is processual throughout. We do, as a matter of fact, believe that there are compelling reasons to interpret the physical world more generally in terms of processes as well; and, if that is right, it is hard to see where we should expect to find exceptions to a universal processualism. But the argument must be made piecemeal.

We proceed as follows. We begin by reviewing what we take to be the major milestones in the history of process philosophy; we then consider some early twentieth-century attempts to develop a processual view of the living world that turn out to be of considerable relevance to our project. Following this, we outline our particular conception of process ontology, as well as our understanding of what processes are and how they relate to things, or substances. We then examine some of the key empirical findings of biology that have prompted us to adopt the processualism we defend, and we illustrate the value of taking a processual stance in a number of prominent philosophical debates. Finally, we discuss several important consequences of embracing an ontology of processes in various areas of biology. We conclude by providing a brief overview of the rest of the essays in this volume.

## 2. Historical Background

The opposition between things and processes as the ultimate constituents of reality is of course an ancient one, commonly traced all the way back to the Presocratics. Heraclitus is the patron saint of process philosophy, at least in the western tradition.[1] The Greek dictum *panta rhei* ('everything flows') encapsulates the Heraclitean doctrine of universal flux. Heraclitus not only emphasized the pervasiveness of change, but also signalled the importance of change in explaining stability over time (Graham 2015). The antithesis of this view is the atomism of Leucippus and Democritus. The indivisible and unchanging material atoms of the ancient tradition provided paradigms for the various notions of substance articulated in subsequent centuries.

Parmenides was another early advocate of substantialism. Although his static view of the world proved too extreme for most subsequent philosophers to accept, his conviction that permanence is more fundamental and more real than change became one of the cornerstones of western metaphysics. It was enthusiastically adopted by Plato in his changeless realm of eternal Forms—and also by Aristotle who, while transforming these essential Forms into worldly entities, nonetheless remained committed to their unchanging character. Aristotelian substances, which are the basic entities of his metaphysics, are distinguished as substances of particular kinds by their essence, and this essence sets unbreachable limits on the kinds of changes that an individual substance can undergo. It is difficult to overstate how influential this essentialist view has been. Many substantialists to this day follow Aristotle in asserting that, to be a thing, one must be a thing of a certain kind, and that the kind to which a thing belongs determines what changes it can undergo while still being what it is.

---

[1] For reasons of space as well as of expertise, we restrict ourselves here to the debate within western philosophy. It is worth noting, however, that process-centred views of reality are quite prominent in non-western philosophical traditions, for example in Buddhist thought (see e.g. Carpenter 2014).

It is interesting to observe that Aristotle's own thought is in many respects more congenial to a process perspective than that of his followers, and this surely has to do with his lifelong fascination with biology. Aristotle took organisms as exemplars of his notion of substance and consequently conceived of them teleologically so as to recognize and unify the different stages of the developmental cycle organisms go through. But, despite his awareness of the distinctive dynamicity of living beings, Aristotle is still best described as a substance ontologist.[2] Historically speaking, Aristotelian metaphysics provided the foundation for the substantialist philosophy that was developed by scholastic thinkers during the medieval period.

Although the scientific revolution is often thought of as a revolt from Aristotelianism, it was certainly not a rejection of substantialism. A central reason for this was the revival of atomism by Boyle, Newton, and others. An atom is a thing or a substance, if anything is. The atoms of early modern science were eternal and permanent in their intrinsic properties. Changes experienced in our macroscopic world were attributed to the motions of, and rearrangements of the relations between, underlying atoms, which remained unchanged throughout such interactions. It is true that, for Locke, perhaps the leading exponent of this philosophy, our lack of access to the microscopic world of atoms justified adopting a sceptical stance about the kinds of entities that we encounter at the macroscopic level. As a result, he did not assume that biological kinds, such as cats or dogs, partook of a common underlying essence. However, this did not contradict his substantialist philosophy; it only restricted our ability to know what kind of substance, if any, natural kinds exemplified.

Various philosophers in subsequent centuries can be associated to some degree with process thinking. Leibniz criticized Descartes' conception of material substance for its lack of activity; and he is often considered a process metaphysician, though we are inclined to think that his system of inherently active 'monads' is too idiosyncratic to lend much support to the process cause, at least as we understand it. Hegel's metaphysics of becoming, wherein nature progressively self-differentiates through the operation of a dialectic that continually integrates conflicting opposites into ever new unities, provides something more unequivocally processual. In fact, attempts to apply dialectical materialism to biology—from Engels' *Dialectics of Nature* to *The Dialectical Biologist* (Levins and Lewontin 1985)—have many commonalities with the present project. The American pragmatists, especially James and Dewey, deserve a mention as well, as they formulated a distinctly processual philosophy in order to come to terms with the implications of Darwin's theory of evolution, though, given their pragmatism, more from an epistemological than from an ontological perspective. Nevertheless, the figure that has come to be most closely associated with process thinking in recent times is unquestionably Alfred North Whitehead, who articulated a comprehensive metaphysical system that conceived of the world as a unified, dynamic, and interconnected whole.

---

[2] Accordingly, the deployment of his ideas in modern biological theory raises a number of problems. For example, it seems difficult to see how the essences Aristotle postulated could encompass either the indefinite developmental plasticity now acknowledged to be a characteristic of organisms (West-Eberhard 2003) or the changes that kinds of organisms undergo over the course of evolution.

Nowadays process philosophy has become almost synonymous with Whitehead's work.[3] Without in any way wishing to detract from Whitehead's importance to the development of process thought, for the purposes of our present project we wish to distance ourselves from the association with Whiteheadian metaphysics. One reason for this is that Whitehead's most systematic metaphysical treatise, *Process and Reality* (Whitehead 1929), is generally agreed to be opaque and at times so obscure as to verge on the unintelligible. His system confers unconventional meanings to familiar concepts (e.g. organism, nexus, satisfaction) and introduces a number of neologisms (e.g. prehension, concrescence, superject) and idiosyncratic technical terms (e.g. actual occasion, subjective aim, extensive continuum) that we have not found particularly helpful in developing the ideas that interest us concerning living systems. We also disagree with Whitehead regarding his insistence on conceiving of reality in atomistic terms—a feature of his process metaphysics that we shall return to later. Finally, and perhaps most importantly, the panpsychist foundations of Whitehead's system, not to mention its theological character, are hard to reconcile with the naturalistic perspective we uphold.

For these reasons, we have found that Whitehead is sometimes as much a liability to process thought—associating it with undesirable philosophical baggage and off-putting prose—as he is a valuable exponent of it. In fact, we suspect that process philosophy has not received the attention it deserves partly because of its close association with Whitehead's work.[4] So, while we are happy to acknowledge the significance of Whiteheadian metaphysics, the reader will not find in this essay, or in any of the ensuing chapters, exegetical discussions of the Category of the Ultimate, the eight Categories of Existence, the twenty-seven Categories of Explanation, or the nine Categoreal Obligations enshrined in *Process and Reality*. All things considered, we are more inclined to risk reinventing the wheel than to look for the concepts and theses we want in Whitehead's metaphysical system.

## 3. The Organicist Precedent

Surprisingly, perhaps, we have found Whitehead to be more useful to us through his influence on an important group of early twentieth-century biologists than through direct engagement with his own work. The members of this collective, known as the *organicists*, produced a large body of literature in the philosophy of biology that predates by several decades the publications generally assumed to have given rise to the modern discipline (Nicholson and Gawne 2015). Interestingly, the book by

---

[3] To illustrate, the journal *Process Studies* defines its subject 'as the study of the thought and wide-ranging implications of Alfred North Whitehead (1861–1947) and his intellectual associates' (http://www.ctr4process.org/publications/process-studies, accessed 16 November 2017). The Australasian journal for process thought is named after one of Whitehead's neologisms, *Concrescence*. And the European Society for Process Thought declares in its mission statement that '[t]he society focuses on Alfred North Whitehead's thought in all its aspects' (http://espt.eu/society, accessed 16 November 2017).

[4] There are notable exceptions, however. Recent attempts to extricate process philosophy from White-head's particular version of it can be found in Rescher 1996 and in Seibt 2003. The *Stanford Encyclopedia of Philosophy* entry on 'process philosophy' (Seibt 2012) also discusses the topic in general terms rather than in Whitehead's own specific framework.

Whitehead that exerted the greatest influence on the organicists was not *Process and Reality*, but *Science and the Modern World* (Whitehead 1925). This earlier work, written in more accessible prose and without much of the conceptual apparatus that characterizes his later metaphysical writings, presented a condensed history of scientific ideas and argued that the mechanicist, reductionist, and determinist view of nature that had reigned supreme since the seventeenth century was no longer defensible in light of the revolutionary developments of modern physics.

In place of this obsolete worldview, Whitehead advocated a philosophy of nature that stressed the development, organization, and interdependence of all things; and he recognized that these concepts were far more congenial to biology than to physics. Biology, Whitehead observed, had historically relied for its development on the solid epistemological bedrock of classical physics, but could no longer afford to do so; it had to spring forth as a unified, autonomous science by carefully scrutinizing and rebuilding its conceptual foundations in accordance with its own needs. As he put it, 'the progress of biology ... has probably been checked by the uncritical assumption of half-truths. If science is not to degenerate into a medley of ad hoc hypotheses, it must become philosophical and must enter upon a thorough criticism of its own foundations' (Whitehead 1925: 25). The organicists adopted this as a rallying cry in promoting their cause: to develop a new philosophy of biology that would emancipate their science from both the physico-chemical reductionism of mechanicism and the obscurantist holism of vitalism in coming to terms with life's dynamic, systemic, and purposive character.[5]

The noted physiologist John Scott Haldane (father of J. B. S. Haldane) was the first to use the label 'organicism' to describe his own views. Looking back at his writings, it is easy to discern a processual thread running through them. Haldane regarded the organism as an integrated and coordinated whole exhibiting a 'delicate regulation [that] is maintained, day after day, and year after year, in spite of all kinds of changes in the external environment, and in spite of the metabolic changes constantly occurring in all living tissues' (Haldane 1917: 17). He observed that organisms remain physiologically constant over time even though, from a purely physical perspective, they are highly dynamic eddies of matter. 'They are constantly taking up and giving off material of many sorts, and their "structure" is nothing but the appearance taken by this flow of material through them' (ibid., 90). When we study the living world, according to Haldane, we are not really dealing with material things at all, but with stabilized processes. He even went as far as to remark that '[t]he conception of a "thing", or material unit, is ... useless in the interpretation of distinctively biological facts' (Haldane 1919: 125).

Another prominent organicist was Edward Stuart Russell, who is probably best known today for his historical treatise *Form and Function* (Russell 1916). In his later, more philosophical works, Russell repeatedly emphasized the purposive character of

---

[5] Many of the British organicists came together in what came to be known as the Theoretical Biology Club, a group of biological thinkers interested in interdisciplinary approaches to the problem of organization that held regular meetings during the 1930s (see Abir-Am 1987; Peterson 2016).

organisms, which are always striving 'actively towards an end, whether of self-development, self-maintenance or the continuance of the race' (Russell 1924: 56). Underlying this was a deeply processual understanding of the organism, which Russell described as 'essentially an activity in course of passage, changing from one form to another, always developing or regressing, but never standing still' (ibid.). Like most other organicists, Russell criticized the machine conception of the organism for neglecting the inherent dynamicity of life, and asserted that '[t]he organism is *not*, like a machine, a static construction, but a constantly changing organization of functional activities' (Russell 1930: 169). Russell also drew attention to the temporal character of the organism, which 'at any one moment of its history must be regarded as merely a phase of a life-cycle', insisting that '[i]t is the whole cycle that is the life of the individual' (ibid., 171). As he put it in a subsequent discussion, '[i]t is as a life-cycle progression and not as a static organisation that the living thing is ultimately to be conceived' (Russell 1945: 186).

Joseph Henry Woodger was a further exponent of organicism. Although today he is mostly remembered—and derided—for his attempts to formalize biological theories, Woodger published a number of non-formal philosophical works in which he articulated a processual view of life (see Nicholson and Gawne 2014). Like Russell, he felt that '[t]here is an urgent necessity for a consideration of *temporal* relations' (Woodger 1929a: 299); and he was perhaps the first philosopher of biology to examine the issue in a systematic way. The following quotation, taken from his *Biological Principles*, offers a flavour of his discussions of the topic:

An organism, whatever else it may be, is an event—something happening. It is temporally as well as spatially extended. It has temporal as well as spatial parts. Your pet dog to-day and your pet dog yesterday are two *different* temporal parts of the same dog, just as his head and his tail are two different *spatial* parts of the same dog. It is in virtue of the particular kind of continuity of the dog yesterday and the dog to-day that we call it the 'same', and this seems to be the proper sense of the term. But it can no more be taken for granted that to-day's temporal part is the same as yesterday's than it can be taken for granted that one spatial part, e.g. the head, is the same as another, e.g. the tail. We know, in fact, that they are not the same. Organisms are temporally as well as spatially differentiated. (Woodger 1929a: 219)

Conrad Hal Waddington is probably the most familiar of the organicists nowadays, as his work has for some time been an inspiration to philosophers of biology sceptical of various aspects of the modern synthesis.[6] Waddington remained a committed processualist throughout his career (see Waddington 1969). 'The fundamental characteristics of the organism', he wrote in *The Strategy of the Genes*, 'are time-extended properties' (Waddington 1957: 189). Biology does not study things; it studies processes occurring at various timescales. According to Waddington, to fully understand an organism, one has to consider how it is affected by four distinct types of temporal

---

[6] The phrase 'modern synthesis' refers to the general theoretical framework of evolutionary biology articulated in the mid-twentieth century, which combined Darwinian natural selection with Mendelian genetics in the form of population genetics and was used to bring together many aspects of comparative anatomy, systematics, ecology, and palaeontology under a common set of explanatory principles.

change—studied, in turn, by four different branches of biology—all of which are proceeding simultaneously and continuously, at various rates:

> An animal functions from minute to minute or from hour to hour, in feeding, digesting, respiring, using its muscles, nerves, glands and so on. These processes of *physiological* functioning may be repeated within periods of time which are short in comparison with the lifetime of an individual animal. But there is an equally important set of processes, of a slower tempo, which require appreciable fractions of the life-history and are repeated only a few times, if at all, during one life-cycle; these constitute *development*. Still longer-term processes are those of heredity, which can only be realised during the passage of at least a few generations and which form the province of *genetics*. And finally, no full picture of an animal can be given without taking account of the still slower processes of *evolution*, which unfold themselves only in the course of many life-times. (Waddington 1956: 3–4, emphasis added)

But it was not just the British organicists who developed a processual understanding of life. The Viennese organicists Ludwig von Bertalanffy and Paul Alfred Weiss, for instance, also defended a processual ontology in their thinking about biology. Bertalanffy (1952: 134) regarded the organism as 'the expression of an everlasting, orderly process' consisting of 'a continuous stream of component materials' (ibid., 133) that flow through it and at the same time constitute it. 'What is described in morphology as organic form', Bertalanffy argued, 'is in reality a momentary cross-section through a spatio-temporal pattern' (ibid., 134). Weiss was equally thorough in his processualism, claiming that biological processes can only be understood in terms of more basic processes, and not in terms of more basic things: 'Life is a dynamic *process*. Logically, the elements of a process can be only elementary *processes*, and not elementary *particles* or any other static units' (Weiss 1962: 3). One of the key implications that Weiss drew from this is that a cell 'can never be defined in terms of a static inventory of compounds, however detailed, but only in terms of their interactions' (ibid.).

Overall, the ontology of organicism was distinctly processual, and this is in no small part a reflection of the influence exerted by Whitehead, both in terms of his timely diagnosis of the collapse of mechanism in physics, with its exciting implications for biology, and in terms of his insightful examination of time, dynamics, and wholeness in the books he wrote prior to *Process and Reality* (see Whitehead 1919, 1920, and especially 1925).[7] We think it is quite significant that the organicists were able to develop a comprehensive processual perspective in biology without availing themselves of the grand metaphysical system Whitehead presented in that book, widely regarded today as his masterpiece. In this respect, it is perhaps more appropriate to describe organicism as a philosophy of biology that was inspired by Whitehead than

---

[7] Russell, for example, described Whitehead's *Science and the Modern World* (Whitehead 1925) as 'the most valuable philosophical contribution of recent years...which is of the highest interest and importance to biology' (Russell 1930: 179). Waddington blamed that same book for his own decision to abandon geology for biology in the mid-1920s (Peterson 2011: 306). And Woodger described his own *Biological Principles*, which he wrote before the publication of *Process and Reality*, as an 'attempt to extend Prof. Whitehead's views to biology' (Woodger 1929b: 345).

as a genuinely Whiteheadian philosophy of biology.[8] Be that as it may, what is relevant for our present purposes is that the organicists showed the way in which one might articulate a processual account of the living world that is naturalistically grounded and empirically informed. We regard the organicists as kindred spirits and consider our project to be continuous with the earlier tradition in the philosophy of biology to which they belong. Of course, a great deal of progress has occurred in biology in the decades since the organicist tradition went into decline; so, while many of their core ideas have stood the test of time, their re-evaluation in the light of these more recent developments is long overdue. About a third of the chapters in this volume feature discussions of, or references to, organicist authors.

## 4. Processes and Things

This essay, and the book more generally, defends the thesis that the right way to understand the living world at all levels is as a hierarchy of processes rather than of things. Philosophically, this is a radical thesis: as we have already seen, an ontology of things, or Aristotelian substances, has dominated western philosophy since the Greeks. As a result, it is generally assumed that substance ontology provides the most 'natural' articulation of our common-sense intuitions about the world.[9] Johanna Seibt, in her contribution to this volume and elsewhere (e.g. Seibt 1996), refers to this as the 'myth of substance'. This pervasive bias towards things is reflected in our everyday language, and it has a direct effect on how scientific research is conducted and on how its results are interpreted. The chapters in this collection illustrate many of the problems that arise from taking the ontological primacy of things as a given in the particular context of the life sciences. Though not every contributor to this book is a fully signed-up process metaphysician, all are exploring topics that point in various ways to the advantages of this alternative position.

What is the difference between a thing, or a substance, and a process? In large part, of course, this is the question that this book, and to a lesser extent the present chapter, are supposed to help to answer. A starting point, however, is the following. Processes are extended in time: they have temporal parts. Whether things have temporal parts is a debated issue. Many philosophers hold that this is not the case. As it is sometimes expressed, a thing is wholly present at any moment when it is present at all. Often this position is combined with presentism, the view that only the present exists at all (Bourne 2006). Both of these theses are contested by four-dimensionalists, who think

---

[8] Indeed, it is important to distinguish the organicists from authors who have sought to develop a processual understanding of biology on the basis of the panpsychist metaphysics laid out in *Process and Reality*. Early attempts to do so include *A Contribution to the Theory of the Living Organism* (Agar 1943) and *General Biology and Philosophy of Organism* (Lillie 1945). Two multicontributor volumes with a similar agenda have appeared more recently: *Mind in Nature* (Cobb and Griffin 1977) and *Life and Process* (Koutroufinis 2014).

[9] Of course, we realize that the concept of substance has lent itself to a wide range of interpretations throughout the history of philosophy, not all of which are mutually compatible. Accordingly, we recognize that substance ontology is not so much a single, distinct doctrine as it is an umbrella term for a collection of philosophical positions that share a commitment to ontologically prioritizing substances (however these may be construed) over processes.

of things as 'space–time worms', extending through time for as long as they exist (Sider 2001). On such a view, only a tiny part of a temporally extended thing is present at any instant. We shall not say any more about this debate here, except to note that four-dimensionalism is already halfway towards a full-blown process ontology. Both populate the world with temporally extended entities with diverse temporal parts. To see where these positions may still differ will require further attention to what it is that makes the temporal parts of an entity parts of the same entity—a question to which we shall return later and that will also be discussed in subsequent chapters, especially those in Part II of the book.

Equally central to the concept of process is the idea of change. A process depends on change for its occurrence. Traditionally, change has been construed as something that happens to things, or substances, typically conceived of as durable, integrated entities that are not dependent on external relations for their existence. Things, in this view, are the subjects of change, and processes merely track modifications in the properties of things over time, or describe means by which things interact with one another. This understanding leads to the assumption that processes always involve the doings of things. Processes therefore necessarily presuppose the prior existence of things.

One problem with this view is that many processes do not in fact belong to particular subjects. Rain, wind, electricity, and light are all commonplace examples of subjectless or 'unowned' processes—processes that are not the actions of individual things. There are numerous subjectless processes in the living world as well: osmosis, fermentation, adaptive radiation, and so on. As Johanna Seibt shows in her chapter, the autonomous existence of subjectless processes is not compromised by the fact that they lack many of the features of concrete particulars, such as determinate boundaries or, for that matter, a unique or specific spatio-temporal location.

A more fundamental problem is that even entities that appear to be the subjects of activities can themselves be construed as specific temporal stages of stable processes; they do not have to be understood as things. Though many processes are defined in terms of the concrete particulars that undergo them, the notion that only things or substances are qualified to count as concrete particulars is nothing more than a prejudice. Many processes are *bona fide* individuals—they are concrete, countable, and persistent units. Non-biological examples include whirlpools, flames, tornadoes, and laser beams. In biology, processes are, as we have already mentioned, dynamically stabilized at vastly different timescales: a matter of minutes for a messenger RNA molecule, a few months for a red blood cell, many decades for a human being, and up to several millennia for a giant sequoia tree. This stabilization can make it easy to mistake them for static things. But, more importantly, it allows them to play the role traditionally attributed to things that undergo processes in substance ontology. The only condition is that the relevant processes must be sufficiently stable on the timescale of the further processes that they in turn undergo. Enzymes can be treated as things because they are stable on the timescale of catalysis. Similarly, white blood cells are stable on the timescale of phagocytosis, alveoli are stable on the timescale of respiration, animals are stable on the timescale of reproduction, and so on.

We believe, then, that it is a mistake to suppose that processes require underlying things, or substances. This commonly held belief corresponds, unsurprisingly, to the original meaning of the term 'substance', which derives from the Latin word *substantia*—literally, that which stands under. In opposition to this view, we take nature to be constituted by processes *all the way down*.[10] This represents a reversal of the substantialist position described above. Instead of thinking of processes as belonging to things, we should think of things as being derived from processes. This does not mean that things do not exist, even less that thing-concepts cannot be extremely useful or illuminating. What it does imply is that things cannot be regarded as the basic building blocks of reality. What we identify as things are no more than transient patterns of stability in the surrounding flux, temporary eddies in the continuous flow of process.

The thoroughgoing processualism we uphold regards change, or better dynamicity,[11] as fundamental or primitive. This dynamicity is extended in time and, like time itself, it is continuous. It is therefore inappropriate to regard it—or any of the myriad processes that constitute it—as a sequence of particular events. To conceive of processes as series of discrete temporal episodes is to overlook the very dynamicity that process philosophy is intended to emphasize. As we noted earlier, this is one of the key reasons why we do not situate our project in the Whiteheadian tradition. Whitehead construed reality in atomistic terms, as being ultimately made up of indivisible units, which he called 'actual occasions', out of which all larger processes are composed. In place of a discontinuous view of the world as a complex aggregation of ultimate elements, we prefer to think of it as a manifold of nested and interrelated processes that collectively constitute a dynamic continuum. This understanding of nature forces us to rethink a number of traditional problems in philosophy, such as the nature of causation, which is precisely what Rani Lill Anjum and Stephen Mumford set out to do in their chapter.

Processes come in many forms, shapes, and sizes. The spatio-temporal organization of a process and its spatio-temporal and causal relations to other processes determine its persistence and stability, and are also what grounds its properties and causal powers. As we have already noted, processes can be 'pure' dynamic activities, or they can be individuals exhibiting most of the characteristics typically attributed to things. Whereas things can generally be individuated by their spatio-temporal locations—things typically exclude other things from the regions of space–time they occupy—this is not typically the case for processes. Many processes have boundaries that are fuzzy or indeterminate, a feature with implications that we shall explore later. Processes are individuated not so much by where they are as by what they do. A series of activities constitute an individual process when they are causally interconnected or when they come together in a coordinated fashion to bring about a particular end. Many of the processes found in the living world, moreover, exhibit a degree of cohesion that demarcates them from their environment and thereby allows us to identify them as distinct, integrated systems—as entities in

---

[10] If we were to use the concept of substance in its original Latin sense, we would say that processes, not things, are the real substances of the world, the ultimate constituents of reality.

[11] Dynamicity appears to be a more suitable concept for our purposes, given that change can carry undesirable substantialist connotations. As we indicated above, if change is described as the alteration of the properties of preexisting substances, then it is not necessary to consider it a basic ontological category in its own right.

their own right. In his chapter, James DiFrisco considers the extent to which the cohesion criterion, first proposed by John Collier (1988, 2004), can provide an effective means of individuating biological processes.

The transition we are urging from a substance ontology to a process ontology has one very important epistemological implication. In any scientific enquiry it is necessary to distinguish what requires explanation from what is background, taken for granted. The orthodox substantialist position of modern science typically takes this background to involve stability: if nothing changes, then nothing requires explanation. This is because the default mode of existence of a thing is stasis and consequently the need for explanation only arises when changes happen to it. For a process, however, change is the norm, and it is its relative stability that takes priority in the explanatory order. If the living realm is indeed processual, then we should consider the central explanandum of biology to be not change but stability—or, more precisely, stability achieved by activity, that is, by change. Take physiology as an illustrative example. Physiology is largely concerned with understanding the multitude of internal processes that enable an organism to stay alive by maintaining its organization over time within a relatively narrow range of parameters. It is quite clear that, in the context of physiological enquiry (which encompasses many more specialized areas of biology), the persistence of an organism is not a background assumption at all, but the very phenomenon that cries out most loudly for explanation. The same can be said of enquiries in development or in immunology, and even (or perhaps especially) in medicine, as illustrated by the discussion of cancer in the chapter co-authored by one of us (Dupré) with Marta Bertolaso.

## 5. Empirical Motivations

In this section we shall examine some well-established scientific facts about life that reveal the unsuitability of traditional substance metaphysics for representing biological reality and which have compelled us to adopt a processual stance towards the living world.

Although this book is primarily concerned with the life sciences, we do not think we should proceed without at least mentioning that the physical sciences already provide powerful motivations of their own for endorsing a process ontology. Nicholas Rescher (1996: 97) has quipped that modern physics 'puts money in the process philosopher's bank account', and it is easy to see why. The advent of quantum mechanics at the turn of the twentieth century led to the dematerialization of physical matter, as atoms could no longer be construed as Rutherfordian planetary systems of particle-like objects. This resulted in the demise of the classical corpuscular ontology of Newtonian physics, which had been one of the pillars of substance metaphysics since the scientific revolution. What had hitherto been conceived of as the ultimate bits of matter became reconceptualized as statistical patterns, or stability waves, in a sea of background activity.[12]

---

[12] Interestingly, the organicists were well aware of the shift towards process that was contemporaneously taking place in physics. Bertalanffy, for example, explicitly compared the processual ontologies of physics and biology, declaring that, just '[a]s in modern physics there is no matter in the sense of rigid and

The subsequent development of quantum field theory, which was articulated to reconcile quantum mechanics with special relativity, has lent further support to the process cause. Quantum fields, which are dynamic organizations of energy distributed in space–time, appear to have purged classical notions of elementary particles from the ontological picture. Although contemporary physicists still routinely speak of 'particles', this term no longer refers to solid microentities or tiny impenetrable granules, but to quantized excitations of particular fields. Quantum fields, in other words, are primary, and the various kinds of particles that physicists refer to are derivative entities, appearing only after quantization. Thus, what contemporary physics seems to be telling us—if we understand the equations realistically—is that the basic ontological constituents of the universe are not elementary particles, understood as minuscule things, but fields extended in space–time. Though we are not entirely sure whether fields are either processes or things, they do appear to be more like the former than like the latter.[13]

If physics directs us towards process metaphysics, and if there are additional reasons for thinking that chemistry is, likewise, amenable to a process ontological interpretation (see e.g. Stein 2004 and the chapter here by Stephan Guttinger), it would be surprising to find that biology pushes us in the opposite direction, towards an ontology of substances. As it turns out, there are good independent reasons for embracing process metaphysics in biology as well, as we shall now see.

*5.1. Metabolic turnover*

One of the strongest motivations for adopting a process ontology in biology stems from a very familiar fact about life, namely that organisms have to eat to stay alive. We can express this in more technical terms using the parlance of thermodynamics by stating that organisms are open systems that must constantly exchange energy and matter with their surroundings in order to keep themselves far from equilibrium. The persistence of an organism is dependent on its ability to maintain continuously a low entropic 'steady state' in which there is a perfectly balanced import and export of materials. When this exchange ceases, the steady state is irretrievably lost and the organism succumbs to equilibrium—and dies.

Because of their particular thermodynamic characteristics, organisms find themselves in the existential predicament of needing to act to continue to exist. Although a car cannot function without fuel, its existence (i.e. its structural integrity) is not compromised when it is deprived of fuel. An organism, by contrast, is *always* acting (or working, in the thermodynamic sense), as it must remain permanently displaced from equilibrium if it is to stay alive. You can leave your typewriter in an empty loft and return a decade later and start using it again. But if you accidentally leave your hamster in the loft, you will not have a hamster for very long.

---

inert particles, but rather atoms are node-points of a wave dynamics, so in biology there is no rigid organic form as a bearer of the processes of life; rather there is a flow of processes, manifesting itself in apparently persistent forms' (Bertalanffy 1952: 139).

[13] The argument we have sketched here to the effect that modern physics motivates process metaphysics draws from a number of sources, including David Bohm's classic *Wholeness and the Implicate Order* (Bohm 1980) and Richard Campbell's more recent *The Metaphysics of Emergence* (Campbell 2015).

This continuous activity, which is truly indispensable for life, is known in biology as the process of *metabolism*. Metabolism encompasses the means by which organisms break down the materials they take in from their environment in order to acquire the energy they need to rebuild their constituents and maintain themselves in a steady state far from thermodynamic equilibrium. Metabolism also includes the processes by which organisms dissipate energy and excrete material wastes back into their environment, thereby conforming to the second law of thermodynamics. This constant metabolic turnover takes place at every level of biological organization. As we indicated at the start of this essay, an organism is not organized as a hierarchy of structures (as a machine is), but as a hierarchy of processes. The lower we go down the biological hierarchy, the faster the rate of material exchange. The stability of each process in the hierarchy is secured by the constant metabolic turnover of components that takes place at the lower level. Accordingly, the stability of a multicellular organism as a whole derives from the continuous regeneration of its tissues, which are themselves maintained by the incessant renewal of their cells, which are in turn stabilized by the ongoing replenishment of their molecular constituents.

The appearance of stasis in biology can be deceptive at any level of organization. Subcellular formations such as the mitotic spindle or the Golgi apparatus seem well-defined structures within a short timescale, but when we consider them for longer temporal intervals it becomes evident that they are but fleeting manifestations of ongoing processes of material exchange. These organelles, like many other molecular assemblies in the cell, do not exist as fixed microstructures but as quasi-stationary patterns—partly fluid, partly consolidated—that persist for a time, before undergoing transformations or disappearing altogether (Kirschner et al. 2000; Misteli 2001). Exactly the same is true for the cells that compose a tissue, for the tissues that constitute an organ, and for the organs that make up an organism. Any given cell or any given tissue instantiates a dynamic steady state, only the form of which persists, while its material constitution is constantly being turned over by metabolic events.

Ultimately, this also applies to a multicellular organism as a whole. This is harder to appreciate because the material regeneration of the form of most macroscopic organisms is so slow that it is not easily perceived by the human eye. But consider the following thought experiment. Imagine an extraterrestrial humanoid life form whose mode of visual recognition was based on the enumeration of the material components that make up particular tokens of general types, rather than on the identification of the general types that are instantiated by particular tokens. Imagine, further, that this alien lands on Earth at a particular location and encounters two dogs: a living dog and a robotic dog. The alien scans the two dogs, catalogues their material constitution for future identification, and returns home. A few years later, the alien returns to Earth to the same location and faces the two dogs it encountered in its first trip. Despite being in the presence of the same two dogs, the alien's cognitive apparatus is such that he is only able to identify the robotic dog and not the living one. From the alien's perspective, the living dog of the first trip has faded out of existence, and an entirely different living dog has taken its place. What this admittedly fanciful thought experiment is meant to illustrate is that, if one focuses on matter rather than on form and allows for a sufficiently extended period of time, the stream-like nature of macroscopic organisms becomes perfectly evident. The fact that this does not happen to be easily perceptible to us does not make it any less true or important.

From a metabolic perspective, it is simply a matter of fact that, in an organism, *everything flows*. Of course, this is not to say that everything flows *at the same rate*. In the human body, for instance, each type of tissue has its own turnover time, depending—at least in part—on the workload endured by its cells. To illustrate: the cells lining our stomach only last around five days; cells of our epidermis are renewed every two weeks; our red blood cells are replenished after four months; our liver as a whole is regenerated on a yearly basis; and our entire skeleton is replaced each decade (Wade 2005). When we consider the turnover time of molecules in our body, the rate of replacement is several orders of magnitude faster. For example, the protein turnover rate in an adult is roughly 8 per cent *per day*, and virtually all the protein molecules in our body are replaced during the course of a year (Haynie 2008).[14] This should not be totally surprising. A cell persists for far longer than any of its molecular constituents, and the same can be said of the lifespan of a tissue by comparison with that of its component cells. In general, none of the parts of an organism is as old as the organism itself. What this implies, as far as you—the reader—is concerned, is that your physical body is several times younger than your actual age.[15] Sadly, however, wrinkles and grey hair renew themselves just as faithfully as smooth skin and more youthfully tinted tresses.

Overall, the reality of metabolism forces us to recognize that organisms, despite their apparent fixity and solidity, are not material things but fluid processes; they are metabolic streams of matter and energy that exhibit dynamic stabilities relative to particular timescales. As processes, and unlike things or substances, organisms have to undergo constant change to continue to be the entities that they are. The chapter by one of us (Nicholson) examines the metabolic character of organisms and its grounding in thermodynamics in greater depth, and explores some of the ontological implications of transitioning from a machine-like to a stream-like conception of the organism. The chapter by Denis Walsh also takes the metabolic aspect of life as a point of departure in arguing that organisms are agents rather than objects and in exploring the consequences of this ontological shift for our understanding of evolution. And the

---

[14] One might be tempted to object that the DNA in our cells constitutes an exception, as it remains unchanged. But what remains unchanged is not the DNA molecules themselves, but their nucleic acid sequence. The actual molecules undergo change when they are replicated during cell division, and indeed the extraordinary precision of replication is achieved only through an elaborate set of editing processes that respond to inaccuracies. It is also worth noting that even the nucleic acid sequence does not stay exactly the same, given that replication errors do occasionally occur, resulting in minute variations that may or may not result in phenotypic changes.

[15] Physiologists, and even members of the general public, have long been aware of this bewildering fact. In the 1905 edition of his novel *The Irrational Knot*, George Bernard Shaw wrote: 'At present, of course, I am not the author of *The Irrational Knot*. Physiologists inform us that the substance of our bodies (and consequently of our souls) is shed and renewed at such a rate that no part of us lasts longer than eight years: I am therefore not now in any atom of me the person who wrote *The Irrational Knot* in 1880. The last of that author perished in 1888; and two of his successors have since joined the majority. Fourth of his line, I cannot be expected to take any very lively interest in the novels of my literary great-grandfather' (Shaw 1905: xvii).

18  JOHN DUPRÉ AND DANIEL J. NICHOLSON

last chapter in the book, by Anne Sophie Meincke, appeals to the processual nature of metabolism in order to deal with the puzzle of personal identity.

## 5.2. Life cycles

Another equally uncontroversial fact about life that similarly compels us to embrace a process perspective is that all organisms undergo a characteristic series of morphological and behavioural changes over the course of their lifetime; they do not stay the same from the moment they come into existence but rather *develop* progressively over time, acquiring certain properties and capacities and losing others along the way. This is commonly referred to as the process of *ontogeny*, and the precise nature, order, and timing of the changes it comprises vary enormously from one species to another.

Let us take a frog as an example. It begins its life as a fertilized egg, which divides and develops into an embryo. When the egg hatches, it leaves its gelatinous enclosure and attaches itself to a floating weed or blade of grass. It then becomes a tadpole with a cartilaginous skeleton, gills for respiration (external gills at first, internal gills later), and a tail for swimming. Its tail keeps on growing and its hind legs gradually appear, followed by its forelegs. The tadpole's lungs also begin to develop at this point, to prepare it for its life on land. Many other morphological changes take place during this time: its nervous system becomes adapted for its eventual life on land, its head becomes more distinct through the repositioning of its eyes, its ear organs begin to form internally and externally, its lower jaw transforms into a big mandible, and its gills are gradually grown over by skin until they disappear. As its tail progressively shortens, the tadpole appears ever more frog-like, until it emerges from the water as a froglet. Eventually, as an adult frog, it finds a mate—if it is lucky—and the entire process starts anew (see Figure 1.1).

**Figure 1.1** Schematic representation of the various stages in the life cycle of a frog

This familiar example showcases some of the problems with the traditional understanding of organisms as things, or substances. When considering a particular organism, there is a general tendency to privilege or prioritize the adult stage of its life cycle (for instance, in the context of taxonomic discussions), as this is the period during which the organism most closely resembles a thing, by virtue of its relative stability. But we should not forget that the organism encompasses the entire life cycle; indeed, it is the life cycle itself that constitutes the organism. Strictly speaking, it is incorrect to speak of an egg developing into a frog, as the egg is really a temporal part of the developmental trajectory that *is* the frog. Organisms cannot be separated from their history. What we perceive as an organism (e.g. a frog) at any given moment represents only a cross section, or time slice, in the unfolding of the persistent process it instantiates. It is important to realize that these time slices do not reflect real discontinuities in the process from which they derive. Although an adult frog is undoubtedly very different from a tadpole, the developmental progression from tadpole to frog is smooth and gradual; there is no sharp boundary demarcating them. Knowing what we know about frogs, it simply makes more sense to think of them as processes rather than as things.

Now one could attempt to retain the commitment to substance ontology by arguing that, despite the developmental transformations that organisms undergo, they nevertheless remain the same thing—or substance—throughout. The problem with this line of argument is that it is surprisingly difficult to specify what stays the same throughout the life cycle of an organism. To refer back to our example, when we consider a fertilized egg, an embryo, a tadpole, a froglet, and an adult frog, it is not clear what properties they all share beyond being temporal stages of the same individual process. In fact, there may well be no interesting properties shared by all.[16]

Before moving on, we wish to emphasize that the idea of a life cycle is broader and more inclusive than the concept of ontogeny. Multicellular organisms undergoing embryological differentiation are not the only biological entities with life cycles. Cells have life cycles as well, which typically involve a growth phase that includes DNA replication followed by mitosis and cytokinesis. Accordingly, the same problems we have discussed in the context of multicellular organisms also apply to individual cells. Even viruses have life cycles. In fact, a wide range of conceptual difficulties disappear when we view viruses as processes rather than as things, as one of us has recently argued (Dupré and Guttinger 2016). Viruses pass through an intricate sequence of stages as part of their life cycle. Some of these stages are highly stable (for instance, the virion stage, which is what most people have in mind when they think of viruses). Crucially, however, the very existence of these stable states can only be accounted for by referring to their role in the larger cyclical process that *is* the virus. Finally, we should also mention that, just as developmental cycles are not the only kinds of life cycles, life cycles are not the only kinds of cycles that living systems undergo.

---

[16] Again, one may be tempted to nominate the DNA sequence for this role. The problem is that genomes are highly dynamic entities that are subject to many kinds of changes during development (Barnes and Dupré 2008). Moreover, it is not even clear that there is any unique entity that qualifies as the genome of an organism (Dupré 2010b).

The field of chronobiology investigates the wide range of cyclic phenomena exhibited by organisms, such as circadian rhythms.

The fact that most biological entities (not just organisms) exist as temporally extended and temporally differentiated life cycles provides strong grounds for endorsing a processual view of the living world. The significance of life cycles for biological ontology is examined in the chapters by James DiFrisco, Paul Griffiths and Karola Stotz, and Flavia Fabris.

## 5.3. Ecological interdependence

A third reason for taking a processual stance in biology concerns two features that are often regarded as defining characteristics of a thing or substance: the first is that it should have boundaries, even if these are sometimes vague, more or less determined by its being the kind of thing that it is; and the second is that it should exhibit a certain autonomy or independence, that its dependence on anything external to it should be, at most, contingent. Neither of these characteristics is easy to reconcile with the well-known fact that organisms do not exist in nature as isolated, or even independent, entities but rather live in densely interconnected communities that provide many of the conditions of existence that enable the survival of their individual members. Indeed, ecology tells us that the environment in which each organism finds itself is partially constituted by the complex network of reciprocal interactions that the organism in question maintains with other organisms. Some of these interactions are so intimate and so fundamental to the survival of the organism that it has been hotly debated whether the interacting entities are distinct at all, or whether they should rather be understood as constituting a single life.

The nature of these interactions varies widely: they can be specific or generic, permanent or transitory, obligate or facultative. They can occur among individuals belonging to the same species or can involve organisms of many different species. Traditionally, there has been a tendency to regard conspecific relations as more intimate and significant. Extreme examples are the highly organized collectives formed by eusocial insects such as ants or honeybees, which are so tightly interwoven that they are often described as 'superorganisms'. However, recent research in various areas of biology has revealed that inter-species associations, generally referred to as *symbiotic* relations, are equally (if not more) fundamental, and are also more pervasive. Indeed, it is becoming increasingly apparent that symbiosis is the rule rather than the exception in the biological realm (Margulis 1998; Gilbert and Epel 2015).

There are various kinds of symbiotic partnerships. Some are mutualistic (i.e. beneficial to both parties), such as the relationship between sea anemones and clownfish, in which the anemones protect the fish from predators (which cannot tolerate the stings of the anemone) and the fish defend the anemones against butterflyfish (which eat anemones). Others are commensal (i.e. beneficial to one party and neutral to the other), such as the relationship between epithytic plants (e.g. mosses, orchids, ferns) and the trees on which they grow in order to increase their exposure to sunlight. And yet others are parasitic (i.e. beneficial to one party but harmful to the other), such as the ticks, fleas, lice, and leeches that feed on the blood of warm-blooded animals. But the extent of symbiosis goes well beyond associations between multicellular organisms (Dupré and O'Malley 2009). Every multicellular organism is itself engaged in an array of symbiotic relationships of all three types with vast numbers of microorganisms. Large organisms ('macrobes') are actually multi-species collectives, or *holobionts*, composed of many different kinds of microbial symbionts—bacteria, archaea, viruses, protists, fungi, and microscopic metazoans such as nematodes—which live in symbiotic associations with their macroscopic eukaryotic hosts. Importantly, these microbial symbionts play crucial roles in the survival, reproduction, and evolution of their hosts (Moran 2006; Gilbert et al. 2012). Even the microbes themselves do not live independently, preferring instead to live in complex communal organizations, or *biofilms*, which often consist of multiple species. A familiar example is dental plaque, in which over six hundred distinct microbial taxa have been distinguished (Marsh 2006). Biofilms, just like organisms, have their own distinctive life cycles (Ereshefsky and Pedroso 2013).

Ecological interdependence, then, is one of the most characteristic aspects of the living world, and it poses major problems for an ontology of things or substances. One reason is that this ontology typically regards relations as external to things. A thing is taken to be what it is independently of the relations it enters in. A precondition of this independence is that it has relatively clear boundaries that enable its objective individuation as a discrete entity. Moreover, the properties (often taken to be essential ones) that determine both its boundaries and its continued existence are grounded in features that lie entirely within those boundaries. The interconnectedness of life challenges all of these substantialist assumptions. Organisms persist by virtue of the intricate webs of relations they maintain with one another, which in part endow them with their distinctive properties, capacities, and behaviour. In the light of ecology in general and of symbiosis in particular, organisms need to be viewed as processes rather than as things. As processes, and in contrast to things, organisms are fundamentally relational entities that affect and are affected by their environment, within which they are firmly embedded, and which is itself constituted by numerous other processes.

Ecological communities or consortia, such as biofilms, holobionts, and super-organisms, are not collections of relatively autonomous things but deeply entangled meshes of interdependent processes. This entanglement can make it extremely difficult to establish unequivocally the boundaries of a biological individual, or even to determine how many individuals we are dealing with in a particular situation. This is why ecological relations are best seen as an intertwining of processes. The biological realm presents us with a seamless spectrum of degrees of intertwining. This spectrum ranges from facultative forms of mutualism or commensalism among multicellular organisms to instances of fully obligate endosymbiosis, as exemplified by mitochondria, which are found in virtually all eukaryotic cells. We are generally comfortable with asserting that in cases of the former type we are dealing with two individuals, whereas in cases of the latter type we are dealing with one individual, as we take the mitochondrion to be fully a part of the cell that contains it. However, for most other forms of ecological association that lie in between these two extremes, it is far less clear that we can make such definite estimations.

How intimate must an association become for two individuals to count as one? Think, for example, of *Buchnera aphidicola*, bacteria that live in specialized cells within aphids and synthesize nutrients that the aphid cannot otherwise produce.

These bacteria have greatly reduced genomes, a fact that makes clear that they could not survive outside the aphid. Does this mean that neither the bacteria nor the aphid can be seen as individuals? One of the merits of an ontology of processes—by comparison with one of substances—is that it makes it possible to embrace the vagueness naturally suggested by such cases. It appears that distinguishing discrete individuals within the deeply entangled flux of interdependent living processes is often a matter of conceptual decision motivated by specific theoretical or practical interests. We shall examine the implications of this in the next section.

Although the early twentieth-century organicists appealed to metabolic turnover and life cycles to motivate their defence of a process philosophy of biology (as can be appreciated from the quotations we offered earlier), they did not really consider ecological interdependence. But it is interesting to note that Whitehead himself recognized its significance.[17] More recently, Charles Birch and John Cobb have used ecological interdependence as the basis of the particular brand of process ontology (which they call 'event ontology') defended in their book *The Liberation of Life* (Birch and Cobb 1981). On their view, relations rather than intrinsic properties are the distinguishing features of entities of all kinds. Stephan Guttinger's chapter builds on Birch and Cobb's ecological model to develop a processual account of biological macromolecules. The chapters by Thomas Pradeu and Frédéric Bouchard consider the significance of symbiosis for philosophical discussions of biological identity and of biological individuality respectively. And Argyris Arnellos presents in his chapter a process-based organizational account of life that recognizes and attempts to integrate its metabolic and ecological dimensions.

## 6. Philosophical Payoffs

Process ontology, as we have seen, is far more attuned to and concordant with the understanding of the living world provided by the findings of contemporary biology than its substantialist rival. But apart from the critical question of empirical correspondence, from a pragmatic perspective process ontology also proves to be the more attractive position to adopt in a number of different contexts. One of its great virtues, as we shall see in this section, is that it is able to offer a cogent metaphysical justification for several important critiques that have recently been put forward in the philosophy of biology. Specifically, in what follows we shall discuss

---

[17] In *Science and the Modern World*, Whitehead illustrated the interconnectedness of his worldview with an ecological discussion of the Brazilian rainforest: 'The trees in a Brazilian forest depend upon the association of various species of organisms, each of which is mutually dependent on the other species. A single tree by itself is dependent upon all the adverse chances of shifting circumstances. The wind stunts it: the variations in temperature check its foliage: the rains denude its soil: its leaves are blown away and are lost for the purpose of fertilisation. You may obtain individual specimens of fine trees either in exceptional circumstances, or where human cultivation has intervened. But in nature the normal way in which trees flourish is by their association in a forest. Each tree may lose something of its individual perfection of growth, but they mutually assist each other in preserving the conditions for survival. The soil is preserved and shaded; and the microbes necessary for its fertility are neither scorched, nor frozen, nor washed away. A forest is the triumph of the organisation of mutually dependent species' (Whitehead 1925: 206). Whitehead's remarks seem remarkably prescient, especially in the light of recent discussions regarding the so-called Wood Wide Web (Giovannetti et al. 2006).

how process ontology serves to ground critiques of essentialism, reductionism, and mechanicism.

## 6.1. Grounding critiques of essentialism

In our view, the ability of substance ontology to answer central philosophical questions is deeply connected to its invocation of essentialism. Assuming that there is some essential property of a thing of a certain kind answers two sorts of questions.[18] First, it tells us what kind of thing we are concerned with and hence, more generally, it allows us to address questions about the nature of classification. If we ask what kinds of things there are in the world and how they are distinguished from one another, we can answer that things are of different kinds if and only if they have different essential properties. Second, the essentialist assumption allows us to answer questions about the persistence of things, for instance what it is for the same thing or substance to continue to exist through time. These questions will be familiar to those concerned with the puzzle of personal identity, which considers how it is possible for someone to be the same person as the one they were decades earlier despite having undergone massive change. Essentialism's answer to these questions is that a thing persists for just so long as it possesses the relevant essential property.

As it happens, however, essentialism—at least as far as the living world is concerned—is untenable, or so one of us has argued for many years (Dupré 1993, 2002; see also Hull 1965; Sober 1980). The questions it aims to answer are badly posed. The assumption that there is some unique natural kind to which a given organism belongs is false (Dupré 2002; Ereshefsky 1992). In actual scientific practice, different theoretical interests (e.g. ecological role, phylogenetic history) dictate different and multiply overlapping ways of dividing biological entities into kinds. A classificatory pluralism that follows from this observation has become quite widely accepted among philosophers of biology.[19] And the failure of essentialism with regards to classification is one of the factors that make the process perspective so attractive. Indeed, it is possible to see the processual character of biological entities as providing a deep explanation of why a multiplicity of ways of classifying such entities is precisely what we should expect to find. Promiscuous realism, as one of us has denominated this pluralism of classifications (Dupré 1993), thus finds a metaphysical justification in process ontology (Dupré forthcoming).

But the pluralist implications of process ontology do not end there. In addition to explaining why we should be pluralists about classification, it also explains why we should be pluralists about individuation. Biology suggests not only that there is no single way of classifying living entities into kinds but that even the division of biological reality into distinct individuals may not be a thoroughly determinate and unambiguous matter. As we argued in our discussion of ecological interdependence,

---

[18] There are philosophical accounts that postulate distinctive individual essences (e.g. Kripke 1980), which could answer the second of the questions we discuss here. We shall ignore these for present purposes.

[19] There are still some defenders of essentialism (e.g. Devitt 2008), as well as philosophers who defend essentialism generally, without much interest in the practical obstacles it faces in the life sciences (e.g. Ellis 2001). At least among philosophers interested in biology (a category that does include Devitt), they are in a distinct minority.

the ubiquity of symbiosis can severely complicate the task of defining the boundaries of a biological entity, as well as that of ascertaining whether particular entities are distinct individuals or parts of another entity. Take, for instance, the trillions of microbes that make up the human microbiome, and without which the human host would rapidly become sick and die. Are these parts of the human organism, or rather just a large consortium of cooperating others? For some practical purposes, for example when describing the conditions for a healthy human life, it will be natural to treat the object of study as one whole system; for others, for example when tracing the various evolutionary lineages to which the various collaborating cells belong, it may be more natural to treat the same object as many. Why should we suppose that there is a single unequivocal answer to this question, rather than many different ones, depending on the issues we are interested in addressing?

By analogy to the promiscuous realism one of us has long advocated, the thesis that there are multiple ways of carving biological entities into distinct individuals can be described as 'promiscuous individualism' (Dupré 2012). What we wish to emphasize here is that this latter thesis, like the former, can be viewed as a direct implication of an ontology that takes processes as fundamental. To illustrate this, consider the processes made famous by Heraclitus: rivers. A river is a part of a widely distributed process of water flowing to the sea, but the division of this process into discrete parts is a matter of human convenience, not an objective fact. The Missouri River is the longest river in North America, and much longer than the Upper Mississippi, but we call the former a tributary of the combination made up of the latter and the (still shorter) Lower Mississippi. Is this a discovery, or is it a mistake? It is surely obvious that this is a convention that resulted from a contingent decision in the past. Similarly, on the question of classification, distinctions between rivers, streams, brooks, lakes, ponds, and so on are rough and convenient ways of dividing up processes of water flow; they hardly reflect sharp and objective distinctions between natural kinds.

Beyond matters of classification and individuation, substance ontology has traditionally called upon essentialism to answer questions concerning persistence. Things persist, the argument goes, by virtue of their continued possession of certain essential properties, which make those things what they are and which remain unchanged over time. The problem for substance ontology has always been that it is extraordinarily difficult to specify any such change-exempt descriptive properties that permanently characterize the essence of things. Indeed, all of the empirical facts that we marshalled in support of the process cause in the previous section cast serious doubts on the existence of essential properties. First, as a consequence of constant metabolic turnover, a biological individual is never materially identical from one moment to the next. Second, because of its life cycle, it undergoes massive morphological changes as it progresses through its various ontogenic phases. And, third, as a result of its ecological interrelations, the symbiotic associations that compose and maintain it change considerably over its lifetime.[20] Overall, substance ontology is not able to deal with the problem of persistence because the constituents of the world—of

---

[20] In the human example we mentioned above, for instance, the symbionts that make up the microbiome of a neonate are very different from those of an adult.

the living world, at the very least—are not endowed with essential properties. The question is: does process ontology actually fare any better? One of the classic objections to process ontology is that it is incapable of accounting for persistent entities or continuants, which anchor our representation of the world (Strawson 1953; Wiggins 2016). Is this correct, and, if so, where does it leave us?

We offer here some thoughts regarding the persistence of processes, which is a recurring topic throughout the book (see especially the chapters by Peter Simons, James DiFrisco, Thomas Pradeu, Daniel Nicholson, and Anne Sophie Meincke). For a start, we do not think that we are dealing with an insoluble philosophical problem, as there is a wide range of uncontroversial physical processes to which we clearly and unproblematically do apply criteria of identity over time. Think of hurricanes, streams, and vortices. A paradigm somewhere between the first and last of these is the Great Red Spot on Jupiter—a storm that has persisted for at least as long as we have had the instruments to observe it, a period of centuries. The mode of persistence of an organism is in many respects quite similar to that of the Great Red Spot. Just as the latter persists by drawing in matter and energy from the violent winds that surround and shape it, so the former persists by securing from its environment the matter and energy it requires to maintain its organization far from thermodynamic equilibrium. The persistence of the Great Red Spot is not based on the mere persistence of any of its individual properties or constituents; it is rather something it *does*—a continuous activity. Even its identity as an individual cannot be taken as a given; it is instead a task that it must constantly accomplish. By means of its continuous activity, the Great Red Spot demarcates itself from the flux of its surroundings. All of these considerations, we believe, are equally applicable to organisms.

So what is it that makes the various temporal stages of the Great Red Spot temporal stages of the same process? We take it that the answer must be sought in various forms of causal connection and continuity between the relevant stages. The momentum of the material that makes up the Great Red Spot at $t_1$ explains the momentum of the material at $t_2$, and the external winds that help to maintain the structure of the Great Red Spot at $t_1$ continue to do so at $t_2$. Indeed, a similar kind of story could be told about those surrounding winds themselves. We suspect that this is the general form of explanation that is needed to account for the persistence of a process. An obvious advantage of this way of explaining persistence is that it poses no restrictions on the amount of change that can take place between the properties of a process at different moments in time.

However, as we illustrated with the example of a river, the ability to track individual processes over time definitively is limited. Rivers split and merge, open up into lakes or deltas, and so on. What counts as the same river is to some degree a matter of convention. Nevertheless, we regard this inherent vagueness as a strength of the process perspective rather than a weakness. We need to come to terms with the fact that many organisms—including all unicellular organisms—are liable to split; those that do not, frequently intertwine and even intermingle as a consequence of their ecological interrelations. The reproduction of a lichen, for instance, requires the independent reproduction of a fungus and a photosynthetic bacterium or alga, and the subsequent merger of the various offspring with suitable new partners (not necessarily of exactly the same kind as their parents) to form new lichens.

More generally, as we have already emphasized, the near universality of symbiosis makes the delineation of biological individuals to some degree indeterminate. Given all this, we should certainly not expect the tracking of entities over time to be a fully determinate matter. This may strike us as a problem, but the truth is that it is only a problem if we already assume that it *should* be possible to perfectly track entities through time in the first place, in accordance with the essentialist stance generally associated with substance ontology. An ontology of processes, besides conforming to what biological research actually tells us about the living world, liberates us from the burden of this expectation.[21]

## 6.2. Grounding critiques of reductionism

Process ontology also helps us to account for the problems generated by reductionism in biology. One of us has been arguing for decades against the temptations of reductionism, temptations that have been exacerbated in recent times by the successes of molecular biology (Dupré 1993, 2010a; Powell and Dupré 2009). Despite these successes, it has become increasingly clear to many biologists, as well as to philosophers, that reductionism is at best a severely limited approach to understanding living systems. While much progress has been made in understanding wholes such as cells in terms of their organelles, macromolecules, and other subcellular assemblies, a consensus is emerging that understanding these entities cannot be fully accomplished without a vital reference to the systems of which they are parts (e.g. Boogerd et al. 2007; Noble 2012).

In the context of substance ontology, however, it is difficult to understand where the failure of reductionism comes from. What could a whole possibly be, beyond the set of constituents that compose it and the physical relations among them—that is, spatial locations and connecting forces? A classic argument of this kind is due to Kim (2005), who has convinced many that it is incoherent to attribute to a physical whole any behaviours that are not ultimately either deductive consequences of the behaviours of the parts that make up that whole or causally impotent epiphenomena. This conclusion is often expressed via the widespread assertion that higher levels must at least supervene on lower levels. If the state of the lower level is fully specified, then everything at the higher level must be fully determined at the same time. Any other possibility is taken to be logically incoherent.

Such considerations have their basis in the conventional structural hierarchy of substance ontology, which views the world in terms of successive levels of organization: elementary particles, atoms, molecules, cells, organisms, and so on. At each level of this hierarchy, it is in principle possible to determine the intrinsic properties of the relevant things. From the properties of things at one level and from the relations between them, we can infer the properties of things at the higher level. Thus the properties of all things are consequences of the properties of their constituents plus the relations between their constituents, down to the most elementary level

---

[21] Strawson (1953) argued convincingly that we could make no sense of a world without persistent individuals that we could reidentify at different times. Note that nothing we have said contradicts this requirement. It is just that the selection of reidentifiable individuals is less objectively determined than Strawson supposed. Or so we claim.

(assuming that there is one). It is difficult to see what else there could be that might contribute to determining a thing's properties.

Matters look quite different when we view the biological world as being organized not as a structural hierarchy of things, but as a dynamic hierarchy of processes, stabilized at different timescales. At no level in the biological hierarchy do we find entities with hard boundaries and a fixed repertoire of properties. Instead, both organisms and their parts are exquisitely regulated conglomerates of nested streams of matter and energy. The processes that make up the biological hierarchy not only compose one another but also provide many of the enabling conditions for the persistence of other processes in the hierarchy, at both higher and lower levels. In other words, the visible and tangible entities at each level are not simply given (as they are in a structural hierarchy of things), but are rather dynamically maintained by continuous activity taking place at higher and lower levels in the same hierarchy.

One of the most significant consequences of the processual hierarchy of the living world, then, is that it makes the physicalist dream of absolute reductionism impossible. The complex web of causal dependencies between the various levels means that we cannot fully specify the nature of an entity merely by listing the properties of its constituents and their spatial relations. It also means that we cannot pick out any level in the hierarchy as ontologically or causally primary. Whereas a substance ontology that presupposes a structural hierarchy of things only allows bottom-up causal influences, a process ontology has no trouble in recognizing that causal influences can flow in different directions. Once we transition from an ontology of substances to an ontology of processes, it is no longer incoherent or mysterious to assert that the properties of the parts are partially determined by the properties of the whole—a claim, by the way, that biologists (especially physiologists and embryologists) have been making for centuries on the basis of their empirical investigations. This point has important philosophical ramifications. The 'downward causation' that naturally results when a process at one level in the hierarchy is stabilized from a level above it has often seemed metaphysically grotesque to philosophers tacitly or explicitly working within a substance ontology. To a process philosopher, on the other hand, such notions present no special problem.

In a world of processes, reductionism makes little sense. We already anticipated this claim in our discussion of ecological interdependence. Processes are inherently relational entities. They influence and are influenced by their surroundings. These surroundings, in turn, are made up of even further processes. In abstraction from their surroundings, processes are nothing at all; they have no independent existence. One needs only to think of waves in an ocean or of gusts of wind in a storm. A process has the properties it does in no small part because of its relations with other processes. Consequently, it cannot be fully explained independently of these relations. A necessary condition for reductionism is that it must be possible to treat entities independently from one another and to consider their structure and constitution independently of the context in which they exist. But this is precisely what an ontology of processes denies. When we accept that the living world is a process world, we are able to understand why reductionism in biology, despite its countless limited successes in local and fixed contexts, can never fully succeed as a global explanatory enterprise, even in principle.

## 6.3. Grounding critiques of mechanism

Ever since the scientific revolution, substance ontology has been associated with mechanicism, the view that nature, together with everything in it, is a machine that operates in a regular and predictable manner and that can be fully explained in mechanical terms. Mechanicism is, of course, a natural expression of substance thought, as machines instantiate many of the properties traditionally attributed to substances: they are fixed material entities with clearly defined boundaries and exist independently of the activities they engage in and of the relations they maintain with other entities. It is not surprising, then, that mechanicism is perfectly consistent with both essentialism and reductionism; a machine is taken to belong to a particular kind because it exhibits certain unchanging properties, and its organization and operation can be completely accounted for in terms of its component parts and their interactions.

Although physics emancipated itself from the mechanicist worldview at the turn of the twentieth century (which is partly what led Whitehead to embrace process metaphysics, as we indicated earlier), mechanicism never really lost its grip on biology; on the contrary, it tightened it after the rise of molecular biology. Nowadays it is not uncommon to regard protein complexes as ingeniously designed molecular machines, cells as intricate sets of circuits that can be partially re-engineered to function in accordance with our needs, development as the computable execution of a deterministic genetic program, or the products of evolutionary change as analogous with optimally designed artefacts. However, it is becoming ever more apparent that all of these views are deeply problematic, as both of us have argued in earlier work (Dupré 2008; Nicholson 2013, 2014).

One very appealing aspect of the process perspective is that it makes the ontological inadequacy of the machine conception of the organism explicit. If organisms are processes rather than substances, then conceiving of them as machines inevitably leads to a distorted understanding of them. Looking back once more at the three empirical motivations we discussed for adopting a process ontology, it is significant that all of them also constitute reasons for resisting the ontological assimilation of organisms to machines. First, the reality of metabolic turnover means that the very structure of every organism, unlike that of any machine, is wholly and continuously reconstituted as a result of its operation. Second, the life cycle an organism undergoes is unlike anything a machine ever experiences. Machines do not develop, nor do they reproduce; their configuration is fixed upon their manufacture, as opposed to that of organisms. And, third, as a consequence of ecological interdependence, no organism can function, or even persist, independently of the entangled web of interrelations it maintains with other organisms. By contrast, the persistence or operation of machines does not depend on their capacity to maintain relations with other machines. Your microwave does not require the presence of other microwaves to heat your food, let alone to persist in your kitchen. By adopting a process ontology we become far less likely to be tempted by the machine conception of the organism, which for many biologists (and for some philosophers) still constitutes the default, and often tacit, ontological understanding of living systems.

But process thinking allows us to do even more. In addition to legitimating and encouraging the recourse to machine analogies in the living world, mechanicism has permeated modern biology in another, more subtle way, namely by popularizing the view that living systems can be fully explained by describing the causal mechanisms that are said to be operating within them. The word 'mechanism' of course is etymologically related to the word 'machine', but today the term refers more loosely to a set of components that causally interact in a regular fashion to produce a phenomenon of interest. The recent realization that biologists frequently invoke mechanisms in their explanations of phenomena has led to an explosion of interest in the philosophy of science, as a growing number of philosophers have tried to come to terms with the nature and scope of so-called mechanistic explanations (e.g. Machamer et al. 2000; Glennan 2002; Bechtel 2006; Craver 2007; Craver and Darden 2013). Although it is undeniable that the elucidation of biological phenomena in terms of mechanisms has proven to be an enormously productive scientific strategy, mechanistic explanations are inherently limited in what they can tell us about living systems, or so we have claimed in previous publications (Nicholson 2012; Dupré 2013). The value of the process perspective in this context is that it helps us to explain these limits.

The view that mechanistic explanations are in principle capable of providing a complete understanding of living systems is problematic because it presupposes an ontology in which substances at least play the central role.[22] The ontological picture that typically motivates appeals to mechanisms is one in which living processes ultimately derive from, and can therefore be explained by, the systematic rearrangement of the preexisting thing-like entities that compose them. But, on the view we have been developing, the components of an organism, be it multicellular or unicellular, are just as much processes as the organism itself. The constituents of a higher-level process do not suffice to explain it, because they themselves—or, more specifically, their stability and activity—cannot be fully understood without reference to that same process, as well as to those above it. The mereological, bottom-up character of mechanistic explanations means that their deployment in a world of processes is inevitably subject to the same limits as more traditional forms of reductionism.

Why, then, have mechanistic explanations turned out to be as successful in biology as they undoubtedly have been? A process perspective allows us to answer this question as well. The reason why mechanistic explanations provide insights (to the extent that they do) is that the components of the mechanisms being described are sufficiently stable on the timescale of the phenomena under investigation. For example, it is possible to explain the phenomenon of muscle contraction in mechanistic terms because the entities primarily responsible for it, namely the actin and myosin filaments in the muscle fibre, are sufficiently stable during the temporal interval in which contraction occurs to be treated as thing-like components of a mechanism. However, this does not mean that actin and myosin filaments really are inherently stable constituents of muscles, for if we change our research question and

---

[22] We do not say that they presuppose a substance ontology *tout court* because, following the canonical statement by Machamer et al. (2000), some philosophical accounts of mechanisms are explicitly dualistic, endorsing both 'entities' (i.e. substances) and 'activities' (i.e. processes). Nevertheless, processes in mechanistic accounts tend to play a role similar to that of properties in standard substance ontology.

inquire instead into the growth of muscular tissue during development—a phenomenon that takes place over a much longer timescale—it is not possible to discern any specific muscle fibres, let alone any actin and myosin filaments within them, that persist for the entire duration of development. In this epistemic context we can no longer characterize actin and myosin filaments as thing-like components of a mechanism, as they themselves now appear to be dynamic and transient entities. Mechanism descriptions, therefore, are accurate only on the particular timescales of the phenomena they are called upon to explain.

A key implication that we take to follow from this is that the mechanisms described in mechanistic explanations should not be treated, in accordance with the assumptions of the older mechanicist tradition, as real things that ontologically make up organisms (in the way they might be supposed to make up machines). Rather, mechanisms in biology are more appropriately understood as heuristic explanatory devices—as idealized spatio-temporal cross sections of living systems that conveniently abstract away the complexity and dynamicity of their biotic and abiotic surroundings and pick out only the causal relations that are taken to be most relevant for controlling and manipulating the phenomena under investigation (Nicholson 2012; Dupré 2013). To suppose that mechanisms are the ontological building blocks of living systems is to commit what Whitehead famously called 'the fallacy of misplaced concreteness' (Whitehead 1925: 52), which occurs every time we confuse our conceptual schemas, models, and analogies with the way things really are. What biologists describe as mechanisms (and their components) are actually manifestations of specific patterns of stability of different processes that unfold concurrently in living systems. Process ontology enables us to understand the limits of mechanistic explanations in biology while simultaneously recognizing and accounting for the scope of their effectiveness.

## 7. Biological Consequences

The move from substance ontology to process ontology has many interesting biological consequences. We have examined some of them over the course of this essay, and many more are explored by other authors in the remainder of this book. Some of these consequences are far from obvious. Ann-Sophie Barwich, for instance, shows in her chapter how a process perspective on perception results in a far more contextual understanding of olfaction than has been traditionally supposed. Moreover, not all of the consequences discussed in the chapters of this book are purely theoretical in character; quite a few of them are closely tied to practical issues. For example, Eric Bapteste and Gemma Anderson argue in their chapter that an increasing appreciation of the evolutionary role of biological processes requires new ways of representing them for analytical purposes.

Before we conclude, we wish to discuss a few rather significant consequences of embracing an ontology of processes in biology that are not directly covered in subsequent chapters. We present these reflections in order to give the reader a fuller sense of the wide-ranging implications of our project, as well as to offer a glimpse of the work that still remains to be done. Below we consider some specific consequences for physiology, genetics, evolution, and medicine.

## 7.1. Physiology

Physiology has traditionally been concerned with the study of function. This emphasis on function in turn has tended to direct physiological research towards the analysis of structure. The reason for this is that it has generally been assumed, in accordance with substance ontology, that structures somehow ground the capacities that constitute functions. Structures are taken to be more or less fixed, and ontologically prior to the functions associated with them. This substantialist understanding pervades the longstanding philosophical debate on function, where, despite major disagreements, the function of an entity—whether or not it is tied to the explanation of its selective advantage—is seen as that entity's contribution to a larger system; and this contribution is made possible by the entity's structural properties. One consequence of the process perspective in biology is that it calls into question this familiar and widespread view.

The various structures that an organism exhibits are not really fixed, but are instead continuously maintained by a large number of carefully regulated processes, which endow them with their relative stability. These structures cannot be taken for granted in our physiological explanations, as they must themselves be accounted for by the various functional activities that enable them to persist through time. It is therefore incorrect to assume that structures are prior to functions, or that functions are determined by structures. The processal nature of organisms means that changes in their functional demands will tend to result in changes in how they maintain and regenerate their respective structures. In biology the relation between structure and function is not linear and unidirectional, as is often supposed, but circular and symmetrical. Neither of the two can be privileged over the other, or even be understood without appealing to the other.

This key insight was already recognized by the organicists working to develop a processual biology in the early decades of the twentieth century. Most of them explicitly argued that structure and function are mutually interdependent features of organisms. As Haldane elegantly expressed it, '[s]tructure and functional relation to environment cannot be separated in the serious scientific study of life, since structure expresses the maintenance of function, and function expresses the maintenance of structure' (Haldane 1931: 22). Structures are not simply given, but instead reflect the stability of functional activities, which are themselves maintained by stable structures. In the last analysis, it makes no sense to separate structure and function, as the two represent different yet complementary 'ways of seeing' the processual reality of living systems, one emphasizing stability and the other emphasizing dynamicity. According to Bertalanffy, '[t]he old contrast between "structure" and "function" is to be reduced to the relative speed of processes within the organism. Structures are extended, slow processes; functions are transitory, rapid processes' (Bertalanffy 1941: 251).

We are inclined to think of structure and function in biology as alternative forms of abstraction from the continuous flow of underlying processes. Structural descriptions abstract away the temporal dimension, as well as selecting non-arbitrary but underdetermined spatial limits for the entities of interest. Functional descriptions bring back the temporal dimension, but they do so at the cost of focusing on a highly specific set of properties of the entities under consideration. More needs to be said, probably, about how structural and functional characterizations abstract.[23] The relevant point for now is that, even though linear explanations of function in terms of structure serve valuable purposes in biological research, we need to keep in mind that they provide limited perspectives on biological phenomena. It is always possible, it seems to us, and ultimately even necessary, to treat biological structures as explananda as well as explanantia.

## 7.2. Genetics

Adopting a process ontology in biology also has an impact on how we think about inheritance. Genetics is generally thought to have originated with Mendel's famous experiments on peas, which concerned dichotomous characteristics that eventually came to be explained in terms of underlying invisible factors called genes. The consequent view that all the heritable information transmitted from parent to offspring resides in material particles, namely genes, which replicate during reproduction and trigger the developmental construction of the new individual, has proved remarkably resilient even as its empirical basis has been increasingly eroded. This understanding fits naturally with a substance ontology, as it regards what is inherited as a collection of things—genes—together with their defining properties, namely the dispositions to cause phenotypic effects. A process perspective brings to the foreground two major problems with this substantialist picture. The first has to do with the supposed atomistic nature of genes, and the second concerns the view of reproduction as the transmission of a set of things, an act construed as a sort of passing of genetic batons.

When genes were first imagined as discrete entities carrying phenotypic information, their ontological status was entirely hypothetical. Although the reality of these entities became progressively accepted, no consensus was reached on their nature until the famous elucidation of the structure of DNA in 1953. As the implications of that discovery became apparent, the view emerged that genes are stretches of DNA defined by a specific sequence of nucleotides, the four chemical structures that alternate in the DNA molecule and provide the information necessary to generate a protein, conceived of as the paradigmatic functional molecule. Genes for all kinds of phenotypic traits—from fur colour to sexual preference—were conceived of as sequences of nucleotides that have the power to cause or alter these developmental outcomes.

This traditional picture has, however, been very widely rejected (see e.g. Keller 2000). The development of most traits is now understood to involve features widely distributed across the genome as well as influences from many aspects of the external environment. This has led to growing doubts about the ontological significance of genes as discrete components of genomes, which in turn has prompted a more direct focus on the genome itself (Barnes and Dupré 2008). The genome, far from being a fixed source of developmental information, is now increasingly seen as a thoroughly processual entity. The stability of the nucleotide sequence, which is so crucial to its functioning, is dependent on multiple processes of correction and editing that

---

[23] It would appear—we note in passing—that distinguishing biological mechanisms involves abstractions of both kinds.

drastically reduce the error rate in its replication. Moreover, the functioning of the genome involves constant changes in its physical conformation that allow the appropriate parts to be more or less accessible for transcription. These changes are in turn partly controlled by certain molecules' attachment to and detachment from particular points on the genome, a process referred to as the *epigenetic* modification of the genome. Epigenetic effects are at the end of a causal chain that can begin far away, for example with maternal interaction with an infant organism (Champagne et al. 2006). Importantly, such causal processes allow environmental factors to affect the functioning of the genome.

Taking a broader view, what these genomic activities indicate is that, far from there being a one-way control of the cell by the genome, as is still sometimes imagined, the genome is in constant two-way interaction with its cellular context and beyond. The persistence of both, in fact, depends on their interrelations. In short, where once we saw a genome as a set of discrete units or things mechanically controlling their wider environment, now we see interactions of a complex dynamic entity with its even more complex surroundings.

Genetic replication is another vital process ripe for reconsideration from a process perspective. Whereas in the context of substance ontology it may seem natural to conceive of replication as simply the copying and production of one thing on the basis of another—something akin to making a photocopy (see Dawkins 1982)—the generation of one process by another is a rather different matter. As James Griesemer (2000, 2005a, 2005b) has emphasized in a series of papers, biological reproduction— unlike photocopying, in which only a re-production of information is required— involves a degree of material continuity between the original and the descendant. Indeed, in the replication of DNA, new double helices are partially constituted from the material of the old double helices. Treating replication as copying has the inadvertent consequence of diverting attention from the causal process that generates it. DNA does not 'self-replicate', as it is sometimes claimed, but is completely dependent for its replication on the participation of an intricate molecular 'machinery'. To think of replication, and by implication of reproduction, as analogous with copying is to abstract away the causality and materiality of the connection between parent and offspring, reducing this connection to an essentially informational relation.

While mechanicist biologists convinced of the machine-like nature of living systems will be comfortable comparing the role of replication in reproduction with the digital duplication of software systems, a biologist committed to a processual understanding of the living world is far less likely to be seduced by such analogies. When parents are recognized as self-maintaining metabolic processes from which offspring somehow branch off, it becomes clear that reproduction must involve material overlap between the two. Accounts of reproduction that fail to include this aspect neglect one of its essential features. Classical models of transmission genetics, Weismann's famous separation of germ and soma, and the more contemporary notion of the genetic program have enabled geneticists to study inheritance as a mere cluster of relations between things, but at the cost of abstracting it from the actual processes of reproduction and development that generate these relations. Although such models have proven to be useful heuristics for scientific research, an adequate theory of inheritance must ultimately include these fundamental processes.

In the end, a process ontology forces us to expand our understanding of inheritance itself. Recognizing the material overlap between parent and offspring in reproduction makes it obvious that what gets transmitted is much more than just the DNA. The material (i.e. cytoplasmic) continuum that exists between parent and offspring includes many molecular systems that can be inherited apart from the genome, such as macromolecular steady states and self-sustaining metabolic loops (see Jablonka and Lamb 2005). Because these epigenetic inheritance systems typically depend on chemical diffusion and molecular transport processes, they exhibit far lower degrees of fidelity than the nucleic acid coding mechanism characteristic of genetic replication. But this does not mean that they are not capable of affecting developmental outcomes; they are, and they do. Once we adopt a processual point of view, the detection of epigenetically inherited traits ceases to be a baffling discovery and becomes something we would actually expect to find.

*7.3. Evolution*

While it is self-evident that evolution is a process, it is less clear how we should think about the nature of the entities that participate in this process. The traditional substantialist stance has been to regard them as things. Process ontology, of course, leads us to understand them as processes. If organisms are developmental cycles, then we should regard these developmental cycles, rather than the thing-like time slices that we abstract from them, as the actual entities that compose the evolutionary process. In fact, this is the view of evolution that developmental systems theorists have been advocating for many years, with the vital corollary that evolution can be driven by changes to any of the factors that contribute to reproducing the developmental cycle (see Griffiths and Gray 1994 and the chapter in this volume by Griffiths and Stotz).

We have to be careful here, however, with the unusual grammar of the word 'evolve'. Although humans evolve, no particular human ever evolves. What evolves is populations. Even the concept of population is not quite appropriate, as it lacks temporal extension. Populations are temporal cross sections of lineages, so it really is the lineages that undergo evolutionary change. To say that humans evolve is to say that humans existing at one stage of the human lineage differ in some systematic way from those existing at an earlier stage. The evolving lineages generally considered to be the central subjects of evolutionary change are the entities we usually refer to as species.[24]

One of the earliest debates in modern philosophy of biology concerned the metaphysical status of species. Because species have historically been treated as paradigmatic classificatory concepts, they were long supposed to designate kinds. But then Michael Ghiselin (1974) and David Hull (1976, 1978) advanced the thesis

---

[24] This is, again, not quite right, as species typically consist of a number of populations—they are often described as 'metapopulations'—that may not be connected to one another reproductively or otherwise. Consequently, the unit we really want as the locus of evolutionary change is that of which the metapopulation is a cross section, something for which, as far as we know, there is no standard term (unless, of course, we count 'species'!). For present purposes, we can overlook this complication.

that species are not kinds but individuals. Although the species-as-individuals view has become widely accepted, it continues to encounter some resistance on the grounds that it is puzzling and counterintuitive. It is objected for example that, as the members of a species are discrete and relatively independent, it is not clear how they can be identified as parts of an individual. Another common complaint is that species have very fuzzy boundaries, whereas individuals typically do not. Interestingly, once we see species not just as individuals, but as individual *processes*, we are able to address both of these concerns. Causal relations between the temporal stages of a lineage, and between the spatial parts of these temporal stages, are responsible for providing that lineage with whatever integrity it has as an individual process. And the problem of vague boundaries turns out not to be a problem at all, as processes tend to lack clearly defined boundaries to begin with (think of the boundaries of a thunderstorm, for instance). The expectation that individuals should have clear boundaries is merely a prejudice of substance thought, which we bypass when we assume a process ontology.

If species or lineages are individual processes, how do they manage to maintain their coherence and stability over time? It would seem that, just as organisms persist by renewing the cells that compose them through constant metabolic turnover, so lineages persist by replacing the organisms that make them up through continuous cycles of reproduction. Now recall that reproduction, properly understood, comprises not only inheritance processes (such as genetic replication) but also developmental ones; and these in turn draw on a highly heterogeneous range of causal factors (Oyama et al. 2001), which means that all of the latter contribute, too, to the stabilization of a lineage. In addition, a lineage persists by virtue of the reciprocal interactions that its members maintain with their environment. Members of a lineage not only adapt to their environment but also modify it, as a consequence of their activities and in accordance with their needs. This process is known as niche construction (Odling-Smee et al. 2003).[25] Last but by no means least, the persistence of a lineage is dependent on natural selection, understood here not as a cause of change but as a stabilizing force. Stabilizing selection leads to the continued production of very similar phenotypes in a lineage (namely the most adaptive ones), and this helps maintain its stability over long periods of time.

Of course, despite these numerous forms of stabilization, lineages do gradually change, and this results in their evolution. A key implication of the process perspective is that it encourages us to embrace a much more pluralistic understanding of evolutionary change than the one assumed by orthodox neo-Darwinism, which regards it almost exclusively in terms of natural selection acting on different alleles in a population. Selection cannot take place unless some other process has already provided the pertinent variants, and the modern synthesis assumption

---

[25] It is important not to confuse niche construction with the notion of the extended phenotype (Dawkins 1982), as they imply opposite ontologies. The former can be seen as an implication of process ontology, whereas the latter is indicative of substance ontology. The extended phenotype extends the boundaries of things (i.e. organisms) beyond their material bodies and onto their environments, but these boundaries remain fully determined by the intrinsic properties of the things themselves. Niche construction, in contrast, suggests that organisms and their environments are causally intertwined and mutually constitute one another.

that random genetic mutations alone can fulfil this role is increasingly coming under attack (Pigliucci and Muller 2010). Lineages are sustained by a wide range of processes, such as reproduction, niche construction, and stabilizing selection, as we have already noted, but also including parental effects, conspecific interactions, and symbiotic associations. It is conceivable that changes in, or disruptions of, any of these stabilizing activities might affect the temporal trajectory of a lineage, and we should therefore consider including them in our explanations of evolutionary change.[26]

We offer one last and more speculative suggestion. Processual systems such as organisms and cells respond to their environment in ways that are conducive to their persistence. A growing number of theorists maintain that such responses are not merely automatic reactions elicited by environmental stimuli, but rather reflect purposive actions performed by organisms on their own behalf and prompted by emerging challenges and opportunities in the environment. It seems uncontentious to suggest that natural selection should favour organisms that do what is most conducive to their survival over those that produce fixed responses to a determinate range of environmental stimuli, though traditionally it was supposed that results of the former kind were achievable only through high degrees of intelligence. In her chapter, Flavia Fabris shows that Waddington believed that such adaptive strategies can be found much more widely (a view that Waddington shared with many of his organicist contemporaries), and that such systemic responses could be explained in terms of the adaptive deployment of hidden genetic variation. Perhaps, as Denis Walsh suggests in his provocative chapter, we should see evolutionary change as resulting primarily from such agent-like capacities of organisms and other processual systems.

*7.4. Medicine*

The process perspective also has interesting consequences for medicine, specifically for how we think about the concept of disease, as has been recently argued by Pierre-Olivier Méthot and Samuel Alizon (2015).[27] The history of medical thought reveals that the understanding of disease has oscillated back and forth between two opposing conceptions. According to the so-called physiological conception, diseases result from disturbances in the functional equilibrium of the body, and their cure reflects the harmonious restoration of this equilibrium. In contrast, the so-called ontological conception views diseases as foreign entities that enter the body, and their cure implies the expulsion of the intruders. The ontological conception is aligned with substance ontology, as it regards diseases as particular things (or properties of things) that are discrete and exist independently of the body they infect, whereas the physiological conception is more congenial to process ontology, as it views diseases

---

[26] One of us has recently elaborated upon these claims in more detail elsewhere (Dupré 2017).

[27] Pierre-Olivier Méthot participated in the workshop at the University of Exeter where many of the papers collected in this volume were first presented, but unfortunately was not able to contribute a chapter. The ensuing discussion considers some of the ideas he presented, many of which are featured in the aforementioned paper.

as temporally extended disruptions in the carefully regulated meshwork of interconnected processes that constitutes the body.

With the rise of medical microbiology in the late nineteenth century the ontological conception became the dominant theory of disease, and so it has remained, more or less, to the present day. The modern notion of a 'pathogen' is clearly derived from it, which accounts for why pathogens have long been considered a discrete category, distinguished from other microbes by their inherent capacity to cause disease in appropriate hosts. Conceiving of the biological realm in processual terms leads us to question the adequacy of such a strongly substantialist understanding of the aetiology of disease. Taking a processual stance today does not mean returning to the old physiological conception of disease. But it does suggest that pathogenicity may not be an intrinsic property of a microbe at all, but rather, as Méthot and Alizon (2015) argue, a contingent property afforded by the particular ecological context in which the microbe finds itself and by the complex and ever changing symbiotic relationship it maintains with its host.

Remarkably, this is precisely the view that appears to be emerging from recent microbiological research. The traditional question 'Is this microbe a pathogen?' is gradually giving way to the question 'Under what ecological conditions is this microbe likely to become a pathogen?'. This shift is partly being prompted by the discovery that microbes thought to be engaged in commensal or mutualistic relations with their host can become pathogenic (i.e. parasitic) as a result of changes in the host environment (this is the case for the microbes that make up the normal microbiota of the human gut, for instance). Conversely, microbes that are usually pathogenic can end up protecting their host against more virulent parasites. Virulence itself, which refers to the degree of damage a pathogen is capable of inflicting on its host, is not a permanent property of the pathogen; it is arguably not even a property *of* the pathogen, but rather the outcome of a specific kind of interaction between the pathogen and its host.

It is becoming increasingly clear that there is nothing about pathogens, as far as their structure is concerned, that fundamentally sets them apart from non-pathogens. There is also growing evidence that infections are frequently caused by more than one type of microbe. Viewing pathogenicity as a property of an individual microbe can therefore be misleading, as it is often a collective property that emerges from the interactions of various kinds of microbes with a host. In addition, it needs to be kept in mind that the microbes themselves are not unchanging things but are constantly evolving—and microbes can evolve very rapidly! Sometimes they evolve the capacity to be pathogens. Indeed, an infection that is initially harmless to one host can become pathogenic in subsequent infections.[28]

The above considerations strongly suggest that there are no definite criteria by which we can unequivocally classify a particular microbe as a pathogen. The more empirical research is conducted, the further away we seem to be from being able to make such classifications, and for good reason. Attempts to classify microbes in this

---

[28] The potentially great rapidity of this process can be attributed to the lateral transfer of so-called virulence factors, which are often packaged as mobile genetic units, such as plasmids, that are transferable to other cells within the microbiome.

way are misguided because they seek to attribute to the microbe a property that is in reality a function of the host, of the microbe, and of their interactions. To attribute pathogenicity to an individual microbe is to commit an indefensible act of abstraction, as it implies ignoring the complex processual context that makes it possible. A microbe is not a thing or a substance, and pathogenicity is not an unchanging property that the microbe carries with it like an essence; it is rather a process, and as such it is to be expected that its characteristics will be found to be transient, context-dependent, historically contingent, and ever subject to ecological and evolutionary changes.[29]

## 8. Conclusions

This essay has concerned itself with a number of issues pertaining to the metaphysics of science. It has shown, we hope, that these issues are of such fundamental importance that they cannot be avoided by anyone seriously concerned with science, be it through philosophical reflection or through empirical investigation. Scientists as well as philosophers are inevitably committed to certain metaphysical views, regardless of whether they are aware of them or not. These views, as Waddington recognized, 'are not mere epiphenomena, but have a definite and ascertainable influence on the work [a scientist] produces' (Waddington 1969: 72). We believe that being explicit about the metaphysical stance one takes to be right for biology is crucial if we are to prevent what Whitehead described as the 'canaliz[ation of] thought and observation within predetermined limits, based upon inadequate metaphysical assumptions dogmatically assumed' (Whitehead 1933: 151). Our aim in this essay has been to defend the metaphysical thesis that a process ontology is the right ontology for the living world.

We are well aware that previous attempts to defend process ontology have often been met with considerable scepticism, if not downright hostility. One reason for this, in our view, is that process philosophers have frequently felt the need to introduce a new lexicon in order to come to terms with the processual nature of existence. Whitehead, of course, is the most notorious example, and his influence on modern process philosophy, including on how it is perceived by those who oppose it, has been enormous. However, as we hope this essay has demonstrated, it is not necessary to appeal to neologisms or resort to opaque prose to make the case for process. Thing-locutions, despite their pervasiveness, do not have to be taken at face value. After all, our linguistic conventions are not always aligned with our ontological convictions; which is why, for instance, we continue to speak of 'sunsets' and 'sunrises' even centuries after the Copernican revolution. It suffices that we realize that English grammar, like that of other Indo-European languages, exhibits a clear bias towards substances, which may well be rooted, at least in part, in our cognitive dispositions.

---

[29] Further implications of process ontology for evolutionary microbiology are explored in Bapteste and Dupré 2013.

Beyond any such inherent bias, we surmise that the widespread prevalence of substance ontology also reflects the fact that in many circumstances it does the job sufficiently well. The relation between substance and process ontology is not completely unlike the relation between classical and modern physics. Just as classical physics provides a convenient approximation of middle-sized physical entities moving at relatively slow speeds but does not constitute an accurate description of physical reality, so substance ontology provides a serviceable characterization of biological entities, especially when considered over short temporal intervals, despite being a fundamentally inappropriate description of the living world. But, although it might seem more intuitive to regard organisms as things than as processes, the situation quickly begins to reverse when we start giving due consideration to time. Many of the methods used to study living entities abstract away from the temporal dimension to facilitate their investigation—one only needs to think of conventional anatomical techniques such as desiccating, pickling, staining, fixing, and freezing. It may be that these methods have led biologists to think in substance terms, or perhaps these methods rather reflect their awareness of the difficulty of dealing adequately with dynamic material and therefore highlight their appreciation of the processual nature of life. Yet, even if the latter is the correct diagnosis, it seems to us that the reliance on such techniques has concealed the deficiencies of theorizing grounded in substance ontology (this claim is further developed by Laura Nuño de la Rosa in her chapter).

Our argument in this essay has been that process ontology is far more concordant with the understanding of the living world provided by contemporary biology than its substantialist rival. The more we learn about life, the more necessary a process perspective becomes. This is particularly the case with regard to the increasing realization of the omnipresence of symbiosis, which directly challenges deeply entrenched substantialist assumptions about the living world. Thus the empirical findings of biology are inexorably driving us towards processualism, even if it is less intuitive than substantialism. It is interesting to observe that physics, which has traditionally been regarded as the more advanced science, was pushed towards process ontology about a century ago (as was argued by Whitehead and others), and now biology—if we and the other contributors to this volume are correct—is following suit. Might this perhaps be an indication that the shift from substantialism to processualism is just something that all sciences go through as they develop? This intriguing possibility was recently suggested by Mark Bickhard:

Every science has progressed beyond an initial conception of its phenomena in substance terms to understanding that they are in fact process phenomena. Fire is no longer modeled in terms of the substance phlogiston, but instead in terms of the process of combustion; heat no longer in terms of caloric, but in terms of random kinetic processes; life no longer in terms of vital fluids, but in terms of special kinds of far from thermodynamic equilibrium processes. And so on. (Bickhard 2009: 553)

Be that as it may, what is evident is that an ontology of processes is more consistent with the facts of biology than is an ontology of substances. Metabolic turnover, life cycles, and ecological interdependence—to mention the three phenomena we have discussed here in most detail—all provide compelling empirical motivations for adopting a process ontology in biology. Philosophers sceptical of our naturalistic approach to metaphysics will doubt whether any amount of empirical evidence could settle, or even be relevant to, the debate between substantialism and processualism. To them, we can at least offer our arguments that processualism can deal more effectively than substantialism with a wide range of philosophical issues, including classification, individuation, persistence, explanation, and abstraction. And, for those who share our discomfort with the interconnected doctrines of essentialism, reductionism, and mechanicism, processualism supplies resources for grounding this discomfort—resources that, as we have seen, are not available to the substantialist. If the value of an ontological position is to be measured, at least in part, by the ability it gives us to solve problems, then processualism can draw support from its application both to biology and to philosophy.

Finally, this essay has illustrated how a processual stance serves as an invaluable heuristic guide in scientific research. It has a number of interesting and sometimes unexpected consequences for fields as diverse as physiology, genetics, evolution, and medicine, where it forces us to question deeply engrained assumptions and to revise basic theoretical tenets. Processualism sheds new light on old problems and even encourages us to change our research questions, and it therefore has the potential to open up novel avenues of empirical investigation. After reading the rest of the essays collected in this volume, we hope that the reader will be convinced at least of the usefulness, and we hope even of the truthfulness, of a processual philosophy of biology.

## 9. Overview of Contributions

Although we have already alluded to the seventeen other contributions to this volume, it might be useful to end by saying something about how the book as a whole is organized. After this extensive introduction, the chapters that follow are classified into four thematic clusters: 'Metaphysics', 'Organisms', 'Development and Evolution', and 'Implications and Applications'.

The five chapters in Part II deal with general metaphysical issues related to the goal of developing a process ontology for biology. Peter Simons argues that a fundamentally processual understanding of the living world requires that we reconceptualize the continuant things that biologists study as secondary 'precipitates' of primary processes. Rani Lill Anjum and Stephen Mumford suggest that biology is better served by a dispositionalist theory of causation than by more traditional Humean accounts, given that only the former can do justice to the dynamicity, continuity, and context-sensitivity of biological phenomena. James DiFrisco attempts to strengthen the case for a process ontology in biology by providing causal—and suitably processual—accounts of individuation and persistence in terms of cohesion and genidentity respectively. Thomas Pradeu explores in more detail the fertility of the concept of genidentity in biology, showing how it leads us to prioritize processes over things. Finally, Johanna Seibt considers how a systematic ontological framework of subjectless processes she has developed over many years, called 'general process theory', can be used to address questions of individuality, composition, and emergence in the philosophy of biology.

The four chapters in Part III concern themselves with the concept of organism—that most central of biological categories—from a processual perspective. Daniel Nicholson calls for a shift in how we think about what organisms are, replacing the conventional machine-like conception with a stream-like one that better captures their processual nature. Denis Walsh claims that recognizing the unique character of organisms as processual agents requires an agential theory to make sense of how they evolve. Frédéric Bouchard maintains that evolutionary individuality comes in degrees, as it reflects various levels of functional integration between intersecting processes. Finally, Argyris Arnellos proposes a process-based organizational ontology for biology in order to account for the integrity of individual organisms while simultaneously explaining their collaborative dimension.

The four chapters in Part IV adopt a process ontology in the specific contexts of development and evolution. Paul Griffiths and Karola Stotz show that developmental systems theory is deeply committed to a processual view of life and offer a number of reasons why processes need to be taken as fundamental in both development and evolution. Flavia Fabris reappraises the core concepts of Waddington's theory of epigenetics and suggests that the genetic assimilation of acquired characters is best explained in processual terms. Laura Nuño de la Rosa looks at how new microscopy, molecular, and computer technologies for modelling biological processes are themselves contributing to a more processual understanding of development. Finally, Eric Bapteste and Gemma Anderson discuss how an increasing appreciation of the evolutionary role of intersecting biological processes requires new ways of representing these processes for analytical purposes.

Lastly, the four chapters in Part V explore broader implications and applications of a processual philosophy of biology. Stephan Guttinger makes the case for a processual understanding of macromolecules, drawing on ecological ideas to argue that only a process view can elucidate their fundamentally relational character. Marta Bertolaso and John Dupré articulate a processual perspective on cancer, arguing that it reflects a failure in the highly complex regulatory system that stabilizes a multicellular organism over time. Ann-Sophie Barwich submits that a process perspective on perception—motivated by recent findings in cognitive neuroscience—results in a far more flexible and contextual understanding of olfaction than has been traditionally supposed. Finally, Anne Sophie Meincke defends the claim that the philosophical dilemma of personal identity can be overcome by conceiving of persons as biological processes.

# References

Abir-Am, P. G. (1987). The Biotheoretical Gathering, Transdisciplinary Authority and the Incipient Legitimation of Molecular Biology in the 1930s: New Perspective on the Historical Sociology of Science. *History of Science* 25: 1–70.

Agar, W. E. (1943). *A Contribution to the Theory of the Living Organism*. Carlton: Melbourne University Press.

Bapteste, E. and Dupré, J. (2013). Towards a Processual Microbial Ontology. *Biology & Philosophy* 28: 379–404.

Barnes, J. and Dupré, J. (2008). *Genomes and What to Make of Them*. Chicago: Chicago University Press.

Bechtel, W. (2006). *Discovering Cell Mechanisms: The Creation of Modern Cell Biology*. Cambridge: Cambridge University Press.

Bertalanffy, L. von. (1941). Die organismische Auffassung und ihre Auswirkungen. *Biologie* 10: 247–58 and 337–45.

Bertalanffy, L. von. (1952). *Problems of Life: An Evaluation of Modern Biological and Scientific Thought*. New York: Harper & Brothers.

Bickhard, M. (2009). The Interactivist Model. *Synthese* 166: 547–91.

Birch, C. and Cobb, J. B. (1981). *The Liberation of Life. From the Cell to the Community*. Cambridge: Cambridge University Press.

Bohm, D. (1980). *Wholeness and the Implicate Order*. London: Routledge.

Boogerd, F. C., Bruggeman, F. J., Hofmeyr, J.-H. S., and Westerhoff, H. V. (2007). *Systems Biology: Philosophical Foundations*. Amsterdam: Elsevier.

Bourne, C. (2006). *A Future for Presentism*. Oxford: Oxford University Press.

Campbell, R. (2015). *The Metaphysics of Emergence*. New York: Palgrave Macmillan.

Carpenter, A. D. (2014). *Indian Buddhist Philosophy*. Durham: Acumen.

Champagne, F. A., Weaver, I. C., Diorio, J., Dymov, S., Szyf, M., and Meaney, M. J. (2006). Maternal Care Associated with Methylation of the Estrogen Receptor-Alpha1b Promoter and Estrogen Receptor-Alpha Expression in the Medial Preoptic Area of Female Offspring. *Endocrinology* 147: 2909–15.

Cobb, J. B. and Griffin, D. R. (1977). *Mind in Nature: Essays on the Interface of Science and Philosophy*. Washington, DC: University Press of America.

Collier, J. (1988). Supervenience and Reduction in Biological Hierarchies. *Canadian Journal of Philosophy* 14: 209–34.

Collier, J. (2004). Self-Organization, Individuation and Identity. *Revue Internationale de Philosophie* 58: 151–72.

Craver, C. F. (2007). *Explaining the Brain: Mechanisms and the Mosaic Unity of Neuroscience*. Oxford: Oxford University Press.

Craver, C. F. and Darden, L. (2013). *In Search of Mechanisms: Discoveries across the Life Sciences*. Chicago: Chicago University Press.

Dawkins, R. (1982). *The Extended Phenotype: The Long Reach of the Gene*. Oxford: Oxford University Press.

Devitt, M. (2008). Resurrecting Biological Essentialism. *Philosophy of Science* 75: 344–82.

Dupré, J. (1993). *The Disorder of Things: Metaphysical Foundations of the Disunity of Science*. Cambridge, MA: Harvard University Press.

Dupré, J. (2002). *Humans and Other Animals*. Oxford: Oxford University Press.

Dupré, J. (2008). *The Constituents of Life*. Amsterdam: Van Gorcum.

Dupré, J. (2010a). It Is Not Possible to Reduce Biological Explanations to Explanations in Chemistry and/or Physics. In J. Ayala and R. Arp (eds), *Contemporary Debates in Philosophy of Biology* (pp. 32–47). Oxford: Wiley Blackwell.

Dupré, J. (2010b). The Polygenomic Organism. In S. Parry and J. Dupré (eds), *Nature after the Genome* (pp. 19–31). Oxford: Wiley Blackwell.

Dupré, J. (2012). *Processes of Life: Essays in the Philosophy of Biology*. Oxford: Oxford University Press.

Dupré, J. (2013). Living Causes. *Proceedings of the Aristotelian Society* 87: 19–38.

Dupré, J. (2017). The Metaphysics of Evolution. *Interface Focus* 7(5). doi: 10.1098/rsfs.2016.0148.

Dupré, J. (forthcoming). Processes, Organisms, Kinds and the Inevitability of Pluralism. In O. Bueno, R.-L. Chen, and M. B. Fagan (eds), *Individuation across Experimental and Theoretical Sciences*. Oxford: Oxford University Press.

Dupré, J. and Guttinger, S. (2016). Viruses as Living Processes. *Studies in History and Philosophy of Biological and Biomedical Sciences* 59: 109–16.

Dupré, J. and O'Malley, M. (2009). Varieties of Living Things: Life at the Intersection of Lineage and Metabolism. *Philosophy and Theory in Biology* 1: e003.

Ellis, B. (2001). *Scientific Essentialism*. Cambridge: Cambridge University Press.

Ereshefsky, M. (1992). Eliminative Pluralism. *Philosophy of Science* 59: 671–90.

Ereshefsky, M. and Pedroso, M. (2013). Biological Individuality: The Case of Biofilms. *Biology & Philosophy* 28: 331–49.

Ghiselin, M. (1974). A Radical Solution to the Species Problem. *Systematic Zoology* 23: 536–44.

Gilbert, S. F. and Epel, D. (2015). *Ecological Developmental Biology: The Environmental Regulation of Development, Health, and Evolution*, 2nd edn. Sunderland: Sinauer Associates.

Gilbert, S. F., Sapp, J., and Tauber, A. I. (2012). A Symbiotic View of Life: We Have Never Been Individuals. *Quarterly Review of Biology* 87: 325–41.

Giovannetti, M., Avio, L., Fortuna, P., Pellegrino, E., Sbrana, C., and Strani, P. (2006). At the Root of the Wood Wide Web: Self Recognition and Non-Self Incompatibility in Mycorrhizal Networks. *Plant Signaling & Behavior* 1: 1–5.

Glennan, S. (2002). Rethinking Mechanistic Explanation. *Philosophy of Science* 69: S342–53.

Graham, D. W. (2015). Heraclitus. In E. N. Zalta (ed.), *The Stanford Encyclopedia of Philosophy*. http://plato.stanford.edu/entries/heraclitus.

Griesemer, J. (2000). Reproduction and the Reduction of Genetics. In P. Beurton, R. Falk, and H.-J. Rheinberger (eds), *The Concept of the Gene in Development and Evolution: Historical and Epistemological Perspectives* (pp. 240–85). Cambridge: Cambridge University Press.

Griesemer, J. (2005a). Genetics from an Evolutionary Process Perspective. In E. M. Neumann-Held and C. Rehmann-Sutter (eds), *Genes in Development: Re-Reading the Molecular Paradigm* (pp. 343–75). Chapel Hill: Duke University Press.

Griesemer, J. (2005b). The Informational Gene and the Substantial Body: On the Generalization of Evolutionary Theory by Abstraction. In M. R. Jones and N. Cartwright (eds), *Idealization XII: Correcting the Model, Idealization and Abstraction in the Sciences* (pp. 59–115). Amsterdam: Rodopi.

Griffiths, P. E. and Gray, R. D. (1994). Developmental Systems and Evolutionary Explanation. *Journal of Philosophy* 91: 277–304.

Haldane, J. S. (1917). *Organism and Environment, as Illustrated by the Physiology of Breathing*. New Haven: Yale University Press.

Haldane, J. S. (1919). *The New Physiology and Other Addresses*. London: Charles Griffin.

Haldane, J. S. (1931). *The Philosophical Basis of Biology*. London: Hodder & Stoughton.

Haynie, D. T. (2008). *Biological Thermodynamics*. Cambridge: Cambridge University Press.

Hull, D. L. (1965). The Effects of Essentialism on Taxonomy: Two Thousand Years of Stasis. *British Journal of Philosophy of Science* 15: 314–26 and 16: 1–18.

Hull, D. L. (1976). Are Species Really Individuals? *Systematic Zoology* 25: 174–91.

Hull, D. L. (1978). A Matter of Individuality. *Philosophy of Science* 45: 335–60.

Jablonka, E. and Lamb, M. J. (2005). *Evolution in Four Dimensions: Genetic, Epigenetic, Behavioral, and Symbolic Variation in the History of Life*. Cambridge, MA: MIT Press.

Keller, E. F. (2000). *The Century of the Gene*. Cambridge, MA: Harvard University Press.

Kim, J. (2005). *Physicalism, or Something Near Enough*. Princeton: Princeton University Press.

Kirschner, M., Gerhart, M., and Mitchison, T. (2000). Molecular 'Vitalism'. *Cell* 100: 79–88.

Koutroufinis, S. A. (2014). *Life and Process: Towards a New Biophilosophy*. Berlin: De Gruyter.

Kripke, S. (1980). *Naming and Necessity*. Cambridge, MA: Harvard University Press.

Levins, R. and Lewontin, R. (1985). *The Dialectical Biologist*. Cambridge, MA: Harvard University Press.

Lillie, R. S. (1945). *General Biology and Philosophy of Organism.* Chicago: Chicago University Press.
Machamer, P., Darden, L., and Craver, C. F. (2000). Thinking about Mechanisms. *Philosophy of Science* 67: 1–25.
Margulis, L. (1998). *The Symbiotic Planet: A New Look at Evolution.* London: Weidenfeld & Nicolson.
Marsh, P. D. (2006). Dental Plaque as a Biofilm and a Microbial Community–Implications for Health and Disease. *BMC Oral Health* 6. doi: 10.1186/1472-6831-6-S1-14.
Méthot, P.-O. and Alizon, S. (2015). What is a Pathogen? Toward a Process View of Host-Pathogen Interactions. *Virulence* 5: 775–85.
Misteli, T. (2001). The Concept of Self-Organization in Cellular Architecture. *Journal of Cellular Biology* 155: 181–5.
Moran, N. A. (2006). Symbiosis. *Current Biology* 16: R866–R871.
Nicholson, D. J. (2012). The Concept of Mechanism in Biology. *Studies in History and Philosophy of Biological and Biomedical Sciences* 43: 152–63.
Nicholson, D. J. (2013). Organisms ≠ Machines. *Studies in History and Philosophy of Biological and Biomedical Sciences* 44: 669–78.
Nicholson, D. J. (2014). The Machine Conception of the Organism in Development and Evolution: A Critical Analysis. *Studies in History and Philosophy of Biological and Biomedical Sciences* 48: 162–74.
Nicholson, D. J. and Gawne, R. (2014). Rethinking Woodger's Legacy in the Philosophy of Biology. *Journal of the History of Biology* 47: 243–92.
Nicholson, D. J. and Gawne, R. (2015). Neither Logical Empiricism nor Vitalism, but Organicism: What the Philosophy of Biology Was. *History and Philosophy of the Life Sciences* 37: 345–81.
Noble, D. (2012). A Theory of Biological Relativity: No Privileged Level of Causation. *Interface Focus* 2: 55–64.
Odling-Smee, J., Laland, L., and Feldman, M. (2003). *Niche Construction: The Neglected Process in Evolution.* Princeton: Princeton University Press.
Oyama, S., Griffiths, P. E., and Gray, R. D. (2001). *Cycles of Contingency: Developmental Systems and Evolution.* Cambridge, MA: MIT Press.
Peterson, E. L. (2011). The Excluded Philosophy of Evo-Devo? Revisiting Waddington's Failed Attempt to Embed Alfred North Whitehead's 'Organicism' in Evolutionary Biology. *History and Philosophy of the Life Sciences* 33: 301–32.
Peterson, E. L. (2016). *The Life Organic: The Theoretical Biology Club and the Roots of Epigenetics.* Pittsburgh: Pittsburgh University Press.
Pigliucci, M. and Muller, G. B. (2010). *Evolution: The Extended Synthesis.* Cambridge, MA: MIT Press.
Powell, A. and Dupré, J. (2009). From Molecules to Systems: The Importance of Looking Both Ways. *Studies in History and Philosophy of Biological and Biomedical Sciences* 40: 54–64.
Rescher, N. (1996). *Process Metaphysics: An Introduction to Process Philosophy.* Albany: SUNY Press.
Russell, E. S. (1916). *Form and Function: A Contribution to the History of Animal Morphology.* London: John Murray.
Russell, E. S. (1924). *The Study of Living Things: Prolegomena to a Functional Biology.* London: Methuen.
Russell, E. S. (1930). *The Interpretation of Development and Heredity: A Study in Biological Method.* Oxford: Clarendon.
Russell, E. S. (1945). *The Directiveness of Organic Activities.* Cambridge: Cambridge University Press.

Seibt, J. (1996). The Myth of Substance and the Fallacy of Misplaced Concreteness. *Acta Analytica* 15: 119–39.
Seibt, J. (2003). *Process Theories: Crossdisciplinary Studies in Dynamic Categories.* Dordrecht: Springer.
Seibt, J. (2012). Process Philosophy. In E. N. Zalta (ed.), *The Stanford Encyclopedia of Philosophy.* https://plato.stanford.edu/entries/process-philosophy.
Shaw, G. B. (1905). *The Irrational Knot.* London: Constable.
Sider, T. (2001). *Four-Dimensionalism: An Ontology of Persistence and Time.* Oxford: Oxford University Press.
Sober, E. (1980). Evolution, Population Thinking, and Essentialism. *Philosophy of Science* 47: 350–83.
Stein, R. L. (2004). Towards a Process Philosophy of Chemistry. *HYLE: International Journal for Philosophy of Chemistry* 10: 5–22.
Strawson, P. F. (1953). *Individuals: An Essay in Descriptive Metaphysics.* London: Methuen.
Waddington, C. H. (1956). *Principles of Embryology.* London: George Allen and Unwind.
Waddington, C. H. (1957). *The Strategy of the Genes: A Discussion of Some Aspects of Theoretical Biology.* London: George Allen and Unwind.
Waddington, C. H. (1969). The Practical Consequences of Metaphysical Beliefs on a Biologist's Work: An Autobiographical Note. In C. H. Waddington (ed.), *Towards a Theoretical Biology*, vol. 2: *Sketches* (pp. 72–81). Edinburgh: Edinburgh University Press.
Wade, N. (2005). Your Body Is Younger than You Think. *New York Times*, 2 August.
Weiss, P. A. (1962). From Cell to Molecule. In J. M. Allen (ed.), *The Molecular Control of Cellular Activity* (pp. 1–72). Toronto: McGraw Hill.
West-Eberhard, M. J. (2003). *Developmental Plasticity and Evolution.* New York: Oxford University Press.
Whitehead, A. N. (1919). *An Enquiry Concerning the Principles of Natural Knowledge.* Cambridge: Cambridge University Press.
Whitehead, A. N. (1920). *The Concept of Nature.* Cambridge: Cambridge University Press.
Whitehead, A. N. (1925). *Science and the Modern World.* Cambridge: Cambridge University Press.
Whitehead, A. N. (1929). *Process and Reality: An Essay in Cosmology.* New York: Macmillan.
Whitehead, A. N. (1933). *Adventures of Ideas.* Cambridge: Cambridge University Press.
Wiggins, D. (2016). Activity, Process, Continuant, Substance, Organism. *Philosophy* 91: 269–80.
Woodger, J. H. (1929a). *Biological Principles: A Critical Study.* London: Routledge & Kegan Paul.
Woodger, J. H. (1929b). Some Aspects of Biological Methodology. *Proceedings of the Aristotelian Society* (n.s.) 29: 351–8.

# PART II
# Metaphysics

# 2
# Processes and Precipitates

*Peter Simons*

> *Ordinary men live so completely within the house of the Stagyrite that whatever they see out of the windows appears to them incomprehensible and metaphysical.*
>
> —C. S. Peirce (2000: 168)

## 1. Introduction

In this chapter I note the pervasive distinction among objects in time, including all objects of interest to biology, between continuants or substance-like entities on the one hand and occurrents or process-like entities on the other. I shall endeavour to make the distinction a little more precise, and argue for the distinctness and real existence of objects of both sorts, while upholding the metaphysical priority of processes. The nature of the relationship between the two sorts turns on a type of causal relationship known as 'genidentity', and the cognitive operation whereby we recognize continuants amid occurrents will be taken as a species of abstraction. The resulting metaphysics prioritizes processes over substances but does not throw substances out completely.

## 2. The Continuant/Occurrent Duality

There is a pervasive and basic duality in the way in which we speak about objects in the real (spatio-temporal–causal) world. Some things are in time in such a way that they grow through the accumulation of parts—*temporal parts*—as time goes by. These are events, processes, and states. The other sort of things are those traditionally called *substances*: people and other animals and organisms, houses, cars and other artefacts, mountains, rivers, planets and stars, and other natural objects. Unlike the other basic sort, they do not add temporal parts but are said to remain identical or the same at different times. It is the same person, cat, car, or river it was yesterday, notwithstanding changes to it in the meantime. I propose in this chapter to use the terminology invented by W. E. Johnson a century ago[1] and call the second

---

[1] Johnson first used the term 'continuant' in print in Moore et al. 1916: 431. 'Occurrent' as contrasted with 'continuant' occurs in Johnson 1921: 199, but was in frequent use as a synonym for 'event' or 'occurrence' in the sixteenth and seventeenth centuries.

group *continuants* and the first group *occurrents* (as in Simons 1987: 118 *et passim*). More recently the terms 'endurant' and 'perdurant' have been used for continuants and occurrents respectively (Lewis 1986: 203 ff.). These have two disadvantages: they are very alike and initially confusing, and they postdate Johnson's coinages by several decades.

*Prima facie* examples of continuants are legion: organisms like ourselves, animals, plants and so on, artefacts like houses, chairs, and ships, geographical features like lakes and valleys, astronomical objects like stars and planets, are all continuants in this sense. For present purposes, the most important types of occurrent are processes. Examples of processes are breathing, growing, flowing, cooling, contracting, singing, and orbiting. Both living and non-living continuants engage in or are involved in or participate in processes and may have processes going on in them. There is one borderline case of objects in time where the continuant/occurrent duality breaks down, and that is for instantaneous objects, which by definition lack temporal parts. I shall not be concerned with such cases.

The continuant/occurrent duality is firmly anchored in our ways of thinking and speaking. Continuants are typically designated by nouns, while occurrents are typically indicated by verbs. The *American Heritage Dictionary* indeed defines a verb as '[t]he part of speech that expresses existence, action, or occurrence in most languages'. The noun/verb distinction was recognized as such by Plato and Aristotle and is universal in human languages. I said that verbs *indicate* occurrents rather than designating or denoting them, because a verb in use predicates rather than names: in 'John is breathing' only the subject term denotes, while the predicate says something of or predicates something of John. If we wish to talk *about* an occurrent, we typically nominalize a verb and form a derivative noun or noun phrase: *John's snoring last night, Luciano's rendering of 'Nessun dorma' in Madison Square Garden in 1987, Vesuvius's eruption in AD 79* all designate occurrents. We are extremely adept at coining and using such nominalizations in both impromptu and routine ways.

## 3. Specification

David Lewis introduced his distinction between enduring (said of continuants) and perduring (said of occurrents) as one among entities all of which *persist*, that is, exist in time and last from one time to some later times. I shall use the word 'persist' in this neutral way.

A *persistent* is an entity that exists in space and time, exists for—or at—more than an instant (and so persists), and at any time at which it exists has a spatial location. Its locations at the times at which it exists sum to a spatio-temporal region that I call its *locus*. For it to be a persistent, there must be parts of its locus that are in time-like separation.[2]

An *occurrent* (perdurant) is a persistent that has disjoint parts that are in time-like separation, such that, at a given time when it exists, there is a maximal part of it, all of

---

[2] An entity, all of whose sub-loci are in space-like separation from one another, can be taken to exist instantaneously in some reference frame, and so not to persist from one time to a later time in that frame.

whose parts exist wholly within that time, so that such maximal parts for disjoint times are disjoint. These are called *temporal parts* of the occurrent. In the limiting case, the time is an instant and the temporal part is called a *phase* or an instantaneous time slice of the occurrent.

A *continuant* (endurant) is a persistent such that, at any instant at which it exists, it is identical with the maximal part of it. Hence, at different instants at which it exists, its maximal parts that exist are identical with one another and with it. By this characterization, a continuant cannot have temporal parts, since an occurrent's maximal parts for distinct instants are disjoint and so not identical, whereas a continuant's maximal parts at distinct instants are it, and so are identical. It further follows that a continuant may have a part at one instant that it does not have at another.

We assume that the non-instantaneous temporal parts of occurrents are themselves occurrents and the parts that a continuant has for a time are themselves continuants. In other words, the formal property of being a continuant or an occurrent is *dissective* (see Leonard and Goodman 1940: 55) or propagates down to parts, instantaneous parts excepted.

Persistents are usually not wholly static but vary with time: a plant grows from a seedling into a mighty tree; the net flow of water in a tidal river is now upstream, now downstream. When a continuant varies in such a way, we say that it changes: it has first one characteristic, then later another characteristic incompatible with the first. The continuant itself is taken to survive the change. When an occurrent varies, we may ascribe the variation to differences of characteristics among its temporal parts: this part of the flow is upstream, that later part is downstream. The whole occurrent, of which these are temporal parts, inherits this variation from its parts in an analogous way to that in which a tiger's coat is variegated because different parts of it have different colours. In the case of a continuant, however, there are no temporal parts from which it can inherit the variation, so it is the continuant itself that changes. Strictly speaking, then, it is incorrect to say that an occurrent changes.[3]

Continuants and occurrents do not merely sit side by side in the metaphysical inventory: they are intimately involved with one another. When a leopard moves through the bush, its movement is an occurrent—as is its breathing, heart beating, tail flicking, and so on. These occurrents in turn involve other continuants: the heart, lungs, and tail of the leopard, the leaves and air molecules it brushes aside. The events and processes in which an organism or other continuant is involved throughout the time it exists constitute what we may call its *life*.[4] How the continuant and its life are related is something the metaphysical story of continuants and occurrents needs to make clear. Continuants and the occurrents in which they are involved seem in many cases to occupy the same spatio-temporal region, albeit in different ways. Again, this is something that needs explaining (see Simons 2014).

---

[3] See Dretske 1967. Dretske's point applies more widely to all variation, not just to motion.
[4] In this special sense, inorganic continuants like rivers and stars also have lives. If this terminology is felt to be unacceptable, another term, such as 'history', could be employed, though this is then ambiguous between events and the description thereof.

## 4. The Priority Question

In the history of philosophy there have been those who regarded continuants as metaphysically more basic than occurrents, for example Democritus, Plato, Aristotle, Aquinas, Descartes, Kant, Brentano, Geach, Strawson and modern neo-Aristotelians such as Jonathan Lowe. They have generally held the basic things to be substances, and in western philosophy they have been the majority. They take processes and events to be changes in or among substances or other continuants. On the other side are those, a minority, who regard processes as more basic than substances, for example Heraclitus, Bergson, Whitehead, Rescher, and Dupré. Occasionally philosophers treat both sides of the duality as equally basic, as does David Wiggins. After many years as a Wigginsian dualist, I now side with those who believe in the primacy of processes.

The metaphysical dispute between three-dimensionalism (endurantism) and four-dimensionalism (perdurantism) is partly skew to the question of priority. Endurantists believe that continuants are (typically) three-dimensional and lack temporal parts. Perdurantists on the contrary believe either that they have temporal parts and are extended in time (worm theory) or that they are instantaneous but stack up in succession (stage theory) (see Sider 2001; Hawley 2001). Continuants, as we conceive and think of them, lack temporal parts, and this is reflected in our explication in the previous section. For this reason, those perdurantists who say that organisms, artefacts, and so on *are* processes are best interpreted as saying that what we *thought* were continuants are in fact occurrents: that, contrary to what we might think, people, cats, trees, and mountains do have temporal parts. Whether they are metaphysically prior to, posterior to, or coeval with perdurants is a different issue. I hold that both continuants and occurrents exist and are differentiated as outlined, but that occurrents, processes foremost, are metaphysically prior.

So, total scepticism aside, there appear to be five possible metaphysical positions with regard to continuants and occurrents and which ones are basic. Here they are, with some recent representatives:

(a) There are only continuants: Brentano (1981).
(b) There are only occurrents and no continuants: Lewis (1986), Seibt (2004),[5] Bapteste and Dupré (2013).
(c) There are both and continuants are prior: Strawson (1959),[6] Lowe (1998).[7]
(d) There are both and they are equally basic: Wiggins (2001).
(e) There are both and occurrents are prior: Whitehead (1978), Rescher (2001), Simons.

---

[5] Seibt is perhaps the most radical contemporary process metaphysician, envisaging as she does a category of general processes that rejects even the endurance/perdurance distinction, as well as the particular/universal distinction: see Seibt 2008 and chapter 6 here.

[6] Strawson (1959) treats bodies as 'basic particulars' and events as secondary. Strictly, however, the asymmetry Strawson detects is one of reference and epistemology rather than metaphysics. Bodily substances could be referentially basic and still be ontologically derivative.

[7] For Lowe, substances are continuants and other things are identity-dependent on substances, but substances are (by definition) not identity-dependent on anything else.

From here on I shall overlook the distinction between events and processes and describe all occurrents as 'processes'. This allows me to smooth out what would otherwise be a small but irritating wrinkle in the history. The man widely (and correctly) regarded as the foremost modern process metaphysician, namely Alfred North Whitehead, did not in fact use the term 'process' as I use it, for a worldly denizen extended in time (see Simons 2015). For Whitehead, somewhat ironically, 'process', as in the title of the *chef d'oeuvre* of process philosophy, *Process and Reality*, stands for what he calls the 'concrescence' or becoming of an individual event; something he (again misleadingly) calls *genetic division*. This, he states explicitly, does not unfold or occur in time (Whitehead 1978: 238). The events themselves, which in his middle philosophy all have proper parts but in his later philosophy have no proper parts at all, are therefore atomic and are redubbed 'actual occasions'. They are in time but do not unfold, since they are atomic, even though they occupy or 'enjoy' a small extended quantum of space–time that comes into existence with them. The mereological analysis of regions is called by contrast *coordinate division*. This dualism of Whitehead's is unnecessary and obfuscating. By calling Whitehead's atomic events and the bigger ensembles they compose 'processes', we are able to continue describing him straightforwardly as a process metaphysician, and we can then safely ignore the mysterious non-temporal becoming what he called 'process'. Whether there are or are not atomic events as Whitehead thought can be left aside here. If there are, then processes in the standard sense are composed of such events and are causally strongly connected parts of the four-dimensional tapestry of occurrence.

## 5. Reasons to Take Processes as Fundamental

As common sense is only the vestibule to serious metaphysics, we cannot rest on it except as offering data to be properly explained and accounted for. Nevertheless, there are solid scientific and metaphysical reasons to be confident that processes not only exist but are more fundamental than substances or other continuants.

In science, processes of various kinds figure ineliminably: in relativity theory, emission, propagation, and absorption of electromagnetic radiation; in quantum theory, exchange of force-carrying bosons; in both, fluctuations in field values. Astronomy deals with star formation, evolution, and death, the formation of planetary systems, the rotation of galaxies, the occurrence of supernovae and gamma radiation bursts. Geology deals with tectonic plate movement, formation and erosion of mountains, precipitation, river capture, soil formation, earthquakes, and much more. The list can be extended indefinitely; I will mention some biological processes below.

Metaphysically, we need truth makers for propositions stating the existence of a continuant *at a time*. Truth makers have to necessitate what they make true. Both Wellington and Napoleon existed on 18 June 1815, but neither of them *necessitated* his own existence on this date, since either could have died earlier.[8] What necessitates the existence of Wellington and Napoleon on that day is the occurrence of the vital processes in virtue of which each of them was alive then (see Simons 2000b).

---

[8] By contrast, each makes true his own absolute, *untensed* existence statement.

As Ramsey (1927) sketched and Davidson (1967) later spelled out more clearly, many event and action predications are covert existential quantifications over individual events, any of which will serve to make the predication true. 'John drank coffee yesterday' may be true no matter how often John drank coffee yesterday, provided that at least one event (or episode) of John's drinking coffee occurred yesterday.

Not all objects occupy space and time in the same way—obviously, since occurrents are, but continuants are not, extended in time. Maybe universals such as being a dog or having a mass of 1 kg are numerically the same at all places and times at which they are instantiated. To make sense of the variety of occupation relations whereby $A$ occupies location $L$—extensively or intensively—we best start with one form of occupation out of which all the others can be derived by abstraction. This is the mode in which processes occupy spatio-temporal regions by being spread out across them (see Simons 2014). Note that this does not analytically entail that any process has a part corresponding to every subregion of the region it occupies: I leave open the possibility of extended simples.

Finally, and this is perhaps the most important reason, processes are causal. It is not that they have causal powers like a coiled spring: they do actually cause things to happen. Causation is arguably fundamental and irreducible: it is, as Mackie (1980) nicely termed it, the cement of the universe. We therefore need terms of causation, and these are events and processes. This also allows for a causal theory of time, one not relying on modality. There is time because stuff happens, and time is directional (at least above a certain granularity) because the causation that makes stuff happen and moves processes along is irreversible.

Causation or determination is that relational factor or species of relational factor in virtue of which one process (partly or wholly) determines or affects the probability that another process occurs. We allow partial as well as total causes, in case some events occur spontaneously or partly spontaneously. We allow negative as well as positive influence: a positive cause compels or inclines something else to happen (increases its probability, to 1 if compelling). A negative cause prevents or inhibits something from happening (decreases its probability, to 0 if preventing). Causation is not necessarily deterministic: we need to allow for uncaused (spontaneous) or partly caused processes.

Causation is called a *factor* because we do not want to posit an additional item called a causal *relation* over and above the processes. To do so would be to invite a regress: what caused the cause to cause its effect? No: processes happen. Some are caused by others, singly or in concert, determinately or inclinationally.

## 6. (Just a Few) Kinds of Processes in Biology

Biology seems at first sight to be primarily about continuants: organisms and their parts (hence anatomy). It is, however, also replete with talk about processes and their parts, because what makes the difference between living and non-living things is their vital processes, at all levels of organization. There is the molecular level: the intricate workings of cell biochemistry—that is, metabolism—through the cell life cycle, mitosis, meiosis, differentiation, and apoptosis. Then there is the complex physiology of multicellular organisms: respiration, digestion, circulation, movement and other behaviours, growth, and maturation. There is also symbiosis, birth, disease, and death. And there is the tapestry of events at the level of populations: adaptation, speciation, evolution, extinction, and so on.

## 7. Continuants out of Processes

If processes are prior to continuants, in what does their priority and the derivativeness of continuants consist, and how are they related? There is a story to be told, and since the categories of continuant and process are extremely general, it has to be extremely general too. Continuants, according to my view (Simons 2000a), are to be understood as invariant *precipitates* of a species of causal relatedness known, after Lewin (1922), as *genidentity*.[9] Genidentity pertains to the vital processes of a continuant—those in virtue of which it exists and continues to do so (this has a causal component). These processes have phases that succeed one another and, when things are going standardly, do so in an orderly way. In this context, 'orderly' means that genidentity as a relation between process phases is symmetric and transitive, so an equivalence relation. Whitehead, less illuminatingly, refers to 'social order' and 'personal order' among events, which allow us to recognize a continuant constituted by them (Whitehead 1978: 34). But while the means by which we gain cognitive access to continuants, recognizing them as the same in their successive appearances, may be termed a species of abstraction, it does not follow that continuants are abstract entities in the standard sense of being outside space, time, and the causal order. On the contrary, many physical continuants such as those instanced before are among the most paradigmatically real and concrete things in the world we experience. They have causal powers, even though they are not themselves causes. The stone against which I stub a toe cannot be faulted for concreteness; it is precisely because the event of my toe's colliding with it causes pain that I am acutely aware of it as a very solid physical body. For that reason, I prefer to say that continuants are not abstracted from processes but are rather *precipitates* of processes: they are what abides, as certain kinds of processes continue and develop.

If it is correct that there are no continuants without genidentity among certain process phases, then that explains the priority of processes. Continuants supervene on processes, though not all processes constitute or precipitate continuants. A dissipative process such as an explosion does not, since it lacks the relative stability required for a continuant. Even here, though, there may be continuants associated with the process. A physical wave such as a water wave, a sound wave, or, thinking of the explosion case, a shock wave is a continuant, not a process, since it may move and change. A water wave can travel long distances, but the material substratum—the water—in which it exists does not travel like this, but only moves locally and briefly. But, unlike more familiar substance-like material continuants, a wave typically propagates and migrates through successive material substrata. That gives it priority. The exact nature of the relationship between continuants and their vital processes

---

[9] The concept of genidentity is also examined in chapters 4, 5, 7, and 11.

apart from the fact of priority requires more consideration, however, and that is why we must look at abstraction in more detail.

## 8. Abstraction

Abstraction is a species of cognitive operation carried out by us, and probably also to some extent by animals of other species. It is very common and underlies a significant portion of sophisticated linguistic and scientific practice. Having given more detailed accounts of abstraction elsewhere (Simons 1981; 1990; 2012),[10] in this section I will focus on an example to illustrate abstraction at work, and in the next section I will apply the idea to processes.

I start from a given domain of objects, which are often called the *concreta*. Among these a special kind of relation applies: relations that constitute exact similarities in some respect. For example, suppose our *concreta* are people and the relation is *weighs as much as* (say, to the nearest 10 g, rounded down). This relation has the following logical properties: everyone weighs as much as him- or herself; and, if two people each weigh as much as a third, then they weigh as much as each other. When people weigh as much as each other, we say they have the same *weight*, and these weights are not concrete but abstract, they are the *abstracta*. The transition from saying that $x$ weighs as much as $y$ to saying that $x$'s weight is the same as $y$'s is typical, and so commonplace we hardly notice it.

Here are a few samples of such transitions from different areas, where the domain is named, the *concreta* are mentioned in a relational statement to the left of the double arrow, and the *abstracta* are named in the identity statement to the right of the double arrow:

(i) People on a given date: $x$ is as tall as $y$ $\Leftrightarrow$ the height of $x$ = the height of $y$
(ii) Straight lines in space: $x$ is parallel to $y$ $\Leftrightarrow$ the direction of $x$ = the direction of $y$
(iii) Bodies: $x$ is as massive as $y$ $\Leftrightarrow$ the mass of $x$ = the mass of $y$
(iv) Collections: there are as many $a$ as there are $b$ $\Leftrightarrow$ the number of $a$ = the number of $b$
(v) People: $x$ earns as much in a year as $y$ $\Leftrightarrow$ the annual income of $x$ = the annual income of $y$

Relations with the logical properties that sustain such transitions are called equivalence relations, and they typically introduce new abstract terms on the right-hand side, hence the term *abstraction*. Frege described abstraction transitions as 'recarving the content' of the sentences (Frege 1951: § 64). Though in the end he preferred a different way to work with *abstracta*, we shall stay with abstraction transitions.

On their own, these do not get us very far. However, when properties and relations can be found that are *invariant* under the equivalence in that any items equivalent to items having the property or relation themselves have the property or relation,

---

[10] The theory of abstraction follows precedents in Weyl (1949), Lorenzen (1962), and Dummett (1973).

then a vocabulary can be built up to deal with the *abstracta* without necessarily mentioning their concrete basis. For example, *is heavier than* is invariant under *weighs as much as*, so

> Jules is heavier than Jim ⇔ Jules's weight is greater than Jim's
> Jules is 70 times heavier than the standard kilogram ⇔ Jules's weight = 70 kg

It is important that, when we move from the left to the right, there is a shift in sense: Jules's *weight* is not heavier than Jim's, it is Jules who is heavier than Jim; his weight is *greater* than Jim's. And while Jules *weighs* 70 kg, his weight does not weigh anything: it *is* 70 kg. This sense adjustment will be important below. As the example suggests, abstraction and invariance are at the basis of many of those measurable characteristics we call *quantities*.

## 9. Abstracting to Continuants

Consider now the genidentity relation on process phases (temporal parts of processes or process complexes). Then

> Processes and their phases: $x$ is genidentical to $y$ ⇔ the continuant of $x$ = the continuant of $y$

This abstracts from particular spatio-temporal location: different genidentical phases will have different temporal locations and may have different spatial locations. But this does not take continuants out of space and time, because, defining

> for all processes $x$ and $y$: $x$ is a temporal part of $y$ $=_{Df.}$ $x$ is a part of $y$ and any part of $y$ that exists only when $x$ exists is a part of $x$

and,

> if $x$ is genidentical to $y$, then the continuant of $x$ (and $y$) exists when $x$ exists and when $y$ exists, and only exists when some $z$ exists such that $z$ is genidentical to $x$.

A continuant exists when and only when its constitutive processes are going on. Since a non-instantaneous process has disjoint temporal parts, any process constituting a continuant has parts that exist at some times and not others during the course of the whole process. For example, being a Tuesday part (when the process is not confined to Tuesday) is not invariant under genidentity. Therefore continuants do not and cannot have temporal parts, even though their lives are limited by the temporal extent of their constitutive processes.

We can likewise define the spatial location of a continuant at a time as the spatial component of that temporal part of its constitutive processes. In general, and relativizing suitably, a continuant's spatial location will be different at different times. In popular parlance, it will *move*. More generally, if a property or relation of a process that constitutes a continuant is not invariant under genidentity, then the adjusted property or relation will not belong to the continuant. If the property is one of a temporal part of the process, we may ascribe the adjusted property to the continuant only by relativizing or indexing it to the time in question. This is why 'at $t$' locutions are needed for continuants; and it helps to explain what their changing consists in.

Finally, a continuant is causally involved insofar as its constitutive processes and their parts are causes and effects of other processes that are not parts of its constitutive processes. So, while physical continuants are supervenient derivative precipitates, they are not abstract entities.

## 10. Modal Properties

Occurrents are subject to mereological and locational essentialism, that is, where and when they are and what their parts are is all essential to them. This cannot be argued here, but it is metaphysically advantageous because it supports a causal theory of time. Modal identity conditions for continuants, however, are considerably more flexible than those of their actual, constituting processes: their location, lifespan, properties, and relations are much more subject to accident and contingency because their dependence on their constituting processes is generic rather than rigid.[11] So they support genuine transworld identity, whereas when we talk about an occurrent using a non-rigid definite description, such as 'the Battle of Waterloo', to the extent that we allow for contingency, we are talking about counterparts.

## 11. Consequences for Biology

Organisms, as well as their parts and collectives, are metaphysically secondary to the processes that constitute and sustain them. This means *you*, both as a continuant human organism and (even more dramatically) as a person. Humans require sustaining vital processes in order to exist, but in certain pathological cases these are insufficient to sustain personhood, which is a far more fragile status than merely being alive. Neonates, comatose humans, and highly senile humans are not, then, persons (sometimes they are temporarily, sometimes not).

Since the natural grain of our language is towards talking about continuants, we need to work on developing vocabulary, metrics, and data representations for processes, their parts, their features, their relations, and their quantities.[12] Because of this natural grain, we cannot jettison our dualistic vocabulary without loss of information and intelligibility, so it is quite right to carry on talking as now and to design biological and medical databases with dualistic vocabulary.[13] Nevertheless, apart from continuously reminding ourselves of the metaphysical priority of processes, it is good to learn to think more in process terms and to reduce our dependence on the Aristotelian heritage.

---

[11] For the distinction between strong (rigid) and weak (generic) dependence, see ch. 8 in Simons 1987.

[12] This extends even to metrology. The SI unit of mass, the kilogram, is currently defined in terms of a body, a platinum–iridium sphere in Sèvres, France. Proposals for a revision basing the kilogram on Planck's constant, whose units are Joule seconds, the unit of action, are therefore a welcome step in the direction of a dynamic, process-based metrology.

[13] As advocated by Grenon and Smith 2004 and implemented in basic formal ontology (BFO): see Arp et al. 2015.

# References

Arp, R., Smith, B., and Spear, A. D. (2015). *Building Formal Ontologies with Basic Formal Ontology*. Cambridge, MA: MIT Press.
Bapteste, E. and Dupré, J. (2013). Towards a Processual Microbial Ontology. *Biology and Philosophy* 28: 379–404.
Brentano, F. (1981). *The Theory of Categories*, trans. R. M. Chisholm and N. Gutterman. Dordrecht: Nijhoff.
Davidson, D. (1967). The Logical Form of Action Sentences. In N. Rescher (ed.), *The Logic of Decision and Action* (81–95). Pittsburgh: University of Pittsburgh Press. (Reprinted in D. Davidson, 2001, *Essays on Actions and Events*, pp. 105–21 Oxford: Clarendon.)
Dretske, F. (1967). Can Events Move? *Mind* 76: 476–92.
Dummett, M. A. E. (1973). *Frege: Philosophy of Language*. London: Duckworth.
Frege, G. (1951). *The Foundations of Arithmetic*, trans. J. L. Austin. Oxford: Blackwell.
Grenon, P. and Smith, B. (2004). SNAP and SPAN: Towards Dynamic Spatial Ontology. *Spatial Cognition and Computation* 4: 69–103.
Hawley, K. (2001). *How Things Persist*. Oxford: Oxford University Press.
Johnson, W. E. (1921). *Logic*, vol. 1. Cambridge: Cambridge University Press.
Leonard, H. S. and Goodman, N. (1940). The Calculus of Individuals and Its Uses. *Journal of Symbolic Logic* 5: 45–55.
Lewin, K. (1922). *Der Begriff der Genese in Physik, Biologie und Entwicklungsgeschichte: Eine Untersuchung zur vergleichenden Wissenschaftslehre*. Berlin: Bornträger/Springer. (Reprinted in C.-F. Graumann, ed., 1983, *Kurt-Lewin-Werkausgabe*, vol. 2, Bern/Stuttgart: Huber/Klett-Cotta.)
Lewis, D. K. (1986). *On the Plurality of Worlds*. Oxford: Blackwell.
Lorenzen, P. (1962). Equality and Abstraction. *Ratio* 4: 77–81.
Lowe, E. J. (1998). *The Possibility of Metaphysics: Substance, Identity and Time*. Oxford: Oxford University Press.
Mackie, J. L. (1980). *The Cement of the Universe: A Study of Causation*. Oxford: Clarendon.
Moore, G. E., Johnson, W. E., Hicks, G. Dawes, Smith, J. A., and Ward, J. (1916). Symposium: Are the Materials of Sense Affections of the Mind? *Proceedings of the Aristotelian Society* 17 (n.s.): 418–58.
Peirce, C. S. (2000). *The Writings of Charles Sanders Peirce: A Chronological Edition*. Vol. 6: *1886–1890*. Bloomington: Indiana University Press.
Ramsey, F. P. (1927). Facts and Propositions. *Aristotelian Society* 7: 153–70. (Reprinted in F. P. Ramsey, 1990, *Philosophical Papers*, ed. by D. H. Mellor, pp. 34–51, Cambridge: Cambridge University Press.)
Rescher, N. (2001). *Process Philosophy: A Survey of Basic Issues*. Pittsburgh: University of Pittsburgh Press.
Seibt, J. (2004). Free Process Theory: Towards a Typology of Processes. *Axiomathes* 14: 23–57.
Seibt, J. (2008). Beyond Endurance and Perdurance: Recurrent Dynamics. In C. Kanzian (ed.), *Persistence* (pp. 133–65). Frankfurt: Ontos.
Sider, T. (2001). *Four Dimensionalism*. Oxford: Oxford University Press.
Simons, P. M. (1981). Abstraction and Abstract Objects. In E. Morscher, O. Neumaier, and G. Zecha (eds), *Essays in Scientific Philosophy* (pp. 377–94). Bad Reichenhall: Comes.
Simons, P. M. (1987). *Parts: A Study in Ontology*. Oxford: Clarendon.
Simons, P. M. (1990). What Is Abstraction and What Is It Good for? In A. Irvine (ed.), *Physicalism in Mathematics* (pp. 17–40). Dordrecht: Kluwer.
Simons, P. M. (2000a). Continuants and Occurrents. *Aristotelian Society* 124: 78–101.

Simons, P. M. (2000b). How to Exist at a Time When You Have No Temporal Parts. *Monist* 83: 419–36.
Simons, P. M. (2012). Abstraktion ohne abstrakte Gegenstände. *Zeitschrift für philosophische Forschung* 66: 114–29.
Simons, P. M. (2014). Where It's at: Modes of Occupation and Kinds of Occupant. In S. Kleinschmidt (ed.), *Mereology and Location* (pp. 59–68). Oxford: Oxford University Press.
Simons, P. M. (2015). Alfred North Whitehead's *Process and Reality*. *Topoi* 34: 297–305.
Strawson, P. F. (1959). *Individuals: An Essay in Descriptive Metaphysics*. London: Methuen.
Weyl, H. (1949). *Philosophy of Mathematics and Natural Science*. Princeton: Princeton University Press.
Whitehead, A. N. (1978). *Process and Reality*. Glencoe: Free Press.
Wiggins, D. (2001). *Sameness and Substance Renewed*. Cambridge: Cambridge University Press.

# 3

# Dispositionalism
## A Dynamic Theory of Causation

*Rani Lill Anjum and Stephen Mumford*

## 1. A Received Orthodoxy

David Hume's biggest influence on the philosophy of causation was not the notion of constant conjunctions, nor was it the theory of counterfactual dependence. As is well known, these two different accounts of causation are both to be found in the first *Enquiry*—indeed, within the very same paragraph (Hume 2007 [1748]: VI, 56). There is, however, an idea of an even greater generality and at a higher level of abstraction that has pervaded much more post-Humean thinking about causation than simply regularity and difference-making theories. This idea has become so orthodox that in most cases it is simply assumed as the required starting point without any discussion. The unexamined assumption is that causation is a relation that relates two distinct events, objects, or existences.

The linguistic expressions of English seem to reinforce the assumption. The lighting of the fuse caused the explosion; dropping the vase caused it to break; ingestion of arsenic caused the victim's death; colliding with the iceberg caused the ocean liner to sink; and so on. These causal statements are all conducive to the received orthodoxy as it's easy to see within them a form *aRb*. *Prima facie*, they describe one event followed by another where the two events are causally related. The causal relation is clearly not merely temporal succession, for an event preceding another is not always its cause. There is thus a whole philosophy of causation that is about correctly identifying what the correct causal relation is that holds between some but not all ordered pairs of events. Some say the relation is an INUS condition (Mackie 1993 [1965]), others that it is a counterfactual dependence involving similarities between possible worlds (Lewis 1986 [1973]).

The orthodoxy immediately suggests a disunited view of the world: that it consists in a succession of distinct, wholly discrete events. The world might not be that way, though. Suppose that things are more unified, dynamic, and continuous and that change occurs in a smooth and gradual processual way. What if, furthermore, this latter view is the one needed in biology to account for causation within, and affecting, living organisms? A problem, we argue, is that, if one begins from the idea that causation has to be a relation between two discrete and 'static' events, then it may already be impossible to formulate a satisfactory theory of what causation is and how it works. The biological case makes this quite apparent. The assumption splits the world asunder into distinct, self-contained fragments, and then tries to find a relation that would stick them all back together again. It is not clear that any relation can do this job. Even if it could, would the world with which we are left—a Frankenstein-world of stitched together pieces—have all the features we require? Would it really be a world of continuity, fit for living biological processes?

Hume urged us to accept a world that was loose and separate: conjoined in some cases, but never connected (Hume 2007 [1748]: VII, 54). He proposed that causation consisted in a complex relation holding between events that consisted of (i) constant conjunction or regularity, along with (ii) temporal priority of cause over effect and (iii) contiguity (Hume 1888 [1739]: I, iii, 14). This gave a causal relation that was enough to explain why we have a psychological habit of inferring the effect from the cause without allowing anything like necessity, which would make the world more connected and less separate. Others have since opted for a stronger relation between cause and effect (e.g. Mellor 1995). But it is the idea of the unconnected world that, we argue, is already damaging enough, no matter what relation one offers to repair it.

There are many different ways in which this Humean view could be developed, and not every way will have all the features that we describe. But let us outline a plausible candidate for the orthodoxy. In the world there is a series of events or states of affairs (Hume says objects). Some of these occur at the same time and many more occur at different times from each other. A ball is blue, for instance; a metal rod is 1 m long; the same metal rod has a temperature of 20°C; a sugar cube is placed in liquid; the same sugar cube dissolves; a rock is pressed against a pane of glass; cryptosporidium is present within a certain human organism; that same human organism is unwell. A key move within empiricist philosophies is to say that in principle any event can follow any other (Hume 2007 [1748]: IV, i). We cannot know a priori what will follow what, as this is a contingent matter. Between any two distinct events, where one precedes the other, those events may be causally related or they may not be. It is thus an empirical matter to discover whether they are causally related.

What it is for one event to be causally related to another is the business of philosophy to decide. Here is where the details can vary. We have already seen Hume's threefold account. Lewis's (1886 [1973]) theory is that, for event A to be the cause of event B, A and B must both occur; and it's also the case that, had A not occurred, B would not either. This amounts, within to the modal realist account, to its being such that, in all the closest possible worlds (in terms of overall similarity) to the actual world in which A did not occur, B did not occur. Suppes (1970) develops the idea that a cause is something that raises the probability of an effect: it has a positive statistical relevance to it. In Armstrong's view, one fact causes another just in case doing so instantiates a contingent law of nature that facts of the first kind necessitate facts of the second kind (Armstrong 1978: ch. 24). In Salmon's (1984) view, a conserved quantity is passed from the cause to the effect. In these theories, there is nothing in the essence of causally related things that makes them be causally related. Whether or not they are just depends on the contingencies of the world, which it is the business of science to discover. We do not have the space here to do full justice to any of these views, or to consider the many other theories of causality on offer.

This orthodox picture, while widespread, should be challenged, we argue. In the first place, we can question its appeal on empirical grounds, in particular on the issue of how well it suits cases of causation that require continuous change, biology being such a case. Second, we challenge also the metaphysical adequacy of this account. In particular, if one prefers a metaphysics of causal powers to a Humean mosaic of distinct existences, one should have a more dynamic and process-like view of things. As others have noted before, particularly Martin (2008: 46), this would seem to require an overthrow of the old two-event model.

## 2. Causation in Biology

Let us now consider some of the lessons that can be drawn from biology. First, biology is notable for its requirement that organisms be in a state of continual change. Movement, flux, development, and process seem essential to life in that to cease to change means death. An explanation for this may be that we have to account for living organisms in terms of functioning (Rosenberg 2008: 513), which has to involve activity. Where different animals have a heart, for instance, the architecture or physical features of the organ does not matter as much as the active, life-sustaining function that it is to perform. This is not to deny that there are cases of relative stability that are important for life too, such as body temperature for humans being within a narrow range close to 37°C. But this stability is also one that is understood to be caused by underlying functions, including homeostatic processes (sweating to cool down; shivering to warm up). An organism's persistence does not depend on being kept in a single state but on being maintained through numerous processes: the continual beating of the heart and breathing of the lungs, a cycle of dehydration, hydration, and absorption of water, metabolism, the ingestion of food and its conversion to energy and waste product, movement around an environment that provides life-sustaining resources, synthesis of vitamins from a variety of sources (such as the sun), sentient and sapient activities, and so on.

To live is thus always to be active to at least some degree. Are there any exceptions? One might argue that it is possible in principle for an organism to undergo a period of motionlessness, such as being in deep hibernation, and yet still be living. But this is not really an exception. In hibernation, body temperature drops and the heart and breathing rates slow considerably; but they do not stop completely. Metabolic activity remains. Thus there is a lowering of degree of motion, but not a cessation of change. This is true even in the extreme cases of suspended animation or cryogenization.

Life suggests a need for *continuous* change. Even the slightest moment of changelessness results in death. Yet the standard assumptions behind many theories of causation accommodate this poorly. Causation is often depicted as relating *changeless* relata—static events or states of affairs—that are wholly temporally distinct. As Bennett (1974) argued, Hume committed to a discrete view of space and time so that causes and effects would not be connected. On such a view, there are smallest units of time and space, which must be indivisible and point-like. This commitment to space–time points survives in Lewis's metaphysics (Lewis 1986c: ix–x). Absurdities arguably accompany the view, though, as it is hard to see how there can be temporal succession in a series of unextended points unless there are temporal gaps between them. A requirement of temporal gaps would be contrary to the initial analysis, however. Even setting that aside, and granting that a succession of point-like times is possible, the view entails also that any property instantiated at a discrete time must be changeless or static. This is because change requires duration; and a moment of time, on this view, has none. The subsequent theory of causation must therefore posit a relation between static relata.

This gives us an account in which change is supposedly produced somehow by an ordered succession of changeless states, as in perdurantist theories of persistence (Lewis 1986b: 202), which will be discussed further below (see also Mumford 2009). The challenge from biology is how well this sits with its requirement for continuous change. Wouldn't perdurantism tell us that organisms are changeless at each point in time, merely progressing from one changeless state to another and then another? It is arguable that this makes change something of an illusion, like the one created through the rapid succession of projected still images, as in old-fashioned analogue film, which shows twenty-four still frames per second.

In addition to this need for continual change, let us note some of the other features of causation in biology, though we will not dwell on these at length. One is that there is a hierarchy of processes occurring simultaneously but on different timescales. Evolution is a long process compared to a complete life cycle, which in turn is much longer that the causal process in which damaged flesh is repaired. Compared to all these, a sneeze is very sudden. A fast process such as a sneeze can, however, play a significant role in a slower evolutionary process. An ability to rapidly expel irritants from the nose, and which can be passed on to offspring, provides an evolutionary advantage. Can we have a single theory that causally explains such overlapping hierarchical processes?

Next, consider the versatility of causes in biology. Genes can take on multiple roles, which is advantageous to survival as they provide back-up mechanisms that ensure the maintenance of vital processes. A well-known example is a famous study involving the Drosophila, which had its eyeless gene removed and subsequently lost its eyes. Within a few generations Drosophilas got eyes again, though without having the gene back (Webster and Goodwin 1996: 87; Skaftnesmo 2005: 72; see also Leiserson et al. 1995). Other genes had apparently taken on a role that was needed by the organism as a whole.

Notably, no biological system is 'closed', and this is an essential feature of its persistence (see Lewontin 2001 [1983] and chapters 7, 8, and 18 here). An organism might be 'organizationally' closed (Mossio and Moreno 2010), but this does not mean that it is causally closed, receiving no external inputs. Symbiosis is required, for example (see Gilbert 2014 and chapters 1, 5, 9, 10, and 15 here). Bacteria are needed by humans and other animals to perform certain functions such as digestion, and it seems at times that there are no clear boundaries between organisms. If an animal cannot survive without the bacteria in its gut, there is a sense in which the bacteria are part of the animal. An organism as a closed system would be, again, a dead organism. Air, sun, food, and symbionts are all needed for regular functioning. In contrast, some standard theories of causation entertain closed-system idealized worlds, such as where there is just one isolated object that collides with another.

We also find in biology an inapplicability of the mono-causal model, which is where there is one cause for one effect. The two-event philosophical view of causation encourages such a model. Instead, in biology we are used to seeing multiple causes of any change: a feature known in genetics as polygeny (Molnar 2003: 194). This causal complexity is emphasized by complex adaptive systems (CAS) theory (see Holland 1992) and by developmental systems theory (DST) (see Oyama et al. 2001 and chapter 11 here), both of which are critical of overly simplistic one-cause models, especially in genetics.

We could go on and mention features such as causation by absence (for instance, where absence of water can kill a plant or animal), the contest and counterbalancing of contrary causal powers, and the multiple causal explanations that seem available for biological phenomena. But we have said enough already to illustrate the complexity that biology must account for in order to explain how life is sustained through ongoing processes of continual change. The standard two-event model, we claim, does not handle this complexity well, offering us just more and more of the same discrete two-events causally related. It was because of its complexity that we took biology as the paradigmatic science in earlier work (Mumford and Anjum 2011: ch. 10)—instead of physics, for instance, which often offers mathematized descriptions of artificially closed systems. Although this may be useful in some branches of physics, it is difficult to draw general lessons for causation in other sciences.

It is one thing to criticize the standard orthodoxy on causation or claim that it is ill suited to some domain. But what, it may be wondered, is the alternative? If there was no credible option but the standard picture, then it would need to stand, perhaps with some adaptation or amendment. However, there *is* another option: a dynamic theory of causation that is a better fit with the needs of biology. We will now go on to describe one such account, namely *causal dispositionalism*.

## 3. Dispositions and Processes

There is a close connection between dispositional approaches to causation and process metaphysics, though this has not always been highlighted (on process theories in general, see Whitehead 1978 [1929] and more updated views in Sellars 1981, Rescher 1996, and Seibt 2008). One barrier has been that dispositionalists are often working within an Aristotelian tradition, which had a basic substance–attribute ontology. Process metaphysics is often antagonistic to the substance–attribute view (e.g., Bickhard 2011). If we look at Aquinas's account of powers (see Feser 2014: chs 2–3), however, we find something that looks more in sympathy with a process view.

A dispositional account of causation should reject the old stimulus–response model of how causal powers are activated. Such a view comes too close to the two-event model, which we have said should be overturned. Instead, Martin's notion of mutual manifestation (Martin 2008: 48–51) serves us better, though not all dispositionalists have embraced this conceptual innovation. Causation occurs, on Martin's view, when two or more reciprocal disposition partners come together to produce a mutual manifestation. Such a manifestation is an effect produced by powers acting together where none of the powers could have produced the same manifestation by acting alone. Our account of mutual manifestation develops Martin's original presentation in emphasizing the processual nature of the manifestation. Martin originally illustrated mutual manifestation with the example of two triangles that come together to form a square jointly. We think there are a number of disanalogies between Martin's putative illustration and real causation (Mumford and Anjum 2016). More typically, when mutual manifestation partners are together, it takes time for them to have their full effect, as Kant (1929 [1781]: A203) already noted. During this time, there is a continual development of change—that is, a *process*—that eventually results in a final point, unless interrupted. Beyond this final point, there is, then, no further change unless new mutual manifestation partners are added to the situation.

A simple illustration is the inanimate case of sugar dissolving in a liquid, such as water. As soon as a sugar cube and water come together, the gradual process of dissolving begins. It will continue unless it is interrupted in some way, such as if the container is overturned, or the water is rapidly evaporated. Without interruption, the process continues until it reaches its natural end point, which is for the sugar cube to be dissolved and held entirely in solution. On standard models of causation, we are often obliged to view such a case asymmetrically: most typically we might tend to think of the active liquid dissolving the passive sugar. But Martin is right to emphasize the symmetry—the mutuality—of the change. Being in a causal partnership changes the sugar cube; but it also changes the liquid, which is a sweet solution by the end of the process. This seems to cohere with a principle of reciprocity in causation that is attractive on independent grounds: a cause brings about a change in the effect, but there is a reciprocal change in the cause when the effect occurs (for a similar approach, see Ingthorsson 2002).

The dispositional aspect is important in this account. Dispositions are properties of things; arguably they constitute *the* properties of things (Mumford 2008). They are a thing's causal powers towards manifestations, those manifestations being conditional upon the empowered particular meeting the appropriate reciprocal disposition partners. Crucially, however, these powers are not *just* for the end point of the causal transaction, such as the property of being dissolved that is the end point of solubility. The power is for the whole process that takes a sugar cube from being in liquid to being dissolved. It matters, to count as solubility, how the power gets to its end state of being dissolved. If it did so via some deviant causal chain, for instance, we would not be convinced that this was a case of genuine solubility.

Contrast this with the two-event model. We might be persuaded to understand, as the cause, the sugar cube's being placed in water and, as the effect, its being dissolved. This assists the idea that causation is a relation between two relata. But what happens between the first relatum and the second? It is not as if the sugar goes instantly from being in liquid to being dissolved. There is a gradual natural process of continuous dissolution in-between. If we examine the sugar at any stage intermediate between its being first in liquid and its being wholly dissolved, we should find that natural process going on. Vitally, at any such intermediate point the sugar and the liquid are still together, changing and disposing towards the end result. The dispositionality is thus found throughout the whole process, no matter how small a part, as Aquinas (*Summa contra gentiles* III, 2) noted.

The dispositional account also suggests a greater connectivity between cause and effect than the standard model. We saw that the orthodox view begins with an assumption that the world's components are discrete, self-contained units, like the tiles in a mosaic. The search for causation is, then, for something that ties some but not all of these discrete units together, though they remain nevertheless distinct existences. Causes and effects are more closely connected on the dispositional account, in the following sense. The placing of the sugar cube in liquid clearly occurs some time before the cube is entirely dissolved, and this appears to support Hume's idea that causes are temporally prior to, and thus disconnected from, their effects. On his view, there can be no temporal overlap at all between causes and effects. It is not even the case that the last moment of the cause can be simultaneous with the first moment of the effect. In contrast, the dispositional view is favourable to simultaneity between the immediate cause and the effect. As soon as reciprocal disposition partners are together, the process that is their mutual effect begins. It takes time to fully unfold but, until the effect is wholly and finally realized, some part of the cause is still present. Again, we can think of the cause as the being together of the water and the undissolved sugar (the cause is not the sugar's being placed in water, which is just the story of how it got there). This cause is in place as long as there is some undissolved portion of the sugar in the liquid. Once it has all gone into the solution, the cause is gone, the process is ended, and the effect is complete. We thus have a temporally extended causal process where the cause is entirely simultaneous with the effect.

This idea is, again, Aristotelian in origin (Aristotle, *Physics* VII, ch. 2), but associated with a metaphysics of substance causation. It trades on the point that, for the cause to bring about the effect, they must exist at the same time. For, if the cause no longer existed, how would it be productive of the effect? (Russell 1992 [1913]: 200 invokes the same argument for different purposes.) Causation looks non-simultaneous, then, only because we conceptually isolate some initial early segment of the causal process and its end point, as if we had a simple relation between the earlier and the later. The problem of causal connection between existences at different times is problematic for any part of the causal process. How could the early part of the process, if it is over, finished, and gone, produce something at a later time if it no longer exists at that later time? Cause and effect are also more closely connected once one couples the powers account with a dense theory of space–time, which we will mention again later on. Although the argument originated in a substance metaphysics, then, it is nevertheless applicable to the more dynamic view proposed here. So applied, it suggests that, rather than being a relation, causation is best understood as a unified unfolding process in which there is gradually less and less of the producing and more and more of the produced.

Can this dispositional account be applied to cases of causation in biology? It should already be clear, we hope, that it can. Consider the process of fertilization, pregnancy, gestation, and child birth in humans, for instance. The two-event model will set the case up as one where an act of sexual intercourse was a cause of a child's being born nine months later. We can hardly deny that; but this superficial articulation overlooks a great deal of complexity that is essential to the actual causal story. In reality the full nine-month process involves a series of different subsidiary processes, sometimes overlapping or involving thresholds where a subsequent stage in the overall process is triggered. During pregnancy there are a number of bodily changes, such as morning sickness and fatigue as the embryo takes hold in the womb, changes in the breasts as they prepare to be ready for feeding, changes in physiology as the placenta grows and diverts nutrition to nurture the growing foetus, processes in which the cervix is gradually softened, ready for dilation, and so on. Some of these processes are short-lived and are small parts of pregnancy as a whole. Others, such as hormonal changes in the mother and growth in the foetus, may last for virtually the whole duration. It is clear that this is a natural process that can run its course and reach an end point, but one that is also capable of interruption, either naturally or artificially. Not all conceptions result in births. There is a tendency or disposition for them to do so and this would remain so even if statistically only a minority of pregnancies ended in births. We would argue that the notion of disposing is not reducible to mere statistical matters, as some tendencies are weak and thus manifest themselves rarely. Consider a drug that has only a weak disposition towards some side effect, manifesting itself only in 1 out of 10,000 cases. We can also see that the notion of mutual manifestation is crucial. First, it is relevant with respect to conception, which requires a sperm and an egg to come together and be, both, changed by their interaction. Second, we can see the mother and the foetus as undergoing their own extended processes of reciprocal changes.

To understand how there can be multiple, interacting processes that occur within a bigger nine-month complex biological process such as pregnancy, we need an understanding of how individual simultaneous causes and effects are capable of forming a temporally extended chain. We cannot deny, for instance, that conception occurs before birth, so how does this square with the alleged simultaneity of causes and effects? It is easiest to explain this with two figures. In the first (Figure 3.1), where temporal duration is directed from left to right, we see a single process of change where the cause gradually merges into the effect. We move from only a small part of the process being the effect to the whole of it being so. This could fit the example of the sugar dissolving in water, or a biological case such as a hormonal change producing a change in the breasts of a pregnant woman.

The diagonal line we have drawn indicating the division between the part of the process that is the cause and the part that is the effect is a straight one, which can be taken as indicative of a very steady, linear process of change at a stable rate. In reality, many processes can produce changes in irregular and non-linear ways, there being only a small part of the effect manifested until a threshold is reached. This is the case, for instance, if a rise in temperature produces perspiration but does so to a disproportionally greater extent once a certain threshold is passed.

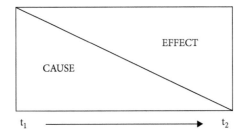

**Figure 3.1** Causation depicted as a gradual process

Given that one such causal process involves a change, that process can result in the realization of new emergent properties (for our preferred account of emergence, see Anjum and Mumford 2017). So one process might be the raising of temperature, where the change is a greater quantity of some measured property. Another process might involve the acquisition of a new property, such as a sugar cube's becoming dissolved or a woman's becoming pregnant. On our account, such properties are to be understood as causal powers (Mumford and Anjum 2011: 124–126). Thus processes can result in the realization of new powers, which are then available to form other mutual manifestation partnerships (of two or more powers). We therefore get temporal duration in causation in two ways. One is simply where a single causal process takes time to fully unfold and develop to its natural end point. The other is where the changes of properties in an unfolding process—either at the end of that process or at some point during it—form mutual manifestation partnerships with other powers and initiate new processes. This is how chains of causes can be connected over time, while each link in the chain involves only simultaneity between cause and effect. Thus an act of sexual intercourse that results in sperm ejaculated into the vagina may be seen as one discrete process. The contact between sperm and egg is the start of another, which then also starts a process of hormonal change within the mother, embryonic development, and so on. We thus have a situation as depicted in Figure 3.2.

This allows us to answer Marmodoro's (2013: 550) concern, which was raised against our account of powers. She says that we take the manifestation of a power to be another power, while she thinks that the manifestation is a different stage of the same, first power. But in a process view it can be both. Given that the manifestation of a power produces a continuous change in properties and that properties are understood to be powerful, the properties produced by causes are both the end point of the original power and a new power, available to form a new reciprocal partnership if circumstances allow. Finally, on this topic of extended chains, we have explained elsewhere how sometimes (though not always) causation travels down such a chain, which makes it technically a non-transitive matter (Mumford and Anjum 2011: 167–73); but we will not repeat the detail of the argument here.

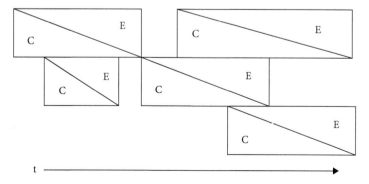

**Figure 3.2** Overlapping causal processes

What of the other features we found to be particularly prominent in biology? We saw that organisms can be regarded as open systems subject to causal interactions with things outside their own bodies, with multiple causal roles for the same genes, complexity, polygeny, with competing and sometimes counteracting powers, and so on. We hope it is clear that the dispositional account can handle such cases.

We allow that mutual manifestation partnerships can consist in any number of reciprocal powers. The two-event model posits one cause for one effect, and it is vital that it does so in order to offer a coherent notion of regularity and/or counterfactual dependence, which are ways the causal relation has been outlined. In the dispositional account, the complexity of causation is embraced, which makes the account compatible with multicausal models in biology such as DST. The biological process of pregnancy is not just a matter of a sperm meeting an egg and fertilizing it. Multiple other powers come into the account. The egg would not develop into a foetus, for instance, without the presence of a hospitable environment, which offers nutrition and the appropriate hormonal triggers. Because the pregnant organism is an open system, further causal powers can enter the picture, either assisting with the pregnancy (folic acid) or hindering it (tobacco smoke inhalation). The relevant operating causal powers can thus be many. Multiple causal roles are allowed for those powers, because it is clear that the same power, when partnered with different others, can produce different effects (what Molnar 2003: 194 calls pleiotropy). Consider the healthy and nutritional powers of the proteins in peanuts, for instance. When those same powers interact with the bodies of peanut allergy sufferers, they can result in a histamine immune response, sometimes producing severe anaphylaxis or even cardiac arrest.

Given that powers can operate in a non-linear way, we see that their operation is also very sensitive to their strength or degree. Thus a specific quantity of a drug such as paracetamol disposes towards pain relief. Taken in a larger quantity—an overdose—the same power can dispose towards pain. The production of dynamic equilibrium states, often so important for proper biological functioning, can also be explained in the theory. It is possible that two powers that are exercised and that tend in opposite directions can cancel each other out. This may involve a homeostatic feedback mechanism. A so-called zero resultant power will thus tend towards stability; that is, towards a situation in which there is no overall change. While none of the component mechanisms are static, their joint action thus produces relative stability (see Dupré 2014: 15 on the stable/static distinction in biology). Body temperature, levels of hydration, vitamin levels, cell renewal, and so on, all have an element of this kind of dynamic equilibrium in their operation. Human absorption of vitamin C (ascorbic acid), for instance, is regulated by a variety of transporters (SVCTs and GLUTs). When intake is high, these transporters lower the rate of absorption; but, when intake is low, absorption increases markedly. The result keeps a body's vitamin levels within a stable range, despite short-term fluctuations in intake.

## 4. Static or Dynamic?

Although the term 'event' is used, and the standard model typically takes events to be the relata of causal relations, those events are occurrences more than changes, in a sense that we will now try to make clear. The suggestion is that a static conception of events is being used (Kim 1976; Lewis 1986a). The standard model is sometimes accompanied by a perdurantist underlying metaphysics where change is something to be explained—perhaps by transitive causal chains—and thus the relata of causal relations have themselves to be changeless (Lewis 1986b: 202–4; Sider 2001; for an overview of the whole debate, see Hawley 2001, especially ch. 3). We interpret this as a discontinuous view of nature, because change is analysed as a mere succession of discrete, self-contained objects, parts, or stages. The perdurantist begins with a metaphysics of static, temporal parts and then tries to explain change through the way in which those parts are arranged. The alternative view is to accept change as basic—something to be found in every segment of a process, no matter how small—and then explain stability, perhaps, as we have already mentioned, as an equilibrium created by counterbalancing powers.

This is the sense in which dispositionalism offers us a dynamic theory of causation. Processes have no motionless parts. Indeed, in no real sense does a process have parts at all; for they can be formed only through the abstraction of a viewer who considers the process. The process is, in reality, an indivisible unity. An event under this conception becomes itself processual. The contrast is with fact-like conceptions, such as those offered by Kim and Lewis. An event, on those views, consists of a property's bearing a particular at a time, such as a ball's being blue at a particular moment. Because no change is needed for this to be the case, it is perhaps better understood as a Tractarian fact or state of affairs (Armstrong 1989). This conception explains why the standard model will usually specify its relata in a non-dynamic kind of way. Causal relations hold between being in water and being dissolved, between being subject to a certain force and taking a new position, between receiving an impact and being broken. These relata do not automatically or obviously involve changes themselves; rather, the change occurs only through the relation of causation that transports us from the cause to the effect.

More naturally, however, we think of events as happenings, and thus as already involving changes. Heating an iron bar causes it to expand, for example, where being heated is what Lombard (1986: 105) calls a dynamic property: one that has to involve change and thus cannot be instantiated at an unextended point in time. However, being heated requires more than just that a certain temperature be realized. What is essential is that there be a direction in the change of temperature—that it be moving from lower to higher temperature rather than vice versa. Thus being heated must involve more than a timeless instant; it must involve a temporally extended period during which there is change. Exactly parallel reasoning applies to the effect, the expansion. Indeed, it is artificial to say that being at a certain temperature causes the iron bar to be a certain length; this makes the relata sound fact-like, and thereby serves a particular metaphysical theory of discrete parts. Instead, we should think of the cause and effect as already being changes, essentially: being heated and expanding. Overall, we find a dynamic account of events more natural. Events are such things as developings and growings; they are flashes and bangs, they are movements of things, they are sneezings, collidings, touchings, breakings, buildings, and so on.

We thus have two very different approaches to the question of causation that are based on opposed metaphysical frameworks. It is the second of these that, we argue, is the more natural one for biology, with its need for continuity. The two frameworks, in summary form, are as follows.

*A. Static relatedness*: The world consists of a succession of static events, as in perdurantism, and then some of those events are 'connected' together by a contingent relation. That relation might be regular succession, counterfactual dependence, probability raising, or something else. Causation is the (unobservable) relation the might hold between some of the static relata; hence the two-event-plus-relation model. Similarly, causal powers are understood according to a conditional analysis that, again, analyses them in terms of a relation holding between events. There are thus no irreducible causal powers in the sense required by the dispositionalist.

This framework is essentially Humean in origin and fits an event/fact ontology. It dispenses with active, powerful particulars and is favoured by empiricist modern philosophy, which is suspicious of strongly modal connections between distinct events. All of Aristotle's four causes are dropped from the account except for efficient causation, which is also given a reductive analysis. Many of the presuppositions are retained even in contemporary anti-Humeanism, simply by making the connection between the causal relata necessary rather than contingent. But this is still merely connecting static parts in an attempt to manufacture change. This model starts with changelessness and then tries to construct change.

*B. Dynamism*: Change is everywhere. It is continuous, in the sense that it does not break down into changeless parts. Causation begins as soon as certain processes are appropriately aligned, which is to say that they form a mutual manifestation partnership with respect to some effect. When they are so aligned, a new process begins and continues dynamically (always in a state of change) until it reaches its end (exhausts itself) or is interrupted, either additively or subtractively.

This framework is roughly Aristotelian in origin, since change is fundamental in the Aristotelian tradition; but it is less committed to substances and more to processes. We can fit the account with Aristotle's four causes in the following way. Particular bearers of powers—*material causes*—can be constructions out of such processes, where what we think of as a single substance is a temporally extended process or a set of processes. Ultimately, it is not that a thing is a subject of change: a thing is constituted by a set of interrelated changes. We can think of a matchstick as an example of material causes, which are bearers of powers. Powers, such as flammability, are akin to the *formal causes*. The *efficient causes* are the so-called stimulus conditions, which we have accounted for here as the coming into being of a mutual manifestation partnership: the striking of the match against a rough surface, for example. The *final cause* is the manifestation type towards which a power is directed—in this case, burning—which gets there through a process of combustion. This final cause is still to be found in any abstracted segment of that process in that the struck match is always directed towards burning until it reaches that end.

Space and time will usually be thought of as dense within this framework, which means that between any two times or places there will always be a third. Space and time will therefore be infinitely divisible. This in itself makes causes and effects connected (Bennett 1974) rather than merely conjoined, as Humeans would have it. If one opted for a discrete view of space–time, then there would always be a gap or a discontinuity between cause and effect, which would always be non-overlapping and temporally separated—something that Hume insists upon. With nothing bridging that gap, it might then seem reasonable to think of causation as nothing more than regular sucession, for instance. With dense space–time, however, Bennett argues that cause and effect can literally be connected, namely through a process in which there are no gaps.

Our concern here has not been directly with the persistence of particulars (on this question, see chapters 1, 2, 4, 5, 6, 7, and 18); but the issue is relevant to our two frameworks simply because perdurantism posits the patchwork of changeless particulars out of which change is constructed. We have argued that a dynamic theory of causation ought to reject such a view. A powerful consideration remains that all the abstracted parts—the segments—of a process remain essential to it on the dispositional view we have outlined. In contrast, it is largely an accidental matter in perdurantism that a temporal part belongs to one particular process rather than another, given that such accounts almost always take the relations between parts to be contingent. As argued, however, on the dispositional view, a process has to go through a particular change or complex of multiple changes in order for it to be the process that it is, and this will be true of all natural causal processes. On the mosaic view, as is clear in Lewis's (1986c: ix) work, it is not essential to any piece of the mosaic that it be situated within the picture it is. All causal and nomic relations will be purely contingent.

## 5. Conclusions

We do not claim that the arguments above constitute a proof in favour of either a dynamic theory of causation or the dispositional variant of it that we prefer. Proofs are rare in metaphysics and philosophy generally. What we have done, however, is advance some considerations that ought be taken into account when selecting the best theory of causation. We have argued that the features of causation found in biology, and not just the features of inanimate causation found in physics, ought to be accommodated. In biology we are confronted with a multilayered complexity, which involves a number of overlapping processes that may all be essential for a proper causal explanation of particular phenomena. This is brushed over in many versions of the standard account, with its two-event model. We have also emphasized the dynamism found in biology, where organisms have to be constantly in motion in order to survive. This, we argued, does not sit well with an ontology in which there is a succession of changeless parts, no matter how those parts are stitched together through some form of relation. A theory of causation suitable to biology must emphasize continuity and connectedness. We added some detail to this, explaining how causes can be temporally extended, but also simultaneous with their effects, and nevertheless also be able to form causal chains of a longer duration.

There are other process theories available. Fair (1979), Salmon (1984: 147), Dowe (2000) and Kistler (2006) all offer alternative theories of causation. However, these are physically reductionist accounts in terms of a transfer of a conserved quantity or some other physical mark. Our account has no such commitment to reductionism, for reasons we cannot detail here (but see Mumford and Anjum 2011: ch. 4). We have been concerned with finding an account of causation fit not just for physics but for the living world as well. We have offered what we think is a plausible dispositional account.

## Acknowledgements

Co-authorship statement: the authors take equal credit for all of their collaborative work, irrespective of credited author order. The authors would like to thank the editors, the anonymous referees, and the audience at Exeter, where an earlier version of this paper was presented. We would also like to thank Elena Rocca, Andrea Raimondi, Flavia Fabris, and Anne Sophie Meincke for further comments. Research for this work was funded by the FRIPRO funding scheme of the Research Council of Norway (NFR).

## References

Anjum, R. L. and Mumford, S. (2017). Emergence and Demergence. In M. Paoletti and F. Orilia (eds), *Philosophical and Scientific Perspectives on Downward Causation*, (pp. 92–109). London: Routledge.
Aristotle, (1996). *Physics*, trans. by R. Waterfield. Oxford: Oxford University Press.
Armstrong, D. M. (1978). *A Theory of Universals*. Cambridge: Cambridge University Press.
Armstrong, D. M. (1989). *A Combinatorial Theory of Possibility*. Cambridge: Cambridge University Press.
Bennett, J. (1974). A Process View of Causality. In R. Whittemore (ed.), *Studies in Process Philosophy*, vol. 1 (pp. 1–12). Springer: Dordrecht.
Bickhard, M. (2011). Some Consequences (and Enablings) of Process Metaphysics. *Axiomathes* 21: 3–32.
Dowe, P. (2000). *Physical Causation*. Cambridge: Cambridge University Press.
Dupré, J. (2014). Animalism and the Persistence of Human Organisms. *Southern Journal of Philosophy* 52: 6–23.
Fair, D. (1979). Causation and the Flow of Energy. *Erkenntnis* 14: 219–50.
Feser, E. (2014). *Scholastic Metaphysics*. Heusenstamm: Editiones Scholasticae.
Gilbert, S. F. (2014). Symbiosis as the Way of Eukaryotic Life: The Dependent Co-Origination of the Body. *Journal of Biosciences* 39: 201–9.
Hawley, K. (2001). *How Things Persist*. Oxford: Oxford University Press.
Holland, J. (1992). *Adaptation in Natural and Artificial Systems*. Cambridge, MA: MIT Press.
Hume, D. (1888 [1739]). *A Treatise of Human Nature*, ed. by L. A. Selby-Bigge. Oxford: Clarendon.
Hume, D. (2007 [1748]). *An Enquiry Concerning Human Understanding*, ed. by P. Millican. Oxford: Oxford University Press.
Ingthorsson, R. (2002). Causal Production as Interaction. *Metaphysica* 3: 87–119.
Kant, I. (1929 [1781]). *Critique of Pure Reason*, ed. by N. Kemp Smith. London: Macmillan.
Kim, J. (1976). Events as Property Exemplifications. In M. Brand and D. Walton (eds), *Action Theory* (pp. 159–77). Dordrecht: Reidel.
Kistler, M. (2006). *Causation and Laws of Nature*. New York: Routledge.
Leiserson, W., Bonini, N. and Benzer, S. (1995). Transvection at the Eyes Absent Gene of Drosophila. *Genetics* 138: 1171-9.
Lewis, D. (1986 [1973]). Causation. In idem, *Philosophical Papers*, vol. 2 (pp. 159–213). Oxford: Oxford University Press.
Lewis, D. (1986a). Events. In idem, *Philosophical Papers*, vol. 2 (pp. 241–69). Oxford: Oxford University Press.
Lewis, D. (1986b). *On the Plurality of Worlds*. Oxford: Blackwell.
Lewis, D. (1986c). *Philosophical Papers*, vol. 2. Oxford: Oxford University Press.

Lewontin, R. (2001 [1983]). Gene, Organism and Environment. In S. Oyama, P. Griffiths, and R. Gray (eds), *Cycles of Contingency* (pp. 59–66). Cambridge, MA: MIT Press.
Lombard, R. (1986). *Events*. London: Routledge.
Mackie, J. L. (1993 [1965]). Causes and Conditions. In E. Sosa and M. Tooley (eds), *Causation* (pp. 33–55). Oxford: Oxford University Press.
Marmodoro, A. (2013). Causes as Powers. *Metascience* 22: 549–54.
Martin, C. B. (2008). *The Mind in Nature*. Oxford: Oxford University Press.
Mellor, D. H. (1995). *The Facts of Causation*. London: Routledge.
Molnar, G. (2003). *Powers: A Study in Metaphysics*, ed. by S. Mumford. Oxford: Oxford University Press.
Mossio, M. and Moreno, A. (2010). Organisational Closure in Biological Organisms. *History and Philosophy of the Life Sciences* 32: 269–88.
Mumford, S. (2008). Powers, Dispositions, Properties, or: A Causal Realist Manifesto. In R. Groff (ed.), *Revitalizing Causality: Realism about Causality in Philosophy and Social Science* (pp. 139–51). London: Routledge.
Mumford, S. (2009). Powers and Persistence. In L. Honnefelder, E. Runggaldier, and B. Schick (eds), *Unity and Time in Metaphysics* (pp. 223–36). Berlin: De Gruyter.
Mumford, S. and Anjum, R. L. (2011). *Getting Causes from Powers*. Oxford: Oxford University Press.
Mumford, S. and Anjum, R. L. (2016). Mutual Manifestation and Martin's Two Triangles. In J. Jacobs (ed.), *Causal Powers* (pp. 74–87). Oxford: Oxford University Press.
Oyama, S., Griffiths, P., and Gray, R. (eds). (2001). *Cycles of Contingency*. Cambridge, MA: MIT Press.
Rescher, N. (1996). *Process Metaphysics: An Introduction to Process Philosophy*. New York: SUNY Press.
Rosenberg, A. (2008). Biology. In S. Psillos and M. Curd (eds), *The Routledge Companion to Philosophy of Science* (pp. 511–19). London: Routledge.
Russell, B. (1992 [1913]). On the Notion of Cause. In *The Collected Papers of Bertrand Russell*, vol. 6 (pp. 193–210). London: Routledge.
Salmon, W. (1984). *Scientific Explanation and the Causal Structure of the World*. Princeton: Princeton University Press.
Seibt, J. (2008). Beyond Endurance and Perdurance: Recurrent Dynamics. In C. Kanzian (ed.), *Persistence* (pp. 133–65). Frankfurt: Ontos-Verlag.
Sellars, W. (1981). Foundations for a Metaphysics of Pure Process. *Monist* 64: 3–90.
Sider, T. (2001). *Four Dimensionalism: An Ontology of Persistence and Time*. Oxford: Oxford University Press.
Skaftnesmo, T. (2005). *Genparadigmets Fall: Sammenbruddet av Det Sentrale Dogmet*. Oslo: Antropos.
Suppes, P. (1970). *A Probabilistic Theory of Causality*. Amsterdam: North Holland.
Thomas Aquinas. (1956). *Summa contra Gentiles*, trans. by V. J. Bourke. New York: Doubleday.
Webster, G. and Goodwin, B. (1996). *Form and Transformation: Generative and Relational Principles in Biology*. Cambridge: Cambridge University Press.
Whitehead, A. N. (1978 [1929]). *Process and Reality: An Essay in Cosmology*, ed. by D. R. Griffin and D. W. Sherbourne. New York: Macmillan.

# 4

# Biological Processes
## Criteria of Identity and Persistence

*James DiFrisco*

## 1. Introduction

The life sciences have long described biological systems as being fundamentally dynamic at all scales, from the evolution of species to the developmental construction of individuals to the continual energetic turnover of metabolism. There has historically been a variety of attempts to motivate a shift in ontology on the basis of the empirically dynamic character of the biological world. Something like 'process ontology of biology' was advanced by many of the organicist philosophers of biology of the twentieth century who came under the influence of Whitehead and Bergson—notably Woodger (1929) and von Bertalanffy (1952); and it has also been suggested by Waddington (1957), by process structuralists like Webster and Goodwin (1996), by the plant morphologist Sattler (1990, 1992), and by Bickhard (2009), Jaeger and Monk (2015), and Dupré (2012)—among others.

Scepticism toward this type of project, such as there is, is less likely to be directed at the biological facts that have inspired it than at the claim to constitute a novel ontology that rivals more traditional ontological frameworks in descriptive richness and explanatory scope. Accordingly, this chapter aims to advance the underdeveloped ontology of individual processes. I focus on explicating what are arguably the two most important categorial features of biological processes—their persistence and their identity or individuation conditions—in contrast to their substance-ontological counterparts. The process ontology I defend is committed to providing a causal account of biological persistence and individuation—a commitment that underlies its advantages as well as its distinctive challenges.

## 2. Ontological Explanation for Scientific Domains

Before examining the aspects of contemporary biology that motivate a processual perspective on life, first it will be necessary to clarify how an ontology can be 'motivated' by a scientific field at all. What basis is there for choosing between a process-based and a substance-based ontology in the context of biology?

Ontological theory choice can be helpfully understood by analogy with scientific theory choice. Scientific theories are typically evaluated on the basis of a cluster of epistemic virtues such as predictive power, explanatory power, unity with other accepted theories, simplicity, and so on. The analogy begins to pull apart on the issue of predictive power, but otherwise it is relatively uncontroversial that one ontology can be simpler than another, more consistent, or more globally unified than another, and that these are desiderata for ontological theories. The epistemic virtue that I highlight, however, which is less well understood, is explanatory power. Rival ontologies can provide better or worse ontological explanations, and it is primarily here that they can come into contact with scientific theories.

To clarify the notion of ontological explanation, we should consider what an ontology for a given scientific domain is supposed to do. The dominant approach to ontology in the analytic tradition, the one initiated by Carnap (1949) and Quine (1939), aims to extract ontological commitments from a language that governs a given domain by examining the wide-ranging structural inferences comprised in the language. Starting from the structure of the language, one can then develop an ontological theory of its domain by providing a systematic account of the truth makers for its sentences. Different ontologies for the same domain differ by offering contrasting descriptions of what it is in virtue of which the true sentences of the language are true.

The sense in which contrasting ontologies can differ as to their explanatory power is characterized by Seibt (2008, 2010, 2015) as follows. In providing a description of truth makers for sentences of a language, an ontology explains why we are justified in drawing certain 'categorial inferences' from these sentences. Consider the following true sentences:

(1) This tree is green.
(2) This is the same tree as the young sapling you saw last year.

An ontology explains, for instance, why one is justified in inferring from (1) that the denotation of 'green' can occur multiply in space whereas the denotation of 'this tree' cannot (Seibt 2008). Many items can be green but only one thing is this tree. An ontology might explain this inference by postulating that a tree is a substance, having the category feature of particularity, that is, having unique spatio-temporal location necessarily. By contrast, 'green' denotes a property or universal that can be multiply instantiated in space and time. Such a theory explains successfully, due to the fact that the theoretical terms 'substance' and 'universal' properly fulfil certain categorial-inferential roles that are fixed by accepted facts about trees—that is, that distinct trees can each be green. Similarly, an ontology explains why a speaker is justified in inferring from (2) that 'young sapling' and 'this tree' are co-referential by categorizing the tree as an individual substance. Invoking the ontological category of substance can explain why the inference is justified, because substances have the category feature of numerical identity over time, so that the young sapling is one and the same thing as the tree presently being observed.

In constructing an ontology specifically for a scientific domain, the categorial-inferential roles are set by the established contents of theories and models in that science. Different categories can fulfil these roles better or worse, and this leads to differences in the explanatory power of the associated ontologies. For example, there is pressure from developmental and evolutionary biology to be able to say that (i) an individual that undergoes substantial changes during ontogeny is still the same individual, whereas (ii) one that reproduces is distinct from any offspring it has. I will argue that the category of substance is too strong to play the inferential role embedded in (i), while the category of process is not. Conversely, the category of process is, like that of substance, *not too weak* to be able to play the inferential role embedded in (ii). This is a case in which a process ontology of individuals has greater explanatory power, because it can explain everything that substance ontology can explain but without occasioning problems from (i). Similarly, aspects of ecology and biochemistry suggest that we should be able to say that certain systems have weak individuality despite being spatially scattered and non-uniquely located. As I will argue in section 4, the category of substance is also too strong here, whereas the category of process is not. If these arguments are right, then process ontology constitutes a better domain ontology than substance ontology due to its greater explanatory power.

The more acute problem for substance ontology in biology is certainly that of persistence despite change; and, as we will see in the following section, it arises due to the connection between substance ontology and essentialism.

## 3. An Argument for Biological Process Ontology

There are, arguably, many reasons to think that contemporary biology favours the adoption of process ontology. I will focus here on one argument for viewing biological individuals as processes, which starts from a familiar story about biological essentialism.

Before the arrival of evolutionary theory, biological species were for the most part still conceived of on the Aristotelian model of fixed kinds, or even as types corresponding to archetypes in the mind of the Creator. Darwin was able to show that the variety of species can be explained by appeal to a long-term evolutionary process of natural selection acting on heritable variations. Not only are the different species we observe today derived from common ancestors, but species kinds are themselves temporary stages in an ongoing evolutionary process. Moreover, the critical importance of variation in this process means that we should not expect intraspecific variation to be confined to variation only in the 'accidental' properties of species. Species are therefore very different from paradigmatic natural kinds such as the elements of the periodic table, because they are continually varying and evolving into new species.

Many philosophers of biology accept the arguments of Ghiselin (1974) and Hull (1978) that species do not belong to the ontological category of classes or kinds at all, but are rather historically extended individuals. On this view, an organism or a population is not an instance of a species in the way in which a piece of gold is an instance of the chemical kind gold, but is rather a part of a species in the way in which one is part of one's family lineage. This conception of species was inspired by the cladistic classification program, which sought to tie biological classifications more closely to actual evolutionary history. On the cladistic view, species are individual segments in the evolutionary tree between phylogenetic branching points. In fact matters are more complicated than a simple contrast between kinds and individuals might suggest. If species were *merely* genealogically defined segments of the phylogenetic tree, it is doubtful that the cladistic classification of an organism would explain very much about it besides its position in the genealogical nexus. Like traditional natural kinds, cladistic classifications are used to identify underlying patterns of resemblance between members of groups, despite the fact that these patterns are spatio-temporally restricted in the way individuals are (see Griffiths 1994: 211). The 'species as individuals thesis' certainly does not resolve all the problems surrounding biological species, but the least one can say is that it recasts the concept of species in a more dynamic light than before. It is also an interesting example of a categorial claim that is similar in form to the one suggested here. The claim is that, within the domain ontology for evolutionary biology and systematics, species lack the category features of generality and unrestricted instantiability and are characterized instead by historical individuality and concreteness.

These kinds of considerations about species have led many philosophers of biology to reject 'biological essentialism'. To be precise, the form of essentialism being rejected is qualitative essentialism, or the view that a kind $K$ has the essential property $P$ if all and only members of $K$ have $P$, where $P$ is an intrinsic property or quality that is in some sense central to explaining what it means to be a $K$. Qualitative essentialism about biological kinds is recognized to be inconsistent with the fact that classification into biological kinds such as species tends to be based not on intrinsic properties (even if it is often informative about intrinsic properties), but on genealogical or historical-evolutionary relations (see Okasha 2002). In the present context, the important point to note is the following: if a species is a class defined by shared qualitative properties among its members, then the existence of significant qualitative variation within the species creates problems for the unity of the class. By contrast, if an organism's being part of a species is a matter of causal and genealogical relations, then there is no such problem. In principle, this ontology imposes no limits on the qualitative variation between different organisms in a species, either synchronically or diachronically.

Similar considerations motivate the categorial claim that biological individuals should be conceptualized as individual processes rather than as substances. As in the case of species, the reason has to do with qualitative variation: that is, qualitative variation between successive stages in the life of a biological system can make it difficult to specify an identical subject that passes through these stages. It is easier to simply drop the requirement that persistence is identity over time, and instead to construe it as a mereological relationship among the stages themselves. But this means discarding the category of substance as the paradigm for individuality in favour of the weaker and more flexible category of process, with its distinctive category features. I defend the view that processes lack the categorial features of numerical identity over time, persisting instead by having temporal parts, and they are individuated primarily by causal relations rather than by location. These are the features that make processes uniquely well suited for the categorial roles licensed by biological theory.

The argument can be made more concrete with an example. The class *Trematoda*, comprised of parasitic flatworms, is characterized by remarkably complex life cycles involving at least two hosts and several morphologically distinct developmental stages. One genus of trematodes, *Fasciola*, is well researched due to its significant economic impact as a cause of disease in ruminants (and sometimes in humans). The life cycle of *Fasciola gigantica*, the giant liver fluke, begins with eggs laid in mammalian gut excrement that hatch under aquatic conditions to release ciliated larvae (*miracidia*) (see Figure 4.1). The free-swimming larvae then penetrate an intermediary host (typically, snails from the family *Lymnaeidae*), where they develop through a number of larval stages. These are the oval sporocyst—a tightly packed ball of germinal cells (C–D in Figure 4.1); the cylindrical redia, which reproduces asexually (E–F in Figure 4.1); and the disc-shaped cercaria, with long swimming tails (G in Figure 4.1). Free-swimming cercariae then leave the intermediate host to encyst on vegetation that will be eaten by its definitive hosts, typically ruminants like sheep or cattle. When the encysted cercariae (i.e. the metacercariae) are ingested, they quickly excyst and take on the adult form of the liver fluke, eventually lodging themselves in the liver or in the bile duct, where they can increase in size by an order of magnitude over the next several weeks. Adult liver flukes reproduce sexually and lay eggs that find their way into the host's excrement, thereby closing the life cycle (see Phalee et al. 2015).

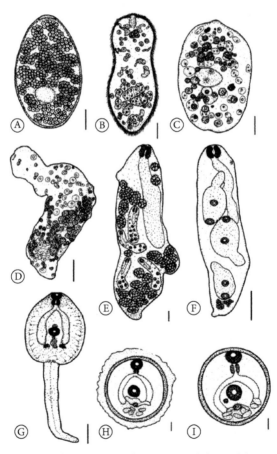

**Figure 4.1** Variation in the life cycle stages of *F. gigantica*: (A) egg, (B) miracidium, (C) young sporocyst, (D) mature sporocyst, (E) mother redia, (F) daughter redia, (G) cercaria, (H) encapsulated metacercaria, and (I) metacercaria. Scale bars (A–D) = 0.03 mm; (E–G) = 0.1 mm; (H–I) = 0.05 mm. Reprinted from Phalee et al. 2015, under a creative commons license.

The development of *Fasciola gigantica* provides a vivid illustration of how different the stages of a life cycle can be. Other examples could be enlisted to highlight variability within ontogenies, as in complete metamorphosis. Intrinsic temporal variability in development is extensive and pervasive enough that attempts to consign the variation between stages to variation in accidental properties in favour of a core of identity would often be biologically unrealistic. This is not to say that no qualities remain the same over time. However, not just any stable qualitative property can provide a basis for diachronic identity. It should in some sense ground and/or explain other important features of the individual, otherwise it could be that a peripheral quality that happens to be shared by all and only members of a kind defines its essence.

Given this explanatory requirement, an initially appealing idea is that diachronic identity in development can be grounded in the genotype—or, better, in the 'extended genotype' formed by gene regulatory networks (GRNs). In contemporary developmental biology, GRNs are frequently assumed to deterministically control development, and thus they could be interpreted as providing the individualized essences of developing organisms (see e.g. Davidson and Erwin 2006; Davidson 2010). The trouble with relying on GRNs for this role is that they often fail to meet the explanatory requirement. It is increasingly recognized that the mapping between GRN and phenotype can be highly non-linear, indeterministic, and degenerate owing to phenomena like developmental system drift (True and Haag 2001), structural robustness and canalization (Huang and Kauffman 2009), alternative splicing, post-translational modification, intrinsically disordered protein domains, and gene expression noise (Niklas et al. 2015). GRNs are no doubt crucial for understanding and explaining developmental processes, but not so far as to define a qualitative essence for a developing organism. And, if this is true for GRNs, it is all the more so for other candidate qualitative properties. The radical but effective solution in this situation is to discard the very requirement for a qualitative essence or diachronic identity criterion and take the developmental process as a whole to be ontologically primary (see chapter 11 in this volume).

This idea draws further support from evolutionary perspectives on life cycle variation. The descriptive study of life cycles has, in recent decades, increasingly come under the purview of life history theory, which uses optimality principles to explain the variation in adaptive traits at different stages of individual lifetimes (see Stearns 1992; Roff 1992; Flatt and Heyland 2011). Principal life history traits include size at birth, growth pattern, age and size at maturity, number, size, and sex ratio of offspring, age- and size-specific reproductive investments and mortality schedules, and longevity (Stearns 1992). These traits often constitute the major fitness components of an individual and are typically constrained by complex trade-offs with one another that demand significant intra-lifetime variation. Applying this framework leads to the important insight that life histories have a more fundamental status for evolution than organisms per se: given that fitness integrates over the entire reproductive performance of a lifetime, selection acts to maximize the fitness of a life history as a whole (Flatt and Heyland 2011; see also Fusco 2001).

A similar kind of intrinsic temporal variation that is found in the life histories of multicellular organisms and in the evolution of species is also found in metabolic processes. Recent research is increasingly highlighting the biological importance of circadian and ultradian variation in cells (see O'Neill and Feeney 2014). Virtually all important cellular events are parts of non-overlapping cycles involving intrinsic variation. For example, the primary metabolism of cells involves continuous oscillation between the chemically incompatible oxidative and reductive phases. It is not as though the contradictory properties instantiated by these stages can be reinterpreted as 'accidental' features, as this kind of variation lies at the heart of what is happening in cellular metabolism. Such variation creates serious strain for the conception of persistence as identity over time. To quote a review of the literature on circadian variation in hepatocytes (i.e. liver cells), this phenomenon: 'leaves one with the overwhelming impression that the humble hepatocyte is simply not the same cell between day and night' (ibid., 2971). Perhaps one should say that, indeed, it is not the same cell, and that the 'cells' are successive stages of an ongoing process.

These examples of intrinsic qualitative variation are relevant to ontology in the following way: the more qualitative variation there is between successive states of a persisting entity, the more difficult it will be to provide an ontological explanation of persistence in terms of qualitative identity or resemblance. This is what is required, however, for the enduring-substance conception of persistence, as I will show in section 4. But the problem is avoidable: without an identical subject that passes through each stage, one can interpret the stages as temporal parts of an extended process. Of course, stages and processes must also have identifying features; but, as we will see, without the requirement of numerical identity through time, defining these features can be much less demanding.

## 4. Criteria of Identity and the Individuation of Processes

It is sometimes thought that any category or kind $K$ that we can coherently talk about—for example, the category of processes—must be associated with a *criterion of identity* for its instances, on the intuition that otherwise $K$ is not theoretically well defined and cannot be actually applied. Some have even considered the absence of a criterion of identity for $Ks$ to license the inference that there are no $Ks$. Quine, for instance, made an inference of this type from the lack of specifiable criteria of identity for intensions (see Horsten 2010: 412). Criteria of identity also appear in one popular view according to which ontological categories are only distinct if they are associated with distinct criteria of identity (Thomasson 2007; Lowe 2006). These considerations seem to press the conclusion that a process-ontological framework must be able to provide a distinctive criterion of identity for processes.

A criterion of identity is a rule specifying the conditions under which items that belong to the same kind are identical. The two most common logical forms for criteria of identity are the following:

(I) $\forall x \forall y \in K : x = y \leftrightarrow R(x, y)$

(II) $\forall x \forall y \in D : f(x) = f(y) \leftrightarrow R(x, y)$

A 'one-level' criterion (I) states that, for objects $x$ and $y$ of kind $K$, $x$ and $y$ are identical iff they stand in relation $R$. The paradigmatic example here is the axiom of extensionality as a criterion of identity for sets, according to which sets $x$ and $y$ are identical iff they have the same members ($R$). For a 'two-level' criterion (II), the right-hand side of the equation is the same as before, but the identity relation is between functions of objects $f(x)$ and $f(y)$, where $x$ and $y$ belong to the same domain $D$ over which functions can be defined to constitute a kind $f(D)$ (Horsten 2010: 414). In Frege's well-known example of a two-level criterion, the criterion of identity for the direction of lines $f(x)$ is given by the relation of parallelism ($R$).

A candidate criterion of identity is generally supposed to meet a few requirements: (a) $R$ must be an equivalence relation; (b) it must be informative, in the sense of excluding at least some interpretations of the non-logical vocabulary on the right-hand side, and thus cannot be a merely logical truth like Leibniz's Law (hereafter LL); (c) the criterion cannot admit of counterexamples; and, more controversially, (d) it cannot be circular or 'impredicative' (see ibid.). Because (II) requires defining a function over a domain of objects, it may be restricted to special classes of singular terms formed by means of functional expressions like 'the direction of' (Lowe 1989: 4). Criteria of identity for processes, if there are any to be found, should accordingly be one-level. Many putative one-level criteria turn out to be circular, however, and whether this afflicts the criteria for processes remains to be seen.

Before proposing a criterion of identity for processes, it will be instructive to consider first an example of (I) that is thought to be viciously circular: Davidson's criterion of identity for events. According to Davidson's criterion, events $x$ and $y$ are identical iff they have the same causes and effects (Davidson 1970: 306). The trouble arises when we recognize that causes and effects are themselves events, and thus we would need a criterion of identity for these cause events and effect events in order for events $x$ and $y$ to be identical, and so on for their causes and effects. Davidson's criterion can therefore individuate events only if events are already individuated.

Seeking to avoid the impredicativity of defining $R$ for events in terms presupposing event identity, Quine (1985) proposed that events are identical iff they occupy the same spatio-temporal region. This criterion successfully escapes circularity, since it not based on a conception of event identity but rather of region identity. Underlying this difference with Davidson is a conception of events not as nodes in the causal nexus but as material contents of regions of space–time. However, Davidson (1985) raised a key counterexample to Quine's criterion. Imagine that a ball is simultaneously rotating and heating up. On the Quinean view, the ball's rotating and its heating up are the same event, because they occupy the same spatio-temporal region. If the ball's heating up causes the surrounding air to heat up, then the undesirable consequence follows that the ball's rotating is also what causes the surrounding air to heat up. Davidson's criterion, by contrast, can distinguish between the rotating and the heating up, but only if other events such as the heating up of the surrounding air are already individuated.

How does the matter stand with processes, then? Although events and processes may be distinguished in different ways (see Simons 2003; Steward 2013), they are relevantly similar in that both are extended in time and intimately bound up with causal relations. If processes are to be individuated causally in some sense, I take the above counterexample to show that sameness of spatio-temporal region alone will not be suitable as a criterion of identity. Something more, as in Davidson's criterion, is needed—but without the same worrisome circularity.

I propose that the right approach to understanding biological processes is to be found in the *causal cohesion* account of individuation developed in a series of papers by Collier and colleagues (Collier and Muller 1998; Collier 2003, 2004). The cohesion account is based on dynamical systems theory and, though not intended specifically for events or processes, it extends over dynamical entities generally—in other words concrete entities that change, that are subject to forces, that are composed of physical parts at some level, and that can be described in terms of successive states. An entity is *cohesive* in the most general sense when it is demarcated by an interaction gradient—that is, when the causal interactions among its parts are stronger than the causal interactions between these parts and their environment, which grants it stability against perturbations (see Collier 2004). The probabilistic perturbation conditions under which a cohesive entity remains stable define its *cohesion profile* (ibid.). 'Stability' in this context refers to the recurrence of a *cohesion regime*, or to the specific interaction structure among parts that constitutes them as a unit. As an example, a Bénard convection cell is a cohesive entity because the macroscopic coordinated interaction among its microscopic parts is stronger than the interactions of these same parts with their environment. The cohesion profile of a convection cell includes its stability against a certain range of temperature changes, but not against perturbations like vigorous stirring. And we say that a Bénard cell is stable due to the recurrence of physical convection—as opposed to turbulence—as its cohesion regime.

Cohesion can take different forms for different types of causal interactions at different scales, and can obtain due to first-order or higher-order dynamic properties. A rock is a cohesive individual, and it is individuated by the strong molecular bonds its parts have with one another but not with the environment. The cohesion of living cells is more complex and dynamic and of a higher order: it occurs by means of the recursive organization of chemical reactions whereby the material constraints on the dynamics of these cells (e.g. membranes, enzymes) are continually regenerated by those dynamics (Mossio and Moreno 2010; see also chapter 10 here). Higher-level and spatially diffuse systems can also be cohesive, for example a biological population, where the relevant interactions are primarily reproductive. According to Mayr's (1942) biological species concept, for example, a population is individuated by the reproductive interactions that are stronger among its parts than they are with parts of other populations. Even flocks of birds and schools of fish are weakly cohesive systems, in virtue of the sensorimotor feedback mechanisms that actively maintain a certain distance between the organisms. A cohesive individual may be considered 'robust' to the extent that it retains stability against a greater number of independent perturbations, and in this sense a single bird will be a more robust cohesive individual than the flock of birds it is part of.[1]

---

[1] For more on the notion of robustness as a criterion for (belief in) the reality of an entity, see Wimsatt (2007: 43 ff.) and Eronen (2015).

Cohesion is a functional property describing the condition under which an arbitrarily selected system is a genuine dynamical individual. Being a functional property, cohesion must be 'filled in' with the relevant interaction type for a given system. When this is done, and when the interactions of the relevant type meet the cohesion condition, we get access to a dynamical explanation grounding the *sortal* for the individual in question. Sortals are terms for sorts or kinds bearing determinate individuation criteria, which allow one to single out individuals of a kind as objects of unambiguous reference. The sortal 'cell', for example, is normally determinate enough to permit counting how many cells there are in a given region, or discriminating where one cell ends and another begins. It is only due to the cohesive interactions of metabolic processes, however, that there are individual cells to be singled out at all. Specifying the cohesion conditions for cells provides a causal explanation of the individuality of cells, and this can also be done for the other dynamical systems studied in biology.

On the basis of this conception of cohesion, the following criterion of identity for dynamical entities such as processes can be put forward: processes are identical iff they have all of the same cohesive properties, including cohesion profile and cohesion regime, and they occupy the same spatio-temporal region. 'Cohesive properties' are to be understood as the dynamical properties a process has that make it cohesive.

How does the cohesion account fare with respect to the requirements on criteria of identity? It satisfies (a), given that sameness of cohesive properties and of region ($R$) are equivalence relations. The criterion is also informative, and thus satisfies (b); despite appearances, it is not merely a reformulation of LL, because any dynamical entity will display both cohesive properties and properties irrelevant to its cohesion (Collier 2003: 106). As for (c), which concerns counterexamples, a criterion of identity based purely on sameness of cohesive properties would face the troublesome counterexample that two spatio-temporally distinct individuals might instantiate exactly the same cohesive properties. The conjunction of cohesive properties and spatio-temporal regions disposes of this as well as of any counterexamples of spatio-temporally coincident but causally distinct processes.

Finally, does the cohesion criterion satisfy (d)—is it circular or impredicative? Arguably, it is not viciously circular like Davidson's criterion of identity; for, whereas Davidson defines $R$ for events in terms of events, here $R$ for processes is defined in terms of cohesive properties and regions rather than processes. However, the general definition of cohesion appears to be formally impredicative. When a cohesive individual is defined as one in which the interactions among *its* parts are stronger than the interactions between these parts and the environment, the definiendum appears in the definiens. Impredicativity could be avoided if one could individuate a cohesive entity solely by examining the interaction structure of a set of putative parts without having to decide in advance which parts are parts of the entity. In many cases, though, it will be necessary to rely on hypothetical demarcations as well as on top-down functional criteria to determine the relevant interaction types for assessing cohesion in specific cases. In determining where one ecosystem ends and another begins, for example, it is necessary to have some idea of where to look as well as of what kinds of interactions to look for (nutrient cycling, trophic exchanges, etc.). Starting from a hypothetical ecosystem, one can decompose the system to examine the interaction structure of a wider set of parts, then use that information to recompose and potentially revise the initial demarcations. There is a sense in which this procedure is circular but, unlike Davidson's criterion, this circularity does not make individuation epistemically inoperable. The fact of starting from some initial system of interest would not seem to pose a problem so long as the revision criteria allow one to eventually converge on the right cohesive individuals, even if the initial system is arbitrary.

One consequence of adopting the cohesion account for processes is that the notion of individuality becomes weaker and more permissive than the individuality associated with substances. Traditionally, substances are characterized by the categorial features of particularity, countability, concreteness, independence, unity, and endurance—among other features (see Seibt 2010 as well as chapter 6 here). Cohesive individuals, however, need not have the feature of particularity, that is, a necessarily *unique* spatio-temporal location; examples would be processes like heat flow, air flow, or chemical reaction networks like the Krebs cycle, which can spatially overlap with other biochemical processes in cells. The parts of a cohesive individual can also be spatially scattered rather than connected—as are for example flocks of birds, populations, or some ecosystems. The 'weak individuality' of cohesive systems may be an advantage in cases where biological models represent causal interactions between non-conventional individuals, such as Lotka-Volterra population interactions. In such cases, weak individuality fulfils the categorial roles for biological individuality better than the strong notion of substance does.

This conception of processes as concrete independent individuals evidently doesn't track exactly with English usage in the case of the word 'process'. This noun is, in fact, ambiguous: sometimes it denotes a concrete occurrence and sometimes an instantiable pattern. The phrase 'the process of evolution' can denote the entire historical chain of developments from the last universal common ancestor to present-day life on Earth, which is a concrete and non-instantiable occurrence. But it can also denote a pattern, or a regular distribution in the properties of some individuals over time, which is instantiable at different times and places in the universe. Concrete processes can be what instantiates these patterns, but the latter must be instantiated in order to exist, because they are abstract and dependent. The focus of process ontology, in my view, ought to be on the development of the category of concrete independent processes rather than on process patterns, as only the former have the right categorial features to replace the category of individual substances.

## 5. Process Persistence: Identity versus Composition

The most salient categorial difference between substances and processes concerns their persistence in time: substances persist by being numerically identical through time, whereas processes persist by having temporal parts at different times. As I have argued, this is also the primary difference that gives a process-based representation of biological systems greater explanatory power in biological contexts, as it has no problem handling phenomena of intrinsic temporal variation (on this issue, see chapter 13). Before concluding in favour of the process view, however, it is necessary to examine in greater depth why exactly the substance view is committed to a problematic form of essentialism and whether the alternative model of persistence is actually coherent.

According to the account of ontological explanation outlined in section 2, evaluating different models of persistence as to their explanatory power will require clarifying the categorial roles for persistence. These can be expressed in ordinary statements about change (A) and reidentification (B) (see Seibt 2008):

(A) This liver fluke was ciliated, but now it is not.
(B) This is the same liver fluke as the one you saw yesterday.

More specific categorial roles can be added for specific domains, for example roles for distinguishing biological parents and offspring. But a model of persistence for *any* class of concrete entities must fulfil categorial roles for both change and reidentification over time, thereby explaining how statements like (A) and (B) can both be true. As might be expected, substance ontology tends to have greater difficulty with (A), whereas process ontology tends to have greater difficulty with (B).

## 5.1. The endurance of substances

Putative biological substances—organisms, organs, cells, and so on—are said to persist by enduring or being 'wholly present' whenever they exist. The easiest way to clarify the notion of being wholly present is to define it negatively, so that it consists in the denial that objects thought to be substances have temporal parts. This requires defining a minimal notion of temporal parthood (TP) at a time (see Sider 1997: 205):

TP: $x$ is a temporal part of $y$ at $t_i =_{df}$ (i) $x$ exists at $t_i$ and only at $t_i$, (ii) $x$ is part of $y$ at $t_i$, and (iii) $x$ overlaps at $t_i$ with everything that is part of $y$ at $t_i$.

I leave it open whether $t_i$ is an instant or an extended interval of time. Temporal parts or stages should be conceived of as the smallest units of difference over time for a given process, relative to a given granularity or resolution. Say $y$ is a whole life history and $x$ is a certain stage in the life history, such as the stage of reaching maturity. The core idea of the above definition is that this stage is temporally located at a certain time, and that everything of the life history that exists at that time is just stage $x$.

When enduring things are said to be wholly present (WP) whenever they exist, this means that they lack temporal parts.

WP: $y$ is wholly present at $t_i =_{df}$ (i) $y$ exists at $t_i$, (ii) there is no $x$ such that $x$ is a proper temporal part (TP) of $y$ at $t_i$.[2]

---

[2] Note that, on this definition, if $y$ is instantaneous, then there can be an $x$ that is an *improper* temporal part of $y$ at $t_i$—i.e., $x$ is the *only* temporal part of $y$. In this case we get the seemingly strange result that, if $y$ is instantaneous, it can both be wholly present and have *a* temporal part. As a consequence, the statements '$y$ endures' and '$y$ perdures' only become distinct if $y$ is non-instantaneous. While this may pose a problem for exdurance views (Sider 1996, 1997, 2001), it is not a problem for other views of persistence in which, if $y$ is instantaneous, then by definition it does not persist.

For an organism $y$, what exists of $y$ at any instant when $y$ exists is $y$ itself, and not just a part of $y$. If $y$ persists or exists at multiple times $t_1, t_2 \ldots t_n$, this means that $y$ at $t_1$ must be *identical* with $y$ at $t_2$ and with $y$ at $t_n$. This is the sense in which endurance is identity through time.

The endurantist view quickly encounters the problem of change, however, as expressed in (A) above. If a liver fluke is ciliated at $t_1$ and then not ciliated at $t_2$, the fluke at $t_1$ cannot be identical with the fluke at $t_2$. Initially one might try to resolve this tension by distinguishing between qualitative identity and numerical identity. A statement about change like (A) would then only pertain to the qualities of the fluke—say, its having cilia or not—while its numerical identity would remain unaffected by change. Similarly, the reidentification expressed in (B) would only pertain to the numerical identity of the fluke, and not to its qualitative features. This distinction between qualitative and numerical identity becomes untenable, however, once they are logically connected by means of LL (see Seibt 2008: 136), which states that there is no numerical identity without qualitative identity.

LL: $x = y \leftrightarrow \forall F \, (Fx \leftrightarrow Fy)$[3]

In effect, if change is qualitative difference over time, then an entity cannot change and remain numerically identical. Endurance would be impossible.

The classical strategy for dealing with this problem is to restrict the scope of LL so that it only quantifies over *essential* qualitative properties. An enduring thing could then change in its accidental qualitative properties while remaining numerically identical over time; but, if it changed in its essential qualitative properties, it would become a numerically distinct individual. This is the sense in which endurantism must be tied to some form of essentialism, because otherwise persistence as numerical identity despite change would not be possible.[4]

As for what the essential properties are, these are often taken to be determined by the sortal that the entity falls under, which specifies identity conditions for the kind of entity in question (Lowe 2009; Wiggins 2001). Substances are of course not the only items that are classified into kinds. There are also kinds of events and processes, and individual events and processes will have essential properties to the extent that they are individuals of certain kinds. The important difference, however, is that substances must fall under sortals that specify synchronic *and* diachronic identity conditions. At every time at which a substance exists, there must be an intrinsic property or set of intrinsic properties whose qualitative sameness grounds its numerical identity through time. Processes, by contrast, can belong to kinds in virtue of essential *relational* properties, or in virtue of intrinsic properties that are instantiated by

---

[3] The principle of the identity of indiscernibles and the converse principle of the indiscernibility of identicals are here combined.

[4] It is possible for a 'continuant' to have a relational rather than qualitative essence, whereas this wouldn't be possible for substances, as long as they have the categorial feature of independence. However, because continuants lack other categorial features besides enduring, they do not comprise a robust ontological category, and so the ontological problems with the category of substance cannot be avoided simply by switching to the more non-committal notion of continuants.

*different* temporal parts. Such is the case, for example, with the process kind 'the life history of *Fasciola gigantica*'.

A different endurantist solution to the problem of change is to hold that all properties instantiated by a substance are time-indexed. It may be contradictory for a substance to be both ciliated and not ciliated simpliciter, but not for it to be both ciliated-at-$t_1$ and not-ciliated-at-$t_2$. The problem here is not just that all intrinsic properties become relations to times, as Lewis (1986) pointed out. Given that the continuant maintains its numerical identity over time, it now has all of its time-indexed properties *at all times*. At $t_1$, the liver fluke has the properties of being ciliated-at-$t_1$ *and* being not-ciliated-at-$t_2$. It has these same properties at $t_2$. Arguably, this fails to meet the explanatory requirements for change as expressed in (A), which says simply that the individual's properties at $t_1$ are different from its properties at $t_2$.

It is worth noting that the problem of change does not arise for the temporal parts view, because the latter does not conceptualize persistence as identity over time— numerical or qualitative. Hence there is no problem in saying that a stage $x_1$ at $t_1$ is ciliated and another stage $x_2$ at $t_2$ is not ciliated, and that stages have their intrinsic properties simpliciter.

There are several other problems with endurantism that may be relevant to onto-logical issues in biology, including the need to introduce artificial discontinuities when a continuant begins or ceases to exist. On closer inspection and viewed with a finer grain, discontinuities such as birth, death, or speciation are based on continuous developments. Similarly, phenomena of fusion, fission, or overlap are more easily handled through the concept of temporal parts. Just as two distinct substances can share spatial parts, processes can share temporal parts without this impinging upon their individuality. The capacity of processes to easily accommodate continuity and overlap is an advantage in biology, where these phenomena are common. Reproductive processes, for example, are thought to involve the offspring and the parent's sharing (some) material parts, and temporal parts of those material parts (see Griesemer 2000).

The strongest argument against endurantism as a general ontological thesis is often thought to be the argument from the relativity of inertial frames in special relativity, which implies that there is no privileged present to ground the notion of being wholly present. In biological contexts, however, the strongest argument targets the endurantist commitment to qualitative essentialism. To be sure, in many cases it will be possible to identify non-trivial properties that retain qualitative sameness over the course of a dynamical trajectory, which can even serve to ground the conception of a substance that traverses the trajectory. But, because there will also be cases where this procedure becomes strained, it should not be built in as a requirement on biological ontology.

## 5.2. The perdurance of processes

The core idea of the alternative model—perdurantism—is to deny that persistence is a matter of the numerical identity of entities that exist at different times, and to claim instead that it is a special mereological relationship.[5] From this perspective,

---

[5] Perdurance theories have often been motivated by the development of the theory of relativity, and in particular by the unified concept of space–time, as famously described by Minkowski. Accordingly, many formulations of perdurantism rely on the idea that, since space and time are fundamentally unified in

statements like (A) are made true by the fact that the ciliated and the non-ciliated organism are different temporal parts of a temporal whole, which is the liver fluke's life history, while statements like (B) are made true by the fact that the parts are parts of the same whole. The main attraction of this view in the present context is that there is virtually no restriction on the variability of properties among the temporal parts themselves, except that they must be able to stand in the appropriate composition relationship.

Many of the objections to perdurantism simply express incredulity that it can adequately capture our habitual thinking about change or transtemporal sameness (e.g. Chisholm 1976; Wiggins 2012), but such objections have little force on the model of ontological explanation from section 2. The main objection that should be addressed concerns the capacity of processes to instantiate the properties that substances can instantiate. For the central strategy of perdurantism is to reject the endurantist idea that statements (A) and (B) refer to the same entities. Instead, change statements refer to temporal parts, whereas reidentification statements refer to temporal wholes (see Seibt 2008: 159). However, it would seem that many of the predications we make of items thought to be substances express properties that cannot be exemplified by either instantaneous stages or temporal wholes.

For example, an organism eats, but one of its instantaneous stages cannot eat. Rather, the stage is succeeded by another stage that has different eating properties. Since eating takes some time to occur, it can only be a property of a stage if the stage is temporally extended. However, if a stage is temporally extended, then arguably it must persist through some change. Then either the stage persists through change by perduring, in which case the same problem is repeated for *its* stages, or it endures, and the perdurance thesis fails to obtain after all (see Seibt 2008: 143). Similarly, the entire life history of an organism doesn't have properties like eating or moving, but only properties like occupying a certain spatio-temporal region.

A perdurantist could respond that a time-extended property instance such as moving, eating, digesting, and so on is itself a process or an activity and need not be exemplified by *a* stage but by several stages that, together, have the suitable interrelations. The above argument begs the question against the perdurance view by assuming that these processes are not analysable into smaller units of difference, since this is just what persistence is, on the perdurance view. But the critic might respond that this does not adequately explain persistence. Specifically, it captures the aspect of difference but not the aspect of sameness. The process of eating is internally differentiated into stages, but what makes them stages of the same

---

relativity theory, so temporal parts must be fundamentally similar to spatial parts. But this introduces a number of unnecessary metaphysical commitments about time into the ontology of perdurance, such as eternalism and the B-theory of time. A theory of temporal parts can be explanatorily adequate to (A) and (B) without invoking a constitutive analogy with spatial parts. The reason why a constitutive (rather than a heuristic) reading of the analogy between spatial and temporal parts implies eternalism and/or the B-theory of time is that spatial parts all exist, which implies that future (and past) temporal parts should also exist (i.e. eternalism). The analogy also causes problems when one imagines temporal parts to be discontinuous, as spatial parts are sometimes taken to be. Neither of these associations is necessary, as can be appreciated by examining ordinary talk about occurrents like events and processes—talk that involves no apparent commitments to eternalism or discontinuity.

process? A closer examination of the composition relation for stages is required for resolving this difficulty.

It will be noticed that the analogous question for enduring substances asks for a criterion of identity—a *diachronic* criterion of identity—that takes the following form (see Merricks 1998):

(III)  $\forall x \forall y \forall t : x \text{ at } t_1 = y \text{ at } t_2 \leftrightarrow R(x, y, t_1, t_2)$

Because process persistence is a mereological relation rather than identity relation, however, it takes a different logical form, where $x_n$ are temporal parts, $y$ is a temporal whole and '<' is the parthood relation:

(IV)  $\forall x \forall y : x_n < y \leftrightarrow R(x_n, y)^6$

The question is, then, under what conditions (R) are temporal parts ($x_n$) parts of the same temporal whole ($y$)? Certain answers to this question would be *prima facie* excluded in biological contexts—for instance, that composition occurs under all conditions (mereological universalism) or under no conditions (mereological nihilism). When evolutionary biologists count the number of offspring in order to measure a parent's reproductive fitness, they do not count later stages in the life history of the parent as another offspring. An ontology ought to be able to differentiate between the persistence of an individual and the reproduction of new individuals, and it is not clear how mereological universalism or nihilism about temporal parts could do so in a principled way.

Composition can be plausibly restricted for biological individuals by specifying R in (IV) as a type of *causal continuity*, sometimes also called 'genidentity' (Lewin 1922; Hull 1978; Guay and Pradeu 2015; see also chapters 2, 5, 7, and 11 here). More precisely: the $x_n$ compose $y$ if $y$ is the sum of causally continuous stages $x_n$. On this view, two temporally separated stages in the life cycle of *Fascicola gigantica* are parts of the same life cycle not because they resemble each other, but because they are linked in a continuous causal chain. Of course, allowing that any amount of causal continuity satisfies R would make R overly permissive, since a stage tends to be causally continuous with more than what should be included in a biological individual. The life cycle of deciduous trees includes the loss of leaves, for example. Let $x_1$ be a tree stage with leaves at $t_1$, $x_2$ be the tree stage without leaves at $t_2$, and $z_2$ be the fallen leaves at $t_2$. Why is $x_1$ continuous with $x_2$ rather than with $z_2$? If dead leaves don't count as cohesive biological individuals, the same question can be posed with $z_2$ as offspring. In response, one can distinguish between partial and total continuity: the continuity between $x_1$ and $z_1$ is partial—namely, it is only the leaves that are retained—whereas that between $x_1$ and $x_2$ is near-total, as it is the whole tree (minus its leaves) that is retained. In different contexts of inquiry and for different kinds of entities, different degrees of partial continuity will dictate whether a given $x_i$ is

---

[6] Note that one can differentiate one-level and two-level principles of composition, where the former characterize the composition relation in terms of relations between parts and wholes, and the latter characterize it in terms of relations between other items (e.g. regions, constituting stuffs) that are functionally related to parts and wholes (see Hawley 2006). Here as before, I assume that one-level principles are more germane to the temporal composition of biological processes.

considered a new individual or a continuation of the same individual. In general, however, the causal continuity relation $R$ can be specified as *relatively greater partial continuity* between stages whenever there are multiple stages partially continuous with the same earlier stage.

This interpretation of (IV), where $R$ is relatively greater partial continuity and the $x_n$ are cohesive individual stages, is capable of explaining the truth of (A) and (B) while also distinguishing between parent and offspring processes in a biologically satisfactory manner. (A) is true because the ciliated fluke and the non-ciliated fluke are qualitatively different cohesive stages ($x_n$), and (B) is true because they are related through causal continuity ($R$). When the liver fluke lays eggs in its definitive host, the eggs are stages of a distinct individual process because they have less partial continuity with earlier stages of the parent than the later parent stages do.

## 6. Conclusions

Process ontologies for the life sciences have continued to attract theorists for a variety of biological reasons. I have argued that, if this sort of ontology is empirically well motivated (where motivation is a matter of ontological explanatory power), then it should be an ontology in which the categorial roles for enduring substances are realized by perduring cohesive processes. This shift has the advantage of dropping the substance ontologist's commitment to qualitative essentialism and the too strong equation of individuality with particularity without any serious sacrifice of explanatory power.

The implications of adopting this sort of framework, given its high level of abstraction, are likely to be felt more in philosophical reflections on the life sciences than in the life sciences themselves. However, to the extent that scientific research is guided by theoretical and philosophical presuppositions, the processual perspective might help to overcome some unnecessary ideological obstacles to theoretical progress; for example, the notion that causation can only occur between particulars, or that informative kind classification requires qualitative essentialism, or that diachronic criteria of identity are necessary for distinguishing between biological individuals.

One interesting consequence of the idea of ontological explanation for scientific domains from section 2 is that, as long as a domain is still developing, the domain ontology cannot make a claim to completeness or finality. A process ontology for biology should therefore be viewed as a provisional categorial interpretation of the biological domain that is susceptible to revision as the science changes. In attempting to get clear on what biological processes are, however, at least we will have provided tools for revision.

## Acknowledgements

The author thanks Johanna Seibt, Jan Heylen, John Collier, Argyris Arnellos, Dan Nicholson, John Dupré, and two reviewers for helpful comments and discussion.

# References

Bertalanffy, L. von. (1952). *Problems of Life*. London: Watts & Co.
Bickhard, M. H. (2009). The Interactivist Model. *Synthese* 166: 547–91.
Carnap, R. (1949). *The Logical Syntax of Language*. London: Routledge & Kegan Paul.
Chisholm, R. M. (1976). *Person and Object: A Metaphysical Study*. London: Allen & Unwin.
Collier, J. (2003). Hierarchical Dynamical Information Systems with a Focus on Biology. *Entropy* 5: 100–24.
Collier, J. (2004). Self-Organization, Individuation and Identity. *Revue internationale de philosophie* 228: 151–72.
Collier, J. D. and Muller, S. J. (1998). The Dynamical Basis of Emergence in Natural Hierarchies. In G. Farre and T. Oksala (eds), *Emergence, Complexity, Hierarchy, and Organization: Selected and Edited Papers from the ECHO III Conference, Acta Polytechnica Scandinavica, MA 91* (n.p.). Espoo: Finnish Academy of Technology.
Davidson, D. (1970). The Individuation of Events. In N. Rescher (ed.), *Essays in Honor of Carl G. Hempel* (pp. 216–34). Dordrecht: D. Reidel.
Davidson, D. (1985). Reply to Quine on Events. In E. LePore and B. McLaughlin (eds), *Actions and Events: Perspectives on the Philosophy of Donald Davidson* (pp. 172–6). Oxford: Blackwell.
Davidson, E. H. (2010). Emerging Properties of Animal Gene Regulatory Networks. *Nature* 468: 911–20.
Davidson, E. H. and Erwin, D. H. (2006). Gene Regulatory Networks and the Evolution of Animal Body Plans. *Science* 311: 796–800.
Dupré, J. (2012). *Processes of Life*. Oxford: Oxford University Press.
Eronen, M. I. (2015). Robustness and Reality. *Synthese* 192 (12): 3961–77.
Flatt, T. and Heyland, A. (eds). (2011). *Mechanisms of Life History Evolution*. Oxford: Oxford University Press.
Fusco, G. (2001). How Many Processes Are Responsible for Phenotypic Evolution? *Evolution & Development* 3 (4): 279–86.
Ghiselin, M. (1974). A Radical Solution to the Species Problem. *Systematic Zoology* 23: 536–44.
Griesemer, J. (2000). Development, Culture, and the Units of Inheritance. *Philosophy of Science, Supplement: Proceedings of the 1998 Biennial Meetings of the Philosophy of Science Association* 67: S348–68.
Griffiths, P. E. (1994). Cladistic Classification and Functional Explanation. *Philosophy of Science* 61 (2): 206–27.
Guay, A. and Pradeu, T. (2015). To Be Continued: The Genidentity of Physical and Biological Processes. In A. Guay and T. Pradeu (eds), *Individuals across the Sciences* (pp. 317–47). New York: Oxford University Press.
Hawley, K. (2006). Principles of Composition and Criteria of Identity. *Australasian Journal of Philosophy* 84 (4): 481–93.
Horsten, L. (2010). Impredicative Identity Criteria. *Philosophy and Phenomenological Research* 80 (2): 411–39.
Huang, S. and Kauffman, S. (2009). Complex Gene Regulatory Networks: From Structure to Biological Observables: Cell Fate Determination. In R. A. Meyers (ed.), *Encyclopedia of Complexity and Systems Science* (pp. 527–60). Berlin: Springer.
Hull, D. L. (1978). A Matter of Individuality. *Philosophy of Science* 45: 335–60.
Jaeger, J. and Monk, N. (2015). Everything Flows: A Process Perspective on Life. *EMBO Reports* 16 (9): 1064–7.
Lewin, K. (1922). *Der Begriff der Genese in Physik, Biologie und Entwicklungsgeschichte: Eine Untersuchung zur vergleichenden Wissenschaftlehre*. Berlin: Springer-Verlag.

Lewis, D. K. (1986). *On the Plurality of Worlds*. Oxford: Blackwell.
Lowe, E. J. (1989). What Is a Criterion of Identity? *Philosophical Quarterly* 39 (154): 1–21.
Lowe, E. J. (2006). *The Four-Category Ontology: A Metaphysical Foundation for Natural Science*. Oxford: Clarendon.
Lowe, E. J. (2009). *More Kinds of Being: A Further Study of Individuation, Identity, and the Logic of Sortal Terms*. Oxford: Wiley Blackwell.
Mayr, E. (1942). *Systematics and the Origin of Species*. New York: Columbia University Press.
Merricks, T. (1998). There Are No Criteria of Identity Over Time. *Noûs* 32 (1): 106–24.
Mossio, M. and Moreno, A. (2010). Organisational Closure in Biological Organisms. *History and Philosophy of the Life Sciences* 32: 269–88.
Niklas, K. J., Bondos, S. E., Dunker, A. K., and Newman, S. A. (2015). Rethinking Gene Regulatory Networks in Light of Alternative Splicing, Intrinsically Disordered Protein Domains, and Post-Translational Modifications. *Fronteirs in Cell and Developmental Biology* 3: 1–13.
Okasha, S. (2002). Darwinian Metaphysics: Species and the Question of Essentialism. *Synthese* 131 (2): 191–213.
O'Neill, J. S. and Feeney, K. A. (2014). Circadian Redox and Metabolic Oscillations in Mammalian Systems. *Antioxidants & Redox Signaling* 20 (18): 2966–81.
Phalee, A., Wongsawad, C., Rojanapaibul, A., and Chai, J.-Y. (2015). Experimental Life History and Biological Characteristics of *Fasciola gigantica* (Digenea: Fasciolidae). *Korean Journal of Parasitology* 53 (1): 59–64.
Quine, W. Van Orman. (1939). Designation and Existence. *Journal of Philosophy* 36 (26): 701–9.
Quine, W. Van Orman. (1985). Events and Reification. In E. LePore and B. McLaughlin (eds), *Actions and Events: Perspectives on the Philosophy of Donald Davidson* (pp. 162–71). Oxford: Blackwell.
Roff, D. A. (1992). *The Evolution of Life Histories: Theory and Analysis*. New York: Chapman & Hall.
Sattler, R. (1990). Toward a More Dynamic Plant Morphology. *Acta Biotheoretica* 38: 303–15.
Sattler, R. (1992). Process Morphology: Structural Dynamics in Development and Evolution. *Canadian Journal of Botany* 70: 708–14.
Seibt, J. (2008). Beyond Endurance and Perdurance: Recurrent Dynamics. In C. Kanzian (ed.), *Persistence* (pp. 133–64). Frankfurt: Ontos Verlag.
Seibt, J. (2010). Particulars. In R. Poli and J. Seibt (eds), *Theory and Applications of Ontology: Philosophical Perspectives* (pp. 23–55). Dordrecht: Springer.
Seibt, J. (2015). Ontological Scope and Linguistic Diversity: Are there Universal Categories? *Monist* 98 (3): 318–43.
Sider, T. (1996). All the World's a Stage. *Australasian Journal of Philosophy* 74 (3): 433–53.
Sider, T. (1997). Four-Dimensionalism. *Philosophical Review* 106 (2): 197–231.
Sider, T. (2001). *Four-Dimensionalism: An Ontology of Persistence and Time*. Oxford: Oxford University Press.
Simons, P. (2003). Events. In M. Loux and D. Zimmerman (eds), *The Oxford Handbook of Metaphysics* (pp. 357–85). Oxford: Oxford University Press.
Stearns, S. C. (1992). *The Evolution of Life Histories*. Oxford: Oxford University Press.
Steward, H. (2013). Processes, Continuants, and Individuals. *Mind* 122: 781–812.
Thomasson, A. L. (2007). *Ordinary Objects*. Oxford: Oxford University Press.
True, J. R. and Haag, E. S. (2001). Developmental Systems Drift and Flexibility in Evolutionary Trajectories. *Evolution & Development* 3: 109–19.
Waddington, C. H. (1957). *The Strategy of the Genes*. London: Routledge.

Webster, G. and Goodwin, B. C. (1996). *Form and Transformation: Generative and Relational Principles in Biology*. Cambridge: Cambridge University Press.
Wiggins, D. (2001). *Sameness and Substance Renewed*. Cambridge: Cambridge University Press.
Wiggins, D. (2012). Identity, Individuation, and Substance. *European Journal of Philosophy* 20 (1): 1–25.
Wimsatt, W. (2007). *Re-Engineering Philosophy for Limited Beings. Piecewise Approximations to Reality*. Cambridge, MA: Harvard University Press.
Woodger, J. H. (1929). *Biological Principles: A Critical Study*. London: Routledge & Kegan Paul.

# 5
# Genidentity and Biological Processes

*Thomas Pradeu*

## 1. Introduction

What exactly is a process view of life? Philosophers of biology and biologists who have recently defended such a view (Dupré 2012; Bapteste and Dupré 2013; Dupré 2014) generally oppose *processes* (characterized by constant change) and *things* (characterized by stability and durability). But two major questions should be raised regarding this view:

(i) What kinds of relationships between 'things' and 'processes' are possible? Is it possible to countenance both 'things' and 'processes' as categories that describe the living world, or are these two categories incompatible?
(ii) Can a priority claim about 'things' and 'processes' be made and, if so, on what grounds? Here the question is to determine what comes first—things or processes. Two main types of priority claims can be made: an *ontological* claim (the biological world is actually made of processes, and things are only partial and temporary stabilizations of processes); or one made from an *epistemological* perspective (for us as human beings, the best way to understand the biological world is to get access to it in terms of processes).[1]

In this chapter I would like to show that a preexisting view of the biological world, called the 'genidentity view', helps—especially in the variant defended by David Hull—to clarify what a process view of life might be. More specifically, the genidentity view is useful because it suggests that (a) the notions of processes and things are both needed in biology, but processes are *prior* to things; and (b) the main interest of adopting a process view is epistemological, not ontological. One key underlying objective of the present chapter will be to address a question that seems decisive for anyone who proposes to conceive of the biological world in terms of processes: what difference does it make, in actual practice, to adopt a process view?

---

[1] On processes in general, and on the opposition between processes and substances, see Seibt 2013.

Let me start with a preliminary definition of the notion of 'genidentity'. In a nutshell, the genidentity view, which has been explored in the contexts of psychology, physics, and biology (Lewin 1922; Reichenbach 1956; Hull 1992; Boniolo and Carrara 2004; Pradeu and Carosella 2006b; Guay and Pradeu 2016b), says that the identity through time of an entity X is given by a well-identified series of continuous states of affairs. Of course, this claim is not sufficient in itself; every precise application of the genidentity view requires a clarification of exactly which continuous states are being followed, and why. In the next pages I examine in detail the concept of genidentity, and then I show why it could constitute a decisive building block for the project of developing a process view for biology. After a short reminder of the origins of the concept of genidentity, I describe its centrality in David Hull's reflections on biological identity. Following this, I suggest an extension of Hull's view on the basis of recent data that demonstrate the ubiquity of symbiotic interactions in the living world. Finally, I explain why genidentity prompts us to adopt a multilevel and mainly epistemological view on biological processes.

## 2. What Is Genidentity? And How Can It Be Applied to the Living World?

What constitutes the identity through time of an entity X? For instance, in what sense can I be said to be the 'same' as the child I was, a cat the 'same' as the kitten it was, or a wave the 'same' wave while it is moving through the sea? These questions are particular instances of the more general problem of diachronic identity (or identity through time), undoubtedly one of the most fundamental and most debated problems of all philosophy. The problem has been raised by major philosophers of the past, including Aristotle, Locke, and Leibniz. More recently, metaphysicians (both 'perdurantists' and 'endurantists') have offered important analyses of the same problem, for example David Wiggins (2001), Peter van Inwagen (1990), Theodore Sider (2001), and Katherine Hawley (2004).

One very interesting, though often neglected, way to address the problem of diachronic identity is to resort to the concept of genidentity. According to the genidentity view, the identity through time of an entity X is given by a well-identified series of continuous states of affairs. As will become clear in what follows, this view insists on the continuity of states rather than on the prior existence of objects; it conflicts radically with several forms of substantialism; and it is based on an epistemological (rather than ontological) attitude regarding science.

The concept of genidentity was proposed in 1922 by Kurt Lewin (1890–1947), a leading German American psychologist, as a way to better understand identity through time (Lewin 1922). It was neglected by most philosophers of science, but Hans Reichenbach (1891–1953) took it very seriously and explored it further. Indeed, Reichenbach examined different conceptions of genidentity and applied them to several physical cases (Reichenbach 1956).[2] Today, however, the notion of

---

[2] On the different versions of the genidentity concept for Reichenbach, see Padovani (2013) and Guay and Pradeu (2016b).

genidentity is rarely used in philosophy in general and almost never in the philosophy of science, where the exceptions are few and far between (e.g. Hull 1992; Boniolo and Carrara 2004; Pradeu and Carosella 2006b; Guay and Pradeu 2016b; see also chapters 2, 4, 7, and 11 here).

In its insistence on defining identity as mere continuity, the concept of genidentity echoes in part Locke's conception of identity. Indeed, in the second edition of his *Essay Concerning Human Understanding*, Locke (1975 [1694]) says that the long-sought 'principle of individuation' is to be found in a simple continuity of states. In the case of living things (plants or animals), the identity of a being is, according to Locke, the continuity of one and the same 'life'. This illustrates what has been said above, namely that genidentity views always need to make clear exactly what states are followed and why (here, the continuity of a 'life'). Applied to humans, Locke's view is as follows: 'This also shows wherein the Identity of the same Man consists; viz. in nothing but a participation of the same continued Life, by constantly fleeting Particles of Matter, in succession vitally united to the same organized Body' (Locke 1975 [1694]: §6, 331). Interestingly, in his *New Essays Concerning Human Understanding*, a text that constitutes a systematic response to Locke, Leibniz (1916 [1765]) strongly rejects Locke's conception of identity because, for Leibniz, continuity by itself is insufficient to define identity: 'By itself continuity no more constitutes substance than does multitude or number... *Something* is necessary to be numbered, repeated and continued' (Leibniz 1916 [1765]: 169).

Let us call *substantialism* the view according to which the identity of a thing X must be understood as the identity of a substance identified beforehand, and *continuism* the view according to which the identity of a thing X is given by a mere continuity of states (as defended by Locke). Substantialism is defended by Leibniz, but also by many contemporary philosophers, under different forms. One version of substantialism is *essentialism*, which states that what makes the identity of X through time is the fact that a core constituent or characteristic of X remains constant through time. In the case of living things, genetic essentialism says that a living thing remains the same through time in virtue of the fact that it possesses the same genome throughout (Kripke 1980). Another, significantly different version of substantialism is the 'neo-Aristotelian' view defended by Wiggins (2001). According to this view, the identity of a thing X is given by a sortal concept (a category), which defines a specific principle of activity (for example, it is possible to understand the identity of a given thing only by determining that it is, say, a dog and that what defines the identity of a dog is a certain principle of activity, common to all dogs).

I suggest here that the genidentity view constitutes a particularly interesting and fruitful version of continuism (and therefore a view that stands in contrast with substantialism in general), and that it can shed light on the question of the diachronic identity of living things. More precisely, I shall defend the thesis that genidentity is the best way to understand the diachronic identity of a living thing and that it helps to make the concept of biological process more precise. I first explain the centrality of the notion of genidentity in Hull's thinking about the problem of individuality and then I show how his view can be extended to reflect important findings in recent biology, in particular regarding the phenomenon of *symbiosis*.

## 3. The Inconspicuous Centrality of Genidentity in Hull's Conception of Biological Individuality

David Hull (1935–2010) was undoubtedly one of the most influential philosophers of biology. Hull mentioned the notion of genidentity several times in his writings (Hull 1986, 1992), but these mentions remained largely unnoticed and, intriguingly, most other philosophers of biology did not follow Hull and did not adopt this notion.

How does Hull specifically apply the idea of genidentity? In other words, which states does Hull think one should follow over time in order to understand the identity of a living thing? His answer is that one should follow the continuity of an *internal organization*. Let me try to explain this idea in more detail.

Even before Hull used the notion of genidentity explicitly, the idea behind it was already present in his writings. In particular, that idea underlies the view of identity presented in one of his most famous and influential texts: 'A Matter of Individuality' (Hull 1978). This paper is often seen as a defense of two theses: that species are individuals rather than classes, and that there is no 'human nature', no 'essence' of humanity (indeed, if humans are considered from the species point of view, that is, as tokens of *Homo sapiens*, it is impossible to define what are the necessary and sufficient characteristics that would make a given entity a human). Nevertheless, what in fact constitutes the basis of these two theses is the conception of identity Hull defends; a conception that is also at the heart of what is arguably the most important contribution of that paper, namely the two diagrams drawn by Hull. Let us now see what exactly is the conception of identity held by Hull and why it is important for our argument.

Hull's starting point is that, at least since Aristotle, most philosophers have had a naïve view of biological individuality. Philosophers often use fictitious examples and, when they do actually speak of a living thing, they generally mean an animal, even a higher vertebrate in most cases (a horse, a cat, etc.). In contrast, Hull insists on the importance of using biological examples that are both more realistic and more diverse. According to him, such examples are more interesting, more complex, and in the end more challenging than the fictitious examples and thought experiments favoured by metaphysicians (the same idea is developed in Hull 2001). Taking into account the actual diversity of the living world implies, for Hull, a suspicion towards conceptions of biological individuality based on common sense and intuitive perception. Indeed, common-sense individuation is too strongly biased by our relative size and perception abilities (Hull 1978, 1992). For example, in dealing with many plants, colonial animals, fungi, microbes, and so on, common sense individuation is of no help whatsoever. Some cases have been much discussed in the biological and in the philosophical literature, such as dandelions, aspens, social insects, ascidians, siphonophores, and biofilms (for overviews on these cases, see Pradeu 2012; Bouchard and Huneman 2013; Guay and Pradeu 2016a).

In contrast to common sense and intuitive perception, Hull seeks to offer a biologically precise criterion for the diachronic identity of biological individuals; and he finds this criterion in the idea of continuity of change. According to Hull, organisms and species belong to the same ontological category, as both must be understood as spatio-temporally localized entities. More radically, Hull's thesis is that any organism or any species is a portion of space and time. Every organism has a starting point and an end, and goes through different but continuous states between these two extremes. Exactly the same is true of every species. For Hull, because living things can undergo massive and unpredictable change, retention of substance (the idea that something of X remains through time) and resemblance (the idea that X looks sufficiently like itself) are useless criteria for biological diachronic identity. The only satisfying criterion is continuity of change.

To describe and defend this continuity-based conception of identity, Hull explicitly endorses the notion of 'genidentity' in several texts and grounds it in the idea of a continuous *internal organization*:

> Three traditional criteria for individuality in material bodies are retention of substance, retention of structure, and continuous existence through time (genidentity). If organisms are to count as individuals, then the first two criteria are much too restrictive. In point of fact, many organisms totally exchange their substance several times over while they retain their individuality. Others undergo massive metamorphosis as well, changing their structure markedly. If organisms are paradigm individuals, then retention of neither substance nor structure is either necessary or sufficient for continued identity in material bodies. The idea that comes closest to capturing individuality in organisms and possibly individuals as such is genidentity. As its name implies, this criterion allows for change just as long as it is sufficiently continuous. The overall organization of any entity can change but it cannot be disrupted too abruptly.
>
> (Hull 1992: 182)

This is the very same conception of identity that underlies Hull's (1978) fundamental diagrams; see Figures 5.1 and 5.2. These two figures offer a description of structural patterns of change in the living world that are equally applicable to organisms and species.

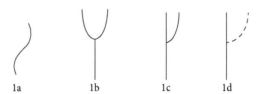

**Figure 5.1** Ontogenetic change and splitting giving rise to the production of new organisms (in the case of organisms), or phylogenetic change and splitting leading to speciation (in the case of species). Based on Hull 1978

**Figure 5.2** Total or partial merging between organisms, or between species. Based on Hull 1978

Figure 5.1 illustrates structural changes associated either with *change* in a living entity or with its *splitting* into two living entities. For Hull, a new entity emerges from a given entity if and only if the internal organization of the original entity is strongly disrupted.[3] This criterion—the *disruption of internal organization*—is not always easy to apply, and by definition the observer often faces a continuum of possible situations, but the examples given by Hull are very helpful. A living entity remains the same (case 1a) even if it undergoes a limited change, or even a radical change, provided that the continuity between these different states can be established (e.g. a caterpillar becoming a butterfly). In contrast, the phenomenon of *splitting* (case 1b) is characterized by a disruption of internal organization: one individual becomes two individuals and the initial individual disappears. Transverse fission in paramecia is an example. In other situations, an individual appears *on* another, preexisting individual, and this new individual becomes progressively autonomous (case 1c). An example is strobilization in certain forms of Scyphozoa (sometimes colloquially called "true jellyfish"). In yet other cases, a small part of an individual gains independence and becomes itself a new individual (case 1d; note that this is a part of an individual—not an individual growing on another, as in case 1c). An example is budding in Hydrozoa (Hydrozoa are Cnidaria that have both a polypoid and a medusoid stage in their life cycles—or at least most of them do). Though classifying all the diversity of real biological phenomena of change and splitting into these four cases would probably prove very difficult, what seems clear and useful is the criterion used by Hull, who asks systematically whether the overall organization of the entity under consideration is disrupted or not. (Importantly, there is transgenerational *material continuity* between a parent and its offspring, but they are characterized by two different *internal organizations*, and it is precisely this criterion that makes the difference between the continuity of one being and the continuity of several beings through reproduction; this issue is addressed in chapter 7.)

Figure 5.2 describes the *merging* of two living entities, or of their parts. To distinguish among the different possible situations, here again Hull uses the disruption of internal organization as a criterion. Two entities can fuse to become one single entity and remain one entity for a significant period of time, so the two initial individuals are lost (case 2a; fusion in amoebas does not constitute an adequate illustration of this case, while the fusion of two germ cells does). In other situations (case 2b), a portion of a first individual becomes a portion of a second individual, the two individuals continue their existence, but both have changed (the first has lost a part, the second has gained a part). Blood transfusion or bacterial conjugation are good examples. In still other situations, a portion of a first individual and a portion of a second individual merge to form a third (new) individual, while the two initial individuals continue their existence (case 2c). Sexual reproduction is a good example. Applied to species rather than organisms, a good example of 2b is introgression and a good example of 2c is speciation by polyploidy (a rather common event in plants, for instance).

---

[3] 'The relevant consideration is how much of the parent organism is lost and its internal organization disrupted' (Hull 1978: 345).

In conclusion, Hull endorsed the genidentity view. For him, what biologists can and must do to account for the continuously changing identity of any living thing is to follow its changes through time, keeping in mind that it remains the same only as long as its internal organization remains the same or changes progressively.

Naturally, one immediate difficulty faced by Hull's account of biological identity is to offer a precise definition of what 'internal organization' (and its disruption) means. Suggestions to move in this direction will be made in section 5; for now, I would like to show why recent biological data about symbiosis strengthen Hull's line of argument.

## 4. Why Cases of Symbiosis Strengthen the Genidentity View

As we have seen, Hull claims that, because living entities can change, merge, and split, it is crucial, in order to understand their diachronic individuality, to be able to actually follow them through time. But how frequent are events of merging and splitting in the living world? Though Hull mentions several important examples of fusion and splitting, symbiosis does not play an important role in his demonstration (apart from the rapid mention of the endosymbiotic event that is at the origin of some organelles). Now, research done on symbiosis in the twenty-first century shows that symbiotic events of fusion and splitting are much more frequent than had traditionally been assumed (McFall-Ngai 2002; McFall-Ngai et al. 2013; Gilbert and Epel 2015). In fact, we will see that the pervasiveness of symbiosis proves that Hull's diagrams, which at first sight might seem to concern only a limited number of biological cases, describe situations that are in fact very frequent in nature. Indeed, by taking into account symbioses, one realizes that living things commonly undergo events like a 2a (fusion) or a 2b (integration with continuation)—or, even more frequently, like an 'inverted 1c' (internalization; see Figure 5.3)—and that they can also split more often than is usually thought.

'Symbiosis' can be understood here in the very broad sense of any close and lasting interaction between two biological entities belonging to two different species. This is in accordance with the traditional definition of Anton de Bary, formulated in 1879 (see e.g. Sapp 1994). Adopting this broad definition is important here, as the

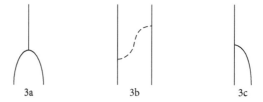

**Figure 5.3** Forms of integration of external biological material, at the organism level or the species level. The first case corresponds to Hull's 2a (fusion), the second to Hull's 2b (integration with continuation), while the third case is an inversion of Hull's 1c. The third case (absent in Hull's analysis) can be called 'internalization', and is described in this chapter as an extremely frequent (though long overlooked) phenomenon in nature

definition can cover cases that range from mutualism (in which the fitness of the two partners is increased by the interaction) to commensalism (a neutral interaction) and to parasitism (in which the fitness of one partner increases while the fitness of the other decreases). Indeed, all these different cases exist among the recently documented examples of symbioses.[4]

Important events of symbiotic fusions occur both at the level of organisms and at the level of species. At the level of organisms, symbioses, long thought to be rather rare, are now considered almost ubiquitous (McFall-Ngai et al. 2013; Gilbert and Epel 2015). Indeed, probably all organisms are hosts of many microorganisms, very often in close and long-lasting associations. Very well documented cases include plants, hydra, cnidarians, sponges, fishes, the squid *Euprymna scolopes*, insects, mice, and humans (Bosch and McFall-Ngai 2011; Nyholm and McFall-Ngai 2014). There is often a co-construction of the host and the microbes, as illustrated by cases such as legume–rhizobia (Oldroyd 2013) or *Euprymna scolopes–Vibrio fischeri* (Nyholm and McFall-Ngai 2004) interactions. In many cases the association is beneficial to one of the partners or to both. In particular, and perhaps counterintuitively, it is frequently the case that interactions with some microbes shape and strengthen the host's immune system (Pradeu and Carosella 2006b; Round and Mazmanian 2009). Very interesting cases are those where some microbes are indispensable for the development of the host (McFall-Ngai 2002; Pradeu 2011). The notion of a *heterogeneous organism* (Pradeu 2010, 2012) captures this very general idea, that all known organisms seem to harbour huge quantities of biological entities belonging to other species and that, in many of them, those biological entities become so integrated into the host's that they can be considered *part* of it. Which of these cases count as merging events, as described by Hull? The decision, for each, will depend, here again, on the degree of disruption of internal organization. For example, the legume–rhizobia symbiosis offers a clear case of merging, and, more precisely, of internalization (case 3c, which can also be called 'inverted 1c'). The plant could live without the bacteria, but the recruitment of the bacteria at the root level helps the plant decisively (the rhizobia create ammonia from nitrogen in the air, which is used by the plant to create amino acids and nucleotides), and the bacteria are very significantly transformed during the process, differentiating into bacteroids. The association between the two partners eventually constitutes a unit that displays a high degree of internal organization. In contrast, the colonization of a host by a microbe (be it pathogenic, commensal, or mutualistic) that would not remain for long in the host, would not be transformed by this interaction, and would not, in turn, have a deep effect on the overall organization of the host. Consequently, it would not count as a case of merging.

Symbiotic events of merging also happen at the species level. A nice example is the obligate symbiotic association between aphids and *Buchnera* symbionts (more precisely, *Acyrthosphion pisum* and *Buchnera aphidicola*). Approximately 160–280 million years ago (Shigenobu and Wilson 2011), an aphid ancestor was infected with a free-living eubacterium, and this eubacterium became established within aphid cells. The host and the *Buchnera* endosymbiont became interdependent and

---

[4] For complementary discussions of symbiosis, see chapters 1, 9, 10, and 15.

unable to survive without each other. The growth of *Buchnera* became integrated with that of the aphids, which acquired the endosymbionts from their mothers before birth. Speciation of host lineages was paralleled by divergence of associated endosymbiont lineages, resulting in parallel evolution of *Buchnera* and aphids (Baumann et al. 1995). Today the aphid–*Buchnera* association (almost all 4,000 extant species of aphid harbour an obligate *Buchnera* symbiont) constitutes one of the best-documented cases of obligate symbiosis, where neither of the partners can survive and reproduce without the other (Shigenobu and Wilson 2011). In such situations the physiological and reproductive integration between the host and the bacterium is so tight that it makes sense to talk about a single unit, constituted by this association.

Merging events can also happen between a virus and a host species (Pradeu 2016). For example, many parasitoid wasps have integrated a polydnavirus into their genome several million years ago. Such polydnaviruses have a beneficial effect on the parasitoid wasps: they enable them to realize their life cycle by laying their eggs into their hosts, where then their offspring grow, often killing the host progressively in the process. In fact, the wasp eggs can survive and develop only because a virus integrated into the wasp's genome actively counters the immune defense of the host larva (Edson et al. 1981; Espagne et al. 2004; Bézier et al. 2009). Many specialists consider that the virus has been so tightly integrated into the host genome that it is no longer possible to regard the virus and the wasp as separate entities (Roossinck 2015; on host–virus mutualisms, see also Virgin et al. 2009).

In many cases, therefore, symbiosis can give rise to new lineages, constituted by the merging of two individuals who belong to different species and subsequently reproduce as new reproductive units. The idea that symbiotic events are crucial in evolution and can even lead to the appearance of new species (i.e. symbiogenesis) is not new (Wallin 1927; Margulis and Fester 1991; Margulis and Sagan 2002), but this phenomenon has recently been illustrated by several examples, including some of those mentioned above. It seems legitimate, in these cases, to use the notions of *heterogeneous species* and *heterogeneous lineages*, on the model of the heterogeneous organism (mentioned above).[5]

What about cases of symbiotic splitting? Taking into account symbioses is likely to lead to a very dynamic view of biological individuality, because many symbiotic interactions change through time. For example, humans are hosts to billions of microbes from birth to old age, but the composition of their microbiome changes significantly through time. Immediately after birth, bacteria colonize the baby upon passage through the birth canal. The microbiota then has a complex history. There is a first period, until approximately the age of one, during which the microbiota has a rather simple composition (*Bifidobacteria* being usually highly abundant in human milk-fed infants); but this changes rapidly and a second period follows during which the microbiota becomes highly diverse (more than 1,000 species, with a clear domination of Firmicutes and Bacteroidetes), indeed unique to each individual,

---

[5] On this issue, see also Dupré and O'Malley 2009 and Bouchard 2010. More generally, on the integration of 'foreign' genetic material, in particular through horizontal gene transfer, see Doolittle and Bapteste 2007 and Bapteste et al. 2012.

and stabilizes (see Candela et al. 2012). This is even clearer in the many cases of parasitic symbioses where the association is transient, for example because the parasite leaves the first host in order to colonize a second, or because the host eliminates the parasite via its immune system. Importantly, even a transient interaction between two living things can lead to very significant changes in their respective internal organizations; hence there is no direct link between the robustness and durability of a symbiotic interaction and the extent to which it impacts the internal organization of the partners.

In conclusion, the double phenomena of merging and splitting happen successively in many instances of symbiotic interactions, probably reflecting complex physiological, ecological, and evolutionary exchanges between the two partners. From all the examples examined here it can be concluded that symbiotic events of merging and splitting are extremely frequent in nature, which makes Hull's analyses and diagrams even more useful than they might have seemed when the paper was published in 1978. The ubiquity of symbiosis decisively strengthens Hull's point that genidentity offers the best way to capture the individuality of biological entities through time. Indeed, it seems clear that using an essentialist account based on genetic homogeneity or an account based on similarity would be entirely inadequate. Only the idea of a continuous change enables us to follow in detail what contributes to the construction of a given living thing; and, here again, the criterion of the degree of disruption of internal organization seems a suitable guide for understanding biological diachronic individuality.

## 5. How Genidentity Helps Define What an Organism Is

The genidentity view seems very useful if you want to understand biological identity. Nevertheless, it faces a series of important challenges, and in fact it is likely that any process-based view of the living world will also have to meet those same challenges. The basic idea at the heart of the genidentity view is to follow a biological process through time. But how does one choose adequately which processes to follow? And how does one follow them in practice?

In my view, these questions are very important, and the answer to them will depend on who asks them and for what purpose. This is where it becomes clear that one of the main interests of the genidentity view is that it places the emphasis on an epistemological, rather than ontological, approach to processes. I do not think that it is possible to prove the ontological claim that the biological world is 'really' made of processes; and, if this is indeed the claim that process philosophers of biology want to make, then they must give an argument for it. However, it is possible to give good arguments in favour of the adoption of an *epistemological* process view and to show that, from this epistemological point of view, the decision to interpret the living world in terms of processes (rather than of already individualized things) makes an important difference to scientific work, because it leads to different perspectives and potentially to different experimental programs.

Biologists will decide which process or processes to follow according to their working questions. For example, one may ask how reproduction is achieved in a given species, or how metabolism is maintained in a cell, or how DNA transcription into RNA occurs. These different processes happen at different levels and involve many different entities. What is crucial is to decide which process will be followed and which criteria can help us consider that we are dealing with *one* continuous process. This is exactly the question raised by Hull. However, he only did so at a very general level, by talking about the maintenance or disruption of the 'internal organization' of an organism or a species. In my view, Hull was on the right track, but the notion of internal organization needs to be defined much more precisely in each specific biological context.

Here I propose to define in precise terms what internal organization is and what its maintenance or disruption entails at the level of an organism (a similar reflection can be produced, and indeed has been produced, at the species level; see e.g. Haber 2016). More specifically, I suggest that immunity helps to offer a more precise conception of genidentity applied to organisms (see also Pradeu 2012: 248-9). In all species (animals, plants, and also prokaryotes), the immune system plays a decisive role in the delineation of the boundaries of the organism because it constitutes a principle of inclusion–exclusion: the immune system is responsible for the rejection or tolerance of any given entity, which means that the immune system determines which entities will be part of the organism and which won't. Importantly, this discrimination mechanism is not based on the traditional 'self' versus 'nonself' distinction, according to which an organism would immunologically reject all foreign entities and would immunologically accept only constituents originating from itself. Actually, every organism harbours huge quantities of genetically foreign entities and triggers, everyday, effector immune responses that target endogenous constituents.[6]

An immunological approach leads to a definition of the organism based on the distinction between two different levels: that of biochemical interactions and that of immune interactions, both necessary to delineate the organism. From that point of view, an organism can be defined as follows:

*Definition*: Organism = a functionally integrated whole, made up of heterogeneous constituents that are locally interconnected by strong biochemical interactions, and controlled by systemic immune interactions.

(Pradeu 2010: 258; see also Pradeu 2012: 243-6)

This definition means that, when entities interact through regular biochemical interactions and are actively tolerated by the continuous action of an immune system, they are part of a higher-level entity, which should be called an 'organism'. Of course, this definition places a strong emphasis on the role of the immune system in the definition of the organism, but it does so on the basis of the argument that the immune system plays a decisive role in delineating the boundaries of any organism.

---

[6] Much more specifically, the discontinuity theory of immunity that I have constructed with immunologists (Pradeu and Carosella 2006a; Pradeu et al. 2013) reflects directly a genidentity perspective. Indeed, a crucial claim of the most elaborate versions of the genidentity view, including Reichenbach's (see Guay and Pradeu 2016b), is that what matters to understanding diachronic identity is not the *degree* of change (i.e. how much it changes) but the *rate* of change (i.e. how fast it changes). The discontinuity theory of immunity is based on the principle that the immune system responds to sudden modifications of the antigenic motifs with which it interacts (for further details, see Pradeu 2012).

This definition rests on the recognition of two layers of interactions (biochemical interactions and immune interactions), which can be seen as a way to make more precise process approaches to the living world. What is suggested here is that there can exist some coalescences of interrelated processes, such as the organism itself, and that following such a coalescence of processes through time might rest on the identification of higher-level processes that control lower-level processes. Thus, an organism can be understood as a local concentration of intertwined biochemical processes under the control of higher-level immunological processes. More generally, it is crucial for a process philosophy to be able to identify not only processes in general, but also 'bundles' of processes (in this case, the organism), as well as to ask how the unity and cohesiveness of these bundles through time is achieved.

The crucial point is that, with the definition of the organism given above, we do not start with a preexisting delineation of the organism and subsequently say that the immune system controls this preexisting delineation. On the contrary, we start with biochemical and immunological interactions and, from the observation of how these interactions work, we deduce what the boundaries of the organism are. In this view, therefore, what comes first is interactions, and the organism 'supervenes' on those interactions. To understand this point fully, it is useful to move away from familiar mammalian examples and to examine more complex cases of biological individuality, in particular colonial organisms. A particularly illuminating case is that of *Botryllus schlosseri* (see Figure 5.4). *Botryllus*, born as a chordate tadpole larva, metamorphs into a sessile, invertebrate juvenile, after which it begins a lifelong, recurring budding process that results in a colony of expanding, asexually derived individuals. The colony is made of genetically identical individuals (zooids) united by a common extracorporeal vasculature. The zooids and the vasculature are embedded in a cellulose-based tunic, and the extracorporeal vasculature ramifies throughout this matrix and at the periphery terminates in finger-shaped projections

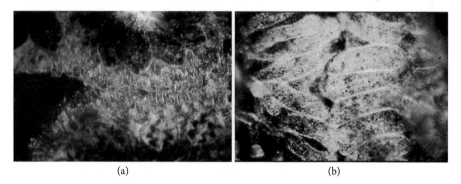

(a)  (b)

**Figure 5.4** Rejection between two colonies of Botryllus schlosseri. Panel A shows an example of rejection between two colonies of Botryllus schlosseri at the level of the colonies themselves; the brown zones at the point of contact between the ampullae (the finger-like structures at the center of the image) show the starting point of the reaction of rejection. Panel B shows an example of rejection between two colonies of Botryllus schlosseri at the much more precise level of the ampullae. Photographs courtesy of Tony De Tomaso, University of California, Santa Barbara

called 'ampullae'. When two colonies meet, an allorecognition reaction occurs, which leads either to vascular fusion or to rejection. Allorecognition is controlled by a single, highly polymorphic locus (the Fu/HC), and the rejection is realized through the triggering of an immune response (Scofield et al. 1982; Nyholm et al. 2006; McKitrick and De Tomaso 2010; McKitrick et al. 2011).

In the case of *Botryllus schlosseri*, as in many other cases of colonial organisms, it is very difficult to say whether what should count as an individual organism is each zooid or the colony as a whole—or perhaps both. In fact, common sense cannot decide between those options. According to the view I have presented, it is the observation of how immune responses occur that tells us what should count as an individual organism. As illustrated in Figure 5.4, immunologically controlled fusion or rejection in *Botryllus schlosseri* occurs at the level of the whole colony, so it is the colony that must count as an individual organism, in accordance with the definition presented above.

The example of *Botryllus* illustrates one very important advantage of the approach presented here: we start with an indistinct, non-individualized reality, about which common sense and perception have little to say, and it is the decision to follow immunological processes that leads us to conclude, in a scientifically precise way, what counts as an individual entity and what its boundaries are.

## 6. Genidentity as a Way to Shed Light on the Notion of Biological Process: 'Priority' as the Central Question

I believe that the multilayered genidentity approach developed here can be useful to the current trend towards a processual biology (Dupré 2012; Bapteste and Dupré 2013; Dupré 2014).

First, it emphasizes the importance of an epistemological approach to processes. Though process proponents often think in ontological terms (this is very explicit, e.g., in Bapteste and Dupré 2013), I do not think that they have hitherto offered compelling arguments for this view. With an epistemological approach, the aim is, more modestly, to show that a process view can make a difference to the actual work of biologists, as a question framed in terms of what processes biologists are interested in and how they should follow those processes through time. (A nice example is the argument offered in Dupré and Guttinger 2016.)

Second, this approach clarifies the idea, often expressed by process philosophers, that we live in a world of change. According to the genidentity view, what is interesting is to study how things change, at which pace, and when they start and cease to exist. Within such a perspective change is pervasive, and what is derivative is not change but the apparent absence of change, that is, regularities and sameness, and therefore even apparent stability must be explained in terms of constantly changing processes.

Third and perhaps most importantly, the genidentity view shows that the crucial claim of a process approach is in fact a claim about *priority*. Indeed, a process approach does not emphasize the importance of change itself (substantialist philosophers perfectly admit that every substance changes constantly) so much as it asks whether the 'thing' or the 'process' comes first; and it answers that things come second, after processes, or derive from them. In other words, the important move of a processual perspective is not, in my view, to go from individuals to processes (by saying, for instance, that individuals are in fact processes), but to go from processes to individuals (as scientists or philosophers, we *decide* to follow some scientifically meaningful processes, and individuals supervene on these processes). This is exactly what I have tried to illustrate in the case of immunology: in this case, it is scientifically identified processes that tell us where the individual lies and what its boundaries are, and not vice versa. We cannot start with the 'thing' *Botryllus schlosseri*, because we just don't know where a *Bottrylus schlosseri* starts and ends; the only solution is to start with processes—namely, as I suggested, biochemical and immune processes—and it is the realization of these processes that tells us where the individual is and what its boundaries are. The example of *Botryllus schlosseri* constitutes in fact a good model for thinking more generally about the individuation of living entities in other species: we cannot assume that we start by identifying an entity and then ask what processes occur in that entity, because we cannot trust our intuitions and perceptions with identifying living entities. So in each case—even in seemingly 'intuitive' cases, such as that of humans—it is necessary to start with scientifically well-defined processes—here, immunological processes; and it is those processes that tell us where the individual is and what its boundaries are.

## 7. Conclusions

The genidentity view seems particularly well suited if we wish to understand the diachronic identity of living things. Given the frequency of events such as extensive structural changes (e.g. through metamorphosis), splitting, and fusion, conceptions of biological identity based on similarity or substance are highly problematic. Hull (1978) perceived this point very well, but we have seen that recent work on symbiosis shows that his view is probably even more compelling today than it was in the 1970s. So the first lesson of the present chapter is that the genidentity view is a very satisfying way to conceive of biological identity.

The second lesson of this chapter is that the genidentity view sheds an important light on process views in biology and philosophy of biology (Dupré 2012; Bapteste and Dupré 2013; Dupré 2014). Indeed, it emphasizes the importance of an epistemological and multilayered approach to processes and suggests that the main claim of a process view is one of priority, namely that processes come first and make it possible to define things, and not the other way around.

Even though the notion of genidentity has not been very popular among philosophers of science, there is still perhaps much that we can learn from it.

## Acknowledgements

I would like to thank Alexandre Guay for our long-lasting mutualistic collaboration. For previous discussions on identity, genidentity, and processes, I thank Giovanni Boniolo, John Dupré, Adam Ferner, Steven French, Peter Godfrey-Smith, Matt Haber, Johanna Seibt, David Wiggins, and Charles Wolfe. Three anonymous referees made very useful comments on the first draft of this chapter. This project has received funding from the European Research Council (ERC) under the European Union's Horizon 2020 research and innovation programme—grant agreement n° 637647—IDEM.

# References

Bapteste, E. and Dupré, J. (2013). Towards a Processual Microbial Ontology. *Biology & Philosophy* 28 (2): 379–404.

Bapteste, E., Lopez, P., Bouchard, F., Baquero, F., McInerney, J. O., and Burian, R. M. (2012). Evolutionary Analyses of Non-Genealogical Bonds Produced by Introgressive Descent. *Proceedings of the National Academy of Sciences* 109 (45): 18266–72.

Baumann, P., Baumann, L., Lai, C. Y., Rouhbakhsh, D., Moran, N. A., and Clark, M. A. (1995). Genetics, Physiology, and Evolutionary Relationships of the Genus Buchnera: Intracellular Symbionts of Aphids. *Annual Review of Microbiology* 49: 55–94.

Bézier, A., Annaheim, M., Herbinière, J., Wetterwald, C., Gyapay, G., Bernard-Samain, S., et al. (2009). Polydnaviruses of Braconid Wasps Derive from an Ancestral Nudivirus. *Science* 323 (5916): 926–30.

Boniolo, G., and Carrara, M. (2004). On Biological Identity. *Biology & Philosophy* 19 (3): 443–57.

Bosch, T. C. G., and McFall-Ngai, M. J. (2011). Metaorganisms as the New Frontier. *Zoology* 114 (4): 185–90.

Bouchard, F. (2010). Symbiosis, Lateral Function Transfer and the (Many) Saplings of Life. *Biology & Philosophy* 25 (4): 623–41.

Bouchard, F., and Huneman, P. (2013). *From Groups to Individuals: Perspectives on Biological Associations and Emerging Individuality*. Cambridge, MA: MIT Press.

Candela, M., Biagi, E., Maccaferri, S., Turroni, S., and Brigidi, P. (2012). Intestinal Microbiota Is a Plastic Factor Responding to Environmental Changes. *Trends in Microbiology* 20 (8): 385–91.

Doolittle, W. F., and Bapteste, E. (2007). Pattern Pluralism and the Tree of Life Hypothesis. *Proceedings of the National Academy of Sciences* 104 (7): 2043–9.

Dupré, J. (2012). *Processes of Life: Essays in the Philosophy of Biology*. Oxford: Oxford University Press. http://www.oxfordscholarship.com/view/10.1093/acprof:oso/9780199691982.001.0001/acprof-9780199691982.

Dupré, J. (2014). Animalism and the Persistence of Human Organisms. *Southern Journal of Philosophy* 52: 6–23.

Dupré, J. and Guttinger, S. (2016). Viruses as Living Processes. *Studies in History and Philosophy of Biological and Biomedical Sciences*.

Dupré, J. and O'Malley, M. (2009). Varieties of Living Things: Life at the Intersection of Lineage and Metabolism. *Philosophy & Theory in Biology* 1: e003.

Edson, K. M., Vinson, S. B., Stoltz, D. B., and Summers, M. D. (1981). Virus in a Parasitoid Wasp: Suppression of the Cellular Immune Response in the Parasitoid's Host. *Science* 211 (4482): 582–3.

Espagne, E., Dupuy, C., Huguet, E., Cattolico, L., Provost, B., Martins, N., et al. (2004). Genome Sequence of a Polydnavirus: Insights into Symbiotic Virus Evolution. *Science* 306 (5694): 286–9.

Gilbert, S. F. and Epel, D. (2015). *Ecological Developmental Biology*, 2nd edn. Sunderland, MA: Sinauer Associates.

Guay, A. and Pradeu, T. (2016a). *Individuals across the Sciences*. New York: Oxford University Press.

Guay, A. and Pradeu, T. (2016b). To Be Continued: The Genidentity of Physical and Biological Processes. In A. Guay and T. Pradeu (eds), *Individuals across the Sciences* (pp. 317–47). New York: Oxford University Press.

Haber, M. H. (2016). The Biological and the Mereological. In A. Guay and T. Pradeu (eds), *Individuals across the Sciences* (pp. 295–316). New York: Oxford University Press. http://www.oxfordscholarship.com/view/10.1093/acprof:oso/9780199382514.001.0001/acprof-9780199382514-chapter-16.

Hawley, K. (2004). *How Things Persist*. Oxford: Oxford University Press.

Hull, D. L. (1978). A Matter of Individuality. *Philosophy of Science* 45 (3): 335–60.

Hull, D. L. (1986). Conceptual Evolution and the Eye of the Octopus. In R. B. Marcus, G. J. W. Dorn, and P. Weingartner (eds), *Logic, Methodology, and Phillosophy of Science* (pp. 643–65). Amsterdam: Elsevier. http://www.sciencedirect.com/science/article/pii/S0049237X09707177.

Hull, D. L. (1992). Individual. In E. F. Keller and E. A. Lloyd (eds), *Keywords in Evolutionary Biology* (pp. 181–7). Cambridge, MA: Harvard University Press.

Hull, D. L. (2001). *Science and Selection: Essays on Biological Evolution and the Philosophy of Science*. Cambridge: Cambridge University Press.

Kripke, S. A. (1980). *Naming and Necessity*, rev. edn. Oxford: Blackwell.

Leibniz, G. W. (1916). *New Essays Concerning Human Understanding*, 2nd edn, ed. by K. Gerhardt, trans. by A. G. Langley. Chicago: Open Court.

Lewin, K. (1922). *Der Begriff der Genese in Physik, Biologie und Entwicklungsgeschichte: eine Untersuchung zur vergleichenden Wissenschaftslehre*. Berlin: Springer.

Locke, J. (1975 [1694]). *An Essay Concerning Human Understanding*, ed. by P. H. Nidditch. Oxford/New York: Clarendon/Oxford University Press.

Margulis, L. and Fester, R. (eds). (1991). *Symbiosis as a Source of Evolutionary Innovation: Speciation and Morphogenesis*. Cambridge, MA: MIT Press.

Margulis, L. and Sagan, D. (2002). *Acquiring Genomes: A Theory of the Origins of Species*. New York: Basic Books.

McFall-Ngai, M. (2002). Unseen Forces: The Influence of Bacteria on Animal Development. *Developmental Biology* 242 (1): 1–14.

McFall-Ngai, M., Hadfield, M. G., Bosch, T. C. G., Carey, H. V., Domazet-Lošo, T., Douglas, A. E., et al. (2013). Animals in a Bacterial World: A New Imperative for the Life Sciences. *Proceedings of the National Academy of Sciences* 110 (9): 3229–36.

McKitrick, T. R. and De Tomaso, A. W. (2010). Molecular Mechanisms of Allorecognition in a Basal Chordate. *Seminars in Immunology* 22 (1): 34–8.

Nyholm, S. V. and McFall-Ngai, M. J. (2004). The Winnowing: Establishing the Squid-Vibrio Symbiosis. *Nature Reviews: Microbiology* 2 (8): 632–42.

McKitrick, T. R., Muscat, C. C., Pierce, J. D., Bhattacharya, D., and De Tomaso, A. W. (2011). Allorecognition in a Basal Chordate Consists of Independent Activating and Inhibitory Pathways. *Immunity* 34 (4): 616–26.

Nyholm, S. V. and McFall-Ngai, M. J. (2014). Animal Development in a Microbial World. In A. Minelli and T. Pradeu (eds), *Towards a Theory of Development* (pp. 260–74). Oxford: Oxford University Press.

Nyholm, S. V., Passegue, E., Ludington, W. B., Voskoboynik, A., Mitchel, K., Weissman, I. L., and De Tomaso, A. W. (2006). Fester, a Candidate Allorecognition Receptor from a Primitive Chordate. *Immunity* 25 (1): 163–73.

Oldroyd, G. E. D. (2013). Speak, Friend, and Enter: Signalling Systems That Promote Beneficial Symbiotic Associations in Plants. *Nature Reviews: Microbiology* 11 (4): 252–63.

Padovani, F. (2013). Genidentity and Topology of Time: Kurt Lewin and Hans Reichenbach. In N. Milkov and V. Peckhaus (eds), *The Berlin Group and the Philosophy of Logical*

*Empiricism* (pp. 97–122). Dordrecht: Springer. http://link.springer.com/chapter/10.1007/978-94-007-5485-0_5.

Pradeu, T. (2010). What Is an Organism? An Immunological Answer. *History and Philosophy of the Life Sciences* 32: 247–68.

Pradeu, T. (2011). A Mixed Self: The Role of Symbiosis in Development. *Biological Theory* 6 (1): 80–8.

Pradeu, T. (2012). *The Limits of the Self: Immunology and Biological Identity*. New York: Oxford University Press.

Pradeu, T. (2016). Mutualistic Viruses and the Heteronomy of Life. *Studies in History and Philosophy of Biological and Biomedical Sciences* 58: 80–8.

Pradeu, T. and Carosella, E. (2006a). On the Definition of a Criterion of Immunogenicity. *Proceedings of the National Academy of Sciences* 103 (47): 17858–61.

Pradeu, T. and Carosella, E. (2006b). The Self Model and the Conception of Biological Identity in Immunology. *Biology & Philosophy* 21 (2): 235–52.

Pradeu, T., Jaeger, S., and Vivier, E. (2013). The Speed of Change: Towards a Discontinuity Theory of Immunity? *Nature Reviews Immunology* 13 (10): 764–9.

Reichenbach, H. (1956). *The Direction of Time*. Berkeley: University of California Press.

Roossinck, M. J. (2015). Move over Bacteria! Viruses Make Their Mark as Mutualistic Microbial Symbionts. *Journal of Virology* 89 (13): 6532–5.

Round, J. L. and Mazmanian, S. K. (2009). The Gut Microbiota Shapes Intestinal Immune Responses during Health and Disease. *Nature Reviews Immunology* 9 (5): 313–23.

Sapp, J. (1994). *Evolution by Association: A History of Symbiosis*. New York: Oxford University Press.

Scofield, V. L., Schlumpberger, J. M., West, L. A., and Weissman, I. L. (1982). Protochordate Allorecognition Is Controlled by a MHC-like Gene System. *Nature* 295 (5849): 499–502.

Seibt, J. (2013). Process Philosophy. In E. N. Zalta (ed.), *The Stanford Encyclopedia of Philosophy*. http://plato.stanford.edu/archives/fall2013/entries/process-philosophy.

Shigenobu, S. and Wilson, A. C. C. (2011). Genomic Revelations of a Mutualism: The Pea Aphid and Its Obligate Bacterial Symbiont. *Cellular and Molecular Life Sciences: CMLS* 68 (8): 1297–1309.

Sider, T. (2001). *Four-Dimensionalism: An Ontology of Persistence and Time*. Oxford: Clarendon.

Van Inwagen, P. (1990). *Material Beings*. Ithaca, NY: Cornell University Press.

Virgin, H. W., Wherry, E. J., and Ahmed, R. (2009). Redefining Chronic Viral Infection. *Cell* 138 (1): 30–50.

Wallin, I. E. (1927). *Symbionticism and the Origin of Species*. Baltimore: Williams & Wilkins.

Wiggins, D. (2001). *Sameness and Substance Renewed*. Cambridge: Cambridge University Press.

# 6
# Ontological Tools for the Process Turn in Biology
## Some Basic Notions of General Process Theory

*Johanna Seibt*

## 1. Introduction

Recently philosophers of biology have presented arguments calling for a reconceptualization of the biological domain that is focused on processes, or even exclusively formulated in terms of processes (Dupré 2012; Falkner and Falkner 2013; Koutroufinis 2014b). Such arguments for a 'process turn' in biology could be strengthened if one could show that recasting biological phenomena in the classificatory terms of a sufficiently precisely formulated process-ontological framework can increase explanatory depth in biology and serve as a heuristic for empirical research. But is there an ontological theory out there that those interested in the process turn in biology could turn to?

From Aristotle onwards, ontology has been under the spell of what I have called the 'myth of substance'—a set of unreflected presuppositions for ontological theory construction that prescribe a focus on static entities, mainly a dualism of particulars and universals, as the most 'natural' way to describe the structure of the world.[1] One good antidote to the myth of substance might appear to be Whitehead's 'philosophy of organism', a metaphysics that presents reality as patterns of events. A small but growing number of philosophers of science are currently exploring Whiteheadian reconceptualizations of the domains of empirical science.[2] The purpose of this chapter is to introduce in outline another candidate process ontology, general process theory (GPT), as an auxiliary conceptual framework for the process turn in biology.

---

[1] On the notion, definition, and detailed studies of the counterproductive effects of the myth of substance or substance paradigm, see e.g. Seibt 1990, 1994, 1995, 1996, 1997, 2005, 2008, 2010, 2014, 2015. I am indebted to discussions with L. B. Puntel and W. Sellars for identifying some of the elements (about twenty) of the substance paradigm; see also Puntel 2002, 2010 and Sellars 1960.

[2] For Whiteheadian process chemistry, see e.g. Earley 2013; for Whiteheadian neurophysiology, see e.g. Brown 2005; for Whiteheadian quantum physics, see e.g. Hättich 2004; Eastman, Epperson, and Griffin 2016. Koutroufinis 2014b collects several proposals for Whiteheadian approaches to biology.

According to GPT, the world is 'the ongoing tissue of goings-on', as Sellars (1981: 57) put it—or, somewhat more precisely, the interacting of more or less generic interaction dynamics.

I will proceed as follows. In section 2 I introduce the new category of the monocategorial framework GPT, called 'general processes' or 'dynamics', and argue that the features of this category actually should be quite familiar to us, since they dovetail with our reasoning about subjectless activities. In section 3 I set out some elements of a non-standard mereology in terms of which relationships between general processes can be formally defined. In section 4 I sketch the dimensions of the classification system of GPT by means of which general processes can be diversified into many types of processes. In particular, I show how the classificatory parameters of GPT can be used to distinguish between processes that have the temporal and logical structure of a 'goal-driven' development, while others have the temporal and logical structure of 'non-developmental' or 'dynamically stable' temporally unbounded activities that persist in time by literal recurrence. In section 5 I offer some ideas on where in the current ontological debate in philosophy of biology the special constructional features of GPT could provide conceptual support for arguments in favour of a process-geared description of the living world. In passing I supply two reasons why Whitehead's philosophy of organism, albeit the most fully worked out process metaphysics to date, may prove an obstacle rather than an aid for the good cause of inviting a revisualization of the biological domain in terms of processes.

Before setting out let me insert two cautionary remarks. First, GPT is a process ontology but not a metaphysics—neither a speculative metaphysics like Whitehead's nor a 'realist' metaphysics, at least not in any of the problematic senses of 'realism' that are currently connected with the recent return to pre-Kantian meta-philosophies in analytical philosophy. Even though GPT lends itself to combinations with a naturalist or scientific realist position in metaphysics, it is itself an ontological domain theory in the Carnapian vein, specifying truth makers for true statements of common sense or science; it merely aspires, as part of an enterprise of philosophy as rational explication, to describe which entities we could rationally take to make true the accepted-as-true sentences $S_i$ of a natural language or scientific theory L.[3] Second, given that I am not a philosopher of biology, the applications of GPT I offer in the last part of this chapter are presumably of heuristic value only. But, since GPT aims to reconstruct the entitative commitments of our everyday common-sense reasoning, and since it is this kind of reasoning that we employ to understand the non-mathematical content of scientific claims, the following outline and some illustrations of possible application paths will make, I hope, for a useful preface to a process ontology of biology.[4]

---

[3] For more details on methodology, see Seibt 1995, 1997, 2000a, 2000c, 2005. On the specific issue of the relation between ontological and scientific theories, see also chapter 4.

[4] The sketch I will present here contains many shortcuts and omissions. For this reason I supply throughout the chapter references to more detailed expositions of single aspects of GPT.

## 2. General Processes or Dynamics: A New Category

GPT is a mono-categorial ontology that postulates only one type of entity, called 'general process' or, to allow for occasional transnumeral references, 'dynamics'. According to GPT, all there is—that is, all the different sorts of entities we speak about in common sense and in science (including in the humanities)—is one variety or other of a general process or dynamics. The main task of GPT is to *differentiate* this basic entity type into subtypes that can form the ontological correlates for true sentences of (part of) some theory T (in common sense or in science).

Like all ontological category terms, the label 'general process' (or 'dynamics') is a theoretical term that receives its meaning and explanatory force from two sources: (i) axiomatically, that is, from the differentiating definitions and principles of GPT; and (ii) from its 'model' or analogical illustration. Since the category of 'general process' is characterized by a feature combination that has not been explored in ontology so far, it will be best to begin by setting out the model of general processes.[5] As we will see, the sort of existence articulated in the category features of general processes may be new in ontology owing to constraints on theory construction introduced by the myth of substance, but it is a familiar element of our common sense and scientific reasoning about the world we live in.

General processes are modeled on 'subjectless' (C. D. Broad) or 'pure' (W. Sellars) activities, as these are denoted by sentences that merely affirm the presence of a dynamic feature, such as 'it is snowing', 'it is itching', 'the fire is spreading', 'photosynthesis occurs everywhere in your garden'. The concept of a subjectless activity consists in its inferential role (in a given language), that is, in the set of inferences that are licensed (and not licensed) by statements about subjectless activities. Subjectless activities qualify as a model for the postulated ontological category of 'general processes' or 'dynamics' to the extent that the inferential role of statements about subjectless activities illustrates the seven category features in terms of which general processes are defined. These seven categorial features are as follows:

(i) Statements about subjectless activities do not license the inference that there is one unique and discrete spatial location or temporal period where the denoted activities occur. A subjectless activity is an entity occurring *somewhere* in space and time; in other words, an entity that is concrete, yet general or non-particular, as these category features are commonly defined. While a particular entity necessarily occurs 'uniquely', that is, in one spatial location at any time of its existence, a general entity may occur 'multiply', that is, in several spatial locations at the same time.

(ii) Non-particular entities cannot be individuated in terms of their space–time location, as particular entities typically are. Instead, as subjectless activities illustrate, they are *individuated in terms of their typical functioning within a*

---

[5] Ontological categories are commonly characterized and coarsely differentiated from each other in terms of theoretical predicates for category features such as *particular, universal, concrete, abstract, discrete, countable, persistent,* etc.; many of these predicates are related to the way in which a type of entity exists in space and time. Note that the term 'space–time' location is used as an abbreviation for 'spatial location at a time", not in the sense of relativity theory.

*dynamic context*. When we say, for example, that 'it is snowing, not raining', 'the fire has stopped, but not the radiation', 'on the West coast there is more wind erosion than water erosion', 'there's water in the fridge but no milk', or 'you can't see the gin in your gin and tonic', we individuate items in terms of what they commonly 'do', engender, or are involved in.

(iii) Subjectless activities are occurrences in their own right rather than modifications of persons or thing-like things. Unlike properties and relations, they are *independent* in the sense that sentences such as 'it is snowing', 'today's rush hour was particularly bad', 'an immune reaction occurred in the sample', or 'high energy radiation was hitting the atmosphere during period T' do not entail either statements about determinate single snowflakes, cars, enzyme molecules, or protons, respectively, or statements about any sort of medium or carrier for these activities.

(iv) Subjectless activities are temporally extended and, like things, they are good illustrations of the category feature of being an *enduring* entity, that is, an entity that persists through time by being 'identical' in time.

(v) Quite unlike things and much like stuffs (water, wood, etc.), subjectless activities are not *countable* in the sense that they do not necessarily occur in space and time pre-packaged into discrete spatio-temporally extended units that afford our common practices of counting. Since subjectless activities are individuated or differentiated in terms of their functional features, we can also count them in this way. For instance, metabolism and photosynthesis are two activities, and so are C3 photosynthesis and C4 photosynthesis.

(vi) In our reasoning about (subjectless) activities we 'zoom in and out'. We accept that there are highly generic and also highly specific activities. For example, we may say that in a tobacco plant the following activities occur: photosynthesis, C3 photosynthesis, C3 photosynthesis in a tobacco plant, C3 photosynthesis in a tobacco plant on a window sill directed eastward, C3 photosysthesis in a tobacco plant on a window sill with spatial coordinates $<x, y, z>$ at time $t'$, and so on. In other words, (subjectless) activities are a good model for entities that are said to be concrete, yet more or less *indeterminate*. Since activities are more or less indeterminate, any activity is possibly multiply recurrent in space and time. This holds even when a space–time location is among its 'determinations' (that is, when it is expressly specified). The activity denoted by the phrase 'photosynthesis extending over the space–time region $<x, y, z, t>$' might de facto not recur in space or time, but this is merely a contingent fact. In other words, the location serves merely to ensure reference to a highly specific and contingently unrepeated yet in principle repeatable activity.

(vii) Subjectless activities are not changes. Constitutive 'phases' of a subjectless activity—for example, the change of place of every single flake that constitutes the dynamicity of the snowing—contribute to the activity's occurring, but not *as* temporal stages or phases. So our reasoning about subjectless activities provides us also with a model for the category feature of being a *dynamic* entity—a category feature rarely used in ontology so far—in a sense where such dynamicity is not immediately associated with the 'telic' or directed dynamicity of developments and with internal temporal differentiation into phases of different kinds.

In sum, then, guided by familiar patterns of common-sense reasoning, we can claim that subjectless activities can serve as a model for an entity that is concrete, non-particular, enduring, more or less indeterminate, dynamic, and individual in the sense of being functionally distinct from others.

## 3. Relationships among General Processes: GPT's 'Levelled Mereology'

GPT is a domain theory that postulates one basic category or entity type and one basic relationship that holds among such entities, namely the relationship of 'being part of' in its most basic sense of 'belonging with', which in everyday speech is used with any entity type.[6] In order to capture the sense of this notion of 'being part of', one cannot, however, resort to any of the standard mereological systems of so-called classical extensional mereology or to their intensional modifications, for these systems axiomatize part–whole relations that are informationally richer (e.g. 'is a spatial part of', 'is a material part of', 'is a functional part of', etc.).[7] Thus the inferential meaning of 'is part of', in its most basic and generic sense of asymmetric 'belonging with', must be captured within a non-standard mereology called 'levelled mereology' (LEM), which operates with an irreflexive, antisymmetric, and non-transitive part relation. In other words, according to the axioms of LEM, when we claim that dynamic D1 is part of dynamic D2, and D2 is part of D3, we speak of parts at different 'levels' of conceptual partition of a phenomenon, and we cannot in all cases conclude that it makes sense to say that D3 is part of D1. Parthood on dynamics does not hold automatically indirectly, that is, across levels, or, technically speaking, parthood on dynamics is not transitive.[8] For this sense of direct parthood—which is, again, our most basic sense of parthood—it holds that no dynamics is part of itself (irreflexivity) and that, if two dynamics seem to be part of each other, they are in fact identical (antisymmetry).

Assume that P is the partition of an entity D, specifying a tree-structure of direct parthood relations with D as the head node. In connection with such a partition of D we can then introduce relationships of indirect parthood relating to the n-th level of D's partition.

---

[6] Consider the following everyday examples, taken straight from the Worldwide Web: 'Blogging is part of life'; 'Russia is part of the West'; 'Music is part of God's Universe'; *All I see is part of me* (book title); 'Learning to negotiate is part of the advocacy process'; 'My heritage is part of who I am'; 'Looking immaculate is part of what I do'; 'Pain is part of running a marathon'; 'Hopping too is part of running'; 'Fab Face is part of Screaming Talent'; 'The concert is part of the 11th Ludwig van Beethoven Easter festival'.

[7] I argue that 'x is part of y' is transitive only if x and y denote spatio-temporal regions. More on this claim and on the details of LEM can be found in Seibt 1990, 2004, 2005, 2008, 2009, 2014.

[8] For example, ionization is part of hydrolysis, which is part of proteolysis, which is part of the adaptive immune reaction by B-cells, which is part of the human immune system. Unless we read 'is part of' in a narrow sense, as 'is a spatio-temporal part of', it is false to say that ionization is part of the human immune system—the functional organization that the latter term identifies normally does not 'reach that far', i.e. it leaves indeterminate how proteolysis occurs.

(Definition of 'n-part'): If P is a (possibly infinite) partition of entity D, let us call the direct parts of D at partition level 1 the 1-parts of D, the direct parts of these at partition level 2 the 2-parts, and so on. In general, the 'n-part' of D in partition P of length m, counting from the top, is any entity at partition level n, where $1 < n \leq m$.

This has a number of advantages. First, one can quantify over the parts in the partition of an entity in a much more differentiated fashion. One can refer to any part above a certain partition level n: 'x is <n-part of y'; at a partition level n: 'x is n-part of y', and below a partition level n: 'x is >n-part of y', as shown in Figure 6.1.

Second, using the basic, not very informative part relation of 'is part of', which supports only weak axioms, it is possible to introduce other, more informative varieties of parthood (spatial parts, material parts, functional parts, morphological parts, etc.) by stipulating additional conditions. In fact, in GPT, each dynamics is defined in terms of a collection of partitions, each created by a parthood relation that either is or is defined in terms of 'is part of'. More concretely, a dynamic D is represented by its base partition (BP) and three or more additional partitions. BP lists all the dynamics that are direct or indirect parts of D, in other words, all the $\leq n$-parts of D, where $n$ is the lowest level of BP (counting levels downwards). In addition to BP, it is useful to specify the spatio-temporal partition SP of D, the material partition MP of D, and the functional partition FP of D (and possibly others); these additional partitions either are included in BP or else include BP. In this way one can refer to the 'parts' of an entity not only with greater precision but also in a less ambiguous and systematically more enlightening way. For example, the question 'How does natural selection work for organisms with hierarchical organization—which parts of the organism are involved?' could be answered by suggesting that only certain first-level and second-level material parts, multicellular subunits and apical meristems, are involved, but not third-level material parts. Moreover, if dynamics are represented in terms of a collection of partitions, one can compare kinds of mereological relationships as they hold on kinds of dynamics. For example, one can compare whether 'is a material part' has the same transitive scope on different kinds of biological dynamics (e.g. on photosynthesis vs mitosis), or whether it has the same transitive scope for biological, chemical, and physical kinds of dynamics.

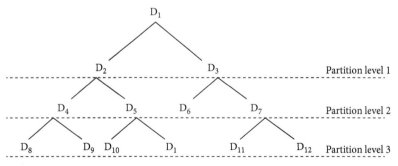

**Figure 6.1** The n-parts of $D_1$ are the dynamics at partition level n. Note that dynamics can appear in several places in a partition—$D_1$ is its own 3-part

A third advantage of operating with a non-transitive parthood relation is that one can interpret the requirement of having 'the same parts' in a more precise and differentiated fashion than usual. In general, it is true that the 'identities' of dynamics are determined on the basis of the so-called 'proper parts principle': names for dynamics $D_1$ and $D_2$ co-refer iff $D_1$ and $D_2$ have the same parts. But in LEM the co-referentiality for dynamics $D_1$ and $D_2$ is defined in terms of 'x is $\leq$ $n$-part of y'. That is, identity is always defined with reference to a given level n of mereological depth in the base partitions of $D_1$ and $D_2$; in order to determine whether $D_1$ and $D_2$, one compares each of the parts that lie on partition levels $1 \leq k \leq n$. By specifying which partition levels are to be taken into account, one can operate with more coarse-grained or more fine-grained requirements for co-referentiality in different contexts. For example, consider the partitions in Figure 6.2.

In some contexts we might want to claim that turf grass and crab grass both grow by making use of 'the same process', namely photosynthesis. Such a statement is made true by the parts of the base partitions of 'photosynthesis in turf grass' and 'photosynthesis in crab grass', if we restrict the level of analysis to partition levels 1 and 2 of these two dynamics, that is, to the 1-parts and 2-parts of photosynthesis in turf grass and photosynthesis in crab grass. In other contexts we want to stress the difference between these two processes, and such a statement is made true by extending the level of analysis to partition level 3, that is, we compare their 1-parts, 2-parts, and 3-parts.

A fourth advantage of LEM is that, unlike standard mereologies with transitive parthood, in this system we can formally represent emergent parts of processes and feedback structures (parthood loops). In particular, as I shall sketch below, it is possible to represent the difference between emergent products of an interaction dynamics without causal role—simple emergence—and emergent products of an interaction dynamics that causally influence the conditions under which the interaction dynamics occurs and is further propagated—generative emergence, as it occurs in self-maintaining systems such as organisms (Seibt 2014; see also sections 5.2 and 5.3 below).

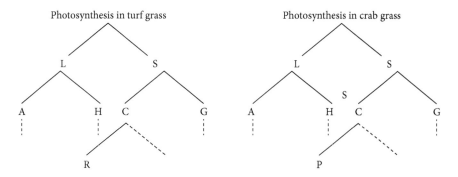

**Figure 6.2** Abbreviations: L: Conversion of light into chemical energy; S: Storage of chemical energy; A: absorption of energy by proteins; H: Formation of hydrogen ions; C: Calvin cycle; G: formation of glucose; R: Carbon fixation with RuBP; P: carbon fixation with PEP. Dashed lines indicate omitted branch regions of the partition

## 4. A Typology of Processes

Since everything there is—that is, everything we refer to, in natural or scientific language, as being in the world—is a general process or dynamics according to GPT, the explanatory power of this framework hinges on the classificatory statements it can deliver in order to describe more precisely what varieties of dynamics there are and how they relate to each other. The classificatory scheme or typology of GPT uses five evaluative 'dimensions': (1) spatio-temporal signature, (2) participant structure, (3) dynamic constitution, (4) dynamic shape, and (5) dynamic context. The classificatory parameters of each dimension are here stated informally, but formal analogues can be defined with the resources of LEM.

### 4.1. Spatio-temporal signature

The most basic concepts of common sense and scientific reasoning, 'thing', 'stuff', 'event', 'state', 'activity', 'field', 'organism', and so on, are associated with characteristic inferential patterns pertaining to how the item in question occurs in space and time. The entity type that an ontological domain theory postulates as ontological correlate or relevant part of a truth maker for a sentence about a thing, some stuff, an event, and so on must be defined in such a way that these inferential patterns are entailed. In GPT this is achieved through a systematic extension of the predicate of *homeomereity*—that is, like-partedness. Aristotle observed that our reasoning about stuffs can be accounted for if we assume that stuffs are 'like-parted (i.e. homeomerous) bodies...composed of [spatial] parts uniform with themselves',[9] and it was noted early on in the debate about the ontological interpretation of 'action types' and verbal aspects that an analogous homeomereity with respect to temporal parts dovetails with our reasoning about activities (Vendler 1957; Kenny 1963; Mourelatos 1978). Just as any spoonful of a puddle of water is like the whole, namely an expanse of water, so any minute of an hour of snowing is like the whole, namely a period of snowing. Thus we can formulate a generalized notion of homeomereity:

> *Like-partedness or homeomereity:* An entity of (proximate) kind K is homeomerous with respect to its spatial extent (temporal extent) iff all of its spatial parts (temporal parts) are of kind K.

Upon closer look, however, our reasoning about stuffs and activities suggests an even more remarkable mereological feature than like-partedness. Since stuffs and activities are purely 'functionally' individuated, it does not make sense to distinguish between a stuff or an activity and its 'nature'—stuffs and activities *are* 'natures', even though they occur concretely. For the ontological correlates of our sentences about stuffs or activities, the following condition holds (where 'all of E occurs in spatio-temporal region R' is to be read as 'all parts of E occur in region R'):

> *(Spatio-temporal) Self-containment or automereity:* An entity E is automerous iff for any spatio-temporal region r (r > 0): if r is a subregion of a spatio-temporal region R in which all of E occurs, then r is a region in which all of E occurs.

---

[9] Aristotle, *History of Animals* 487$^a$2. See Seibt 2017.

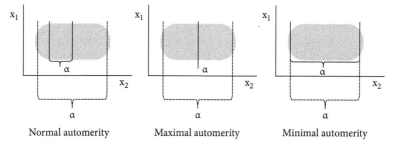

**Figure 6.3** Graphical illustration of degrees of self-partedness (automerity) in space or time, depending on the interpretation of coordinates x1 and x2 (as spatial coordinates or with x1 as spatial and x2 as temporal coordinate, respectively; in the first case one spatial dimension is surpressed, in the latter case two, as usual)

In other words, the entities denoted by sentences about stuffs and activities are not only like-parted in space and time, respectively. They are also literally the same individual; that is, they are *recurrent* in space and time, respectively.

Our reasoning about mixtures (e.g. (a) fruit salad, (a) forest) and about repetitive sequences (e.g. hammering) suggests that on these occasions we refer to entities that are uniformly structured only for a certain 'grain size' of parts. Similarly, since no spatial part of a frog is a frog and no temporal part of a symphony or of a translation process (in gene expression) is a symphony or a translation, our reasoning about things or developmental events requires that we postulate entities for which it holds that there are *no* parts like them or containing them. Thus the predicates of like-partedness and self-containment can be generalized in two respects: first, with respect to dimensionality and, second, with respect to degree (see Figure 6.3 for a graphical illustration):

*Maximal, normal, minimal homeomereity:* An entity D of (proximate) kind K is maximally/normally/minimally like-parted$_K$ in space(/time) iff *all/some but not all/none* of the spatial(/temporal) parts of the spatio-temporal extent of D are of kind K.

*Maximal, normal, minimal automereity:* An entity D is maximally/normally/ minimally self-contained in space(/time) iff a spatio-temporal region in which D exists has *only/some but not all/no* spatial(/temporal) parts in which all of D exists.[10]

---

[10] Generalized homeomereity and automereity in this sense were first introduced in Seibt 1997. In a different terminology, the properties of homeomereity and automereity have also been called 'dissectivity' (a cut in any region occupied by C will yield C) and have been supplemented with the property of 'additivity' (adding C to C will yield more C). Zemach's (1979) little noticed fourfold classification of entity types in terms of 'bound' and 'continuous' occurrence in spatial and temporal dimensions is kindred in spirit. Whether minimal homeomereity always implies minimal automereity depends on whether proximate kinds are restricted to *natural* kinds—if an entity D is defined as sortally amorph, that is, as changing continuously its proximate natural kind in ways that are sometimes connected with the idea of 'flux', then we might say that D is temporally and spatially automerous (kindless occurrence occurs in every temporal or spatial subregion of the spatio-temporal region R in which kindless occurrence occurs), but we might nevertheless allow that an occurrence of D in R* is of the proximate non-natural kind K *occurring in R*,* and thus minimally homeomerous with respect to K. Similarly, maximal automereity mostly, but not always,

The individual entities of the ontological tradition, for example, material objects but also Whiteheadian 'actual occasions', cannot be self-contained or recurrent in a region, since they are conceived of as particular entities; they are individuated in terms of their location, and thus are by definition (i.e. necessarily) non-recurrent. Spatio-temporal self-containment is a coherent concept only for functionally individuated entities and can be coherently defined only in terms of the 'is part of' relation, in the wide sense of asymmetric 'belonging with' described in section 3. For example, breathing is part of walking, and so is moving your legs, lifting and placing your feet, swinging your arms, keeping balance, moving forward; assuming that these five activities are all of what is part of walking, we can coherently say that any hour in which walking exists has temporal parts in which all of (what is part of) walking exists. As I will elaborate briefly below, spatio-temporal self-containment provides a straightforward account of persistence as identity (recurrence across time) as well as a consistent account of generality (recurrence of features in space).

The predicates for different varieties of homeomerity and automereity can be combined to define the 'spatio-temporal signature' of a dynamic.[11] Together with other conditions, these spatio-temporal signatures can be used to define ontological correlates or parts of truth makers of the statements of natural or scientific languages. As I have shown elsewhere, to demonstrate the wide scope of GPT for the interpretation of ontological commitments expressed in natural language, we can define spatio-temporal signatures of entities that reflect the inferential patterns that carry the ten basic types of inferential information in natural languages. In other words, GPT can be used to define ontological correlates that dovetail with the way natural languages guide our reasoning about things, events as developments, events as results, stuffs, activities, states, collectives, sets, sorts, and features.[12] Here I will briefly illustrate six spatio-temporal signatures.[13]

Type-1 dynamics: Dynamics D is the ontological correlate of an *activity* statement iff
(i) D is temporally maximally automerous and
(ii) D is spatially maximally, normally, or minimally automerous.

Condition (ii) highlights that the concept of activity implies spatial occurrence without further specification. The ontological correlates of sentences about activities may be spatially maximally automerous (e.g. 'the water is boiling', 'it is itching', or 'the light is shining') or spatially normally automerous (e.g. 'the vortex is turning', 'all

---

implies maximal homeomerity: the entity that is first a caterpillar and then a butterfly is maximally automerous (along the temporal dimension) but not maximally homeomerous (along the temporal dimension).

[11] More precisely, the 'mereological signature of the spatio-temporal distribution' of a dynamic.

[12] In Seibt (2015a) I argue, on the basis of research in linguistic typology on the verb and noun systems of the world's languages, that GPT is currently the only (western) analytical ontology that can aspire to formulate truth-makers for sentences of a wide range of the languages of the world; otherwise in analytical ontology the peculiar inferential patterns of the noun system of Indo-European languages are simply read into the structure of the ontological domain to create the long-standing fixation on concrete individuals that are particular and countable per se.

[13] I shall retain here the GPT terminology and label these dynamics type-1, type-2, and type-5 through to type-8, instead of using continuous enumeration.

three beehives are swarming'), or spatially minimally automerous (e.g. 'the soccer team is singing', 'the immune system is working normally again').

This is quite similar in the case of type-2 dynamics, that is, the ontological correlates of sentences about developments such as 'the plant grew from 10 cm to 2 m', 'the stickleback moved into the nest', 'the ribosomes disassembled'; the dynamics referred to by sentences like these must be temporally minimally automerous—the 'grain size' of their temporal occurrence is the entire temporal extent of any temporal region in which they occur. But, as the last example of a collectively performed 'accomplishment' shows, a type-2 dynamic may be normally homeomerous and automerous in space.

> Type-2 dynamics: Dynamics D is an ontological correlate for a *development* (accomplishment) statement iff
> (i) D is temporally minimally homeomerous and temporally minimally automerous, and
> (ii) D is spatially minimally automerous (and spatially minimally homeomerous), or spatially normally automerous (and spatially normally homeomerous).

Omitting spatio-temporal signatures for the ontological correlates of sentences about 'results' and 'states', of particular interest for the interpretation of biological claims may be the fact that GPT postulates processes or dynamics also as ontological counterparts of our talk about things or singular objects, collectives, sets, sorts, masses, and features. (To simplify, I will speak about the ontological counterpart of a noun N, in order to abbreviate the more cumbersome formulation 'ontological counterpart of a truth maker for a statement containing a referential use of noun N'.) Process philosophers occasionally say that objects are abstracted from processes. From the point of view of GPT, such a statement is somewhat misleading, since objects *are* a type of process or, more precisely, our talk about objects can be ontologically interpreted as being about a certain type of process.

> Type-5 dynamics: Dynamics D is an ontological counterpart for a *singular object noun* iff
> (i) D is spatially minimally automerous (and minimally homeomerous) and
> (ii) D is spatially normally homeomerous[14] and
> (iii) D is temporally maximally automerous.

The dynamic denoted by 'a cat', for instance, is spatially minimally homeomerous and automerous: no spatial part of the extent occupied by a cat is a region where a cat exists. Minimal automereity warrants unique bounded spatial existence as well as the failure of additivity and dissectivity, as required by our common-sense reasoning about cats. Condition (ii) is added here to establish that D might be either an integrated whole (e.g. a living organism) or a locally homogenous material expanse (e.g. the ontological counterpart of *a/this stone* or *a/this lake*), but not an entirely arbitrary collection of things without any sortal identity or family resemblance

---

[14] An entity D of kind K is (spatially vs temporally) maximally/normally/minimally homoeomerous or 'similar-parted' iff all/some/none of the (spatial vs temporal) parts of the extent of D are of a kind K* that is a necessary condition for being of kind K.

among its members (see Seibt 2000b). Condition (iii) states that type-5 dynamics persist by *literally recurring*—type-5 dynamics are wholly present at any moment of their existence.

The ontological counterparts of sentences about collectives, on the other hand, should fulfil inferential requirements of the following spatio-temporal signature:

> Type-6 dynamics: Dynamics D is an ontological counterpart for a *collective* iff
> (i) D is spatially minimally automerous and
> (ii) D is spatially minimally or normally homeomerous
> (iii) D is temporally maximally or normally automerous.

A dynamics D that fulfils the application conditions of, for example, the English expression '(this) herd' should not be spatially recurrent, since collective nouns are said to denote several entities conceived of as a unit. Dynamics D does not recur as a unit in any of its spatial parts. This does not exclude, however, that the region occupied by D has spatial parts that also fulfil the application conditions of the English expression 'a herd'. This is captured in condition (ii)—some collectives are weakly structured groups, and the same structure may recur within the group (e.g. a large herd may contain smaller herds), while strongly structured groups (e.g. a soccer team, the processes that make up an organism) do not. Condition (iii) accounts for the fact that some collectives (such as herds) persist through recurrence, while others (e.g. the metabolic processes that co-occur in an organism) take time.

Some languages (e.g. Oromo), but not English, have 'set nouns' that refer to entities as 'one or several' of a kind. The spatio-temporal signature for the relevant ontological counterpart is as follows:

> Type-7 dynamics: Dynamics D is an ontological counterpart for a classification with a *set noun* iff
> (i) D is spatially minimally automerous and minimally homeomerous, or
> (ii) D is spatially normally automerous and normally homeomerous, and
> (iii) D is temporally maximally automerous.

While the English noun 'sheep' is not a proper set noun, it might do to adumbrate the idea: the disjunction of clauses (i) and (ii) allows for the ontological counterpart of 'sheep' to be *one or several* entities. I mention type-7 dynamics here only to contrast this type with the following one, type 8, these being two possible candidates for an ontological interpretation of biological species as individuals (see section 5.1). Nouns referring to concrete items can, but do not need to, imply that the items denoted occur in a bounded region or have determinate shapes. While Indo-European languages predominantly use nouns that carry these implications, the so-called 'sort nouns' of classifier languages do not. Consider the mereological counterpart of a dynamic that is indeterminate with respect to shape and boundedness:

> Type-8 dynamics: Dynamics D is an ontological counterpart for a *sort noun* iff
> (i) D is spatially normally automerous and spatially normally homeomerous, and
> (ii) D is temporally maximally, normally, or minimally automerous.

English counterparts of the sort nouns of classifier languages are noun phrases in generic sentences such as 'the African elephant is on the brink of extinction' or terms for natural kinds such as 'humankind'. The ontological counterpart of 'the African elephant' or 'humankind' is normally automerous—some of the parts of regions in which the African elephant or humankind exists are regions in which the African elephant or humankind exists. This entails indeterminate spatial location and accounts for the inhomogeneity of sorts, that is, for the fact that there are minimal portions satisfying the noun 'the African elephant' or 'humankind'. Condition (ii) again acknowledges that sort nouns can imply all varieties of temporal existence.[15]

The spatio-temporal signature specifies the 'mode of spatio-temporal occurrence' of a dynamics. While the relevant inferential patterns of natural languages suggest that there might be ten such modes of occurrence (here selectively illustrated), scientific languages may require additional ones. But there are many scientific kinds that can be characterized in terms of one of the ten modes of occurrence. For example, let the gene for protein P be the statistical relationship between sequences of DNA and coded proteins; the UBE3A gene for the protein ubiquitin ligase may be interpreted as a type-6 process (collection) of type-6 processes (collections) of type-5 processes (DNA sequences and amino acid sequences). Describing dynamics in terms of their spatio-temporal signatures emphasizes that many of them are quite unlike the bounded developmental processes or events (the growth of the oak tree) that the ontological tradition so far saw fit to acknowledge. Dynamics may be gappy, relational, unbounded, and non-countable—they can occur distributed into disconnected spatio-temporal regions and have themselves such distributed processes as their parts. Let me now turn to a brief sketch of the remaining four dimensions of the classificatory system of GPT.

### 4.2. Participant structure

The participant structure of a general process or dynamic states what types of processes are involved in a complex dynamics, and in what role. There are three basic roles, labelled 'agent', 'patient', and 'interagent', but additional ones can be defined recursively. For example, for the ontology of biology, it would be of particular interest to introduce the role of an emergent constituting constrainer (ECC), as a special variety of interagent. Self-maintaining far-from-equilibrium systems like a candle flame, a hurricane, or a biological organism *are* complex dynamics that emerge from, and constrain, certain interactions and thereby constitute the dynamic system (Christensen and Bickhard 2002; Bickhard 2009; Bickhard and Campbell 2002; see also chapters 1, 7 and 10 here).

Differences in the participant structure, that is, in the types and roles of participants in occurrences, are often encoded linguistically. In general, however, since neither the syntax nor the verb semantics can be used as reliable indicators of the

---

[15] To operate again with English counterparts for illustration, the ontological counterparts of 'the African elephant', 'humankind', and 'the completion of the zygote' are temporally maximally, normally, and minimally automerous, respectively.

participant structure of a dynamic, it is useful to articulate the causal composition of a complex dynamic in terms of these simple role concepts.[16]

## 4.3. Dynamic composition

The third dimension of the classificatory system of GPT analyses the partition of a dynamic in terms of various predicates for linear and non-linear composition. In LEM, the sum $D_1$ of two items $D_2$ and $D_3$ may have *n*-parts at (graphically) lower levels that are neither parts of $D_2$ nor parts of $D_3$. How these additional parts of a dynamics emerge and what role they have is reflected in the partition structure. For example, one may contrast the partition structures representing 'weak emergence' (i.e. the emergence of a part without subsequent dynamic role) with those representing 'generative or strong emergence', where emergent parts both facilitate and constrain the dynamics they are emerging from (Seibt 2014).

To illustrate, consider the differences in dynamic composition among (i) a linear mechanism, (ii) a feedback loop, and (iii) a self-maintaining dynamic. To simplify the exposition, let me here switch to the idiom of 'episode' and 'type'. Even though in GPT, where all individuals are non-particular entities, there are no 'tokens' in the traditional sense, given that tokens are particular entities, the occurrence of a dynamic $D_1$ in a determinate location can be treated, in Leibnizian style, as the occurrence of more a specific dynamics $D_2$, which we may call an 'episode of process type' $D_1$. From here on 'the dynamics of *[linguistic expression]*' will be used as shorthand for 'the dynamics denoted by *[linguistic expression]*'.

A mechanism is a linear sequence of process types $D_1 \ldots D_n$ such that it holds for any concrete 'run' of the mechanism that any member in the sequence of episodes of processes from $D_2$ to $D_n$ directly dynamically presupposes its predecessor $D_{n-1}$.[17] Importantly, if a mechanism M is run repeatedly, the episode that occurs at stage $S_i$ of M is each time an episode of the same process type $D_i$ (e.g. an episode of 'this gear's making half a turn'). In contrast, in a simple loop of positive or negative feedback, it is also the case that, after initialization, each episode directly and dynamically presupposes the previous one, but what happens at stage $S_i$ is not with absolute regularity always an episode of the same specific process type; rather, at different times, we find at $S_i$ episodes of process types that are merely *similar*—at each time a different intrinsic efficacious character may be actualized, in dependence of (i.e. dynamically presupposing) a regulatory change to the 'causal signal' upstream.

Finally, consider a system of processes that maintains itself far from the thermodynamic equilibrium, for example a burning candle or a living organism. The component processes of the burning candle or of the organism (e.g. the melting of the wax, the percolation of the wax in the wick, the combustion in the flame, the air convection that adduces oxygen and carries away residues; or respectively

---

[16] Many sentences with intransitive verbs do not express a one-subject process. Some involve implicit references to locations, times, or observers—consider 'he disappeared', 'the wedding took place', 'the game was disappointing'—others express qualifications of dynamic shape or dynamic context of a process—for example 'the feeding frenzy increased', 'the rate of change remained constant', 'this development could not be stopped'.

[17] The notion of direct dynamic presupposition I must leave here undefined; see Seibt 2016.

the processes constituting catabolism, energy transformation, anabolism, immune reaction, and reproduction) not only feed into each other in the way in which this could also be said to hold for a mechanism or feedback loop; they each depend on, or dynamically presuppose, not only each other but also the occurrence of the entire process system (Bickhard 2004; Seibt 2009). Taken in isolation, the process types 'melting of wax' or 'percolating of wax' can have episodes that occur without any candle burning; and the same holds of the process types 'electron flow' or 'carbon fixation'. However, once the process system D of a burning candle or prokaryotic organism is up and running, any episode *in D* of 'melting of wax' or 'carbon fixation' directly and dynamically presupposes not only an episode in D of 'heating' or 'electron flow', respectively, but also an episode of D—for the process system D as a whole and for each of its constituents to occur, nothing but the occurrence of D is required.[18]

As these examples may convey, the predicates used in the formulation of the classificatory parameters in the dimension of dynamic composition (here, 'dynamically presupposes') involve modalities of the possible and necessary, not merely the modality of what occurs normally or for the most part, which is captured in partitions set up with the 'is part of' relation of LEM. Precisely how these strong modalities can be integrated into the formal framework of GPT—which is conceived of as a naturalist and so to speak 'actualist' ontology eschewing possible worlds realism—is currently an open question.

## 4.4. Dynamic shape

The fourth dimension of evaluative parameters of the classificatory system of GPT relates to differences in behaviour that can be geometrically described in terms of predicates of phase space trajectories—for example, whether a process stays in the same regions of its phase space, periodically switches between different attractors, or behaves chaotically. Some of these differences in trajectory can be related to intuitive characterizations of the 'flow' of a process—for example, whether it is slow or fast, or whether it is, metaphorically speaking, a sparse, normal, or rich realization of a process type. Such adverbial modifications of process types have so far received very little attention in ontology, traditional and contemporary.

The relevant classificatory predicates for such modes of performance can be defined in terms of deviations from typical partitions, that is, the partitions of generic dynamics. For example, while the partition of the dynamics D 'reproduction in *Hemidactylus garnotii* (Indo-Pacific house gecko)' will contain the dynamics 'copulating with a member of the species of different sex' and 'sexual reproduction', there are episodes of D whose partitions contain the dynamics 'copulating among female members of the species' and 'parthenogenesis', which could be classified as a sparse realization of dynamic D. Instead of, or in addition to, characterizing modes of performance on the basis of structural features of the partitions of 'types' and

---

[18] In contrast, an episode in a mechanism does not directly dynamically presuppose a previous run of the entire process system; and the same holds for or an episode in simple regulatory feedback system. In both of these types of system, episodes depend on the one hand on the preceding episode (direct dynamic presupposition) and, on the other, on an episode outside the system that initializes the episode at the first stage of the system (indirect dynamic presupposition).

'episodes' (i.e. dynamics and their specifications), one can also resort to dynamical systems theory and correlate the different modes of performance with deviations from the 'normal' shapes of trajectories in the phase space.

For the ontology of biology, it would seem all important to introduce suitable predicates for the modes of performance for processes, since the latter are often mentioned as playing a decisive role in natural selection.

### 4.5. Dynamic context

Finally, processes are differentiated with respect to how they relate to their *dynamic context*. In natural language we often characterize a dynamics D in relation to the dynamics that occurs or occur before or after D, or of which D is a temporal part. This is done, for example, by means of so-called verbal aspects (e.g. perfective, progressive, repetitive, ingressive, egressive). The technical predicates of the fifth classificatory dimension shall capture the relationships expressed by verbal aspects of this kind, as well as corresponding scientific predicates specifying contextual conditions. Other classificatory parameters in this dimension specify how a dynamics affects its dynamic context and is affected by it. Of particular importance for the ontology of biology may be the distinction between dynamics that have *linear causal impact* on their immediate environment (such as disturbances in the air flow around a kite) and dynamics with *non-linear causal impact* (such as changes in ecosystems that alter selection pressures).[19]

In sum, then, in GPT a dynamics D is classified by way of determining parameters associated with five evaluation perspectives, specifying:

(1) how D occurs in space and time, that is, D's mode of spatio-temporal existence or spatio-temporal signature;
(2) what causal (also agentive) roles are performed by which of D's parts, that is, D's participant structure;
(3) how the part processes of D compose (i.e. how they feed into each other, where dynamics 'emerge', and in which sense), that is, D's dynamic composition;
(4) how D is performed in relation to the 'norm' reflected in the partitions of the relevant process type or generic dynamics, that is, D's dynamic shape; and
(5) how D relates to its environment, that is, D's dynamic context.

## 5. Applying GPT in the Philosophy of Biology

GPT was developed in 1990 as a project of paradigm revision in ontology, in order to show that the three big problems of the ontological tradition—the problem of individuation, the problem of universals, and the problem of persistence—are artifacts of a certain tradition of ontological theory construction, namely the substance paradigm or the myth of substance (Seibt 1990, 1994, 1995, 1997, 2004). Accordingly, the applications of GPT have been mainly focused on topics in general ontology, in particular on the problem of persistence (Seibt 1997; 2008). Elsewhere

---

[19] These two examples are due to Bickhard and Campbell (2002).

I have shown that the current main alternatives in the theory of persistence, the perdurance approach—both the 'stage version' and the 'hunk' version—and the endurance approach, incur their familiar problems precisely because they are based on the presupposition that all concrete individuals must be fully determinate particular entities. Once we relinquish this assumption and allow for individuals that are concrete but general (that is, indeterminate with respect to their location and other aspects), we can consider persistence as the *recurrence* (in time) of an individual. In GPT, assertions about persistence, such as 'this is the same cat that we saw in the fall', are made true by a dynamics D—here, say, *catting-F-ly*,[20] where *F-ly* is a way of physical appearance—which recurs identically through time, yet happens not to recur in space. On the other hand, an assertion about change, such as 'but it (the cat we are seeing now) is much thinner now than it (the cat we saw then) was', is made true by a pair of different dynamics $D_2$ and $D_3$—here (*catting-F-ly*)-*thin-ly* and (*catting F-ly*)-*fat-ly*—that are specifications of $D_1$. In other words, once we give up on two elements of the myth of substance, namely the presumption that all individuals are particulars and the presumption that persistence statements and statements about change have the same subject, the puzzle of persistence has a straightforward solution. When we assert change, we are not literally making an assertion about that which we claim to have persisted. We are making an assertion about how that which persists is turning out now in comparison to how it used to turn out, and we say that how it is turning out now or occurring now is a specification of, but is not identical to, how it is occurring all the time.

This analysis of statements about persistence and change has three specific advantages. First, it allows for straightforward temporal and causal continuity (unlike the perdurance view). Second, it can be coherently formulated for either eternalist or presentist conceptions of time. Third, it makes do with concrete individuals only, in other words without having to postulate an additional category (e.g. a universal or a Whiteheadian eternal object) that relates to concrete individuals through an additional relationship (e.g. instantiation or Whiteheadian ingression, respectively).

In short, since GPT relinquishes the traditional fixation on discrete, particular, determinate individuals, it can offer a straightforward approach to an ontological interpretation of statements about general features in space ('x and y are both F') and time ('x-now is the same F as y-then'). In addition, the new category of non-particular, indeterminate concrete individuals might have applications in the ontology of quantum field theory and can be used to devise ontological interpretations of claims of interactions, from the interactivist interpretation of cognition to claims about social interactions (ch. 5 in Seibt 2005; also Seibt 2009 and 2014).

While it may be clear already from the sketch in the preceding sections *that* this framework might be employed in support of a 'process turn' in the philosophy of biology, let me highlight here three aspects of *how* or *where* the constructional ideas of GPT would seem particularly useful.

---

[20] Compare Quine's (1985: 169) playful rendition of 'a white cat faces a dog and bristles' as 'it's catting whitely, bristlingly, and dogwardly'. Ontology is never in the business of linguistic revision, but occasionally some tongue-in-cheek transpositions are useful.

## 5.1. Biological individuality

An ontology like GPT, which is based on a rejection of the classical variety of concrete individuals, can offer a new perspective in the recent debate about individuality in biology.[21] As several authors have argued, given our current state of knowledge about multispecies consortia (e.g. biofilms such as dental plaque) and plants, the standard criteria of individuation used in biology (germ/soma separation, developmental bottleneck, sexual reproduction, genome from one species, genome forming reproductive lineages, physical boundaries, or immune response) are no longer unproblematic tools for an unambiguous demarcation of those bits of organic matter that should be classified as an individual (Clarke 2010, 2011, 2012; Dupré 2012; Ereshefsky and Pedroso 2013, 2015). This situation has been understood as an opportunity to discuss critically the set of criteria and to streamline one of the four main perspectives (evolutionary, topological, physiological, and organizational) that commonly have been guiding definitions of biological individuality or individuation. Thus it has been argued that some of the more generic criteria (e.g. the 'interactor' approach instead of the more specific 'bottleneck' criterion) should be adopted for the notion of evolutionary individuality (Ereshefsky and Pedroso 2015).

To assist this discussion, one could reformulate the candidate criteria in the terminology of GPT so as to make the relevant proposals more precise. Hull's notion of an interactor, for example, has the drawback that is operates with a vague notion of a 'cohesive whole': an interactor is 'an entity that directly interacts as a cohesive whole with its environment in such a way that replication is differential' (Hull 1980: 318). A cohesive whole in Hull's sense has 'reasonably sharp beginnings and endings in time' (ibid., 313) and must 'exist continuously through time and maintain its internal organization' (ibid.). A cohesive whole in this sense is a type-5 dynamic—it is temporally maximally automerous, spatially minimally automerous, and spatially minimally homeomerous. Thus a dynamics D that is an interactor must be a type-5 dynamics; and the fact that an interactor interacts with its environment only as a cohesive whole can be captured by the requirement that the participant structure of D only has D as agent in all interactions with processes outside of the spatial region occupied by the spatial parts of D. Such a translation into the more abstract idiom of GPT would support the general strategy of the interactor approach of defining individuality in terms of an entity's function in the process type of selection: 'in order to be selected...an entity must be an individual. Anything that can be selected the way an organism can, must be the same sort of thing an organism is' (ibid. 326).

An alternative suggestion in the debate has been to determine a joint underlying rationale that can explain the previous success of the standard individuation criteria and to develop more generic rephrasings of individuation criteria in terms of 'mechanisms...whose effect is to constrain the extent to which populations at different compositional scales exhibit heritable variance in fitness' (Clarke 2012: 356). The relevant mechanisms (i.e. the constraining sources of heritable variance or fitness differences) pertain to 'interactions amongst parts' (ibid.) of a collection of

---

[21] Chapters 9 and 10 in this volume also address this debate.

cells, where the notion of 'part' at issue is non-transitive (though not explicitly identified as such) and the interactions referred to are at different partition 'levels'. GPT could be used to represent in more formal terms the central claim of this proposal, namely that individuality is relative to a partition level, that is, relative to mechanisms that 'fix the degree to which units at any particular hierarchical level are individuals' (ibid., 351).

GPT could also be used in the debate about biological individuality to advance a much bolder claim. Our current focus in ontology on classical (i.e. discrete, particular, unified, persistent) individuals stems from Aristotle's attention to biology and from his ontological intuition that 'living things'—typically, animals and non-clonal plants—exemplify what it means *to be*. To Aristotle, such beings seemed to enjoy a distinctive developmental independence and organizational unification that illustrated a 'primary' sense of Being or *ousia* ('being', 'essence'), which he characterized as powerful self-realization. At a time when biological research has uncovered a rather different picture, characterized by macrobe–microbe dependencies, 'genetic commons', and a profusion of dependencies in cycles and networks, it seems puzzling that philosophers of biology still would want to hang on to a notion of the individual that is motivated by obsolete biology. While philosophers of biology have observed that individuality is always relative to (the principle of individuation associated with) kind K (Ereshefsky and Pedroso 2013) and that different individuals must be treated as numerically different, for example as two (Clarke 2012), it has not been questioned, as far as I can see, whether one should continue to endorse the traditional tenet that the principle of individuation must also double as a principle of counting. For some kinds of entities—oak trees, cats, chairs, and tables, which ontologists throughout the centuries have not tired of discussing—the principle of individuating what is of kind K automatically supplies a principle of counting that which exists or occurs of kind K, since what is of kind K exists or occurs in 'prepacked' units. But, for other kinds of entities (e.g. activities and stuffs), the principle of individuation for K leaves open how we bundle into units and count what is of kind K. Briefly put, we have been duped by tradition and the grammar of Indo-European languages when we think that whatever is countable as a kind also must be a kind of countables. For example, we differentiate water and milk, or walking and waltzing, in terms of chemical composition or types of bodily movements, respectively, but this criterion or principle of individuation for stuffs and activities, which surely gives us two stuffs and two activities in each case, leaves open how we should count what exists or occurs of them. So, if the standard criteria of biological individuality are not suitable for receiving a unique and unambiguous count of plants, one might address this problem by specifying new criteria for counting (to use on purpose this contorted expression) 'that which exists of plant individuals'.

On the other hand, one might challenge the assumption that how we individuate a plant of kind K determines how we count that which is of kind K.[22] The plant demographer's dilemma (i.e. should one count gamets or ramets?) can be solved by treating gamets, ramets, and any other unit one might count by analogy with litres

---

[22] This assumption drives especially Clarke's treatment of biological individuality; see Clarke 2012.

and grams: as units one may count by, not as units that individuate the entity of kind K that is being counted. Alternatively (or additionally), one may point out that once we liberate—within a process ontology like GPT—the ontological notion of 'concrete individual' from its clandestine traditional restriction to things and living *things* (i.e., animals or non-clonal plants) understood as determinate, discrete particulars, we may take each of the standard criteria for biological individuality to yield an individual—the dynamics of immune response, the dynamics of sexual reproduction, the dynamics of germ/soma separation, and so on. Instead of discussing whether entities of kind K qualify as proper individuals, we would then be discussing which individuals (dynamics) overlap or co-occur in dynamics of kind K.

## 5.2. Biological composition

In GPT a dynamics is represented by several partitions—the partition of the basic part relation 'is part of' and the partitions of derivative part relations, such as 'is spatio-temporal part', 'is material part', 'is functional part', and so on. How many levels of these partitions identify the dynamics in question may change with a change in communicative context—the biophysicist may countenance more material parts of a frog than the morphologist—and their maximal depth is relative to the current state of inquiry. That a dynamics is represented by several partitions in GPT is in line with Winther's recent analysis of biological part–whole explanations, which emphasizes that 'there are multiple cross-cutting manners of abstracting a system into kinds of parts—i.e., there are multiple partitioning frames' (Winther 2011: 397). Winther's suggestion that 'parts are abstracted through partitioning frames closely linked to explanatory projects' (ibid., 400) nicely fits with the constructive strategies of GPT, namely (i) to begin with the most generic part-relation terms from which specific part relations (functional, structural, or even more specific: morphological, physiological etc.) can be defined; (ii) to operate with a part relation that—unlike the transitive 'part of'—does not impose any implicit domain restrictions and allows, for instance, genes in the role of 'structure parts' and 'activity parts' (see ibid., 412); and (iii) to associate terms with default partitions. That the constructional strategies of GPT in this way match the use of classificatory partitions in biology becomes particularly important when the latter shift focus onto process-based classification:

> The more general point is that classifying a thing as a cheetah identifies a set of processes in which it can be involved. Classifying it in other ways might identify different processes. Such possibilities of multiple, perhaps cross-cutting, classification become more salient as classification becomes less determinate. This will be most clearly the case among the microbes.
>
> (Dupré 2012: 78)

The fact that in GPT the identity of a dynamics can be defined at different depths, according to context, by taking more or fewer levels of the representing partition into account, dovetails with another recurrent argument for the process turn, namely that developments and dynamically stable units can be realized along different paths. To explain, in the GPT formalism, the identity of a dynamic can be unfolded by adding levels to the base partition that represents D, then compressed again, accordion style. In other words, we can move between more coarse-grained and more fine-grained identity conditions. Consider now the statement 'the *same* development or homeostatic unit H can been realized along path A as well as path B'. The truth maker for this statement is a dynamic D that is defined for example by the upper five levels of the base partition BP that represents D, while the truth maker of the sentence 'H can be realized via path A' is D as defined for example by ten levels of the base partition BP representing D (where partition levels 6 through to 10 state the dynamics required for a realization along path A).

*5.3. Emergence*

Do biological process systems exhibit cases of generative strong emergence, that is, of emergent processes with a causal role both in the sense of constraining and in the sense of generating the processes from which they emerge? For reasons of space I must leave matters in the hypothetical mode here: if there were cases of generative emergence, they would surely make for the most compelling argument for a process turn in biology (as well as in cognitive science; see Bickhard 2009b; Boogerd et al. 2005; Seibt 2015). It is important to note, however, that, at this potentially most strategic point, the categories and the basic setup of Whitehead's philosophy of organism do not provide immediate conceptual support. Whitehead's ontology contains the means to describe weak emergence and even emergence with 'downward' influences of emergent parts onto other parts of a process. But in Whitehead's setup the interactions between the processual parts of an entity-constituting process ('concrescing occasion') can never affect the process of 'concrescence' itself. All concrescences point forward, as it were, and cannot affect themselves or any part that constitutes them. In GPT, on the other hand, such forms of generative feedback or literal self-modification can be defined in terms of self-similar partitions. A self-modifying or self-regulating dynamics D—that is, a dynamics with negative or positive feedback such as a self-maintaining dynamics—has a base partition BP with some (graphically speaking, upper) section S consisting of all $\leq$n-parts of D in BP and S is repeated further below in BP at a partition level n + m, 2 < m.[23]

# 6. Conclusions

The purpose of this chapter was to offer a sketch of a process ontology that may contain relevant conceptual resources for philosophers of biology who are promoting a revisualization of the domain of biology in terms of processes. I have introduced the basic concepts and constructional ideas of GPT, an ontology or domain structure theory that operates with only one category, the category of concrete, non-particular, dynamic individuals called general process or dynamics, and a five-dimensional parameter matrix that is used to classify the many different types and subtypes of general processes. By way of three selective illustrations, I have adumbrated how the framework of GPT can be used to define processes as ontological counterparts for the basic kinds of entities that we commit ourselves to in our reasoning practice, both in common-sense judgements and in science (entities such as things, stuffs,

---

[23] For further details, see Seibt 2014. See also Koutroufinis 2014a, where non-Whiteheadian and Whiteheadian explanations of embryogenesis are discussed.

developments, results, activities, states, collectives, sorts, features). Since GPT is constructed on the basis of a systematic rejection of the unreflected presuppositions of traditional substance metaphysics, and in particular of the traditional axiom that concrete individuals must be particulars (i.e. necessarily uniquely located), determinate in all their features, and countable (existing or occuring in discrete units), I have suggested that the process individuals of GPT might be used to reconfigure the debate about biological individuality. GPT uses a special mereological system based on a non-transitive part relation called levelled mereology. This sort of mereology and, more specifically, the fact that dynamics are represented by collections of partitions seem to fit with biological part–whole explanations better than extant mereological systems. Finally, GPT represents differences in process architectures in terms of structures of partitions, and it therefore holds out the prospect of defining types of biological mechanisms, various forms of feedback, and 'emergence' relations—including 'self-referential' processes—in formal terms.

GPT is a reconstructive framework intended to provide what Carnap called a 'rational explication'. Unlike Whitehead's philosophy of organism, GPT is not a speculative metaphysics. For this reason, the conceptual tools of GPT are bound to appear as shallow reformulations of what we know anyway. In contrast, if we reconstruct biological concepts with the tools of Whiteheadian metaphysics, all terms acquire additional semantic content deriving from speculative principles. But precisely the interpretational austerity of GPT is, in my view, an advantage rather than a drawback, especially in a situation where philosophers of science wish to encourage a new way of visualizing a scientific domain. In a communicative situation where the force of habitual categorizations in a science must be broken, it is strategically preferable to refrain from recasting the new view in the—initially rather impenetrable—theoretical terms of Whitehead's speculative metaphysics. To speak of 'dynamics' and their parts and to be able to express differences in process flow and feedback more precisely, in terms of partition structures, may be a useful stage along the way.

## Acknowledgements

I am much indebted to James DiFrisco, John Dupré, and an anonymous reviewer for many most helpful comments.

## References

Andersen, H., Barker, P., and Chen, X. (2006). *The Cognitive Structure of Scientific Revolutions*. Cambridge: Cambridge University Press.

Bickhard, M. H. (2004). Process and Emergence: Normative Function and Representation. *Axiomathes* 14: 121–55.

Bickhard, M. H. (2009a). The Biological Foundations of Cognitive Science. *New Ideas in Psychology* 27 (1): 75–84.

Bickhard, M. H. (2009b). The Interactivist Model. *Synthese* 166 (3): 547–91.

Bickhard, M. H. and Campbell, D. (2002). Emergence. In P. Andesern, C. Emmeche, N. O. Finnemann, and P. Christensen (eds), *Downward Causation* (pp. 322–49). Aarhus: Aarhus University Press.

Boogerd, F. C., Bruggeman, F. J., Richardson, R. C., Stephan, A. and Westerhoff, H. V. (2005). Emergence and Its Place In Nature: A Case Study of Biochemical Networks. *Synthese* 145 (1): 131–64.

Brown, J. W. (2005). *Process and the Authentic Life: Toward a Psychology of Value.* Frankfurt: Ontos Verlag.

Chalmers, D., Manley, D. and Wasserman, R. (2009). *Metametaphysics: New Essays on the Foundations of Ontology.* New York: Oxford University Press.

Christensen, W. D. and Bickhard, M. H. (2002). The Process Dynamics of Normative Function. *Monist* 85 (1): 3–28.

Clarke, E. (2010). The problem of biological individuality. *Biological Theory* 5 (4): 312–25.

Clarke, E. (2011). Plant Individuality and Multilevel Selection Theory. In Brett Calcott and Kim Sterelny (eds), *The Major Transitions in Evolution Revisited* (pp. 227–50). Cambridge, MA: MIT Press.

Clarke, E. (2012). Plant Individuality: A Solution to the Demographer's Dilemma. *Biology & Philosophy* 27 (3): 321–61.

Dupré, J. (2012). *Processes of Life: Essays in the Philosophy of Biology.* Oxford: Oxford University Press.

Earley, J. E. (2013). An Invitation to Chemical Process Philosophy, in J.-P. Llored (ed.), *Epistemology of Chemistry: Roots, Methods and Concepts* (pp. 529–39). Newcastle: Cambridge Scholars.

Eastman, T., Epperson, M., and Griffin, D. (2016). *Physics and Speculative Philosophy: Potentiality in Modern Science.* Berlin: De Gruyter.

Ereshefsky, M. and Pedroso, M. (2013). Biological Individuality: The Case of Biofilms. *Biology & Philosophy* 28 (2): 331–49.

Ereshefsky, M. and Pedroso, M. (2015). Rethinking Evolutionary Individuality. *Proceedings of the National Academy of Sciences* 112 (33): 10126–32.

Falkner, G. and Falkner, R. (2013). The Role of Biological Time in Microbial Self-Organization and Experience of Environmental Alterations. In A. Nicolaidis and W. Achtner (eds), *The Evolution of Time: Studies of Time in Science, Anthropology, Theology* 72–93. Sharjah/ info@benthamscience.org: Bentham Science.

Hättich, F. (2004). *Quantum Processes: A Whiteheadian Interpretation of Quantum Field Theory.* Münster: Agenda Verlag.

Hull, D. L. (1980). Individuality and Selection. *Annual Review of Ecology and Systematics* 11: 311–32.

Kenny, A. (1963). *States, Performances, Activities. Action, Emotion and Will.* New York: Humanities Press.

Koutroufinis, S. A. (2014a). Beyond Systems Theoretical Explanations of an Organism's Becoming: A Process-Philosophical Approach. In S. Koutroufinis (ed.), *Life and Process: Towards a New Biophilosophy* (pp. 99–133). Berlin: De Gruyter.

Koutroufinis, S. A. (2014b). *Life and Process: Towards a New Biophilosophy.* Berlin: De Gruyter.

Mourelatos, A. (1978). Events, Processes, and States. *Linguistics and Philosophy* 2 (3): 415–34.

Puntel, L. B. (2002). The Concept of Ontological Category: A New Approach. In R. Gale (ed.), *The Blackwell Guide to Metaphysics* (pp. 110–30). Oxford: Blackwell.

Puntel, L. B. (2010). *Structure and Being: A Theoretical Framework for a Systematic Philosophy.* University Park, PA: Penn State University Press.

Quine, W. Van O. (1985). Events and Reification. In E. Lepore and P. M. McLaughlin (eds), *Actions and Events: Perspectives on the Philosophy of D. Davidson* (pp. 162–71). Oxford: Blackwell.

Seibt, J. (1990). Towards Process Ontology: A Critical Study in Substance-Ontological Premises. PhD Dissertation, University of Pittsburgh.
Seibt, J. (1994). Der Mythos der Substanz: Eine Studie zur Präsuppositionsstruktur der Identitätsdiskussion. *Konstanzer Berichte: Philosophie der Geistes- und Sozialwissenschaften* 15: 1–35.
Seibt, J. (1995). Individuen als Prozesse: Zur prozeß-ontologischen Revision des Substanzparadigmas. *Logos* 2: 352–84.
Seibt, J. (1996). Non-countable Individuals. *Southwest Philosophy Review* 12 (1): 225–36.
Seibt, J. (1997). Existence in Time: From Substance to Process. In J. Faye, U. Scheffler, and M. Urs (eds), *Perspectives on Time* (pp. 143–82). Dordrecht: Kluwer.
Seibt, J. (2000a). Constitution Theory and Metaphysical Neutrality: A Lesson for Ontology? *Monist* 83: 161–83.
Seibt, J. (2000b). The Dynamic Constitution of Things. *Poznan Studies in the Philosophy of the Sciences and the Humanities* 76: 241–78.
Seibt, J. (2000c). Ontological Categories: The Explanation of Categorial Inference. In D. Greimann and C. Peres (eds), *Wahrheit, Sein, Struktur: Auseinandersetzungen Mit Metaphysik* (pp. 272–98). Hildesheim: Olms Verlag.
Seibt, J. (2004). Free Process Theory: Towards a Typology of Occurrings. *Axiomathes* 15: 23–55.
Seibt, J. (2005). General Processes: A Study in Ontological Category Construction. PhD Dissertation, University of Konstanz.
Seibt, J. (2008). Beyond Endurance and Perdurance: Recurrent Dynamics. In C. Kanzian (ed.), *Persistence* (pp. 121–53). Frankfurt: Ontos Verlag.
Seibt, J. (2009). Forms of Emergent Interaction in General Process Theory. *Synthese* 166 (3): 479–512.
Seibt, J. (2010). Particulars. In R. Poli and J. Seibt (eds), *Theories and Applications of Ontology: Philosophical Perspectives* (pp. 23–55). New York: Springer.
Seibt, J. (2014). Non-Transitive Parthood, Leveled Mereology, and the Representation of Emergent Parts of Processes. *Grazer Philosophische Studien* 91: 165–91.
Seibt, J. (2015). Ontological Scope and Linguistic Diversity: Are There Universal Categories? *Monist* 98 (3): 318–43.
Seibt, J. (2016). How to Naturalize Sensory Consciousness and Intentionality within a Process Monism with Normativity Gradient: A Reading of Sellars. In J. O'Shea (ed.), *Sellars and His Legacy* (pp. 187–221). Oxford: Oxford University Press.
Seibt, J. (2017). Homeomerous and automerous. In H. Burkhardt, J. Seibt, G. Imaguire, and S. Gerogiorgakis (eds), *Handbook of Mereology* (pp. 255–61). Munich: Philosophia Verlag.
Sellars, W. (1960). Grammar and Existence: A Preface to Ontology. *Mind* 69: 499–533.
Sellars, W. (1981). Foundations for a Metaphysics of Pure Process: The Carus Lectures of Wilfrid Sellars. *Monist* 64: 3–90.
Vendler, Z. (1957). Verbs and Times. *Philosophical Review* 66: 143–60.
Winther, R. G. (2011). Part–Whole Science. *Synthese* 178 (3): 397–427.
Zemach, E. M. (1979). Four Ontologies. *Journal of Philosophy* 123: 231–47.

# PART III
# Organisms

# 7
# Reconceptualizing the Organism
## From Complex Machine to Flowing Stream

*Daniel J. Nicholson*

> *The machine analogy has put us on a wrong scent... How long are we to persist in refusing to look sheer hard facts in the face, merely in the interests of a seventeenth-century analogy which by now may well have outgrown its usefulness? Sooner or later biology will have to take account of them if there is to be any theoretical biology.*
>
> —Joseph Henry Woodger (1930: 15–16)

## 1. Introduction

The greatest intellectual revolutions are those that lead to such a profound reorientation in our habits of thought that following their occurrence it becomes almost impossible to comprehend what it was like to think about things in any other way. They transform our understanding so fundamentally that they come to ground and guide our inquiries without themselves ever being directly subject to them. A paradigmatic example is the *mechanization of the world picture* that took place during the scientific revolution (Dijksterhuis 1961). Although there is nothing inevitable about seeing the world as a vast, finely tuned machine (indeed, to Aristotle as to most other ancient thinkers, such a view would have seemed alien and artificial), after the seventeenth century it became difficult to think about nature in any other way. Thereafter, the natural was mechanical and the mechanical was natural.

This radical conceptual transformation, which in many ways precipitated the rise of modern science, is a testament to the power of metaphors. The critical role that metaphors play in the conceptualization of phenomena has not always been appreciated by philosophers. In fact, for much of the twentieth century, metaphors were dismissed as decorative literary devices, of little relevance to scientific understanding. The verifiability principle of logical empiricism rendered any appeals to metaphors meaningless and pushed metaphorical language in general beyond the realm of cognitive significance.[1] Ironically, the idea that metaphors are irrelevant to the

---

[1] Max Black, probably the first analytic philosopher to take metaphors seriously, bitterly complained about his peers' reaction to the use of metaphors: 'To draw attention to a philosopher's metaphors is to

pursuit of generating scientific knowledge emerged during the very same period that witnessed the reorganization of all of natural philosophy around a single metaphor: that of the *clockwork universe* (Collingwood 1945; Dear 2006). Many of the pivotal figures of early modern science and philosophy displayed a dismissive—if not downright hostile—attitude toward metaphors, denouncing them as illegitimate rhetorical devices that compromise the clarity and objectivity of rational discourse.[2] Today such views are rare, as there is widespread recognition of the indispensable roles that metaphors play in scientific theory and practice (e.g. Keller 1995; Maasen et al. 1995; Brown 2003). But out of the endless array of metaphors used in science, it is difficult to think of one that has been more dominant and has exerted a greater influence than the *machine metaphor*, which provided the basic theoretical foundation for mechanicist natural philosophy in both physics and biology.

Although the mechanicist worldview, with its emphasis on reductionism and determinism, collapsed in physics following the quantum revolution of the early decades of the twentieth century, it somehow managed to survive in biology. For a time—especially during the interwar years—it seemed as if biology too would abandon mechanicism, as a collective of biological thinkers known as the *organicists* began to articulate a post-mechanicist philosophical foundation for biology that explicitly rejected the ontological assimilation of organisms to machines (see Nicholson and Gawne 2015). The organicists were inspired by Alfred North Whitehead, who had written in 1925 that '[t]he appeal to mechanism on behalf of biology was in its origin an appeal to the well-attested self-consistent physical concepts as expressing the basis of all natural phenomena. But at present there is no such system of concepts' (Whitehead 1925: 129). Nevertheless, in the end mechanicism not only prevailed but was actually reinvigorated by the meteoric rise of molecular biology (see e.g. Monod 1971). The neo-Darwinian view of evolution that became established during the same period also contributed to the consolidation of mechanicism in biology (see e.g. Williams 1966).

Elsewhere (Nicholson 2013, 2014) I have referred to the central tenet of biological mechanicism—the metaphorical redescription of the organism as a machine—as the *machine conception of the organism* (MCO). The MCO is one of the most pervasive metaphors in modern biology. Part of its success lies in its remarkable plasticity, as it is able to take a variety of different forms, depending on the context. To mention only a few of its contemporary manifestations, in developmental biology it equates the embryo with a computer that executes a predetermined set of operations in accordance with a program encoded in its genes (e.g. Jacob 1973); in evolutionary biology it assimilates organisms to optimally designed artefacts, blindly engineered by natural

---

belittle him—like praising a logician for his beautiful handwriting. Addiction to metaphor is held to be illicit, on the principle that whereof one can speak only metaphorically, thereof one ought not to speak at all... [Let us] not accept the commandment, "Thou shalt not commit metaphor", or assume that metaphor is incompatible with serious thought' (Black 1962: 25).

[2] Thomas Hobbes, for instance, declared that '[m]etaphors, and senslesse and ambiguous words, are like *ignes fatui*; and reasoning upon them, is wandering amongst innumerable absurdities; and their end, contention, and sedition, or contempt' (Hobbes 1996: 36). One cannot help but wonder how Hobbes saw no inconsistency in decrying the usefulness of metaphors by using one to make his point.

selection (e.g. Dawkins 1986); and in molecular biology it identifies the cell as a factory of highly specialized molecular machines (e.g. Alberts 1998).

In recent years, however, there have been growing voices of dissent from the mechanicist orthodoxy, as more biologists and philosophers have begun to question the theoretical legitimacy of the MCO (e.g. Rosen 1991; Lewontin 2000; Kirschner et al. 2000; Henning and Scarfe 2013). It is becoming clear that the MCO offers only a partial and rather distorted view of living systems. Most significantly for the purposes of the present volume, the uncritical—and often tacit—acceptance of the MCO is one of the major reasons for the persistence of substance metaphysics in biology. This should not be surprising, as mechanicism has always served as the main vehicle for substance thinking in science. After all, what are machines if not persistent material things with determinate sets of properties and which exist independently of the activities they engage in? Demonstrating the ontological inadequacy of the MCO is a necessary first step if we are to come to terms with the processual nature of life and lay the foundations for a processual philosophy of biology.

In an earlier paper (Nicholson 2013), I argued that the MCO fails to provide an appropriate understanding of living systems because organisms and machines differ from one another in a number of crucial respects. Most fundamentally, I claimed, organisms are *intrinsically purposive* (in the sense that their activities and internal operations are ultimately directed towards the maintenance of their own organization), whereas machines are *extrinsically purposive* (given that their workings are geared towards fulfilling the functional ends of external agents).[3]

The present chapter presents a totally different argument against the MCO: one based on thermodynamic considerations. As I will show in the next section, thermodynamics supplies a surprisingly effective means of elucidating the ontological distinction between organisms and machines.[4] The thermodynamic character of life is incompatible with the MCO and calls for the adoption of a processual view of the organism, which is exemplified by the Heraclitean metaphor of the stream of life. In section 3 I will examine the intellectual development of this alternative metaphorical conception and consider the extent to which it captures the nature of living systems. I will follow this in section 4 by discussing three specific ontological lessons that we can draw when we reconceptualize the organism from complex machine to flowing stream. The first relates to questions of normativity and agency, the second concerns the problem of persistence, and the third addresses the nature and origins of order. I will conclude by briefly reflecting on the broader consequences of this shift in perspective.

---

[3] Another way of expressing this is that organisms act on their own behalf, while machines serve the interests of their makers or users (for the full argument, see Nicholson 2013; for various illustrations, see Nicholson 2014).

[4] It is a pity that thermodynamics has not played a greater role in the philosophy of biology, especially given that the implications of thermodynamics for the field were explicitly recognized by some of its earliest practitioners (see Needham 1928: 81–5). The only major exception was the debate—particularly prominent during the late 1980s—regarding proposals to extend, modify, or even reformulate the neo-Darwinian theory of evolution in accordance with thermodynamic principles (e.g. Brooks and Wiley 1986; Wicken 1987; Weber et al. 1989).

## 2. Organisms ≠ Machines: The Argument from Thermodynamics

Ironically, thermodynamics initially served to vindicate, rather than undermine, the MCO (understood here as a heat engine rather than a clock). The science of thermodynamics arose from the desire to understand heat engines, particularly the relationship between heat and work, and the presumed conformance of organisms to the first law of thermodynamics—a version of the principle of the conservation of energy—enabled them to be treated as veritable engines. The pioneer in this respect was Antoine Lavoisier, the father of modern chemistry, who in the late eighteenth century famously characterized respiration as a form of combustion. Along with Pierre-Simon Laplace, Lavoisier conducted the first calorimetry experiments, comparing the heat and carbon dioxide produced by a guinea pig with that produced by the combustion of carbon.[5]

During the nineteenth century, the rise of thermodynamics was so intertwined with concurrent developments in physiology—united as these disciplines were by their common interest in 'engines', be they technological or biological—that some of the earliest enunciations of the first law, such as those by Hermann von Helmholtz and Robert von Mayer in the 1840s, were tied to efforts to elucidate the relation between chemical energy and physiological activity. Helmholtz, a physiologist by training, was led to his formulation of the first law of thermodynamics by his demonstration of the equivalence between animal heat and energy, as well as by his discovery that only physico-chemical processes are involved in the generation of animal heat. And Mayer, a practising physician, gave very explicit consideration to the bearing of the first law on organisms:

> In the living body carbon and hydrogen are oxidized and heat and motive power thereby produced. Applied directly to physiology...the oxidative process is the physical condition of the organism's capacity to perform mechanical work and provides as well the numerical relations between [energy] consumption and [physiological] performance.
>
> (Meyer, quoted in Coleman 1977: 123)

At the end of the century, Max Rubner conclusively established that organisms are subject to the first law by showing experimentally that the amount of energy returned to the environment by an organism (for instance, in the form of excretory products and heat) is equivalent to the energy taken in, assuming no change in weight. Although this by itself did not prove that organisms are heat engines—in fact, upon close examination, the original analogy between combustion and respiration turns out to be rather problematic[6]—it appeared to offer little reason to question the MCO.

---

[5] Lavoisier's own formulation of the MCO went like this: 'The animal machine is governed by three main regulators: respiration, which consumes oxygen and carbon and provides heating power; perspiration, which increases or decreases according to whether a great deal of heat has to be transported or not; and finally digestion, which restores to the blood what it loses in breathing and perspiration' (Lavoisier, quoted in Jacob 1973: 43).

[6] In combustion, the surmounting of the energy of activation—which is necessary for the accomplishment of oxidative reactions—is achieved by raising the temperature considerably, whereas in respiration this is not needed. Instead, respiration relies on the enzymatic lowering of the energy of activation. If the

The second law of thermodynamics is a completely different matter. Indeed, it is when we consider how organisms conform to it that the MCO absolutely breaks down. The second law negates the possibility of a perfectly efficient transformation of heat into work. It stipulates that the amount of free energy (i.e. energy capable of performing work) is constantly decreasing, while the amount of dissipated energy (measured in terms of entropy) is correspondingly increasing. Every natural change, whether physical or chemical, exhibits this utterly irreversible tendency—pithily described by Arthur Eddington as the 'arrow of time'—which results in a net, ever growing increase in disorder. Such an inexorable trend towards a uniform distribution of heat and the consequent 'running down' of the universe into a state of dead inertness is diametrically opposed to what we find in the living world, where there is a clear evolutionary tendency for complexity and organization to increase progressively with time. What are we to make of this paradoxical situation?

The founders of thermodynamics were perfectly aware of this paradox, but instead of dealing with it they simply ignored it. William Thomson (later Lord Kelvin), who coined the term 'thermodynamics', explicitly excluded living processes in his formulation of the second law in 1851.[7] Years later, Helmholtz declared that whether or not the second law is violated by 'the fine structure of living organized tissue appears to me still to be an open question, the importance of which in the economy of nature is very obvious' (Helmholtz, quoted in Needham 1928: 81). James Clerk Maxwell attempted to confront the problem by suggesting how the second law might be contravened, but his suggestion required postulating a cunning microscopic being—which Thomson dubbed 'Maxwell's Demon'—capable of sorting molecules according to their speed without the expenditure of work, thereby reducing the overall level of entropy. Given the total lack of evidence for the existence of such a fanciful creature, by the early twentieth century biologists—with the exception of a few neo-vitalists such as James Johnstone (1921)—assumed that organisms do conform to the second law. The question that remained was how.

The eventual resolution of the paradox came with the realization that the second law requires only that the universe *as a whole* exhibits an increase in entropy. Local eddies of order (or 'negative entropy') can be sustained and even propagate, as long as, overall, there is a global entropic increase. This was lucidly pointed out by Erwin Schrödinger in his influential book *What Is Life?* (Schrödinger 1944). Schrödinger explained that an organism stays alive in its highly organized condition by importing matter rich in free energy from outside of itself and degrading it in order to maintain a relatively low entropic state within its boundaries. The organism thus preserves its internal organization—thereby eluding (at least for a time) the inert, time-invariant state of thermodynamic equilibrium we call *death*—at the expense of

---

transformation of energy were to take place in organisms in the same way that it does in heat engines, then, at temperatures at which living systems can exist, the coefficient of their useful activity would fall to an insignificant fraction of 1 per cent (see Oparin 1961).

[7] Thomson's original enunciation of the law was this: 'It is impossible *by means of inanimate material agency*, to derive mechanical effect from any portion of matter by cooling it below the temperature of the coldest of surrounding objects' (Thomson, quoted in Keller 1995: 49, emphasis added).

increasing the entropy (in the form of heat and other waste products) of its external environment.[8]

Understanding how organisms conform to the second law allows us to see why they are fundamentally different from machines. Organisms have to constantly exchange energy and matter with their surroundings in order to maintain themselves *far from thermodynamic equilibrium*. Machines, on the other hand, exist in equilibrium or near-equilibrium conditions, and consequently do not have to constantly exchange energy and matter with their surroundings. Organisms, in other words, are necessarily *open systems*, whereas machines can be open or closed. As a result, they differ in the kind of stability they exhibit. Machines exhibit a static stability, which they attain when they reach an equilibrium state that reflects the cessation of their activity. Organisms, in contrast, exhibit a *dynamic stability*, which is based on their capacity to actively maintain a low-entropic 'steady state' where there is a continuous, perfectly balanced import and export of materials. The stability of machines at equilibrium means that they do not require free energy for their preservation, while the opposite is true for organisms.[9] A further difference is that, whereas the activity of a machine is temporary and its onset is reversible (given that a machine can return to a state of operation after being at rest), the actively maintained steady state of an organism is *fixed* and *irreversible*. An organism must remain permanently displaced from equilibrium; the moment it yields to it, death inevitably and irrevocably ensues.

The theoretical distinction between equilibrium and non-equilibrium systems is of paramount importance. Classical thermodynamics was only ever equipped to deal with equilibrium systems (which is why the aforementioned paradox concerning life and the second law arose). The recognition of non-equilibrium systems led to the development in the mid-twentieth century of what has come to be known as *non-equilibrium thermodynamics*, which concerns itself with steady states, irreversible processes, and non-linear reactions. The subject matter of this relatively new branch of thermodynamics extends beyond the living realm, as organisms are not the only far-from-equilibrium open systems found in nature. Whirlpools, flames, and tornadoes are familiar examples. Less familiar but well-studied cases include Bénard convection cells and oscillating chemical reactions such as the Belousov–Zhabotinsky reaction. Ilya Prigogine, whose foundational work in establishing non-equilibrium thermodynamics earned him a Nobel Prize in 1977, referred to these open systems as *dissipative structures*. Perhaps the most significant achievement of this new field of physics has been to show how self-organization arises in nature—that is, to explain how the macroscopic patterns of order displayed by dissipative structures spontaneously emerge from non-linear interactions and become stabilized in far-from-equilibrium conditions through an ongoing flux of energy and matter (see Nicolis

---

[8] Organisms, then, far from disobeying the second law, help to enact it. Life, with its unassailable tendency to proliferate, actively contributes to the dissipation of energy by leaving large amounts of entropic waste in its wake.

[9] The distinction I draw here between static and dynamic forms of stability corresponds to the distinction some authors have made between 'energy-well' and 'far-from-equilibrium' stability (e.g. Bickhard 2000; Campbell 2009).

and Prigogine 1977). Organisms, from this perspective, are the most stable and complexly differentiated dissipative structures in existence.

In biological theory, the thesis that organisms are dynamically stable open systems was most systematically articulated by Ludwig von Bertalanffy (1932, 1942, 1950, 1952). However, as Evelyn Fox Keller (2008) has shown, its intellectual roots can be traced back to the ideas of a number of nineteenth-century authors, including Herbert Spencer, Gustav Fechner, and especially Claude Bernard, who memorably asserted that organisms persist in time by keeping their internal environment constant in the face of external disturbances. Bernard's conception was the basis for Walter Cannon's famous concept of *homeostasis*, which he coined with the explicit purpose of accounting for the distinctive thermodynamic character of the organism:

> The highly developed living being is an *open system* having many relations to its surroundings… The coordinated physiological reactions which maintain most of the *steady states* in the body are so complex, and are so peculiar to the living organism, that it has been suggested… that a specific designation for these states be employed—*homeostasis*. (Cannon 1929: 400, emphasis added)

The above thermodynamic considerations point to *metabolism* as the single most important characteristic of life. Metabolism refers to the balanced coupling of the energy-releasing processes of catabolism (i.e. the breakdown of organic matter by means of cellular respiration) with the energy-consuming processes of anabolism (i.e. the buildup of the macromolecular constituents of cells) that are continuously going on in an organism. Crucially, metabolism is what maintains the organism in a steady state far from equilibrium, liberating large amounts of free energy for the organism to use while simultaneously 'freeing it from all the entropy it cannot help producing while alive' (Schrödinger 1944: 71). It is this emphasis on metabolism, which the non-equilibrium thermodynamics of organisms prescribes, that enables us to understand the ontological inadequacy of the MCO.[10]

## 2.1. Addressing potential objections to the argument

Let us now consider a series of objections that may be raised against the preceding argument. First of all, while some machines (like clocks and computers) are closed systems, others (like heat engines and water pumps) are decidedly open, as they exchange matter, and not just energy, with their surroundings. Although the difference between them and organisms is obvious when they are at rest, do they not display a dynamic form of stability when they are in operation that is akin to that shown by organisms? Can we not say, in the case of engines, for example, that they 'metabolize' their fuel just as organisms metabolize their food?

The problem of equating fuel with food is that it drastically underestimates the physiological pervasiveness of metabolism. No matter how dynamic a functioning

---

[10] Metabolism, incidentally, is also the basis for Humberto Maturana and Francisco Varela's (1980) seminal theory of autopoiesis (literally self-production), which has been fruitfully elaborated in recent decades by a number of philosophers and theoreticians of biology interested in the idea of autonomy (see, e.g., Moreno and Mossio 2015 and chapter 10 here).

machine may be, it is always possible to distinguish the machine's physical frame—which remains fixed—from the materials that flow through it. The actual structure of the machine does not itself take part in the chemical transformations that the fuel undergoes as it passes through it. Instead, it serves as a channel that facilitates the exchange of materials as fuel is converted into waste. An organism, in contrast, changes wholly and continuously as a result of its metabolizing activity. Organisms are constantly being reconstituted from the matter they import from their surroundings, and consequently it is impossible to maintain the distinction between food materials and bodily constituents. As Hans Jonas phrased it, in an organism '[t]he exchange of matter with the environment is not a peripheral activity engaged by a persistent core: it is the total mode of continuity (self-continuation) of the subject of life itself' (Jonas 1966: 76, n. 13).[11] This is why the fuel–food analogy is so misleading, and why the stability of a machine—despite its apparent dynamicity—ultimately resides in an unchanging material structure. In machines there is a specific 'inflow' and a specific 'outflow'. In organisms *everything flows*.[12]

Even after accepting this, a critic might wish to insist that it is still inappropriate to characterize the form of machines as fixed. Although machines do not exchange their material constitution externally in the way organisms do, many do modify their structure *internally* to suit their purpose. Windmills, catapults, bicycles, and typewriters all change their physical configuration in order to accomplish their function. Why, then, should we not regard these internal structural rearrangements as comparable to the ones that organisms undergo?

The trouble with this (more restricted) comparison is that it overlooks the fact that the physical displacements that the parts of a machine undergo conform rigidly to a precise, predetermined cycle of operations. The successful execution of a function by a machine implies a periodic restoration of the spatial relations among its parts—a 'resetting' of its internal configuration that enables it to perform its function again. So, although a machine may contain parts that move about, this does not mean that it goes through a genuine process of transformation. Its physical architecture is still very much fixed; it is just that this fixity is reflected in a recurrent spatio-temporal pattern rather than in a totally static structure. The only real change that machines experience is the gradual wearing down of their parts, which eventually leads to their irrevocable entropic degradation (unless, of course, an external agent intervenes).

The situation with organisms is completely different. Organisms autonomously modify their structure in response to cues from their environment. When injured, they are usually able to heal themselves and repair the damage—and this is as true for bacteria as it is for complex multicellular organisms like trees and vertebrates. Some organisms (e.g. salamanders) can even regenerate entire body parts, often following the self-amputation of limbs, in order to avoid predation. The astonishing plasticity

---

[11] Jonas' insightful analysis of metabolism is also discussed in chapters 8 and 18.

[12] It is rather striking that the organicists were already aware of this fact. For example, John Scott Haldane, who was heavily influenced by Claude Bernard, wrote in 1917 that '[t]he organs and tissues which regulate the internal environment... are constantly taking up and giving off material of many sorts, and their "structure" is nothing but the appearance taken by this flow of material through them' (Haldane 1917: 90; see also Russell 1924 and Woodger 1929).

of organisms contrasts with the brittleness of machines, which tend to stop working when their parts break or are damaged. Of course, redundancy and self-repair can be built into the design of machines to some extent. Nevertheless, although this can make their operation more robust and more reliable, the inherent limitations of their fixed architecture remain.

Not only do organisms modify themselves, but they do so *adaptively*, in a way that optimizes their physiological performance. Now a critic could still argue that servomechanisms—machines controlled by negative feedback with a certain capacity to self-regulate (e.g. thermostats) or to self-steer (e.g. homing missiles)—also modify their operation adaptively in response to external inputs; in fact, they are often described as 'homeostatic machines'. However, this argument misleadingly conflates the thermodynamic meaning of Cannon's original use of the word 'homeostasis' with the cybernetic meaning the term acquired after the Second World War. The adaptability of servomechanisms (measured in terms of input–output adjustments) is of a very limited kind, its mode and range being defined in advance in accordance to a set design. Servomechanisms are closed, near-equilibrium systems, and consequently they are not capable of truly adaptive self-maintenance.[13]

At this point, the most doggedly persistent of critics may take all of the above considerations on board and simply resort to stretching the concept of machine sufficiently for it to encompass everything that is distinctive about organisms. But in the end this does not really help the critic's defense of the MCO, as it undermines the very properties of machines that make the MCO a heuristically useful idealization in the first place. An instructive illustration of this can be found in a curious paper titled 'Living and Lifeless Machines', in which the distinctive properties of organisms are forcefully shoehorned into the framework of the MCO. This leads the author to make a number of exceptionally strange assertions. For example, he notes that '[t]he living body is analogous to a motor car in which the chassis, brakes, cylinders, pistons, connecting rods, valves and bearings all contained combustible material, some of which was burnt whenever the driver placed his foot on the accelerator' (Kapp 1954: 101). The question we must ask is this: how is such a bizarre imaginary motor car still analogous to an *actual* motor car? Can the causal operation of the latter really be used to shed light on the causal operation of the former? If not, then what is the point of clinging on to the MCO, if the price to be paid is that our understanding of machines has to become completely distorted in order to accommodate the characteristic attributes of organisms?[14]

The far more sensible option is to simply accept the fact that machines are not good models for coming to terms with the ontology of organisms. The MCO, despite its obvious heuristic value in biological research, does not provide an adequate

---

[13] For more detailed critiques of the cyberneticists' efforts to assimilate servomechanisms to organisms, see Taylor 1950, Jonas 1953, Oparin 1961, and Nicholson 2013.

[14] Kapp is by no means the only author to stretch the MCO beyond breaking point. The history of biology is littered with memorable examples, including Julien Offray de La Mettrie's *clock that winds itself* or Karl Ernst von Baer's *machine that constructs itself*, to paraphrase their respective formulations of the MCO. A more recent example is Richard Dawkins' (1998: 17) 'machine that work[s] to keep itself in being, and to reproduce its kind'.

theoretical understanding of the nature of living systems. As we have seen, the thermodynamic character of life demands a *processual* conception of the organism. Whatever else they may be, living systems are highly stabilized flows of energy and matter. Machines may take part in various processes, but organisms are *themselves* processes. This inescapable fact must constitute the starting point for any theory of the organism. In the next section we will examine the history of attempts to develop an alternative conception of living systems that successfully reflects their processual nature.

## 3. The Stream of Life: A Processual Conception of the Organism

The MCO has been the most pervasive view of living systems since the seventeenth century, but it is certainly not the only conception one finds when surveying the history of biological thought. In fact, many of those who criticized the MCO in the past sought alternative metaphors that could highlight the very features of organisms that the MCO conveniently ignores or inadvertently distorts. A good place to start is the work of Bertalanffy, who—as we have already mentioned—was responsible for popularizing the idea that organisms are open systems that maintain themselves in a steady state far from equilibrium.

In his organicist treatise *Problems of Life*, Bertalanffy (1952) illustrated the processual nature of the organism by appealing to the famous aphorism of the Presocratic philosopher Heraclitus that it is impossible to step into the same river twice because fresh water is forever flowing through it. A stream is never the same at two successive temporal points; it is permanently changing. This image encapsulates the Heraclitean worldview, which emphasized the endless movement and change of all things. Bertalanffy argued that '[w]ith this Heraclitean thought we put our finger on a profound characteristic of the living world' (ibid., 124). Like the river, ever changing in its waves yet persisting in its flow, an organism only *appears* to be constant and invariable, but in reality it is the manifestation of a ceaseless current. As Bertalanffy put it, 'living forms are not *in being*, they are *happening*; they are the expression of a perpetual stream of matter and energy which passes the organism and at the same time constitutes it'. He referred to this processual view of the organism as the *stream of life conception* (SLC) and counted it 'among the most important principles of modern biology' (ibid.).

The SLC allows us to grasp, in simple and evocative terms, many of the key characteristics of organisms that were highlighted in the previous section. The external form of a stream is stable only because of the constant flow of water molecules that enter into it and emerge out of it. The moment this flow is interrupted, the stream itself disappears, as its very existence depends on the steady movement of water passing through it. In the same way, the physical form of an organism is merely the visible expression of the constancy of catabolic and anabolic processes going on within it. Its persistence through time is entirely dependent on the extremely intricate balancing of these two opposing kinds of reactions. As metabolism proceeds, with the steady import of nutrients and export of wastes, not much remains at a later time of the matter that once composed the organism. The SLC thus embodies two essential and complementary aspects of organismic dynamics: the continuous exchange of matter that lies at the very heart of the concept of metabolism on the one hand,[15] and the surprising stability of form that is maintained in spite of this material exchange on the other.[16]

Bertalanffy deserves credit for being the first to explicitly recognize that the thermodynamic openness of organisms requires the adoption of a processual perspective. However, he was by no means the first biologist to propose that processual metaphors offer a more accurate portrayal of the ontology of living systems. Throughout the nineteenth and the first half of the twentieth centuries, a wide range of authors independently arrived at the conclusion that organisms are best served by metaphorical conceptions that stress their dynamic, non-equilibrium qualities. We can explore the versatility of the SLC by considering some of its most notable historical formulations.

As early as 1817, Georges Cuvier felt compelled to define life as a *vortex*—that most paradigmatic of dissipative structures—using language that clearly prefigures the thermodynamic observations that Schrödinger and others would make more than a century later:

Life then is a vortex, more or less rapid, more or less complicated, the direction of which is invariable, and which always carries along molecules of similar kinds, but into which individual molecules are continually entering, and from which they are continually departing; so that the *form* of a living body is more essential to it than its *matter*. As long as this motion subsists, the body in which it takes place is living—*it lives*. When it finally ceases, *it dies*.

(Cuvier 1833: 14)

The vortex metaphor enabled Cuvier to make conceptual sense of the persistence of organismic form coupled with the transience of its constituent materials. This idea was picked up some years later by William Whewell, who—in the section of *The Philosophy of the Inductive Sciences* that dealt with what he was the first to call 'the philosophy of biology'—paraphrased Cuvier by asserting that 'life is a constant form of circulating matter' (Whewell 1840: 46). Thomas Henry Huxley identified life more specifically with a vortex of water—a *whirlpool*—remarking that the constituents of an organism 'stand to it in the relation of particles of water to a cascade, or a whirlpool'. Moreover, just as 'the stoppage of a whirlpool destroys nothing but a form, and leaves the molecules of the water, with all their inherent activities intact, so what we call the death...of an animal, or of a plant, is merely the breaking up of the form, or manner of association, of its constituent organic molecules' (Huxley 1870: 402).

By the early twentieth century, advances in the study of biochemical energetics—driven by research into the thermodynamics of living systems—had become difficult

---

[15] The term 'metabolism', we should not forget, derives from the Greek word for change. More explicitly still, the German word for metabolism is *Stoffwechsel*—literally, 'material exchange'.

[16] An interesting implication of the dynamic stability of form is that, as Nicholas Rescher (1996: 52–3) has acutely observed, Heraclitus was only half-right when he declared that we cannot step into the same river twice. We may not be able to step twice into the same *waters*, but we can certainly step twice into the same *river* (that is, of course, assuming *we* also stayed the same!).

to ignore, and this is reflected in the SLC formulations of the time (cf. Gilbert 1982). Lawrence Henderson (1913: 23–4), for instance, argued that '[l]iving things preserve, or tend to preserve, an ideal form, while through them flows a steady stream of energy and matter which is ever changing'. John Scott Haldane (1919: 49) also stressed the energetic and material flux taking place in the organism, declaring that 'organic structure is nothing but a molecular stream'. Charles Sherrington (1940: 82), for his part, described the cell as 'an eddy in a stream of energy' and as 'a stream of movement which has to fulfil a particular pattern in order to maintain itself' (ibid., 83).

While some of those who adopted the SLC drew quite generically on dissipative structures like vortices in the way that Cuvier had done—such as Ralph Stayner Lillie (1945: 28), who emphasized 'the "vortex-life" feature of the vital constitution'—the majority were captivated specifically by the fluidity of *water*, finding in streams, waterfalls, and rivers the most suitable analogues for the organism. Edward Stuart Russell (1924: 6) observed that, '[j]ust as in a stream a ripple of constant shape and position is formed by the water flowing over a pebble, so the apparently static form and composition of organic substance are merely the expression of continuous... activity'. A similar assertion was made by Edmund Sinnott (1955: 117), who wrote that '[a]n organism has a sort of fluid form like a waterfall, through which water ceaselessly is pouring but which keeps in its descent a definite pattern'. Conrad Hal Waddington pointed out in *The Strategy of the Genes* (Waddington 1957: 2) that organismic form 'is more nearly comparable to a river than to a mass of solid rock', and Alexander Oparin (1961: 9) stated—echoing Bertalanffy and alluding to Heraclitus—that '[o]ur bodies flow like rivulets, their material is renewed like water in a stream. This is what the ancient Greek dialectician Heraclitus taught'.

Water, however, was not the only resource available to those wanting to underscore the dynamic stability of living systems. Some biologists resorted to fire in their articulation of the SLC, identifying the organism with a *flame*. John Burdon Sanderson Haldane (son of John Scott Haldane) claimed that 'a man is as much more complicated than a flame as a grand opera is more complicated than a blast on a whistle. Nevertheless, the analogy is real' (Haldane 1940: 57). This is not such a far-fetched analogy as it may seem. When a candle is lit, the flame flares up but almost immediately settles into a stable dynamic form that represents the attainment of a steady state. As long as it is continuously supplied with wax and oxygen, the flame is able to maintain itself far from equilibrium. It achieves this by keeping its temperature above the combustion threshold and vaporizing the wax, which induces convection that pulls in oxygen and removes combustion products. For Haldane, the flame analogy depicted the inherent dynamicity of life in ways that accentuated the inadequacy of the MCO. He indicated, among other things, that 'a flame is like an animal in that you cannot stop it, examine the parts, and start it again, like a machine. Change is part of its very being' (ibid.).[17]

Nevertheless, by the mid-twentieth century the SLC was rapidly losing ground to the MCO. This was partly a consequence of the shift in the biological agenda that took place during this period. The focus on metabolism and energetics that had

---

[17] For additional flame-based formulations of the SLC, see Brillouin 1949 and Bertalanffy 1967. The thermodynamic character of flames and its relevance for understanding organisms is also examined in chapter 10.

shaped physiological and biochemical research in earlier decades gave way to molecular biology's intense preoccupation with the structure and specificity of macromolecules, especially nucleic acids and proteins. As interests shifted from the plasticity and adaptability of biological form to the coding, replication, and expression of genetic information, so did the metaphorical conceptions used to characterize living systems. In the last third of the twentieth century, the MCO regained its place at the centre of biological theory and the SLC almost completely disappeared from the biological discourse.

Only in recent years—as the explanatory limits of molecular biology have become apparent—have we begun to witness the first signs of a revival of the SLC. In an influential article titled 'A New Biology for a New Century', Carl Woese specifically singled out the MCO as one of the major obstacles impeding further progress in biology. In place of the MCO, Woese invoked the SLC as a more appropriate metaphor with which to think about organisms:

If they are not machines, then what are organisms? A metaphor far more to my liking is this. Imagine a child playing in a woodland stream, poking a stick into an eddy in the flowing current, thereby disrupting it. But the eddy quickly reforms. The child disperses it again. Again it reforms, and the fascinating game goes on. There you have it! Organisms are resilient patterns in a turbulent flow—patterns in an energy flow. A simple flow metaphor, of course, fails to capture much of what the organism is. None of our representations of [the] organism capture [sic] it in its entirety. But the flow metaphor does begin to show us the organism's (and biology's) essence. And it is becoming increasingly clear that to understand living systems in any deep sense, we must come to see them not materialistically, as machines, but as (stable) complex, dynamic organization[s]. (Woese 2004: 176)

Woese is, of course, right to point out that the SLC does not capture every aspect of living systems. Even different versions of the SLC vary in their capacity to portray particular features of organisms. For example, a flame depicts metabolism more accurately than a whirlpool in that metabolism is essentially a series of chemical reactions; and, while the steady state of a flame is similarly sustained by continuous chemical changes, the steady state of a whirlpool is not.

At a more general level, however, it is obvious that organisms differ from flames, whirlpools, and other dissipative structures in a number of ways. For a start, organisms exhibit a far greater degree of stability, being able to maintain themselves for much longer periods of time. The key to their extraordinary stability lies in their ability to store energy, which enables them to manage their metabolic needs without having to rely on a constant supply of external energy, like other dissipative structures. In addition, organisms are distinctive in that they are demarcated by a physical boundary—a semi-permeable membrane—which helps regulate the intake and outtake of materials flowing through them. It is also evident that organisms display much higher levels of internal complexity, as they are functionally differentiated and hierarchically organized. Most dissipative structures lack these features because they spontaneously self-organize under appropriate conditions—a phenomenon that Stuart Kauffman (1995) memorably called 'order for free'—and this results in a single ordered macroscopic structure within which it is difficult to differentiate distinct functional contributions to the maintenance of the overall system.

Organisms, on the other hand, do not arise spontaneously but instead derive from previous organisms, and their structure reflects the gradual consolidation, through the eons of evolution, of an intricate 'higher-order self-organizing dynamic among component self-organizing processes' (Haag et al. 2011: 329).

Some authors have seen in these differences an unbridgeable gap separating organisms from all other dissipative structures—a gap that undermines any attempts to elucidate the former by examining the latter. For instance, Alvaro Moreno and Matteo Mossio have recently argued in their book-length treatment of autonomy (which they regard as the defining characteristic of living systems) that non-living dissipative structures 'are not relevant for understanding autonomy, not only because they are "too simple"... but also because they cannot be taken as a "starting point" for the emergence of closure and autonomy' (Moreno and Mossio 2015: 18). This seems a rather excessive (not to mention premature) conclusion. Inanimate self-organizing systems are undoubtedly simpler than animate ones, but there is a clear continuity between the two, as the latter must have emerged from the former at some point in the distant past. So, if there is anything that non-living dissipative structures offer biology, it is precisely a starting point from which one may investigate life's origins.

But, even more importantly, organisms and other dissipative structures are fundamentally isomorphic from a purely physical point of view. As we saw in the previous section, they can be understood by means of the same thermodynamic principles. Even Moreno and Mossio admit that autonomy 'is essentially grounded in thermodynamics' (ibid., 6) and that many of the autonomous features of organisms 'in fact *derive* from the fact that they are thermodynamically open systems, in far-from-equilibrium conditions' (ibid., xxviii). It is therefore difficult to see what exactly the problem is, if there really is one, with invoking the SLC and drawing on simple dissipative structures to shed light on aspects of more complex ones—provided, of course, that one recognizes that knowledge of the former cannot by itself suffice to explain the latter.

## 4. Organisms as Streams: Three Lessons for Biological Ontology

The SLC constitutes a promising point of departure for thinking about the ontology of organisms. In direct contrast to the MCO, it correctly identifies and accurately portrays the specific thermodynamic character of living systems. It supplies a firm physical foundation upon which we can begin to articulate a theory of the organism that does justice to its thoroughly processual nature. Although undoubtedly not sufficient, the SLC does show us a path through which we can ultimately *arrive* at a fully fledged biological understanding of organisms that is simply beyond the reach of the MCO. In this respect, the SLC can act as a ladder that can eventually be kicked away after it has served its purpose. After centuries of dominance of the MCO, the question in front of us is simple: what does biological ontology look like when we reject the mechanical and adopt the processual? In this section I will examine three concrete ontological lessons that we can draw from reconceptualizing the organism from complex machine to flowing stream. The first relates to questions of normativity and agency, the second concerns the problem of persistence, and the third addresses the nature and origins of order.

*4.1. First ontological lesson: 'Activity is a necessary condition for existence'*

Perhaps most evidently, using the SLC to shed light on the ontology of organisms allows us to make sense of their peculiar existential predicament. Owing to their thermodynamic condition, organisms—like all other dissipative structures—can only exist insofar as they are able to maintain themselves in a steady state far from equilibrium, and this requires a constant expenditure of free energy. The existence of a whirlpool, for instance, is a direct consequence of its own unremitting activity, which is what enables it to maintain itself through time. In the same way, if an organism is to stay alive, it has to keep acting ('working', in the thermodynamic sense) to avoid the ever present threat of equilibrium. To stop acting is to stop existing. As far as any living system is concerned, to be *is* to act. We can draw from this our first ontological lesson, which is that, for an organism, *activity is a necessary condition for existence*.

This assertion constitutes a radical departure from the conventional ontological stance of the western philosophical tradition, which is firmly rooted in substance metaphysics. A great deal of philosophical thought throughout history has been tacitly committed to the scholastic principle *operari sequitur esse* (Rescher 1996)— that is, activity is subordinated to being and thus follows from it; there can be no activity if there is no being to begin with. The processual nature of organisms requires that we relinquish this principle, given that—in the biological world, at least—activity and being necessarily presuppose one another. Being is neither ontologically nor temporally prior to activity, as the very existence of a living being is only possible by means of continuous activity.

Unsurprisingly, the MCO perfectly encapsulates the traditional substance-ontological position. The existence of a machine is totally independent of whether or not the machine happens to be performing its function. Machines have two modes of being: they can be active ('on') or they can be at rest ('off'). They can move back and forth between these two states without jeopardizing their structural integrity, and this is due to the fact that they exist in reversible, near-equilibrium conditions. The problem is that organisms, much like waterfalls or tornadoes, do not have an 'off' switch.[18] Metabolic processes can be slowed down, for example during sleep or hibernation, but they cannot be stopped completely. Even organisms in extremely dormant states, such as seeds or spores, exhibit basal levels of metabolic activity. Absolute stasis

---

[18] It should not escape anyone's attention that the ability to study machines successfully when they are turned off is the key to the enormous methodological appeal of the MCO. If organisms are machines, then we are justified in believing that they can be fruitfully investigated in abstraction from time. It is only when we adopt the SLC and are confronted with their processual nature that we realize that studying organisms atemporally is not to study them as they actually exist. It becomes apparent that using methods that strip organisms of their temporal extension (methods such as anatomical techniques that involve the desiccation, pickling, fixing, or freezing of biological samples) means resigning ourselves to characterizing static snapshots of an inherently dynamic reality.

signifies death, not inactivity, as it implies the irreversible attainment of thermodynamic equilibrium.

The indissoluble bond linking existence to activity in organisms and other dissipative structures obtains because their operation is directed *inwardly*, towards the generation and maintenance of their own organization. The operation of machines, by contrast, is directed outwardly, towards the production of something external to themselves. Organisms, unlike machines, are *autopoietic*; they persist as a result of their own activity. This ongoing self-producing activity is not optional—not undergoing constant metabolic regeneration is not a possibility. The thermodynamically grounded fact that organisms *need* to keep acting in order to keep existing helps to account for the emergence of a rudimentary form of *normativity* in nature (cf. Mossio et al. 2009; Christensen 2012). It is because its existence depends on its own activity that an organism *must* act in accordance to the operational norms that enable it to persist through time. If the organism stops following these norms, it ceases to exist. What this means is that it is in principle possible to objectively specify what is intrinsically 'good' or 'bad' for an organism (that is to say, what is and what is not in an organism's 'interest') by evaluating its activities according to the contribution they make towards the preservation of its organization in far-from-equilibrium conditions.[19]

Attributing intrinsic normativity to the behaviour of organisms implies ascribing *agency* to them in some minimal (i.e. non-intentional) sense. Fortunately, the ontological interdependence of activity and existence also helps us come to terms with this elusive notion (cf. Barandiaran et al. 2009; Barham 2012). It is generally assumed that, in order to be an agent, an entity must be able to distinguish itself from its surroundings and, in doing so, delineate an external world with which it can maintain causal interactions. Organisms demarcate themselves from their surroundings through their metabolic activities, by taking in and pumping out energy and matter from outside of themselves. Importantly, the capacity of organisms to individuate themselves and interact with their environment is a direct consequence of their thermodynamic exigency to regulate their exchanges with it in order to ensure their continued viability.

As agents, organisms are inherently *active*, as opposed to machines, which are typically reactive. The former are primary sources of activity, whereas the latter must be activated by external means. A clock has to be wound up, a computer has to be turned on, a car has to be started, the keys of an organ have to be pressed, and so on. This crucial difference—already highlighted by Bertalanffy (1952), among others—reflects yet another respect in which the MCO distorts biological ontology. According to the MCO, the organism is essentially a passive system, being set into action through outside influences. Just as a vending machine, by virtue of an internal mechanism, delivers an article after a coin is inserted, so an organism performs a

---

[19] Of course, it is also possible to make normative claims about the operation of machines, but there is an essential difference. Whereas in the organism the norms of its operation are endogenously generated and are intrinsically relevant to its own continued existence, in the machine they are imposed by an external agent (usually the machine's maker or user), who monitors the machine's operation and evaluates its performance according to his or her own needs or interests.

preset operation upon receiving a stimulus from its environment.[20] In actual fact, because activity is a necessary condition for existence, a stimulus never really triggers the onset of activity in a hitherto inactive organism, but rather modifies the preexisting network of processes that are already occurring within it.[21]

This leads to the important conclusion that, more often than not, it is not so much the nature of the external stimuli as the organism's internal physiological state that determines its reactions and behaviour. And as an organism's current physiological state is a product of the sequence of past events that led to it, it follows that the *history* of an organism fundamentally shapes its behaviour. Again, the contrast with machines is instructive. The operation of a machine is not significantly influenced by its history. After a machine executes its function, its configuration is reset to its default inactive state, and every time this happens the particular historical record of its operation—beyond the gradual wearing down of its parts—is erased. However, because an organism never stops functioning during its lifetime (as we have seen, it *cannot* do so), it never returns to the same exact state. The upshot of all of this is that organisms cannot be fully accounted for without affording careful consideration to their individual historical (i.e. developmental) trajectories.

One last implication of this first lesson is that it explains why organisms are inextricably intertwined with their environment. After all, the incessant activity of organisms that guarantees their continued existence is an activity of exchange *with* their environment. Organisms are totally dependent on their environment for the energy they need to maintain themselves far from equilibrium. In addition, they are quite literally composed of the materials they import from it. It is therefore misleading to presume that an organism can be understood in isolation from the environment in which it is always embedded. This assumption, like so much of what we have discussed, has its basis in the MCO. Although a machine requires energy from outside of itself to carry out its function, its existence does not rely on the fact that it has permanent access to environmental resources. A machine is placed in an environment and reacts to the stimuli it receives from it. An organism, by comparison, is *made* from its environment, and at the same time helps to *construct* it through its activities (which include its metabolic exchanges with it). This process is known as *niche construction*, and it has recently been recognized to have profound consequences for evolution (see Odling-Smee et al. 2003).

## 4.2. Second ontological lesson: 'Persistence is grounded in the continuous self-maintenance of form'

Our second lesson concerns the conundrum of diachronic identity, often referred to as 'the problem of persistence'. As we will see in what follows, taking the SLC as the

---

[20] This problematic conception has found its way not only into the study of physiology and animal behaviour but also into molecular biology. As Robert Rosen wryly remarked, '[g]enetic engineers, who are the molecular biologists turned technologues, habitually regard their favorite organism, *E. coli*, as a simple vending machine; insert the right token, press the right button, and the desired product is automatically delivered, neatly packaged and ready for harvest' (Rosen 1991: 21).

[21] Chapter 17 illustrates this claim in the context of olfaction.

starting point for our biological ontology forces us to rethink how we individuate and reidentify a particular organism over time. Ordinary physical objects are usually reidentifiable by means of their material constitution, which tends to remain invariant. The problem with streams, flames, and hurricanes is that their identity over time does not coincide with the identity of the materials that compose them. In the same way, the material content of an organism is in constant flux throughout its lifetime. As a result, no two 'time slices' of it are materially identical. Only in death, when the organism finally succumbs to thermodynamic equilibrium, does its material constitution stop changing.

The challenge that this poses for understanding biological persistence was carefully examined by Jonas (1966), who concluded, very much in line with the biologists who have advocated the SLC, that in an organism form is *emancipated* from matter. Organismic form exhibits a degree of independence that enables it to continue to exist despite incessant material exchange—indeed *because* of it.[22] For Jonas, organisms invert the ontological relation between matter and form found in inanimate objects like a stone or a lump of iron. Whereas in the latter form is subordinated to matter (as form in such instances reflects nothing more than the contingent spatial configuration of a physical body), in the former it is matter that is subordinated to form (as form here specifies a unified, causally efficacious whole).

Machines, however, are considerably different from stones and lumps of iron in that their form is not merely accidental. As with organisms, their matter is arranged into a specific organization that allows them to perform their function. The relevant difference with regard to organisms, as we noted earlier in the chapter, is that their form is manifested as a fixed structure rather than as a persisting flow. But suppose that a machine kept its form intact despite changes to all of its material parts. What then? The famous 'ship of Theseus' of antiquity serves as a convenient example to explore this scenario. According to Greek legend, for centuries this ship performed an annual voyage from Athens to the island of Delos. Over time, the ship underwent numerous repairs and replacements of worn-out parts, until eventually it contained none of its original planks, ropes, or sails. In spite of this, Athenians continued to regard it as the same venerable ship. One could imagine the same situation applying to a number of machines today, such as a medieval cathedral clock still in operation, or a painstakingly restored steam locomotive. The question is: do such machines—let us call them 'Thesian machines'—exhibit the sort of diachronic identity that characterizes organisms?[23]

---

[22] Note, however, that the independence of form from matter is not absolute. Form never totally transcends the domain of matter altogether, as it can only emancipate itself from a specific material constitution by adopting *another* material constitution. One should therefore be sceptical of claims by proponents of so-called 'artificial life' research that the living form can be logically decoupled from any concrete material instantiation and re-created in a computer medium (for an in-depth examination of this issue, see the essays in Boden 1996).

[23] John Locke, in the second edition of his *Essay Concerning Human Understanding* (published in 1694), already considered a version of this question, noting, in relation to the identity of animals compared to that of machines, that, '[i]f we would suppose [a] Machine one continued Body, all whose organized Parts were repair'd, increas'd or diminish'd, by a constant Addition or Separation of insensible Parts, with one Common Life, we should have something very much like the Body of an Animal' (Locke, quoted in McLaughlin 2001: 177).

On the face of it, organisms and Thesian machines display the same dynamic stability of form. Nevertheless, upon closer examination, we are able to see that they differ in three key respects (cf. Jonas 1968). First, the replacements of the material parts of a Thesian machine are caused from the outside, that is, by an agent other than the machine itself. This is by virtue of the fact that machines are extrinsically purposive. In an organism, on the other hand, the material exchange is caused from within, which means that the organism, being intrinsically purposive, is its own agent of change. Organisms thus persist actively, by maintaining themselves, whereas Thesian machines persist passively, by means of external interventions. Second, the replacement of parts in a Thesian machine does not take place continuously, as a matter of course, but is rather a consequence of contingent events. For example, the ship of Theseus would have required repairs after incurring damage during a thunderstorm or after an accidental collision with rocks near the shore, but it could well have *not* required such repairs, if those particular circumstances had been different. In contrast, the material exchange in an organism is—as we have already discussed—a constant, inviolable feature of its persistence through time; it is neither contingent nor accidental. And, third, the form of an organism does not stay fixed during its lifetime. Organisms grow and develop. Some undergo major morphological changes, such as metamorphosis. Nothing comparable happens to a machine (Thesian or otherwise), given that the fixity of its internal structure is precisely what allows it to perform the function it was designed for.

These differences prevent us from straightforwardly concluding that the maintenance of form, coupled with the exchange of matter, is a sufficient condition for the persistence of organisms—even if it might be for Thesian machines. The maintenance of form in an organism is actually a type of *self*-maintenance. This self-maintenance, moreover, is a *necessary* rather than a contingent occurrence, and it takes place *continuously*, in an uninterrupted sequence of causal events that collectively lead to the progressive modification of the form that is being maintained, as the organism grows and develops. It is clear, then, that, in order to comprehend the persistence of organisms, we must first let go of the MCO in all of its possible manifestations—including hypothetical Thesian ones—and take fully on board their processual nature.

As processes, organisms are extended and differentiated not only in space but also in time. It is wrong to speak of an organism and its history as if the two were somehow separable. Strictly speaking, an organism does not *have* a temporal trajectory; it *is* itself a temporal trajectory. What we perceive as an organism at any point in time represents only a cross section (or a time slice) in the unfolding of the process it instantiates. And it is this entire four-dimensional process, rather than any of its momentary three-dimensional manifestations, that constitutes the actual living entity (cf. Woodger 1930; Torrey 1939).[24] These perdurantist considerations signal

---

[24] From this perspective, the process of metabolism could be said to take ontological precedence over the organism that undergoes it, given that what appears as an organism at a given time derives its existence from the metabolic process it embodies. This is, I think, what Jonas had in mind when he cryptically remarked that 'the organism must appear as a function of metabolism rather than metabolism as a function of the organism' (Jonas 1966: 78).

the importance of *causal continuity* in biological persistence. This criterion of diachronic identity—which indexes the continuous interlinking of temporal states that an entity undergoes—was termed 'genidentity' by Kurt Lewin in the 1920s, and it has recently been defended by several philosophers of biology (see Boniolo and Carrara 2004; Guay and Pradeu 2016; and chapters 2, 4, 5, and 11 here).

Let me now try to come to some sort of conclusion. We have seen that it makes no sense to identify an organism over time with the materials that compose it, given that these are constantly being replenished by the whole. The constituents of an organism at any particular instant are only the temporary realization of the self-producing organizational unity of the whole. Unlike the MCO, the SLC accurately reflects the fact that the matter of an organism is necessarily and continuously exchanged while its form is actively maintained. Bringing together the autonomous maintenance of form with the causal continuity of process that makes up a living entity over time, we can draw our second ontological lesson, which is that, as far as organisms are concerned, *persistence is grounded in the continuous self-maintenance of form*.

Two issues require further clarification in relation to this lesson. The first is that, as we have indicated above, to speak of the maintenance of form is not to say that it remains totally fixed over time; form *does* change gradually as the organism develops. It is rather to emphasize that form is stabilized sufficiently to be reidentifiable as an uninterrupted steady state in spite of the constant turnover of matter that realizes it. The second clarification is that the relentless flux of matter entering and leaving the organism does not prevent us from identifying *parts* within it over time.[25] It is important in this context not to confuse material constituents with architectural components. The former are ephemeral and accidental (as they are constantly being exchanged), whereas the latter are persistent and necessary, inasmuch as they contribute—physiologically and morphologically—to the preservation of the organism as a whole.[26]

Before moving on, I wish to draw attention to one more aspect of the diachronic identity of organisms that the SLC can help elucidate, and that is the issue of what may be called 'cross-generational identity'. If we take seriously the processual idea that the organism is ontologically subsidiary to the self-maintaining metabolic stream that instantiates it at any given moment, then it follows that the identity of this stream can be maintained *across* generations of organisms. Reproduction, in this view, can be reinterpreted as the means by which a self-maintaining metabolic stream perpetuates itself beyond the lifespan of individual organisms (cf. Hardy 1965; Griesemer 2000; Saborido et al. 2011). A key advantage of conceptualizing reproduction in this way is

---

[25] Part decompositions do not lose their explanatory power when we adopt a processual conception of the organism. Making the case for this epistemological claim, however, would take me beyond the scope of the present discussion.

[26] The organicists drew attention to the contrast between the maintenance of form (both of the whole and of the parts) and the fluidity of matter by distinguishing between the 'biological' and the 'physico-chemical' viewpoints of the organism. Haldane, for instance, wrote that 'when we have observed the shape of a friend's nose we can predict from the *biological* standpoint that it will be the same a year hence, though from a *physical* and *chemical* standpoint a very small portion of the same atoms or molecules may be present in the nose after the year' (Haldane 1931: 140–1, emphasis added; similar claims can be found in Ritter 1909, Russell 1924, Woodger 1929, and Bertalanffy 1933).

that it is better able to account for *epigenetic inheritance*, as it assumes a far greater degree of continuity between parent and offspring. It is worth elaborating on this point a little. If reproduction essentially boils down to the replication of the genetic material of the parent(s), which—in accordance with the MCO—supplies the blueprint for the *de novo* programmatic construction of the offspring, then epigenetically inherited traits can only be treated as oddities or anomalies. But, if the offspring is construed—in accordance with the SLC—as the offshoot of a self-maintaining metabolic stream (or, in the case of sexual reproduction, as the 'intersection' of two such streams), then it is evident that the genetic material is only going to constitute a part (albeit a very important part) of what gets transmitted during reproduction. The essential point is that, within the process-ontological framework of the SLC, the detection of epigenetically inherited traits ceases to be a strange and surprising discovery and comes to be something we would actually expect to find.

## 4.3. Third ontological lesson: 'Order does not entail design'

We turn now to a third area of biological ontology that can be illuminated by rejecting the MCO and adopting the SLC: the nature and origins of biological order. Organisms and machines are both, to be sure, highly ordered systems. In abstraction from time, their hierarchical structure is quite comparable, which is one of the reasons why the MCO is such a useful heuristic tool. However, when they are considered in time, it becomes apparent that their mode of organization is fundamentally different. Machines exhibit a static organization, in the sense that their physical architecture—as well as the degrees of freedom of their parts—is fixed upon manufacture. Organisms, on the other hand, exhibit a *dynamic* organization in the sense that their form, as we have discussed above, reflects a stabilized pattern of continuous material exchange with their environment. Organismic organization, is dynamic in a further respect, namely in its capacity to modify itself so as to compensate against external perturbations—a feature we have also discussed.

The ontological chasm separating organisms from machines widens even more when we consider the origins of the order they each display. The order of a machine invariably reflects a particular *design*—a preexisting plan, usually in the form of a blueprint or a diagram, which has been implemented by an external agent (i.e. the machine's creator). The striking thing about the order of all dissipative structures, including organisms, is that it arises in the absence of design. As we noted earlier in the chapter, the recognition that natural systems can spontaneously self-organize from non-linear interactions and become stabilized in far-from-equilibrium conditions through a constant flux of energy and matter is probably the most momentous discovery of non-equilibrium thermodynamics. Nevertheless, it has proven remarkably difficult to incorporate this insight into our biological understanding of organismic order. One reason is undoubtedly that biologists today are more accustomed to thinking about organisms mechanically (in accordance with the MCO) than thermodynamically (in accordance with the SLC).

It is rather ironic that Schrödinger's *What Is Life?*, which helped to shed light on how life conforms to the second law of thermodynamics, was simultaneously responsible for spawning the modern mechanicist conception of organismic order. Schrödinger argued that the source of all biological order is to be found in the chemical structure of a single molecule, the self-replicating chromosome, which he conceived as an 'aperiodic crystal' in order to account for its stability in the face of thermal fluctuations. In this respect, organisms are no different from machines, which also exhibit rigid, solid-state structures capable of withstanding thermal agitation that enable them to operate in a regular, orderly way. In Schrödinger's own words, 'the clue to the understanding of life is that it is based on a pure mechanism, a "clock-work"... [that] also hinges upon a solid—the aperiodic crystal forming the hereditary substance, largely withdrawn from the disorder of heat motion' (Schrödinger 1944: 82, 85).

According to Schrödinger, organisms and machines are both subject to the same preformationist 'order-from-order' principle. Just as the order of a clock derives from the preexisting plan of a clockmaker, so the order of an organism derives from the 'hereditary code-script' contained in its genome, which specifies 'the entire pattern of the individual's future development and... its functioning in the mature state' (ibid., 21). But genes, for Schrödinger, do not just store the information for development; they are also 'instrumental in bringing about the development they foreshadow. They are law-code and executive power—or, to use another simile, they are architect's plan and builder's craft—in one' (ibid., 22). These provocative ideas served as the basis for the metaphor of the *genetic program*, which came to dominate late twentieth-century developmental biology (see Keller 2000). This incarnation of the MCO equates the embryo with a computer that executes a predetermined set of operations in accordance with a program encoded in its genes. The genetic program model is fraught with problems but, since I have examined them in detail elsewhere (Nicholson 2014), I will refrain from doing so again here. Suffice it to say that it has taken biology decades to remove the misconceptions that resulted from it; indeed they still linger.

Ultimately, Schrödinger was right to suppose that genes are material carriers of information. There *really is* a code connecting DNA to RNA to the primary structure of proteins—a fact that has become enshrined in every biology textbook as 'the central dogma of molecular biology'. But he was wrong to localize in the genome all the information required to specify the adult organism, and even more so to invest it with the causal power to initiate, control, and direct the developmental process. Ontogeny, it turns out, is a highly heterogeneous process involving the confluence of numerous intersecting causal factors, only some of which have their physical basis in the DNA. Replacing the MCO with the SLC enables us to make sense of this empirical finding. Once we internalize the notion that the organism is a thermodynamically open dissipative structure, we are able to see that its order cannot possibly derive from any one of its material constituents, and that it must instead be construed as a systemic property emerging from the collective dynamics of the complex web of chemical reactions that underlie it.

Organisms are, of course, rather peculiar dissipative structures in that they do not spontaneously self-organize, like whirlpools or tornadoes. This is precisely where genes come into the picture. Genes can be said to encode a historical record of successful modes of self-organization—a record that liberates organisms from the burden of having to 'reinvent' the metabolic pathways of chemical transformation they need to survive every time they undergo a reproductive cycle (cf. Schneider and Kay 1995; Weber and Depew 2001). From the perspective of the SLC, the role of genes is not to initiate, control, or direct development, but rather to constrain the possible paths of dynamically stable forms of self-organization to those with the highest probability of producing a viable, structurally and functionally differentiated adult. The genome, in this view, constitutes a sort of *catalogue* or *database* of effective self-organization strategies that is transmitted from one generation to the next in a given lineage. What follows from this empirically motivated reconceptualization is the conclusion that biological order does not come preformed in a static 'order-from-order' structure (as Schrödinger famously conjectured), but rather emerges progressively, through an epigenetic 'order-from-disorder' process.[27] By restating this thesis in simpler, more straightforward terms, we arrive at our third and final ontological lesson, which is that, in biology, *order does not entail design*.

## 5. Conclusions

After the scientific revolution, the notion that nature is a well-oiled machine proved irresistible. The machine metaphor conforms to our naïve, pre-theoretical expectations about the world, which are grounded in good old-fashioned substance metaphysics; and the profitable deployment of this metaphor in different areas of scientific inquiry historically served to reaffirm such preconceptions. This, in turn, helped to legitimize the ontological adequacy of the metaphor, and it vindicated the mechanicist conviction that nature is lawful, deterministic, and totally explainable in reductionistic terms. Over the past century, however, we have slowly been coming to the realization that that this view of nature simply does not work (cf. Whitehead 1925; Prigogine and Stengers 1984; Dupré 1993). Physicists first, and biologists more recently, have begun to challenge the substantialist assumptions that underlie the mechanicist worldview, which emerged with the rise of modern science.

As far as the living world is concerned, non-equilibrium thermodynamics demonstrates with piercing clarity that organisms are not fixed things with predefined sets of unchanging properties, but resilient *processes* exhibiting dynamic stabilities relative to particular timescales. What I have sought to convey in this chapter is that the findings of thermodynamics render elaborate philosophical arguments in support of a processual view of life almost unnecessary. The idea that an organism is an open system which must constantly exchange energy and matter with its environment in order to keep itself far from equilibrium is not a metaphysical claim but a scientific fact. Of course, a great deal needs to be said philosophically about what kind of processes organisms are and what exactly follows from their processual nature. But the crucial point is that, if we want an ontology of life that is grounded and informed by natural science, then a processual account is unavoidable. Whatever else

---

[27] Interestingly, Schrödinger actually discussed an 'order from disorder' principle in *What Is Life?*, but he took it to apply exclusively to the sort of order described by statistical mechanics, which arises from the statistical averaging of vast numbers of molecules that, taken together, display regular, law-like patterns of behaviour (e.g. diffusion). The very possibility of order spontaneously emerging in far-from-equilibrium conditions did not even occur to him.

organisms may be, what cannot be denied is that they are stable metabolic flows of energy and matter.

As the MCO is the perfect biological embodiment of the commitments of substance metaphysics, in order to come to terms with the processual nature of life we require a different theoretical conception of the organism. We have seen that the history of biological thought already furnishes us with one such alternative, the SLC, which, although harking back to Heraclitus, only became fully articulated in the nineteenth and twentieth centuries—often in explicit opposition to the MCO. By metaphorically appealing to familiar non-living dissipative structures, the SLC enables us to grasp, in simple and evocative terms, the dynamic, far-from-equilibrium features of organisms that a thermodynamic perspective compels us to consider. And, just as the MCO has an impressive range of incarnations (the organism has been variously construed as a clock, a steam engine, a chemical factory, or a computer, depending on context and historical period), so does the SLC prove to be remarkably versatile in its manifestations, invoking as it does streams, vortices, whirlpools, or flames, depending on the aspects of the organism being highlighted. Of course, organisms are quite different from all of these entities, and consequently their correspondence with them is necessarily incomplete. But the SLC is still a considerable improvement on the MCO, as it accurately portrays the physical conditions of life and provides the foundation for a scientifically grounded understanding of the organism capable of making sense of its processual nature.

Taking the SLC as the cornerstone of our biological ontology has a number of interesting philosophical consequences. We have had the opportunity to explore in some depth three such consequences, which we have formulated as 'lessons' in order to underscore the pedagogical payoff of reconceptualizing the organism from complex machine to flowing stream. Importantly, this process of reconceptualization does not render the MCO useless or irrelevant; on the contrary, it highlights its enormous heuristic value. We should not underestimate the fact that it is only by uncovering how the MCO *fails* to truthfully capture the organism that we have managed to elucidate its processual nature and derive our three ontological lessons. Bertalanffy was quite right to remark that 'we cannot speak of a machine "theory" of the organism, but at most of a machine fiction' (Bertalanffy 1933: 38). It remains, nevertheless, an *extremely useful* fiction.

With regard to the SLC itself, our discussion has shown that it displays a number of features that we tend to look for in a scientific theory: it is able to organize a large body of facts, establish connections between seemingly disparate concepts, and make sense of unexpected empirical findings. By adopting the SLC we have found a way to think naturalistically about normativity and agency, we have grasped the ineliminable role that history plays in shaping biological behaviour, and we have accounted for the inextricable link between organism and environment. The SLC has also given us a new handle on the problem of persistence, and it has allowed us to understand why organismic order needs to be construed as a systemic property. In addition, it has helped us come to terms with certain phenomena, such as niche construction and epigenetic inheritance, which seem perplexing and intractable when viewed from the perspective of the MCO. Finally, at a most general level, we have seen how the SLC brings biological principles into closer contact with physical ones by means of non-equilibrium thermodynamics, thereby paving the way for a non-reductionist, non-mechanicist reconciliation of biology with physics.

## Acknowledgements

I have benefitted greatly from the comments I received on earlier versions of this chapter by John Dupré, David Depew, Alvaro Moreno, James DiFrisco, Stephan Guttinger, Sune Holm, Jan Baedke, Laurent Loison, Richard Gawne, and Andrea Raimondi. I am also grateful to the audiences at the HOPOS meeting in Halifax, the PBUK workshop in Cambridge, the ISHPSSB meeting in Montréal, and the EPSA conference in Dusseldorf for valuable feedback on presentations of this material. Finally, I am pleased to acknowledge financial support from the European Research Council under the European Union's Seventh Framework Program (FP7/2007-2013)/ERC grant agreement n° 324186.

## References

Alberts, B. (1998). The Cell as a Collection of Protein Machines: Preparing the Next Generation of Molecular Biologists. *Cell* 92: 291–4.
Barandiaran, X., Di Paolo, E., and Rohde, M. (2009). Defining Agency: Individuality, Normativity, Asymmetry and Spatio-Temporality in Action. *Journal of Adaptive Behavior* 17: 367–86.
Barham, J. (2012). Normativity, Agency, and Life. *Studies in History and Philosophy of Biological and Biomedical Sciences* 43: 92–103.
Bertalanffy, L. von. (1932). *Theoretische Biologie*, vol. 1: *Allgemeine Theorie, Physikochemie, Aufbau und Entwicklung des Organismus*. Berlin: Gebrüder Borntraeger.
Bertalanffy, L. von. (1933). *Modern Theories of Development: An Introduction to Theoretical Biology*. Oxford: Oxford University Press.
Bertalanffy, L. von. (1942). *Theoretische Biologie*, vol. 2: *Stoffwechsel, Wachstum*. Berlin: Gebrüder Borntraeger.
Bertalanffy, L. von. (1950). The Theory of Open Systems in Physics and Biology. *Science* 111: 23–9.
Bertalanffy, L. von. (1952). *Problems of Life: An Evaluation of Modern Biological and Scientific Thought*. New York: Harper & Brothers.
Bertalanffy, L. von. (1967). *Robots, Men, and Minds: Psychology in the Modern World*. New York: George Braziller.
Bickhard, M. H. (2000). Autonomy, Function and Representation. *Communication and Cognition* (Special issue) 17: 111–31.
Black, M. (1962). *Models and Metaphors: Studies in Language and Philosophy*. Ithaca, NY: Cornell University Press.
Boden, M. A. (ed.). (1996). *The Philosophy of Artificial Life*. Oxford: Oxford University Press.
Boniolo, G. and Carrara, M. (2004). On Biological Identity. *Biology & Philosophy* 19: 443–57.
Brillouin, L. (1949). Life, Thermodynamics, and Cybernetics. *American Scientist* 37: 554–68.
Brooks, D. R. and Wiley, E. O. (1986). *Evolution as Entropy: Toward a Unified Theory of Biology*. Chicago: University of Chicago Press.
Brown, T. L. (2003). *Making Truth: Metaphor in Science*. Urbana: University of Illinois Press.
Campbell, R. J. (2009). A Process-Based Model for an Interactive Ontology. *Synthese* 166: 453–77.
Cannon, W. B. (1929). Organization for Physiological Homeostasis. *Physiological Reviews* 9: 399–431.

Christensen, W. (2012). Natural Sources of Normativity. *Studies in History and Philosophy of Biological and Biomedical Sciences* 43: 104–12.
Coleman, W. (1977). *Biology in the Nineteenth Century: Problems of Form, Function, and Transformation*. Cambridge: Cambridge University Press.
Collingwood, R. G. (1945). *The Idea of Nature*. Oxford: Oxford University Press.
Cuvier, G. (1833). *The Animal Kingdom, Arranged in Conformity with Its Organization*. New York: G. & C. & H. Carvill.
Dawkins, R. (1986). *The Blind Watchmaker*. New York: Norton.
Dawkins, R. (1998). Universal Darwinism. In D. L. Hull and M. Ruse (eds), *The Philosophy of Biology* (pp. 15–37). Oxford: Oxford University Press.
Dear, P. (2006). *The Intelligibility of Nature: How Science Makes Sense of the World*. Chicago: Chicago University Press.
Dijksterhuis, E. J. (1961). *The Mechanization of the World Picture*. New York: Oxford University Press.
Dupré, J. (1993). *The Disorder of Things: Metaphysical Foundations of the Disunity of Science*. Cambridge, MA: Harvard University Press.
Gilbert, S. F. (1982). Intellectual Traditions in the Life Sciences: Molecular Biology and Biochemistry. *Perspectives in Biology and Medicine* 26: 151–62.
Griesemer, J. (2000). Reproduction and the Reduction of Genetics. In P. Beurton, R. Falk, and H.-J. Rheinberger (eds), *The Concept of the Gene in Development and Evolution: Historical and Epistemological Perspectives* (pp. 240–85). Cambridge: Cambridge University Press.
Guay, A. and Pradeu, T. (2016). To Be Continued: The Genidentity of Physical and Biological Processes. In A. Guay and T. Pradeu (eds), *Individuals Across the Sciences* (pp. 317–47). New York: Oxford University Press.
Haag, J. W., Deacon, T. W., and Ogilvy, J. (2011). The Emergence of Self. In J. W. van Huyssteen and E. P. Wiebe (eds), *In Search of Self* (pp. 319–37). Grand Rapids: William B. Eerdmans.
Haldane, J. B. S. (1940). *Keeping Cool and Other Essays*. London: Chatto & Windus.
Haldane, J. S. (1917). *Organism and Environment, as Illustrated by the Physiology of Breathing*. New Haven: Yale University Press.
Haldane, J. S. (1919). *The New Physiology and Other Addresses*. London: Charles Griffin.
Haldane, J. S. (1931). *The Philosophical Basis of Biology*. London: Hodder & Stoughton.
Hardy, A. C. (1965). *The Living Stream: A Restatement of Evolution Theory and Its Relation to the Spirit of Man*. London: Collins.
Henderson, L. J. (1913). *The Fitness of the Environment*. New York: Macmillan.
Henning, B. G. and Scarfe, A. C. (eds). (2013). *Beyond Mechanism: Putting Life Back into Biology*. Lanham: Lexington Books.
Hobbes, T. (1996). *Leviathan*, ed. by R. Tuck. Cambridge: Cambridge University Press.
Huxley, T. H. (1870). Address to the British Association: Liverpool Meeting, 1870. *Nature* 2: 400–6.
Jacob, F. (1973). *The Logic of Life: A History of Heredity*. New York: Pantheon.
Johnstone, J. (1921). *The Mechanism of Life in Relation to Modern Physical Theory*. London: Edward Arnold.
Jonas, H. (1953). A Critique of Cybernetics. *Social Research* 20: 172–92.
Jonas, H. (1966). *The Phenomenon of Life: Toward a Philosophical Biology*. Evanston: Northwestern University Press.
Jonas, H. (1968). Biological Foundations of Individuality. *International Philosophical Quarterly* 8: 231–51.
Kapp, R. O. (1954). Living and Lifeless Machines. *British Journal for the Philosophy of Science* 5: 91–103.

Kauffman, S. (1995). *At Home in the Universe: The Search for Laws of Self-Organization and Complexity*. Oxford: Oxford University Press.
Keller, E. F. (1995). *Refiguring Life: Metaphors of Twentieth-Century Biology*. New York: Columbia University Press.
Keller, E. F. (2000). *The Century of the Gene*. Cambridge, MA: Harvard University Press.
Keller, E. F. (2008). Organisms, Machines, and Thunderstorms: A History of Self-Organization, Part One. *Historical Studies in the Natural Sciences* 38: 45–75.
Kirschner, M., Gerhart, M., and Mitchison, T. (2000). Molecular 'Vitalism'. *Cell* 100: 79–88.
Lewontin, R. C. (2000). *The Triple Helix: Gene, Organism, and Environment*. Cambridge, MA: Harvard University Press.
Lillie, R. S. (1945). *General Biology and Philosophy of Organism*. Chicago: University of Chicago Press.
Maasen, S., Mendelsohn, E., and Weingart, P. (eds). (1995). *Biology as Society, Society as Biology: Metaphors*. Dordrecht: Kluwer.
Maturana, H. R. and Varela, F. J. (1980). *Autopoiesis and Cognition: The Realization of the Living*. Dordrecht: Reidel.
McLaughlin, P. (2001). *What Functions Explain: Functional Explanation and Self-Reproducing Systems*. Cambridge: Cambridge University Press.
Monod, J. (1971). *Chance and Necessity: An Essay on the Natural Philosophy of Modern Biology*. Glasgow: Williams Collins Sons.
Moreno, A. and Mossio, M. (2015). *Biological Autonomy: A Philosophical and Theoretical Enquiry*. Dordrecht: Springer.
Mossio, M., Saborido, C., and Moreno, A. (2009). An Organizational Account of Biological Functions. *British Journal for the Philosophy of Science* 60: 813–41.
Needham, J. (1928). Recent Developments in the Philosophy of Biology. *Quarterly Review of Biology* 3: 77–91.
Nicholson, D. J. (2013). Organisms ≠ Machines. *Studies in History and Philosophy of Biological and Biomedical Sciences* 44: 669–78.
Nicholson, D. J. (2014). The Machine Conception of the Organism in Development and Evolution: A Critical Analysis. *Studies in History and Philosophy of Biological and Biomedical Sciences* 48: 162–74.
Nicholson, D. J. and Gawne, R. (2015). Neither Logical Empiricism nor Vitalism, but Organicism: What the Philosophy of Biology Was. *History and Philosophy of the Life Sciences* 37: 345–81.
Nicolis, G. and Prigogine, I. (1977). *Self-Organization in Non-Equilibrium Systems: From Dissipative Structures to Order through Fluctuations*. New York: J. Wiley & Sons.
Odling-Smee, J., Laland, L., and Feldman, M. (2003). *Niche Construction: The Neglected Process in Evolution*. Princeton: Princeton University Press.
Oparin, A. I. (1961). *Life: Its Nature, Origin and Development*. Edinburgh: Oliver & Boyd.
Prigogine, I. and Stengers, I. (1984). *Order Out of Chaos: Man's New Dialogue with Nature*. Toronto: Bantam Books.
Rescher, N. (1996). *Process Metaphysics: An Introduction to Process Philosophy*. Albany: SUNY Press.
Ritter, W. E. (1909). Life from the Biologist's Standpoint. *Popular Science Monthly* 75: 174–90.
Rosen, R. (1991). *Life Itself: A Comprehensive Inquiry into the Nature, Origin, and Fabrication of Life*. New York: Columbia University Press.
Russell, E. S. (1924). *The Study of Living Things: Prolegomena to a Functional Biology*. London: Methuen.
Saborido, C., Mossio, M., and Moreno, A. (2011). Biological Organization and Cross-Generation Functions. *British Journal for the Philosophy of Science* 62: 583–606.

Schneider, E. D. and Kay, J. J. (1995). Order from Disorder: The Thermodynamics of Complexity in Biology. In M. P. Murphy and L. A. J. O'Neill (eds), *What Is Life? The Next Fifty Years* (pp. 161–73). Cambridge: Cambridge University Press.

Schrödinger, E. (1944). *What Is Life? The Physical Aspect of the Living Cell*. New York: Macmillan.

Sherrington, C. S. (1940). *Man on His Nature*. Cambridge: Cambridge University Press.

Sinnott, E. W. (1955). *The Biology of the Spirit*. New York: Viking Press.

Taylor, R. (1950). Comments on a Mechanistic Conception of Purposefulness. *Philosophy of Science* 17: 310–17.

Torrey, T. W. (1939). Organisms in Time. *Quarterly Review of Biology* 14: 275–88.

Waddington, C. H. (1957). *The Strategy of the Genes*. London: George Allen & Unwin.

Weber, B. H. and Depew, D. J. (2001). Developmental Systems, Darwinian Evolution, and the Unity of Science. In S. Oyama, P. E. Griffiths, and R. D. Gray (eds), *Cycles of Contingency: Developmental Systems and Evolution* (pp. 239–53). Cambridge, MA: MIT Press.

Weber, B. H., Depew, D. J., Dyke, C., Salthe, S. N., Schneider, E. D., Ulanowicz, R. E., and Wicken, J. S. (1989). Evolution in Thermodynamic Perspective: An Ecological Approach. *Biology & Philosophy* 4: 373–405.

Whewell, W. (1840). *The Philosophy of the Inductive Sciences*, vol. 2. London: John W. Parker.

Whitehead, A. N. (1925). *Science and the Modern World*. Cambridge: Cambridge University Press.

Wicken, J. S. (1987). *Evolution, Thermodynamics, and Information: Extending the Darwinian Program*. Oxford: Oxford University Press.

Williams, G. C. (1966). *Adaptation and Natural Selection*. Princeton: Princeton University Press.

Woese, C. R. (2004). A New Biology for a New Century. *Microbiology and Molecular Biology Reviews* 68: 173–86.

Woodger, J. H. (1929). *Biological Principles: A Critical Study*. London: Routledge & Kegan Paul.

Woodger, J. H. (1930). The 'Concept of Organism' and the Relation between Embryology and Genetics, Part I. *The Quarterly Review of Biology* 5: 1–22.

# 8
# Objectcy and Agency
## Towards a Methodological Vitalism

*Denis M. Walsh*

## 1. Introduction

My objective here is to offer a methodological proposal predicated on a metaphysical position—neither of which has much credence or currency in modern philosophy of biology. The metaphysical position is that organisms constitute a special category of entity; they are natural agents. The methodological proposal is that, because organisms are agents, a genuine understanding of the difference they make to the world requires a battery of theoretical concepts and explanatory modes that do not apply to the study of non-living things. Organisms call for a special kind of theory, an agent theory. Most of our familiar scientific theories are not of this sort; they are object theories. The principal difference between agent theories and object theories resides in the way they treat the elements of their respective domains. I introduce two neologisms to mark the distinction: *objectcy* is the role played by the elements in the domain of an object theory, and *agency* is the role played by the elements in the domain of an agent theory. The proper study of organisms, I claim, requires us to take their agency seriously.

I call this agent-centred approach *methodological vitalism*. I am aware that the epithet 'vitalism' trails more than a whiff of odium in its wake. Vitalism, of the sort commonly associated with Hans Driesch and others, is roundly considered to be thoroughly discredited (Garrett 2013; Nicholson and Gawne 2015), even downright daft. In most cases the opprobrium is well earned. Prominent versions of vitalism in the late nineteenth to early twentieth century tended to set living things apart from non-living, on the supposition that they partake of a non-material vital substance, or are propelled or guided by non-material vital forces. I have no truck with this substance or ontological vitalism. But not all vitalisms are of this sort. The British Emergentist C. D. Broad advocates a form of materialism he calls 'emergent vitalism', according to which the behaviours of living matter cannot be adequately accounted for by the sciences of non-living things:

[W]e have no right to suppose that the laws which we have discovered by studying non-living complexes can be carried over without modification to the very different case of living complexes. It may be that the only way to discover the laws according to which the behaviour of the separate constituents combines to produce the behaviour of the whole in a living body is to study living bodies as such. (Broad 1925: 68–9)

E. S. Russell prefers the label 'organicism' but expresses a similar sentiment that living things engender new methods of study.[1]

The living thing can be treated as a physico-chemical system or mechanism of great complexity, and no one would dream of denying the validity and value of biochemical and biophysical research. But such an approach leaves out of account all that is distinctive of life...I try to show that we cannot disregard these unique characteristics of life without losing all hope of building up a unified, coherent and independent biology.   (Russell 1945: viii)

J. S. Turner (2013) credits the French developmental biologist Claude Bernard with a materialist form of vitalism.[2] Erwin Schrödinger, in his landmark essay *What Is Life?*, encapsulates the idea that living things, while material entities, nevertheless make distinct methodological demands on the natural sciences:

[F]rom all we have learnt about the structure of living matter, we must be prepared to find it working in a manner that cannot be reduced to the ordinary laws of physics. And that not on the ground that there is any 'new force' or what not, directing the behaviour of the single atoms within a living organism, but because the construction is different from anything we have yet tested in the physical laboratory.   (Schrödinger 1944: 76)

Methodological vitalism locates itself in this tradition.

Vitalism of this sort pays a dividend to evolutionary theory. It has long been noticed that our best theory of evolution, for better or worse, relegates organisms to a marginal role, opting instead to prioritize genes and changes in ensembles of 'gene ratios'.[3] I maintain that modern synthesis evolutionary biology doesn't recognize the contribution of organisms to evolution for the simple reason that it can't. It is the wrong sort of theory—an object theory. Insofar as it deals with organisms at all, it recognizes them only as objects. But, I claim, organisms participate in evolution as agents. Their contribution to evolution can only be adequately captured by an evolutionary agent theory. An evolutionary agent theory is an instance of methodological vitalism. It holds that the contribution of organisms to evolution demands a set of proprietary concepts and methods that apply exclusively to living things.

## 2. An Ontological Surprise

The primary substances of our commonsense ontology are objects. They are constituted of matter and take their definitive properties from their material constitution. These definitive properties are generally thought of as intrinsic causal dispositions, propensities to behave in certain ways when they encounter certain external conditions (Ellis 2001; Bird 2007). These properties in turn fix the individuation and persistence conditions of ordinary objects. They determine the kind of thing each entity is and the number, degree, and sorts of changes that each can undergo without ceasing to exist. Clearly, such things persist if they undergo no changes in their

---

[1] The various species of vitalism and organicism are nicely surveyed by Nicholson and Gawne 2015 and by essays in Normandin and Wolfe 2013. On organicism, see also chapters 1, 7, 11, 12, and 13 here.

[2] Turner 2013 calls Bernard's position 'process vitalism'.

[3] The reasons for the marginalization of organism are discussed extensively in Walsh 2015.

material constitution. But not all primary substances are like that. Some things do not merely persist through change; they *subsist in change*. That is to say, their individuation and persistence conditions involve the constant exchange of matter and energy with the environment. They cease to exist when their material constitution ceases to change. This is a broad category of beings. It comprises cyclones, convective cells, flames, and much else besides.

Organisms constitute a special class of these *processual* objects. They subsist not merely by exchanging matter and energy with their environments, but through metabolism. Metabolism is the process by which an organism synthesizes the materials of which it is made. Through the exchange of matter and energy, the organism builds order internally while decreasing it in its environs. Such exchange is necessary for them to resist thermodynamic decay. Hans Jonas dubs the precarious mode of existence of organisms their 'thermodynamic predicament'.[4] In coping with their predicament, organisms build themselves, organize themselves, and maintain themselves:

[I]n living things, nature springs an ontological surprise in which the world-accident of terrestrial conditions brings to light an entirely new possibility of being: systems of matter that are unities of a manifold...in virtue of themselves, for the sake of themselves and continually sustained by themselves. (Jonas 1966: 79)

According to Jonas, an organism is not determined by the matter of which it is constituted—not in the way in which an ordinary object might be. Instead, an organism and its constituent matter stand in a dialectical relation of 'needful freedom':

[T]his double aspect shows in terms of metabolism itself: denoting, on the side of freedom, a capacity of organic form, namely to change its matter, metabolism denotes equally the irremissible necessity for it to do so. (Ibid., 83)

In engaging in the metabolic struggle against its thermodynamic predicament, an organism creates and individuates itself. 'The ontological individual, its very existence at any moment...its duration is, then, essentially its own function, its own concern, its own continuous achievement' (ibid., 80). It is this unique capacity that makes organisms natural agents. As Di Paolo tells us, an agent is 'a self-constructed unity that engages the world by actively regulating its exchanges with it for adaptive purposes that are meant to serve its continued viability' (Di Paolo 2005: 442).

Quite how nature manages to spring this ontological surprise is the subject of vigorous investigation. Autonomous systems approaches to the study of complex entities offer a compelling account of how natural agents arise. The nature of natural agents is usually elucidated using a cluster of concepts: *closure, autonomy*, and *coupling*.[5] Varela (1979) defines *organizational closure* as a property of a bounded, unified system in which the component processes comprise a network, each of which depends on other elements of the network for its existence and maintenance. In a metabolically closed system, 'each metabolite or enzyme needed for the maintenance of the system is

---

[4] I am grateful to Alex Djedovic for helpful discussions on this issue. See chapter 7 for a complementary examination of this topic.

[5] For various versions of this approach, see Varela 1979; Thompson 2007; Barandiaran et al. 2009; and Moreno and Mossio 2015. Chapter 10 in this volume also exemplifies this approach.

produced by the system itself' (Di Frisco 2014: 500). For Moreno and Mossio (2015: 23), this sort of closure is a 'general invariant of biological organisation'. They continue: 'Biological individuality, we think, has much to do with organisational closure, to the extent that one may conjecture that closure in fact defines biological individuality'.

Organizationally closed systems are autonomous. This is to say that they have the capacity to promote their own existence and to maintain their own structural and functional integrity across a range of internal and external conditions (Thompson 2007: 44). Because autonomous systems persist by exchanging matter and energy with their environments, they must be coupled with their environments. Coupling is the ability of the system to engage in the kind of reciprocal interactions with its environment that result in its continued viability.

Being an agent in this minimal sense, then, consists in an organizationally closed system's capacity to build and maintain itself through the exchange of matter and energy, to differentiate itself from its environment through this capacity, and to exploit its environment in ways that promote its own continued persistence (Barandiaran et al. 2009; Moreno and Mossio 2015). That is to say, an organism is not merely capable of engaging in these activities; doing so is a condition of its very existence. 'Its "can" is a "must", since its execution is identical with its being. It can, but it cannot cease to do what it can without ceasing to be' (Jonas 1966: 83).

Autonomous systems theory and related disciplines offer something of great value to philosophical naturalists. They give us an account of the place of organisms as agents in the natural world—of organisms as self-making, self-individuating, processual things. Moreover, the account accomplishes this in a way that requires no special methodological pleading. They tell us how the entities and activities that make up a system interact with one another to give rise to the definitive properties of an agent; and they do so in a way that requires only minor emendations to traditional mechanistic approaches to understanding the workings of complex entities (Bechtel 2013). Importantly, they close a gap between living and non-living matter upon which the viability of substance vitalism appeared to depend: 'non-life and life share a huge and biologically significant territory that buffers and makes more complex any account of either' (Dupré and O'Malley 2013: 335). If traditional forms of vitalism raise a challenge to naturalism—that of specifying how arrangements of non-living matter can give rise to living organisms—then autonomous systems theory meets this challenge.

It is worth noting, however, that citing the mechanisms by which agents are realized might not be sufficient to account for the difference that agents make to the world. This has little to do with the special nature of agents, much less with any deficiency in autonomous systems theory. In general, it seems, understanding how complex material entities are realized is seldom sufficient for describing and explaining the ways in which the world is different as a consequence of their existence.

## 3. Phases

The ontological surprise that produces organisms is remarkable, but it is by no means one of a kind. The emergence of new phenomena in complex physical systems is the rule rather than the exception. As physical systems take on new configurations, they inaugurate new, highly distinctive properties, regularities, and relations that do not exist in their absence. Understanding these new phenomena calls, in turn, for special theoretical concepts that are not required to account for domains in which these configurations do not occur. Typically, the new concepts are defined over the macrolevel behaviours of these new configurations of matter, quite independently of the details of their microlevel realizers (Morrison 2015). Examples are not hard to find. Phase transitions provide some of the most vivid cases.

As the early universe cooled and expanded, it underwent a series phase transitions (Gleiser 1998). In baryogenesis, thought to have occurred $10^{-32}$ seconds after the Big Bang, a phase transition breaks the symmetry between baryons and antibaryons, yielding a preponderance of the former (Coles 2000). At this point matter becomes more plentiful in the early universe than antimatter. This, in turn, facilitates the subsequent evolution of a stable material universe, which could not have occurred without a breaking of the baryon–antibaryon symmetry. Having stable matter in the world makes a difference (to say the least) that could not adequately be described without the proprietary concepts that describe the behaviour of matter, for example the concepts 'quark', 'baryon', or 'meson'.

Further expansion and cooling of the early universe facilitated yet another phase transition, nucleosynthesis, which occurred between three and twenty minutes after the Big Bang. This transition saw the prevalence of protons over neutrons, which in turn produced the cocktail of light elements—H, $^3$He—that make up the stars. We have stars and nuclear fusion, and eventually the heavier elements thanks to this phase transition. A world with atomic nuclei behaves differently from one without. In order to account for this behaviour we need concepts that apply exclusively to atomic nuclei, that is, to strong and weak nuclear forces (Coles 2000).

Analogously, if less exotically, the presence of fluids brings forth a whole new range of physical phenomena.[6] Fluids flow. Their flow may be laminar or turbulent. It generates lift and buoyancy. It may transfer heat through organiconvection cells. There are storms, ocean currents, tectonic flow—not to mention diffusion, buoyancy, surface tension, cell membranes, osmosis, erosion, flying, sailing, surfing, music, and beer—because there are fluids. Fluid dynamics, in turn, has its own proprietary theoretical concepts. Viscosity, for example, is essential to the explanation of the behaviour of fluids. Viscosity is realized as collisions between the particles that compose a fluid. Yet viscosity is not conceptually tied in any way to the microscopic conditions of its realization. It is defined in terms of its dynamics (Fulda 2016). Viscosity just *is* the resistance of a fluid to sheer forces.

In general, the explanations we find in fluid dynamics do not depend on citing the mechanical microconditions in which the macrolevel phenomena are realized. In fact it appears that we cannot account for all the phenomena that fluid dynamics explains by attending to the particles of a fluid. In order to explain the formation of droplets, for example, physicists employ an idealizing assumption that fluids are *not* made up of discrete interacting particles (Batterman 2005). Without this continuum assumption, the models of fluid dynamics cannot explain the propensity of fluids

---

[6] See Fulda 2017 for an extended discussion on the example of viscosity. I thank him for his help here.

to form droplets or to undergo phase transitions. Batterman argues that we should take seriously the idea that our models of fluid dynamics correctly identify physical discontinuities that we cannot adequately represent through the concepts of finite discrete particles and their interactions. The reason, in the case of droplets, is that 'the ultimate breakup profile is independent of the microscopic details of the breaking' (Batterman 2005: 242). The lesson can be generalized to all fluid phenomena. In fact, it holds true of an enormously broad and important class of sciences known as 'continuum mechanics', which deal with the macroscopic structural properties properties—tensile strength, conductivity, malleability, magnetism—of complex agglomerations of matter. These sciences all proceed on the assumption that macroscopic materials are continuous and non-particulate.

Typically, the macrolevel behaviours of these configurations have a significant degree of epistemic independence from their micro realizers: 'we need not appeal to the micro phenomena to explain the macro processes' (Morrison 2015: 105). The reason is that many of the macro regularities exhibit a comparable degree of 'metaphysical independence': 'most of the details of the [microscopic] arrangement, are irrelevant' (Batterman 2015: 133) at the scale at which the phenomena of interest are manifest. The upshot is that the account we offer of how these phenomena are realized may not have a particularly close relationship to and may form no real part of the account we provide of the difference they make.[7]

As with stable matter, atomic nuclei, magnets, superconductors, excitable media (Solé and Goodwin 2000), fluids, and tissues, so too, I suggest, with agents. A cascade of metaphysical consequences follows from the appearance of agents in the world. Where there are agents there are observable regularities, modal relations that just do not feature in non-agential worlds. The range of these new agential phenomena is no less interesting than other macrodynamic phenomena such as superconductivity, laminar flow turbulence, convection, magnetism, conductivity, or tensile strength.[8] They issue in explanatory demands that are no less challenging. Agential phenomena call forth a battery of theoretical concepts that are not needed to explain natural phenomena where no agents are involved. These concepts, too, manifest a significant degree of epistemic independence with the concepts we employ to explain how agency is realized in complex material systems. The moral of this methodological digression is that, as physical systems take on new configurations, they bring new phenomena into existence. These new phenomena, in turn, often call for new theoretical concepts.

## 4. Agential Dynamics

Because there are agents, there are goals, means, norms, hypothetical necessity, and a special mode of explanation—teleology. Goals are simply the end states that a goal-directed system tends reliably to attain and would reliably attain across a range of counterfactual circumstances. An agent's pursuit of its goals—its goal-directedness—is,

---

[7] Fulda 2017 beautifully illustrates the way in which naturalizing a phenomenon consists of two distinct stages: (i) locating it in the causal structure of the world; and (ii) constructing a theory of the difference it makes. These two stages may have varying degrees of independence.

[8] I am also suggesting that they are no more arcane.

in turn, an observable feature of its gross behaviour. It consists in the agent's capacity to marshal its causal resources in a manner that brings about the reliable attainment and maintenance of an end state. So goals are an objective, natural feature of a world that contains agents. It is often thought that hypostatizing goals is antithetical to naturalism. Some insist that it commits us to intrinsically evaluable states of affairs (Bedau 1998). Others suppose that to be a goal is to be an object of thought, to be desired by a cognitive agent, or to be represented under the 'guise of the good' (Boyle and Lavin 2010). But, if agency is a kind of observable activity and goals are its end states, then the natural, non-psychological status of agency and goals is as unimpeachable as that of fluidity and viscosity.

If goals are natural, then so are means. Means are simply those elements of an agent's repertoire that are conducive to the attainment of its goals. Given the existence of goals and means, there is a special pair of modal relations between them. The relation that holds between goals and their means is hypothetical necessity (Fulda 2016). An agent will implement an element of its repertoire (often enough) because that action is necessary, under the circumstances, for the attainment of that agent's goal. It holds whenever, in a set of circumstances, the goal would not occur unless the means did. Hypothetical necessity is not a causal relation—goals don't cause their means, they hypothetically necessitate them—but it is a natural one nevertheless. Hypothetical necessity entails that, without the action in question, the goal would not have occurred and, with it, the goal it occurs reliably. Its dual is the relation of conducing. Whereas ends hypothetically necessitate their means, means conduce to their ends. Conducing is not the same as causing; $m$ conduces to $e$ only if $e$ is a goal and, under the circumstances, $m$ would reliably cause $e$ across a range of counterfactual conditions.

Means occur *because* they conduce to agents' goals. Where there are agents, certain events occur reliably and predictably—because they are goals or means to them—that would otherwise occur only rarely and by chance. This robust counterfactual relation between goals and the activities of agents can be exploited in explaining why agents do what they do (Walsh 2012a). Explanations that cite goals in this way are teleological (Walsh 2008).

An agent has a repertoire, a range of activities that it can undertake in a given set of circumstances. On occasion, some elements of the agent's repertoire may be more conducive to the attainment of its goals than others. Hence it is possible to assess an agent's actions in respect of their appropriateness. Responses are appropriate if they are conducive to the agent's goals (or if they are hypothetically necessary).[9] In this way agency also issues in a form of natural normativity. Agents are normatively required to bring about those states of affairs that are hypothetically necessary for the attainment of their goals (Broome 1999).[10] So, where there are agents, there are natural norms too.[11]

---

[9] I take it that conduciveness is only sufficient for appropriateness.

[10] I contrast this approach to natural normativity with that offered by Barandiaran et al. 2009. These authors only acknowledge the normative requirement for an agent to promote its own persistence. But agents are generally capable of pursuing a range of goals, each of which normatively requires its means. Again, I am indebted to Alex Djedovic here.

[11] Philosophical folklore has it that you can't derive an 'ought' from an 'is'. Maybe so. Nor can you derive continuum mechanics from the finite arrangement of discrete particles, but that does not impugn the naturalness of either continuum mechanics or normativity.

The relationship between an agent and its conditions of existence is not like the relation between a run-of-the-mill object and its environment. The two are in a sense intimate and non-separable. What the agent experiences and responds to in pursuit of its goals is a set of relational properties that have salience for the agent. These features, in turn, depend jointly on the features of the environment and on the goals and the capacities of agent. Kurt Goldstein captures the idea:

> The environment of the organisms is by no means something definite and static but is continuously forming commensurably with the organism's development and activity. One could say that the environment emerges from the world through the being or actualization of the organism.... Environment first arises from the world only when there is an ordered organism. (Goldstein 1995: 85)[12]

The relation that Goldstein points to is an ecological one. An agent's environment presents opportunities for, or impediments to, the attainment of its goals. In a word, agents experience and respond to their conditions as *affordances*. Affordance is a theoretical concept borrowed from J. J. Gibson's ecological theory of perception.

> The affordances of the environment are what it offers the animal, what it *provides* or *furnishes*, for good or ill...I mean by it something that refers to both the environment and the animal... It implies the complementarity of the animal and the environment. (Gibson 1979: 127)

Affordances are not environments. They are emergent phenomena that, once again, only exist where there are agents. To be an agent is to respond to one's conditions as promoting or impeding the pursuit of goals; to be an affordance is to be a set of conditions that are salient to an agent's pursuit of its goals. 'Affordances are opportunities for action; they are properties of the animal–environment system that determine what can be done' (Stoffregen 2003: 124).

There is a relation of reciprocal constitution between an agent's abilities and its affordances. Affordances determine what an agent can and should do, given its goals and its repertoire. Conversely, the goals and repertoire of an agent determine what its conditions of existence afford. Moreover, as organisms change in response to their affordances, so too do the affordances. In turn, a change in affordances alters what the agent can do. This sort of constitutive reciprocity between an entity and its conditions (i.e. affordances) exists only where there are agents.[13]

All in all, the presence of agents in the world makes quite a difference. Agents behave in a wholly distinctive way, and in so doing they introduce a range of new phenomena, relations, and regularities that do not figure in the ontology of an agent-free world. Consequently we need a battery of theoretical concepts and methods to describe this range of facts: *goal, means, affordance, repertoire, salience, reciprocal constitution, normative requirement, hypothetical necessity, teleology*. Note that these concepts that describe the ontological consequences of agents are defined in terms of agents' gross behaviour. They are not defined in terms of the microscopic realizations of agency. This is the reason why autonomous systems theory (and related fields),

---

[12] The original German language version, *Aufbau des Organismus*, appeared in 1935.
[13] These ideas are developed in more detail in Walsh 2012b and Walsh 2013.

while giving us a compelling mechanistic account of how agents are realized in the natural world, do not provide an account of the difference they make.

In this respect, the concepts we need in order to capture the differences that agents make are of a piece with the theoretical concepts of viscosity or weak and strong nuclear forces, excitable media, or superconductivity. They pick out macrolevel phenomena that enjoy a degree of epistemic independence over the details of their realization. There is an important difference, however, between (say) fluid dynamics and agent dynamics. The concepts we invoke to describe the dynamics of agents involve us in a non-standard kind of scientific theory. This will need a little explaining.

## 5. Object Theories and Agent Theories

I began this chapter with the claim that the elements of our common-sense ontology are objects, defined and individuated by their material constitution. Scientific theories by and large are structured expressly to deal with them; they are object theories. Organisms are fundamentally different kinds of things; they are agents. In this section I want to suggest that this metaphysical difference raises a methodological problem for any science that seeks to encompass the difference that organisms make to the world. In particular, an object theory encounters a specific kind of difficulty in articulating the contribution that organisms make to evolution. Object theories are not aptly suited to doing so. For that we need a different kind of theory, an agent theory.

### 5.1. Objectcy

An object theory seeks to describe and explain the changes in a domain of objects by setting out a space of possible alternatives for those objects—a state space—and by articulating principles that account for the possible trajectories of the objects through the state space. Objects play a specific role in object theories; I shall call it 'objectcy'. Objectcy consists in the fact that the elements of the domain remain unaltered (with respect to the theory's conserved quantities) unless they are influenced by external sources of change. Objects do not initiate their own changes in the state space. The principles we call upon to explain their changes—for example laws of nature, initial conditions, the space of possible configurations—are exogenous to the objects. In an object theory, the laws of nature and the state space remain constant as the objects traverse the space. The physicist Lee Smolin (2013) has associated this type of theory to what he calls the 'Newtonian paradigm'. According to Smolin, theories in the Newtonian paradigm pose two simple questions: '(i) what are the possible configurations of the system? and (ii) what are the forces that the system is subject to in each configuration?' (Smolin 2013: 44).

Object theories are marked by a kind of transcendence of the explanatory principles over the objects in the domain. The laws, the initial conditions, and the state space exist independently of the objects. They are 'givens'. They remain constant as the objects change. This in turn introduces an explanatory asymmetry between the principles and the objects. The principles explain the changes to the objects in the domain, but the objects do not explain the principles. We cannot, for example, look to the motions of the planets to explain why the laws of gravitation are as they are. Nor can we cite the structure of atoms to explain why the strong, weak, and electromagnetic forces are as they are.

## 5.2. Agency

In an agent theory, the elements of the domain take on a much different role. For want of a better word, I shall call it 'agency'. Agency consists in the fact that the elements of the domain (the agents) initiate their own changes. Agents and the principles we call upon to explain their behaviour have a particularly intimate relation. Agents initiate changes in the state space in response to their affordances, which are jointly constituted of the agents' goals, capacities, and their external circumstances. Agents and affordances are in this sense 'commingled'. The range of possibilities open to an agent (the state space) is itself determined jointly by that agent's condition and capacity to respond to them. Moreover, the conditions, the possibilities, and the capacities of agents co-evolve. As agents change in response to their conditions, so do the conditions. And, as the conditions change, so does the range of possibilities open to the agent (i.e. the state space).

Whereas object theories are characterized by transcendence and explanatory asymmetry, agent theories are characterized by what I shall call 'immanence' and 'explanatory reciprocity'. An agent's conditions and its capacities to act are immanent in the agent's engagement with its environment. The conditions that agents experience and their capacities to respond to them are interpenetrating and interdefining; each partially constitutes the other. Because the conditions and agents constitute one another and co-evolve, each can be (partially) explained by appeal to the other. The activities of the agent can be explained as a response to its conditions and, reciprocally, the change in conditions can be explained as a consequence of the activities of the agent.

Good examples of agent theories are a little thin on the ground, but Lee Smolin (2013) has recently made a quite surprising proposal. Smolin argues that the Newtonian paradigm has failed to generate a theory that governs all of the physical world, because it delivers the wrong kind of theory. The best that such a theory can do is tell us that, *given* the laws and some initial conditions, the universe should evolve in such and such a way. But it could not tell us *why* these initial conditions and these laws obtain. This is a deficiency; these questions presumably have answers. However, the answers do not fall within the ambit of the theory. Smolin finds this unsatisfactory from a theory of everything: 'Nothing outside the universe should be required to explain anything inside the universe' (Smolin 2013: 121–2). The complete theory of the universe must abandon the Newtonian paradigm; 'the remedy must be radical, not just the invention of a new theory but...a new type of theory' (Smolin 2013: 250). The new type of theory Smolin envisages is one in which the laws of nature and the principles that explain how the universe changes evolve as the universe does, in such a way that each explains the other. As the universe evolves, as it grows and complexifies, the laws, the conditions, and the space of possible states will also evolve (Smolin 1997). Smolin is calling for an agent theory of the universe.[14]

---

[14] Clearly an entity does not need to be an 'agent' in the sense outlined by autonomous systems theory to be the element of an agent theory. My supposition, however, is that the fact that organisms are agents (in the latter sense) necessitates an agent theory to describe them. I thank Lee Smolin for a helpful exchange around this view.

Whether or not an agent theory of everything is in prospect, the emergence of organisms as agents in the natural world at least provides the opportunity for the development of a more modest example. The purposive behaviour of organisms and their relation to their conditions of existence exhibits the sort of immanence and reciprocity that call for an agent theory.

As indicated above, the concepts and methods we need in order to articulate an agent theory of organisms are already to hand: *goal, means, affordance, repertoire, salience, reciprocal constitution, normative requirement, hypothetical necessity, teleology*. These theoretical tools exhibit the distinctive marks of an agent theory: immanence, reciprocity, and co-evolution. For example, we do not fully understand the response of an agent to its conditions unless we understand what those conditions afford the agent. But we do not understand what an agent's situation affords the agent unless we know the agent's goals and its repertoire. Nor do we understand the evolution of an agent's actions over time, unless we grasp how these actions alter its affordances, which in turn structures its range of possibilities. Further, a successful explanation of the actions of a well-functioning agent will need to show that the conditions, repertoire, and goals of the agent didn't merely *cause* the response, they normatively required it. It will need to show us that the agent ought to have done what it did, in order to achieve its goal. Explanations of agency qua *agency* are, thus, teleological.

We have got this far. Agents make an ontological difference to the world. They usher in a range of phenomena—goals, means, normative requirements, and so on—and call for the sort of explanations that do not feature in agent-free worlds. In order to capture the difference that agents make, we need an agent theory. Organisms are agents by their nature. If what makes organisms organisms makes a difference to evolution, then it would seem that we could only fully account for their contribution with an agent theory of evolution.

## 6. Evolution and Agency

Modern synthesis evolutionary theory is an object theory. It conforms nicely to the Newtonian paradigm. Its objective is, inter alia, to explain the presence and prevalence of organismal traits. Organisms occupy the objectcy role. The space of alternatives is traditionally represented as a fixed landscape of phenotypes and their fitnesses (McGhee 2007). Populations of organisms are propelled through this space by exogenous forces of selection and drift (Sober 1984). Organisms occupy a marginal place in modern synthesis thinking. They are the products of the activities of more fundamental entities—replicators; and the victims of their own external conditions—the environment. As Richard Lewontin puts it, modern evolutionary theory

> is a theory of the organism as the *object*, not the subject, of evolutionary forces. Variation among organisms arises as a consequence of internal forces that are autonomous and alienated from the organism as a whole. The organism is the object of these internal forces, which operate independently of its functional needs or of its relations to the outer world.
>
> (Lewontin 1985: 87; emphasis added)[15]

---

[15] I thank Jonathan Kaplan for pointing out quite how germane this passage is.

The resonances with the Newtonian paradigm are clear, as Lewontin emphasizes. There is nothing about the objects of the domain—organisms—that answers Smolin's two questions: (i) 'what are the possible configurations?'; and (ii) 'what are the forces?'. The possible configurations comprise the set of phenotypes made by gene combinations and their fitnesses as determined by their environments. The forces are selection, drift, and mutation, all things that *happen to* organisms.[16] Organisms initiate no evolutionary changes of their own. Lewontin continues:

> Thus classical Darwinism places the organism at the nexus of internal and external forces, each with its own laws, independent of each other and of the organisms that is their creation.... The organism is merely the medium by which the external forces of the environment confront the internal forces that produce variation. (Lewontin 1985: 88)

The environment is wholly autonomous of organisms. It has the capacity to mould form so as to meet its exigencies while remaining unaffected by organisms themselves.[17]

> Organisms respond to the environment, but the environment is largely autonomous with respect to the organisms. The environment is seen as either stable (as far as the time scale of the evolutionary process in question is concerned) or else as changing according to its own intrinsic dynamics. (Godfrey-Smith 2001: 254)

The autonomy of the environment and the explanatory asymmetry of environment over form are hallmarks of the objectcy of organisms in the modern synthesis.

There is much about the process of evolution that the modern synthesis, like any theory in the Newtonian paradigm, leaves unexplained. The range of variants available to selection is fixed by random mutation, not by the properties of organisms. The good locations in fitness space are determined by the environment (McGhee 2007). These are 'givens' that fall beyond the purview of evolutionary theory.[18]

Lewontin has long been an outspoken critic of these defining features of the modern synthesis. He rejects the detachment (the 'alienation') of organism from environment that is so central to the modern synthesis. He objects to the modern synthesis portrayal of organisms as simply effects of the activities of genes, subject to the vicissitudes of their environments. Furthermore, he rejects the autonomy and asymmetry of the organism–environment relationship:

> First, it is not true that the development of an individual organism is an unfolding or unrolling of an internal program.... Second, it is not true that the life and death and reproduction of an organism are a consequence of the way in which a living being is acted on by an autonomous environment. (Lewontin 1978: 89)

Lewontin counters that the conditions to which organismal form evolves are strictly underdetermined by the features of the external environment. The reason is that the

---

[16] The other putative force, migration, is evidently something that organisms do.

[17] Niche construction theory (Odling-Smee et al. 2003) amply demonstrates the implausibility of the supposition of environmental autonomy over biological form. But niche construction theory retains the explanatory externalism of the Newtonian paradigm (see Walsh 2012b).

[18] See Wagner 2014 for a compelling argument that the origin of evolutionary variants should not be treated as a primitive given by evolutionary theorists.

external environment underdetermines the way in which the organism *experiences* the environment. The difference between the external environment and the 'experienced environment' arises from the contribution of organisms. Lewontin stresses that organisms actively participate in creating the conditions to which biological form evolves: 'the environments of organisms are made by organisms themselves as a consequence of their own life activities' (ibid., 64). This co-constitution of organism and its conditions has implications for natural selection:

> Natural selection is not a consequence of how well the organism solves a set of fixed problems posed by the environment; on the contrary, the environment and the organisms actively co-determine each other. (Ibid., 89)

In light of this, Lewontin has repeatedly called for a revision of the modern synthesis conception of organism–environment relations. His own version is remarkably reminiscent of Gibson's, and indeed of Goldschmidt's (both quoted above):

> There is no organism without an environment, but there is no environment without an organism. There is a physical world outside of organisms and that world undergoes certain transformations that are autonomous.... But the physical world is not an environment, only the circumstances from which environments can be made. (Ibid., 86)

Organisms contribute to the conditions under which they evolve in myriad ways. Their own size, structure, behaviour, physiology, and development determine the ways in which environmental features impact on organisms. They actively select which features of the environment are relevant for their survival. They change the features of their external environments (Odling Smee et al. 2003).

While Lewontin's critique emphasizes the capacity of organisms to influence their experience of the external environment, it must also be noted that organisms have an influence on their own adaptive repertoires. They achieve this through a variety of means. Organisms respond to environmental stresses by regulating the genome's structure and function, for example through adaptive DNA methylation (Dowen et al. 2012; Herman and Sultan 2016). One particularly prominent form of genome regulation is found in intracellular genetic engineering processes (Shapiro 2013). Organisms reconstruct their genomes in response to their conditions. The single-celled eukaryote *Oxytrichia trifallax*, for example, excises over 90 per cent of its somatic genome and reorganizes the rest (Chen et al. 2014).[19] This is an extreme example, but not an especially exotic one. The engineering of the genome by the organism is commonplace. Cells actively cut, transpose, copy and fix their genomes. They do so in highly sensitive, adaptive ways:

> Cells operate under changing conditions and are continually modifying themselves by genome inscriptions... Research dating back to the 1930s has shown that genetic change is the result of cell-mediated processes, not simply accidents or damage to the DNA. This cell-active view of genome change applies to all scales of DNA sequence variation, from point mutations to large-scale genome rearrangements and whole genome duplications. (Shapiro 2013: 287)

---

[19] I thank Greg Rupik for drawing my attention to this example.

The understanding of genome function is itself shifting (Barnes and Dupré 2008). The genome is no longer seen as embodying a program for building an organism. Rather, it is increasingly considered an 'organ', under the control of the cell and of the entire organism:

> [I]t is more accurate to think of a cell's DNA as a standing resource on which a cell can draw for survival and reproduction, a resource it can deploy in many different ways, a resource so rich as to enable it to respond to its changing environment with immense subtlety and variety.
> (Keller 2013: 41)

The control that the organism exerts over the capacities of the genome is an important part of the organism's adaptive, purposive response to its conditions. Even individual cells manifest this adaptive agency:

> A major assertion of many traditional thinkers about evolution... is that living cells cannot make specific, adaptive use of their natural genetic engineering capacities. They make this assertion to protect their view of evolution as the product of random, undirected genome change. But their position is philosophical, not scientific, nor is it based on empirical observations. (Shapiro 2011: 55–56)

Organisms adaptively regulate their own repertoires in other ways too. One vivid example is found in the plasticity of development.[20] Plasticity achieves a number of functions in evolution. It initiates new forms. 'Responsive phenotype structure is the primary source of novel phenotypes' (West-Eberhard 2003: 503). By permitting the development of the organism to accommodate to its circumstances, plasticity buffers the organism against the deleterious effects of perturbations. 'Phenotypic accommodation reduces the amount of functional disruption occasioned by developmental novelty' (ibid., 147). Phenotypic plasticity orchestrates the development of complex adaptations. The evolution of complex adaptations requires coordination between an organism's various developmental systems. For example, the adaptive evolution of tetrapod limb structures requires coordination between the development of bone, muscle, nervous, circulatory, and integumentary systems (at least). If each system had to wait for a fortuitous mutation in order to produce the appropriate accommodation, complex evolutionary adaptations might never arise (Pfennig et al. 2007). These are all ways in which organisms regulate their repertoires. By altering their capacities in response to their conditions, of course, they also further change their affordances.

One of the dominant themes of twenty-first-century evolutionary biology has been the discovery of the active role of organisms in evolution. Yet we are lumbered with a theory of evolution, the gene-centred modern synthesis, that gained its enormous influence under the supposition that the contribution of organisms to evolutionary dynamics is negligible (Walsh 2007).[21] Modern synthesis evolutionary thinking typically excludes developmental plasticity, learning, cultural transmission, behaviour (Bateson and Gluckman 2011; Vane-Wright 2014; Bateson 2014; Corning 2014) and

---

[20] The plasticity of development is discussed at length in chapter 12.
[21] It was not intended to be that way by those who forged the modern synthesis. Mayr and Dobzhansky, for their part, insisted that individual organisms make a substantive contribution to adaptive evolution. See Depew 2017 for an enlightening discussion.

ecological engineering (Turner 2000) from its roster of evolutionary processes. It is imperative that we incorporate these factors into our account of evolution (Laland et al. 2014).

Perhaps it is possible to extend or amend the modern synthesis so as to make it accommodate these insights (Pigliucci and Müller 2010). Perhaps it can happily assimilate the activities of organisms without undergoing any major rejigging (Wray et al. 2014). But the very structure of the modern synthesis suggests otherwise. It is an object theory. Object theories, as we have seen, do not represent agency *as agency*. Yet the contributions that organisms make to evolution are consequences of their agency. I suggest that the modern synthesis has consistently failed to assimilate organisms into evolutionary thinking because it is constitutionally incapable of doing so. It is the wrong kind of theory. Perhaps what is needed, as Smolin suggests for cosmology, is a radically new kind of theory of evolution, an agent theory.

An agent theory would represent adaptive evolution as following from the adaptive, purposive engagement of organisms with their affordances. An evolutionary agent theory would emphasize the endogenous source of changes in form. In doing so, it would encompass the 'immanence' and 'reciprocity' of the relation between form and affordance. It would also acknowledge the role of organisms in securing the high-fidelity inheritance of characters. It would be sensitive to the ways in which the affordances that impinge on an organism are not imposed on it exclusively by exogenous factors but are rather the joint product of the organism's own capacities and the features of its setting. It would also underscore the co-evolution of form and affordance. In responding adaptively to conditions of existence, organisms alter their affordance landscapes. These altered affordances, in turn, redound to organisms. The conditions to which adaptive evolution responds explain the evolution of form, and changes in form explain the evolution of the conditions. The organism's contribution to evolution consists in its capacity to respond to perturbations, to maintain its viability, and to innovate. The novelties that provide the raw materials of evolution, the conditions to which the evolution of form responds, the possible trajectories through state space—these are not *given*. They are *constructed* by organisms' purposeful engagement with the world. They are manifestations of the agency of organisms. It is the objective of an agent theory of evolution to capture the contribution of this ecological dynamics to evolution.

## 7. Conclusions

Darwin's theory of descent with modification established that those 'endless forms most beautiful and most wonderful' are the consequence of the 'struggle for existence'. The simple, elegant idea is that evolution happens because of what organisms do. Yet this insight has been comprehensively lost from the modern synthesis theory of evolution. Evolution, on the modern synthesis view, happens because of what genes do. According to the modern synthesis, organisms are objects of evolutionary forces, middlemen built by genes and selected by, and alienated from, their environments. They are passive with respect to the genuinely evolutionary processes. Recently, however, the significance of Darwin's insight that evolution happens because of what organisms do is beginning to receive renewed attention.

What organisms do is quite unlike what any other natural entities do. Organisms constitute a distinct ontological category. They are a special kind of processual thing; they are agents. The existence of agents ushers in a range of natural phenomena, regularities, and modal relations that are absent from an agent-free world. These include goals, means, affordances, norms, and hypothetical necessity. Agency is an ecological phenomenon. It is the process by which a goal-directed system marshals the resources of its adaptive repertoire in response to the affordances it both experiences and makes. In responding to their affordances, organisms create the conditions under which they evolve. In this way organisms enact evolution. The proper representation of this ecological dynamics requires a special kind of theory, an agent theory of evolution. That theory, in turn, deploys a battery of concepts and methods that have no place in the study of the non-living world. Methodological vitalism is the view that evolution should be studied from the perspective of the distinctive role that agents play in enacting evolution.

## Acknowledgements

I am happy to acknowledge help from Fermín Fulda and Alex Djedovic and from the Philosophy of Biology discussion group at IHPST, University of Toronto. An earlier version was delivered at the '30 Years of Dialectical Biology' conference in Bordeaux. I thank the organizers and the attendees there, especially Sonia Sultan, for the valuable discussion. Dan Nicholson and John Dupré and three anonymous referees provided very helpful comments.

## References

Barandiaran, X., Di Paolo, E., and Rohde, M. (2009). Defining Agency: Individuality, Normativity, Asymmetry and Spatio-Temporality in Action. *Journal of Adaptive Behavior* 1: 1–13.

Barnes, B. and Dupré, J. (2008). *Genomes and What to Make of Them*. Chicago: Chicago University Press.

Bateson, P. (2014). New Thinking about Biological Evolution. *Biological Journal of the Linnean Society* 112: 268–75.

Bateson, P. and Gluckman, P. (2011). *Plasticity, Robustness, Development and Evolution*. Cambridge: Cambridge University Press.

Batterman, R. (2005). Critical Phenomena and Breaking Drops: Infinite Idealizations in Physics. *Studies in the History and Philosophy of Modern Physics* 36: 225–44.

Batterman, R. (2015). Autonomy and Scales. In B. Falkenburg and M. Morrison (eds), *Why More Is Different: Philosophical Issues in Condensed Matter Physics and Complex Systems* (pp. 115–35). Dordrecht: Springer.

Bechtel, R. (2013). Addressing the Vitalist's Challenge to Mechanistic Science: Dynamic Mechanistic Explanation. In S. Normandin and C. Wolfe (eds), *Vitalism and the Scientific Image in Post-Enlightenment Life Science, 1800–2010* (pp. 345–70). Dordrecht: Springer.

Bedau, M. (1998). Where's the Good in Teleology? In C. Allen, M. Bekoff, and G. Lauder (eds), *Nature's Purposes: Analyses of Function and Design in Biology* (pp. 261–91). Cambridge, MA: MIT Press.

Bird, A. (2007). *Nature Metaphysics: Laws and Properties*. Oxford: Oxford University Press.

Boyle, B. and Lavin, D. (2010). Goodness and Desire. In S. Tenenbaum (ed.), *Desire, Practical Reason, and the Good* (pp. 202–33). Oxford: Oxford University Press.

Broad, C. D. (1925). *Mind and Its World*. London: Routledge & Kegan Paul.

Broome, J. (1999). Normative Requirements. *Ratio* 12: 398–419.
Chen, X., Bracht, J. R., Goldman, A. D., Dolzhenko, E., Clay, D. M., Swart, E. C., et al. (2014). The Architecture of a Scrambled Genome Reveals Massive Levels of Genomic Rearrangement during Development. *Cell* 158: 1187–98.
Coles, P. (2000). *Cosmology: A Very Short Introduction*. Oxford: Oxford University Press.
Corning, P. (2014). Evolution 'On Purpose': How Behaviour Has Shaped the Evolutionary Process. *Biological Journal of the Linnean Society* 112: 242–60.
Depew, D. (2017). Natural Selection, Adaptation, and the Recovery of Development. In P. Huneman and D. Walsh (eds), *Challenging the Modern Synthesis: Adaptation, Inheritance, and Development* (pp. 37–67). Oxford: Oxford University Press.
Di Frisco, P. (2014). Hylomorpohism and the Metabolic Closure Conception of Life. *Acta Biotheoretica* 62: 499–525.
Di Paolo, E. (2005). Autopoiesis, Adaptivity, Teleology, Agency. *Phenomenology and the Cognitive Sciences* 4: 429–52.
Dowen, R. H., Pelizzola, M., Schmitz, R. J., Lister, R., Dowen, J. M., and Nery, J. R. (2012). Widespread Dynamic DNA Methylation in Response to Biotic Stress. *Proceedings of the National Academy of Sciences* 109 (32): E2183–E2191.
Dupré, J. and O'Malley, M. (2013). Varieties of Living Things: Life at the Intersection of Lineage and Metabolism. In S. Normandin and C. Wolfe (eds), *Vitalism and the Scientific Image in Post-Enlightenment Life Science, 1800–2010* (pp. 311–44). Dordrecht: Springer.
Ellis, B. (2001). *Scientific Essentialism*. Cambridge: Cambridge University Press.
Fisher, R. A. (1930). *The Genetical Theory of Natural Selection*. Oxford: Clarendon.
Fulda, F. (2016). *Natural Agency: An Ecological Approach*. PhD Dissertation, University of Toronto.
Fulda, F. (2017). Natural Agency: The Case of Bacterial Cognition. Unpublished manuscript.
Garrett, B. (2013). Vitalism versus Emergent Materialism. In S. Normandin and C. Wolfe (eds), *Vitalism and the Scientific Image in Post-Enlightenment Life Science, 1800–2010* (pp. 127–54). Dordrecht: Springer.
Gibson, J. J. (1979). *The Ecological Approach to Visual Perception*. Boston: Houghton Mifflin.
Gleiser, M. (1998). Phase Transitions in the Universe. *Contemporary Physics* 39: 239–53.
Godfrey-Smith, P. (2001). Organism, Environment and Dialectics. In R. Singh, C. Krimbas, D. Paul, and J. Beatty (eds), *Thinking about Evolution* (pp. 253–66). Cambridge: Cambridge University Press.
Goldstein, K. (1995). *The Organism: A Holistic Approach to Biology Derived from Pathological Data in Man*. New York: Zone Books.
Herman, J. J. and Sultan, E. E. (2016). DNA Methylation Mediates Genetic Variation for Adaptive Transgenerational Plasticity. *Proceedings of the Royal Society B* 283. doi: 10.1098/rspb.2016.0988.
Jonas, H. (1966). *The Phenomenon of Life*. Evanston: Northwestern University Press.
Keller, E. F. (2013). Genes as Difference Makers. In S. Krimsky and J. Gruber (eds), *Genetic Explanations: Sense and Nonsense* (pp. 329–45). Cambridge, MA: Harvard University Press.
Laland, K., Uller, T., Feldman, M., Sterelny, L., Müller, G. B., Moczek, A., et al. (2014). Does Evolutionary Theory Need a Rethink? Yes: Urgently. *Nature* 514: 161–64.
Lewontin, R. C. (1978). Adaptation. *Scientific American* 239: 212–30.
Lewontin, R. C. (1985). The Organism as Subject and Object of Evolution. In R. Levins and R. Lewontin, *The Dialectical Biologist* (pp. 85–106). Cambridge, MA: Harvard University Press.
Lewontin, R. C. (2001). *The Tripe Helix: Gene, Organism, and Environment*. Oxford: Oxford University Press.

Mareno, A. and Mossio, M. (2015). *Biological Autonomy: A Philosophical and Theoretical Enquiry.* Dordrecht: Springer.

McGhee, G. (2007). *The Geometry of Evolution: Adaptive Landscapes and Theoretical Morphospaces.* Cambridge: Cambridge University Press.

Morrison, M. (2015). Why Is More Different? In B. Falkenburg and M. Morrison (eds), *Why More Is Different: Philosophical Issues in Condensed Matter Physics and Complex Systems* (pp. 91–114). Dordrecht: Springer.

Nicholson, D. J. and Gawne, R. (2015). Neither Logical Empiricism nor Vitalism, but Organicism: What the Philosophy of Biology Was. *History and Philosophy of the Life Sciences*, 37: 345–81.

Normandin, S. and Wolfe, C. (eds). (2013). *Vitalism and the Scientific Image in Post-Enlightenment Life Science, 1800–2010.* Dordrecht: Springer.

Odling-Smee, F. J., Laland, K., and Feldman, M. (2003). *Niche Construction: The Neglected Process in Evolution.* Princeton: Princeton University Press.

Pfennig, D. W., Wund, M., Snell-Rood, E., Cruickshank, T., Ciliberti, S., Martin, O. C., and Wagner, A. (2007). Innovation and Robustness in Complex Regulatory Gene Networks. *Proceedings of the National Academy of Sciences* 104 (34): 13591–6.

Pigliucci, M. and Müller, G. (eds). (2010). *Evolution: The Extended Synthesis.* Cambridge, MA: MIT Press.

Russell, E. S. (1945). *The Directiveness of Organic Activities.* Cambridge: Cambridge University Press.

Schrödinger, E. (1944). *What Is Life?* New York: Dover.

Shapiro, J. (2011). *Evolution: A View from the 21st Century Perspective.* Upper Saddle River: FT Press Science.

Shapiro, J. (2013). How Life Changes Itself: The Read-Write (RW) Genome. *Physics of Life Reviews* 10: 287–323.

Smolin, L. (1997). *Life of the Cosmos.* Oxford: Oxford University Press.

Smolin, L. (2013). *Time Reborn: From the Crisis in Physics to the Future of the Universe.* New York: Houghton Mifflin Harcourt.

Sober, E. (1984). *The Nature of Selection.* Cambridge, MA: MIT Press.

Solé, R. and Goodwin, B. (2000). *Signs of Life: How Complexity Pervades Biology.* New York: Basic Books.

Stoffregen, T. (2003). Affordances As Properties of the Animal-Environment System. *Ecological Psychology* 15: 115–34.

Thompson, E. (2007). *Mind in Life: Biology, Phenomenology and the Sciences of Mind.* Cambridge, MA: Harvard University Press.

Turner, J. S. (2000). *The Extended Organism: The Physiology of Animal-Built Structures* Cambridge, MA: Harvard University Press.

Turner, J. S. (2013). Homeostasis and the Forgotten Vitalist Roots of Adaptation. In S. Normandin and C. Wolfe (eds), *Vitalism and the Scientific Image in Post-Enlightenment Life Science, 1800–2010* (pp. 271–92). Dordrecht: Springer.

Vane-Wright, D. (2014). What Is Life? And What Might Be Said of the Role of Behaviour in Its Evolution? *Biological Journal of the Linnean Society* 112: 219–41.

Varela, F. J. (1979). *Principles of Biological Autonomy.* New York: Elsevier.

Wagner, A. (2014). *The Arrival of the Fittest: Solving Evolution's Greatest Puzzle.* New York: Current Books.

Walsh, D. M. (2007). Development: Three Grades of Ontogenetic Involvement. In M. Matthen and C. Stephens (eds), *Handbook of the Philosophy of Science* (pp. 179–99). Amsterdam: Elsevier.

Walsh, D. M. (2008). Teleology. In M. Ruse (ed.), *Oxford Handbook of the Philosophy of Biology* (pp. 113–37). Oxford: Oxford University Press.

Walsh, D. M. (2012a). Mechanism and Purpose: A Case for Natural Teleology. *Studies in the History and Philosophy of Biology and the Biomedical Sciences* 43: 173–81.

Walsh, D. M. (2012b). Situated Adaptationism. In W. Kabesenche, M. O'Rourke, and M. Slater (eds), *The Environment: Philosophy, Science, Ethics* (pp. 89–116). Cambridge, MA: MIT Press.

Walsh, D. M. (2013). Adaptation and the Affordance Landscape: The Spatial Metaphors of Evolution. In G. Barker, E. Desjardins, and T. Pearce (eds), *Entangled Life* (pp. 213–36). Dordrecht: Springer.

Walsh, D. M. (2015). *Organisms, Agency, and Evolution.* Cambridge: Cambridge University Press.

West-Eberhard, M. J. (2003). *Developmental Plasticity and Evolution.* Oxford: Oxford University Press.

Wray, G. A., Hoekster, H., Futuyma, D., Lenski, R., Mackay, T., Schluter, D., and Strassman, J. E. (2014). Does Evolutionary Theory Need a Rethink? No: Everything Is Fine. *Nature* 514: 161–4.

# 9

# Symbiosis, Transient Biological Individuality, and Evolutionary Processes

*Frédéric Bouchard*

## 1. Introduction

The debate about how to define biological individuality has a long tradition, both in philosophy and in biology. From the Greeks onward there has been a recognition that our intuitions about the metaphysical primacy of individual organisms is at best programmatic and at worse unjustified. Most common notions of individuality either fail for some paradigmatic individuals (e.g. we may assume that individuality demands some sort of autonomy, and yet a human baby is not autonomous) or apply to biological systems that, to many, do not seem to merit to be identified as individuals (e.g. an ant colony as an emergent individual, also known as a 'superorganism'). This philosophical problem has gained renewed currency in contemporary philosophy of biology in response to debates about group selection and whether groups exist with their own group-level properties (see Sober and Wilson 1998; Wilson and Sober 1989; and Okasha 2006 concerning group selection). Much theoretical work has been conducted to better buttress our metaphysical understanding of biological individuals and how they fit into various biological explanations (see e.g. Huneman 2014a and 2014b for how theoretical frameworks of weak and strong individuality fit into evolutionary and ecosystem biology; Pradeu 2012 for immunology; and Haber 2013 for a discussion of what he calls the problem of the paradigm and how it muddles our attempts to make sense of individuality). Although much of this work has focused on operational problems—that is, on questions such as 'How do individuals actually fit into scientific explanations?'—many philosophers have been interested in the intersection between the metaphysical question concerning the nature of individuals in general (e.g. a chair as an individual) and how this issue is similar to and different from understanding biological individuals (e.g. Dupré 1993; O'Malley and Dupré 2007; Wilson 1999, 2004a, 2004b, 2008).

In this chapter I wish to address the issue of how to define biological individuals beyond common intuitions we have about individual organisms. I will argue that biological individuals are defined via the processes they are involved in. Processes make up the individuals. But if individuals are in some sense process-laden, couldn't we dispense with individuals altogether and consider only processes in the same way? If the focus of biological individuality is the process that generates it (in this case, the process of natural selection), couldn't we forget about individuals and focus the explanation solely on processes? I will argue that doing so would prove too epistemically onerous because of the way evolutionary explanations are structured. Individuals and processes should be co-occurring in our explanations.

John Dupré in recent publications has toyed with this shift to a process-based ontology. Harking back to Heraclitus and Whitehead, Dupré writes:

> Most philosophers, if asked what were the most basic constituents of their ontology, would probably name things and properties.... There is, however, an alternative ontology, one generally attributed in antiquity to Heraclitus that takes things themselves to be only temporary manifestations of something more fundamental: change or process.... I do want to claim that an ontology of processes is better suited to understanding the nature of life and the living.
> (Dupré 2014: 81)

This proposition is similar to discussions about the metaphysics of causality where Wesley Salmon (1984) among others suggested that causality is not about individual events but about continuous processes intersecting through time. Salmon's theory of causal process was geared towards providing a novel theory of explanation (a discussion of which is beyond the scope of this chapter). A better understanding of causal *processes* (instead of events) enabled Salmon to offer a Humean theory of scientific explanation where causality was both objective and contingent and allowed for indeterminism in the natural world (see Dowe 1992 for analysis). Salmon's account was influential in general philosophy of science but was also very influential for certain philosophers of biology—notably Brandon (1990), who imported Salmon's account of propensities into his understanding of evolutionary processes and individuals. Whereas Salmon (and Brandon in a different context) was interested in the *continuous* aspect of processes, in this chapter we will see how the debate between an individual-based ontology and a process-based ontology highlights the *transient* nature of biological organization and function. We can better understand biological individuals by paying closer attention to processes.

Bapteste and Dupré (2013) argue that, in our biological ontology, processes are prior to individuals but that individuals can remain in the scientific picture. While I agree with many of the points they make, in this chapter I emphasize why individuals *must* remain in the scientific picture, especially when adaptations are concerned. The upshot is that this will give us a better understanding of how evolution works on vastly different but intersecting temporal scales for entities with various levels of functional integration. In some sense I will argue that, for evolutionary explanations, there is no dilemma between individual-based and process-based ontologies: they are both part of the same picture.

To get there, I will need to reprise parts of arguments offered elsewhere (Bouchard 2009, 2010, 2011). Our focus shall be on the question of individuality that arises in certain cases of symbiosis. We shall see how a revised notion of biological individuality is necessary on the basis of actual biological cases that need better accounts of their ontological status. Specifically, I will discuss the case of the Hawaiian Bobtail squid and its interaction with *Vibrio fisheri* (here my exposition is based mainly on

McFall-Ngai and Montgomery 1990, Jones and Nishiguchi 2004, and Bouchard 2010). This will lead me to defend a definition of biological individuality that can apply to 'difficult' and exotic cases of associations of individual organisms (sometimes from different species) that generate emergent individuals (i.e. novel individuals that are caused by the interactions of underlying lower-level individuals).

Wilson and Sober's notion of biological individuality in terms of functional integration and common fate (Wilson and Sober 1989) can offer a guide to the perplexed about individuality. To paraphrase their proposal, we get the following notion of biological individuality:

> A biological individual is a functionally integrated entity whose integration is linked to the common fate of the system when faced with selective pressures from the environment.

I have argued elsewhere (Bouchard 2010, 2013) that this definition of biological individuality is fundamentally about how the process of natural selection generates and maintains biological individuals. The proposed definition of individuality is intended both as an operational definition that helps to make sense of certain scientific projects and as a metaphysical claim about what inhabits the universe independently of our needs and uses for certain types of scientific explanations. What I haven't discussed before, however, is the implication of this account for our assessment of the importance of processes in our understanding of nature.

At the end of the chapter I will argue that, although our metaphysical account could focus on processes and minimize the role that individuals play in explanations, this would be a pyrrhic victory, for it would be difficult to offer an operationally equivalent definition of these processes in an evolutionary explanation. Biological individuals may be temporary 'eddies in the constant flux of process' (Dupré 2015: 81), but viewing them as genuine individuals is necessary for evolutionary explanations nonetheless. In a naturalistic metaphysics, this may be an indication that individuals are real enough and should retain a prominent role in our explanations.

## 2. Beyond Replicators

Evolutionary theory allows for the explanation of the 'perfect' organs that organisms seem to have. They have traits that appear well suited to their circumstances. Dawkins (1976) codified adaptation as a relationship between 'replicators' (usually genes) and 'vehicles'—or, to use Hull's (1980) better coinage, 'interactors'—which, most of the time for Dawkins, are individual organisms.

You have variation among replicators, which translates into variation expressed by interactors. Interactors are differentially selected according to their properties by various selective pressures and, if these features are heritable, this will in turn translate into differential representation among replicators in the following generation. This differential representation will appear both among the interactors and among the replicators that generate them. In such an interplay, differential fitness or differential evolutionary success is in terms of survival and reproduction, or rather in terms of differential propensities to survive and reproduce (see Brandon 1990, 2008).

In this view of evolution, the explanatory burden is on the replicator, for it is the only common denominator across generations. Interactors are explicitly defined in relation to replicators. The adaptive signal is about replicators, and the phenotype is at best the repeater or at worse the noise across generations. The distinction between interactor and replicator is a powerful explanatory factor, because it allows adaptive explanations via the tracking of changes in populations of interactors or replicators and the interaction (via heritability) of the two explanatory levels. Well-adapted interactors thrive; less well-adapted interactors do not. This allows replicators to thrive through the generations. If the environmental conditions are right, these small victories accumulate through the ages to form the adaptations that we observe in nature.

For simple organisms, this interplay between interactors and replicators seems explanatorily satisfactory. This assumes a strong link between the interactor population and the replicator population. But what if some biological systems decouple (to various degrees) interactor populations and replicator populations? I have argued (in Bouchard 2009, 2010, 2013) that this is what occurs in cases of symbiosis.

Take an adaptation—say, the ability to digest cellulose. In most simple organisms we assume that this feature of a given interactor is related to a given set of replicators. But if this ability emerges from the interaction of organisms of different species, the claim that replicator success explains interactor success (or vice versa, if you are not a gene centrist) relatively to a specific adaptive trait loses some of its explanatory appeal. In Bouchard 2014, I have explained, using the comparison between a termite whose cellulose digestion does not depend on another species (*Macrotermes michaelseni*) and one whose cellulose digestion does (*M. natalensis*), that traditional co-evolutionary accounts have difficulty comparing symbiotic communities with autonomous organisms. Moreover, the fact that the symbiont (in this case *Termitomyces*) is ecologically acquired raises the prospect that one will not always have neat intergenerational linkages between interactor populations and replicator populations in relation to a specific adaptation (in this case cellulose digestion).

If the bearers of adaptation (or, to use the term in Lloyd 2012, the 'manifestors' of adaptation) are complex multispecies assemblages, we should expect our description of the adaptive process to be complex as well. Multispecies assemblages put stress on the idea that replicator populations and interactor populations are always coupled. If that is the case, then one should expect individuality to become more complicated as well. Below I will briefly discuss an example I examined in more detail in Bouchard 2010, to show just how much the transiency of these interactions between species raises metaphysical and operational problems about individuality and processes.

## 3. Seeing the Light

The Hawaiian Bobtail squid (*Euprymna scolopes*) interacts with the bacteria *Vibrio fischeri* in a way that generates light (this phenomenon is called 'bioluminescence'). This interaction is the 'new' poster child of symbiosis research. McFall-Ngai has contributed a great deal to the detailed study of this symbiotic community; most of the description offered here is inspired by her work and that of other colleagues (see e.g. McFall-Ngai 1999, 2014; McFall-Ngai and Montgomery 1990; Jones and

Nishiguchi 2004). 'How does it glow' and, more importantly, 'Why does it glow?' or 'What is this bioluminescence good for?' are the traditional adaptationist questions. A plausible hypothesis is that bioluminescence allows Hawaiian Bobtail squids, through counterillumination, to avoid predation (see Jones and Nishiguchi 2004). A predator identifies its prey by the shadow the latter casts when swimming between the predator and the light above itself. If you glow and, moreover, if your glow matches other sources of illumination in the environment, you don't cast a shadow. That is to say, when the squid glows, it is unrecognizable as a prey to its predator.

Bioluminescence is obviously a wondrous trait demanding explanation, and natural selection seems to provide it. But things get murkier when one asks questions about the bearer of the adaptation. Who is the individual bearing the adaptation? For whom is the glowing good? What is it that glows? To answer these questions, one needs to understand how bioluminescence is triggered.

The *Vibrio* colonizes the apical surfaces of epithelial cells at each squid generation. The *Vibrio* then reproduces inside individual squids. When it achieves high enough density, quorum sensing triggers a chemical reaction that generates light. This light is oriented by the squid thanks to developmental changes (a lens-like structure) that are initiated by the colonization itself. In other words, the bacteria both trigger developmental changes in the squid and produce the light that these changes take advantage of. Each day, the squid flushes out much of the *Vibrio*, bacteria density falls, and bioluminescence is temporally 'turned off'. The remaining *Vibrio* then starts reproducing again inside the squid, until a high enough density is achieved once more. Quorum sensing triggers the bioluminescence and, as a result, the confused predator keeps looking unsuccessfully for the squid prey. The squid alone cannot glow, and the *Vibrio*, although it could reproduce to high enough densities *sans* squid and generate quorum sensing and bioluminescence, does not actually generate bioluminescence in the wild outside the squid. Bioluminescence is a transient, intermittent, yet recurrent state of a *community*. What glows is a temporary assemblage of species interacting in the right way.

As I have argued in more detail in Bouchard 2010, the question of individuality is not trivial in this case: how many individuals and how many types of individuals are there? Using Wilson and Sober's notion of individuality (which is based on functional integration and common fate), I offered a few possible scenarios (ibid., 632):

(a) Considering that the squid can survive without the bacteria, do we have 1 squid and a multitude ($10^9$) of *V. fischeri* = 1 billion and 1 individuals?
(b) Considering that the bioluminescence, because of the quorum sensing, is a collective property of the bacteria, do we have 1 squid + 1 *Vibrio* superorganism = 2 individuals?
(c) Considering that the symbiotic community has its own additional survival potential (i.e. its own emergent common fate), should we say that we have 1 squid + 1 billion *Vibrio* + 1 *Vibrio* superorganism + an emergent squid/colony superorganism = 1 billion and 3 individuals?

These scenarios highlight that, because functional integration and common fate can be achieved at various levels of organization, one can have overlapping individuals operating at different temporal scales and with different levels of transiency and continuity. In the following section I will examine this issue in more detail, to see how individual-based and process-based ontologies fare in relation to it.

## 4. Transient and Intermittent Individuals

Traditional intuitions about biological individuality are tailored to account for the common individual organisms that we encounter as human beings and that conform to our perceived experiences of our own individuality: animals. And yet, as we have seen, many biological systems at levels of organization below and above animals form systems that deserve to be understood as genuine emergent biological individuals. Such emergent individuals force us to consider accounts of biological individuality that depend less on structure—on the basis of the common origin of the parts of the system, which in many respects is how we conceive of organisms as the expression of shared DNA and as belonging to a single lineage—and more on the functional integration of parts in a larger functional whole. How does this functional integration obtain, and how is it maintained? Via the process of natural selection and adaptation.[1] An adaptation is a trait that *functions* in a way that increases the fitness of its bearer. To be functionally integrated is to have one's parts work in a way that maintains the system as a cohesive whole, as an adaptation bearer. Here my understanding of functional integration is inspired by McShea and Venit's (2002) discussion of *connectedness* between zooids in a colony.[2] Functional integration is a matter of degree. But because the parts, in this account, do not have to be related, they will not have commensurable differential reproductive success.[3]

In the specific case of symbiosis that we have discussed here, there is reproduction at the *Vibrio* level, and at the squid level, but not at the community level. And yet the community is the bearer of the adaptation (i.e. bioluminescence) that results from the temporary but repeated ecological interaction between organisms of different species. The squid–*Vibrio* assemblage persists better with the bioluminescence than without it. But it would be inaccurate to say that this assemblage reproduces. There is repetition of this association and its emergent properties, which has lead me to argue that, instead of focusing on differential reproductive success of individual organisms, one should focus on the differential persistence of lineages (broadly construed).

---

[1] The reader may compare this account of functional integration to the one offered in chapter 10.

[2] '*Connectedness*. The assumption here is that connectedness reflects the degree to which lower-level entities can share resources and function in a coordinated fashion, and therefore the degree to which the colony operates as a unified whole. Connectedness can take a variety of forms, including physical attachment; sharing of a gut, coelom, vascular system, or nervous system (as in some colonial invertebrates); and behavioral interactions mediated by pheromones, sound, or physical contact (as in social insects and vertebrates)' (McShea and Venit 2002: 311).

[3] This was my reason for proposing a different understanding of evolutionary success in terms of differential persistence. I have argued extensively elsewhere (Bouchard 2004, 2008, 2011, 2014) that, by understanding evolutionary success in terms of differential persistence instead of differential reproduction, one can account for many species and associations of species that cannot be readily accommodated in reproduction accounts.

The lack of reproduction should not jettison the possibility of an evolutionary explanation. From clonal species favouring growth over reproduction to associations of organisms from same or different species forming emergent individuals, the biological world is rife with systems for which evolutionary success is more about increasing the potential for 'survival' (or 'persistence', to be more precise) than about generating more copies of itself. Survival is related to the 'common fate' identified by Wilson and Sober as a necessary property of biological individuals, and this common fate is made possible via the functional integration of the parts. Individuals have always been understood in relation to their integration as wholes. For our purposes here, the point is that biological individuals are defined by the functional integration of their parts (or by their connectedness, in accordance with McShea and Venit 2002) and by the common fate of the whole system. In the case of biological individuals, both of these features are the result of the process of evolution through natural selection: functional integration is the result of adaptive processes; and the common fate is success (or failure) when faced with pressures from the environment (i.e. selective pressures).

This example and the treatment offered here evoke other traditional metaphysical problems about individuality that will concern us for the remainder of this chapter. First, it becomes obvious that individuality is a matter of degree: if functional integration is the principle of individuality, then we must accept that there are degrees of individuality. Functional integration and common fate are never absolute, in part because evolutionary success is never absolute and depends to a large extent on the external environment in which the individual operates. A hostile environment may weaken the functional integration of an individual, while other environmental conditions may strengthen the functional relationships between the parts of a system. Our strong intuition that animals are paradigmatic individuals may be vindicated by the fact that the functional integration or connectedness of their parts is higher in most conditions than, say, that of a herd of bisons (see Clarke 2010 and Godfrey-Smith 2009 for a defence of this idea). This is not a novel result, but focusing on multispecies associations makes this issue more salient. Looking at the functional integration of the parts of an individual organism, one also recognizes that the degree to which a part is functionally integrated into the whole will vary from one part to another: my heart is more functionally integrated to me as a whole (and to my fate) than my toes are. Functional integration is a question of degree among the parts themselves.

This should not be surprising, given that the environment dictates the conditions of emergence and persistence of a system, and this environment will put varying stress on different parts of a given individual. This raises a second classic ontological problem for biological individuality: identity through time.[4] If we accept that individuality is about functional integration and that functional integration is a question of degree, we must also accept that an individual's degree of individuality may fluctuate over time. How do we conceive of the permanence of individuals with fluctuating degrees of functional integration? A functional

---

[4] On the issue of diachronic identity, see also chapters 1, 2, 4, 5, 6, 7, 11, and 18.

account of individuality based on evolutionary processes has to take into account the transiency of biological individuals.

At one level, it is trivial to say that individuals are transient. Our lifespans are transient compared to the permanence of Mount Everest or even to that of a protected artefact in a museum. But focusing on fluctuating functional integration raises the spectre that individuals are not only *transient* but also *intermittent*, because some of the parts appear and disappear on vastly different temporal scales.

One symbiont may not go through intergenerational change, while another symbiont may have much shorter generational time and therefore see much more intergenerational change. This leads us to a view of individuality where the parts of a given individual are not synchronized. Parts of the same whole (in the case of multispecies individuals) do not belong to the same temporal scales. While the squid may persist for X years, the individual *Vibrio* may survive less than a day, and yet be part of a colony that surpasses the lifespan of the squid temporarily hosting it.

If functional integration obtains via evolutionary processes, it is the interactions between species that become the locus of individuality: the individual that is the bioluminescent *Vibrio*–squid community *exists only when* both species interact in the right way. This integrated ecological interaction obtains once a day. But how are we to make sense of the periods in which the *Vibrio* density is not high enough (after the daily purge)? How are we to interpret the succession of bioluminescent events? How are we to understand the relation between a single squid individual and a single *Vibrio* colony, when the former interacts with the latter periodically but intermittently and the latter is formed by ever-changing individual *Vibrio* bacteria? Do we have a single, temporally discontinuous, functional individual, or a succession of very short-lived (transient) distinct *Vibrio*–squid emergent individuals?

Take an individual squid—call it Bob—and an individual *Vibrio* colony—call it Kala. Bob and Kala are well integrated individuals persisting over similar temporal scales. But how is Kala formed and maintained? The answer is: via the succession and aggregation (accomplished in the right way) of myriads of individual *Vibrio* bacteria (i.e. Vib, Viba, Vibo, etc). Once a day, Bob and Kala form an emergent functional individual (the glowing squid–*Vibrio* community). How many emergent individuals are there?

(1) Do we have a single individual appearing and disappearing once a day (i.e. the same KaBo blinking in and out of existence every time quorum sensing is achieved)?

OR

(2) Do we have a succession of communities leading to distinct multispecies individuals (i.e. KaBo, BoKa, etc.)?

Integrating a process-based approach to biological explanations, one can see why the former should be favoured. If we are to focus on functional integration, it seems perverse to multiply individuals when, after all, it is a given set of intersecting processes that leads to the periodic blinking of Kabo in and out existence.

While individuals may exist by virtue of their functional integration, it is the continuity of process intersections that leads to their identity through time, even if their degree of individuality fluctuates.

Note that the issue is not novel or especially applicable to my account and examples: Hull famously defined species as spatio-temporally located individuals, and yet the individual organisms constituting the species are a succession of transient parts. Even in cases where one gets overlapping generations, one has the succession of distinct individual parts. The additional difficulty with multispecies associations is that in most cases the parts spend much of their lifespan outside of the emergent individual. For complex emergent individuals, the functional integration is not only transient; it is intermittent. The issue is whether this intermittency disaggregates the whole into a succession of distinct wholes or merely reflects a fluctuation of integration that leads to fluctuation in the degree of shared individuality of the whole. A squid+*Vibrio* is, both literally and figuratively, a flickering individual.

I suggest that, when we are faced with such transient and intermittent individuals, a process perspective allows us to identify (i.e. specify the identity of) an individual through time, however discontinuous that individual may be. Shouldn't we, then, entertain the possibility that, in the biological world, processes and not individuals are primitive, and that individuals can be demoted in our explanations? This is how Bapteste and Dupré (2013: 380) propose to redefine not only the microbial world but biology in general: 'We understand living things to be most fundamentally the consequences of numerous interweaving (occasionally nested) processes. Although it is common to describe the domain of biology as consisting of things, for example organisms, cells, genes, and so on, we understand even these as ultimately processual.'

The appeal of processual explanations is evident in evolutionary biology: adaptation unfolds over time, traditionally across generations of individual organisms. At first glance, a process-based ontology seems to be compatible with such a view of life—a view that life unfolds over many generations. The real question is whether the centrality of process should displace individuals altogether. Here the answer depends on whether you demand that individuality reveals a substance (in an Aristotelian sense) or whether you allow that individuals can be defined as the result of processes in interaction.

Bapteste and Dupré (2013: 381) write: 'For these reasons, the ontology we aim to describe is an ontology of processes. A processual ontology should characterize entities in terms of how they emerge, are maintained and are stabilized.' Although they offer compelling reasons for thinking of processes as a necessary condition for biological interactions (and for associated explanations), they do not explain why a process-based ontology could truly overshadow an individual-based ontology. They offer good reasons for entertaining a process-based ontology, but it is less clear what role is left for individuals to play: the reader may infer that Bapteste and Dupré prescribe forgetting about individuals. To frame this in a language that has probably never been used to characterize Dupré's work, on one reading Bapteste and Dupré are reductionists ('individuals are ultimately about processes'), but on another they are eliminativists ('individuals hinder our explanations, we should forget about them, and focus on processes instead'). Based as it is on Dupré's well-established promiscuous realism, the first reading is probably closer to the intended project.

Ellen Clarke diagnoses the importance of individuals in our evolutionary explanations:

It is hard to overemphasize the importance of individuals within the Modern Synthesis. They are central to the inner logic of evolution by natural selection, according to which evolution occurs because of the differential survival and reproduction of individuals. Even in its most abstract minimal formulations, the action of a selection process requires that there be a multiplicity of objects that are sufficiently separate from one another that they can be differentially deleted or copied.   (Clarke 2010: 313)

Arguably, evolutionary explanations are individual-based. Rightfully highlighting the centrality of process does nothing to reduce the dependency of process on individuals. The explanandum in the case of bioluminescence was the trait that seemed to confer an adaptive advantage to an individual. What would the explanandum and the explanans be if we got rid of entities and of their relative success?

Cases such as the symbiosis discussed here show how much processes need to be taken seriously, both in our scientific explanations and in our metaphysics; but they also show how processes underpin a revised notion of individuality instead of displacing or minimizing individuals.[5]

## 5. Conclusions

Historically, biological ontology has focused on individual organisms, their parts (organs and cells), and the aggregation of individuals (groups, species, etc.). In the second half of the twentieth century, the development of molecular biology tied to a neo-Darwinian account of evolutionary processes raised the possibility that the core level of organization may be the gene (or a gene complex), its parts (e.g. nucleotides), and the aggregation of genes and their expression (from genomes to cells, to organisms, and to species). With developments in metagenomics and a better understanding of symbiotic associations, we are now entertaining the possibility of multispecies assemblages that form emergent individuals. In this chapter I have discussed some of the implications of such developments for our understanding of evolution and individuality.

The goal here was not (and could not be) to offer an exhaustive account of this new biological ontology, but rather to examine one key implication of this story: if biological individuals emerge through a progressive functional integration that arises from and is maintained by selective processes, this entails that biological individuality is a matter of degree: a degree that increases or decreases over time. Being an individual is not a binary affair. If being an individual is not an absolute and if 'weak' individuals are as important to evolutionary explanation as 'strong' individuals (i.e. strongly integrated individuals with a clear common fate), then why not shift the focus away from individuals altogether in favour of focusing on the processes that generate them?

One must not forget what was the explanandum in the first place: adaptive traits. How do we explain bioluminescence? How do we explain the ability to digest cellulose? These traits are what forced us to consider these complex explanations in the first place. It is easier to make sense of these adaptive traits if they are properties of an individual

---

[5] The implications of symbiosis for biological ontology are also considered in chapters 1, 5, 10, and 15.

(whatever its level of organization: gene, organism, population, or species) that faces a selective environment (two individuals interact, generating a new individual with emergent traits) than if they are merely a causal nexus of intersecting processes (the squid process interacting with the *Vibrio* process generating a bioluminescence process). Or, to be more precise, we know how to think about competing individuals with differential evolutionary success, but it is less clear how we should think of competing bundles of processes and what would count as evolutionary success for these bundles. We have to wonder about whether a process-based ontology (in its eliminativist form, and possibly in its reductionist form) can accommodate Darwinian explanations of adaptations in any shape or form. Adaptation has always been both a process and a result (i.e. the trait; see Brandon 1990); but, more importantly, adaptations *as the traits/ properties of individuals* were the original explanandum of evolutionary theory. Any thorough attempt to introduce a process-based ontology will have to accommodate this fact or to demonstrate the explanatory benefit of minimizing the role that individuals play in our explanations.

The appeal of a revival of a process-based ontology for the biological world must not overshadow the potential cost of making Darwinian explanations incompatible with said ontology. If adaptations remain wondrous features of the natural world, and if such features are to be explained in a Darwinian framework (however broadly construed), then the bearers of adaptation will be individuals (at some level of organization or another). Adaptations explain the thriving of some entities that bear them, and these traits emerge and unfold through a process of adaptation to natural selection. Natural selection and adaptation are the processes, but individuals remain in the picture as the entities generated and transformed through these processes. Notable proponents of a process-based ontology, such as Bapteste and Dupré, are right to indicate the severe blind spots of substance-based ontologies, especially in microbiology. But new adherents of a process ontology for biology should not take Bapteste and Dupré further than what they intended: from a process-based ontology it does not follow that there is an explanatory gain from getting rid of individuals altogether. If adaptations are a genuine question (and they are), processes are necessary to our biological explanations, but they are not sufficient. Individuals will need to remain in the picture as well.

## Acknowledgements

Many thanks to the helpful comments of the reviewers as well as to the discussants at the University of Exeter workshop 'Process Philosophy of Biology'. Thanks also to John Dupré, François-Joseph Lapointe, and Eric Bapteste for discussions on this topic. Their questions and comments have greatly improved my thinking on these issues.

## References

Bapteste, E. and Dupré, J. (2013). Towards a Processual Microbial Ontology. *Biology & Philosophy* 28 (2): 379–404.

Bapteste, E., Lopez, P., Bouchard, F., Baquero, F., McInerney, J. O., and Burian, R. M. (2012). Evolutionary Analyses of Non-Genealogical Bonds Produced by Introgressive Descent. *Proceedings of the National Academy of Sciences* 109 (45): 18266–72.

Bouchard, F. (2004). *Evolution, Fitness and the Struggle for Persistence*, ed. by A. Rosenberg. Durham, NC: Duke University Press.

Bouchard, F. (2008). Causal Processes, Fitness and the Differential Persistence of Lineages. *Philosophy of Science* 75 (5): 560–70.

Bouchard, F. (2009). Understanding Colonial Traits Using Symbiosis Research and Ecosystem Ecology. *Biological Theory* 4 (2): 240–6.

Bouchard, F. (2010). Symbiosis, Lateral Function Transfer and the (Many) Saplings of Life. *Biology & Philosophy* 25: 623–41.

Bouchard, F. (2011). Darwinism without Populations: A More Inclusive Understanding of the 'Survival of the Fittest'. *Studies in History and Philosophy of Science Part C: Studies in History and Philosophy of Biological and Biomedical Sciences* 42 (1): 106–14.

Bouchard, F. (2013). What Is a Symbiotic Superindividual and How Do You Measure Its Fitness? In F. Bouchard and P. Huneman (eds), *From Groups to Individuals: Perspectives on Biological Associations and Emerging Individuality* (pp. 243–64). Cambridge, MA: MIT Press.

Bouchard, F. (2014). Ecosystem Evolution Is About Variation and Persistence, Not Populations and Reproduction. *Biological Theory* 9 (4): 382–91.

Brandon, R. (1970). *Adaptation and Environment*. Princeton: Princeton University Press.

Brandon, R. (2008). Natural Selection. In E. N. Zalta (ed.), *The Stanford Encyclopedia of Philosophy*. http://plato.stanford.edu/archives/win2008/entries/natural-selection.

Clarke, E. (2010). The Problem of Biological Individuality. *Biological Theory* 5 (4): 312–25.

Dawkins, R. (1976). *The Selfish Gene*. New York: Oxford University Press.

Dowe, P. (1992). Wesley Salmon's Process Theory of Causality and the Conserved Quantity Theory. *Philosophy of Science* 59 (2): 195–216.

Dupré, J. (1993). *The Disorder of Things: Metaphysical Foundations of the Disunity of Science*. Cambridge, MA: Harvard University Press.

Dupré, J. (2014). A Process Ontology for Biology. *The Philosophers' Magazine*: 81–8.

Dupré, J. (2015). A Process Ontology for Biology. *Physiology News* 100: 33–4.

Dupré, J. and O'Malley, M. A. 2009. Varieties of Living Things: Life at the Intersection of Lineage and Metabolism. *Philosophy & Theory in Biology* 1. http://quod.lib.umich.edu/cgi/t/text/text-idx?c=ptb;view=text;rgn=main;idno=6959004.0001.003.

Godfrey-Smith, P. (2009). *Darwinian Populations and Natural Selection*. London: Oxford University Press.

Haber, M. (2013). Colonies Are Individuals: Revisiting the Superorganism Revival. In F. Bouchard and P. Huneman (eds), *From Groups to Individuals: Perspectives on Biological Associations and Emerging Individuality* (pp. 195–217). Cambridge, MA: MIT Press.

Hull, D. L. (1980). Individuality and Selection. *Annual Review of Ecology and Systematics* 11: 311–32.

Huneman, P. (2014a). Individuality as a Theoretical Scheme. I: Formal and Material Concepts of Individuality. *Biological Theory* 9 (4): 361–73.

Huneman, P. (2014b). Individuality as a Theoretical Scheme. II. About the Weak Individuality of Organisms and Ecosystems. *Biological Theory* 9 (4): 374–81.

Jones, B. W. and Nishiguchi, M. K. (2004). Counterillumination in the Hawaiian Bobtail Squid, Euprymna Scolopes Berry (Mollusca: Cephalopoda). *Marine Biology* 144 (6): 1151–5.

Levins, R., and Lewontin, R. (1985). *The Dialectical Biologist*. Cambridge, MA: Harvard University Press.

Lloyd, E. (2012). Units and Levels of Selection. In E. N. Zalta (ed.), *The Stanford Encyclopedia of Philosophy*. http://plato.stanford.edu/archives/spr2012/entries/selection-units. Palo Alto, Stanford University.

McFall-Ngai, M. J. (1994). Animal–Bacterial Interactions in the Early Life History of Marine Invertebrates: The Euprymna scolopes/Vibrio Fischeri Symbiosis. *American Zoologist* 34: 554–61.

McFall-Ngai, M. J. (1999). Consequences of Evolving with Bacterial Symbionts: Insights from the Squid-Vibrio Associations. *Annual Review of Ecology and Systematics* 30 (1): 235–56.

McFall-Ngai, M. J. (2014). The Importance of Microbes in Animal Development: Lessons from the Squid-Vibrio Symbiosis. *Annual Review of Microbiology* 68: 177–94.

McFall-Ngai, M. J. and Montgomery, M. K. (1990). The Anatomy and Morphology of the Adult Bacterial Light Organ of Euprymna Scolopes Berry (Cephalopoda:Sepiolidae). *Biological Bulletin* 179 (3): 332–9.

McShea, D. W. and Venit, E. P. (2002). Testing for Bias in the Evolution of Coloniality: A Demonstration in Cyclostome Bryosoans. *Paleobiology* 28 (3): 308–27.

Millstein, R. L. (2006). Natural Selection as a Population-Level Causal Process. *British Journal for the Philosophy of Science* 57: 627–53.

Millstein, R. L. (2009). Populations as Individuals. *Biological Theory* 4 (3): 267–73.

Okasha, S. (2006). *Evolution and the Levels of Selection*. New York: Oxford University Press.

O'Malley, M. A. and Dupré, J. (2007). Size Doesn't Matter: Towards a More Inclusive Philosophy of Biology. *Biology & Philosophy* 22 (2): 155–91.

Pradeu, T. (2012). *The Limits of the Self: Immunology and Biological Identity*. Oxford: Oxford University Press.

Queller, D. C. and Strassmann, J. (2009). Beyond Society: The Evolution of Organismality. *Philosophical Transactions of the Royal Society B: Biological Sciences* 364 (1533): 3143–55.

Rosenberg, E. and Zilber-Rosenberg, I. (2011). Symbiosis and Development: The Hologenome Concept. *Birth Defects Research Part C: Embryo Today: Reviews* 93 (1): 56–66.

Salmon, W. C. (1984). *Scientific Explanation and the Causal Structure of the World*. Princeton: Princeton University Press.

Seibt, J. (2013). Process Philosophy. In E. N. Zalta (ed.), *The Stanford Encyclopedia of Philosophy*. http://plato.stanford.edu/archives/fall2013/entries/process-philosophy. Palo Alto, Stanford University.

Sober, E, and Wilson, D. S. (1998). *Unto Others the Evolution and Psychology of Unselfish Behavior*. Cambridge, MA: Harvard University Press.

Wilson, D. S. and Sober, E. (1989). Reviving the Superorganism. *Journal of Theoretical Biology* 136 (3): 337–56.

Wilson, J. (1999). *Biological Individuality the Identity and Persistence of Living Entities*. Cambridge: Cambridge University Press.

Wilson, R. A. (2004a). *Boundaries of the Mind: The Individual in the Fragile Sciences*. Cambridge: Cambridge University Press. http://amazon.com/o/ASIN/0521544947.

Wilson, R. A. (2004b). *Genes and the Agents of Life: The Individual in the Fragile Sciences Biology*. Cambridge: Cambridge University Press. http://amazon.com/o/ASIN/0521544955.

Wilson, R. A. (2008). The Biological Notion of Individual. In E. N. Zalta (ed.), *The Stanford Encyclopedia of Philosophy*. http://plato.stanford.edu/archives/fall2008/entries/biology-individual. Alto, Stanford University.

# 10

# From Organizations of Processes to Organisms and Other Biological Individuals

*Argyris Arnellos*

## 1. Introduction

Even though our world is continuously changing, contemporary thought is still dominated by a 'substance-' or 'particle-based' metaphysics, according to which static individuals composed by basic bits of matter constitute the world. Process philosophy opposes this view by suggesting that all things have to be conceived fundamentally as processes of various scales and complexity and with an inherent causal efficacy. Several thinkers (e.g. Christensen and Bickhard 2002; Ulanowicz 2009, 2013; Bickhard 2011a, 2011b; Dupré 2012) suggest that there are strong reasons for treating process as prior to substance in biology. Most of these suggestions are implicitly introduced as challenges to traditional assumptions. Issues of boundaries and individuation are among the first to be challenged. In a process ontology in biology, openness is the default. As Campbell (2009: 464) points out, the survival of a living system is due to the stability of its far-from-equilibrium processes as 'a function of their being *necessarily open* processes'. Considering biological entities as open processes that participate in various causal networks and life as a hierarchy of such processes, Dupré and O'Malley (2012b: 225) have suggested that *collaboration* should be reconsidered as a central characteristic of life and its evolution.

The consideration of open processes as the basic ontological category implies that living systems are abstracted manifestations of the continuous underlying substratum of biological processes. Consequently, there are no definite answers to questions such as what the boundary of fungi or of molds is, or to questions like how many individuals we count in a field of crab grass, because the answers to these questions depend on the presupposed criteria of individuation and of boundaries—if there are any (Bickhard 2011a; Dupré 2012). It follows from this that there are various ways of dividing living systems into biological individuals—a position that Dupré (2012: 241) has called 'promiscuous individualism'.

However, an excessive emphasis on collaboration, combined with promiscuous individualism, suggests something much stronger than the thesis that there is more than one way to define a biological individual. Ultimately, all attempts at individuation may turn out to be completely futile, since, 'overall, deep and extensive collaborations between biological entities blur—at the very least—any distinction between so-called individual organisms and these larger organismal groupings of which they are parts' (ibid., 222). In light of these claims, the central contention of this chapter is that the emphasis on the collaborative and collective dimensions of the living world runs the risk of overlooking the importance of individual biological organizations as the very conditions of possibility for the subsequent build-up of more complex collaborations in the course of evolution.[1]

A process-based metaphysics implies that living systems (such as cells, multicellular systems, ecosystems, organisms) are just temporary phenomena—the products of the dynamics of some processes. Nevertheless, regardless of whether one takes these manifestations as classes of 'topological knot persistences' with respect to various kinds of dynamical properties (Bickhard 2011b) or, more simply, as transient intersections (i.e. causal loci) of multiple processes that operate on different timescales (Dupré 2012), such manifestations are processes that show some temporal stability. Consequently, what a process-based metaphysics indicates is that the becoming and persistence of such stabilities should be explicated rather than be taken for granted (Bickhard 2011b). Accordingly, my main aim is to propose *an organizational account of the various biological stabilities* that arise from the nexus of continuous change of biological processes. Once this is in place, the living realm can be described in a way that considers the diverse characters of different types of biological organizations as a result of different collaborative schemes, but without necessarily having to subscribe to the pluralism implied by promiscuous (biological) individualism.

I begin in section 2 by arguing that the main implications of the thesis of promiscuous individualism combined with an excessively collaborative view of life are a vague definition of organisms and a blurred position with respect to (a) the distinction between organisms and biological individuals, (b) the consideration of microbial communities as organisms, and (c) the distinction between life and non-life. In section 3 I outline a process-based organizational ontology for biology in which biological individuality is grounded on a special type of highly integrated far-from-equilibrium self-maintaining organization of self-generating and self-reinforcing processes. In section 4 I apply this ontology to describe different schemes of collaboration in multicellularity and to argue that the form of unicellular (organismal) integration is not exported to all types of multicellular collaboration. In section 5 I draw on the findings of sections 3 and 4 to critically revisit the implications presented in section 2. I offer some concluding remarks in section 6.

---

[1] There should be no doubt that, as Dupré and O'Malley (2012b) have stressed, collaboration is one of the main dimensions of life and its evolution. Actually, even the concept of autonomy in biology (which places emphasis primarily on individual organisms) is theoretically and philosophically elaborated as entailing interdependence (through interactive openness) between an autonomous entity and its environment (see Rosslenbroich 2014; Moreno and Mossio 2015; Arnellos 2016). Therefore, my objective here is not to undermine the importance of the concept of collaboration in the description of the biological realm, but to suggest alternatives to some of its (in my view) undesirably pluralistic implications.

## 2. Some Implications of the Promiscuous Individualism Thesis Combined with a Hypercollaborative View of Life

Dupré (2012) has suggested that a process ontology is the most appropriate kind of ontology for thinking about life, organisms, and the organization of the biological world in general.[2] The main claim is that the biological realm consists of processes rather than things. The entities that we commonly represent or model as thing-like, such as organisms and other biological individuals, are just particular time slices of their life cycles, which Dupré takes to be a more basic reality (ibid., 2). Each one of those individuals is just an abstraction from the underlying causal nexus of biological processes. They are not necessarily bounded and, when they are, they can have different boundaries that result from the very dynamics of the related processes.

In a way, such 'processual' biological systems are inherently open (Campbell 2009), in that they are prone to *collaboration* (in a sense that entails both selfish/competitive and cooperative interactions) with other systems. Dupré argues that the omnipresence of collaboration between microbes (through lateral gene transfer) and between microbes and multicellular systems (through symbiosis) threatens the traditional view of an organism as a monogenomic biological individual and as the only, or even the central, unit of selection. Instead, one should consider biological entities as processes collaborating in various causal networks, and life as a hierarchy of such processes (Dupré 2012: 223, 225, 227).

Dupré also advocates a pluralism with respect to drawing individual boundaries; he argues that 'drawing boundaries round biological objects is to an important extent a matter of human decision driven by particular human goals, practical or theoretical' (ibid., 241). There are various ways of dividing living systems into organisms, according to the goals of the observer—a position Dupré calls 'promiscuous individualism' (ibid.). These ideas have several wider implications.

The first implication is that the position results in a rather vague definition of organisms, as well as in a blurred position with respect to the distinction between organisms and biological individuals. Dupré (ibid., ch. 13) stresses the inadequacy of the 'monogenomic differentiated cell lineage' (MDCL) view of organisms. In accordance with his thesis of promiscuous individualism, Dupré suggests that the correct answer to the question of what an organism is 'requires seeing that there is a great variety of ways in which cells, sometimes genomically homogeneous, sometimes not, combine to form integrated biological wholes' (ibid., 88). Some might say that this is not very informative, since there is no definite list of concrete criteria that the concept of an organism should satisfy. But this is exactly the reason why Dupré and O'Malley (2012a) urge that the classical concept of organism is a problematic biological

---

[2] The issues discussed in this section are based on the work of Dupré (2012), who has so far offered the most inclusive discussion of the conceptual implications of the adoption of a process view in biology.

category. And although in several different cases Dupré suggests, and even favors, a 'functional organism' concept (see e.g. Dupré 2012: 124–5), he ultimately stays consistent with promiscuous individualism and does not provide a unique definition. For Dupré, 'the omnipresence of symbiosis should be seen as undermining the project of dividing living systems unequivocally into unique organisms' (ibid., 8). This is the reason why 'what is an organism, and whether something is a part of an organism or not, are not questions that admit of definitive answers' (ibid., 153). Due to extensive collaboration in the biological world, there are various ways to draw individual boundaries that reflect real biological aspects of the multiply symbiotic systems that make up the biological realm. This is not just an epistemic view. For Dupré, 'ontological boundaries are relative to the issues with which we are concerned, which is a central part of the reason why there is no unique ontology' (ibid., 97). The analogy with Dupré's promiscuous realism makes this also an ontological view (see also Wilson 2012). Accordingly, since there are multiple biological individuals out there to classify, the question of how many biological individuals are out there neither has nor requires a definite answer. In all, Dupré's definitions of an organism are so broad that there is no firm basis for distinguishing organisms from other biological individuals.

A second implication is the consideration of microbial communities as multicellular organisms. This is also implicitly related to the inadequacy of the MDCL point of view of organisms. O'Malley and Dupré (2007) argue that the MDCL conception prevents us from identifying microbial communities as *bona fide* organisms due to the polygenomic character of such communities. Moreover, for Dupré and O'Malley biofilms are much more than simple aggregations of individuals; they are self-organizing entities that operate as functional units. This being the case, they suggest that the only reason for not considering microbial communities as multicellular organisms is that the classic understanding of multicellularity is exclusively based on knowledge of multicellular eukaryotes.

A third implication is the blurring between life and non-life, mainly due to the highly important role of collaboration in life and its definition. Dupré and O'Malley (2012b) see a tension between two widely discussed criteria for life: replicating lineages and metabolic self-sustainability. They suggest that this tension can be overcome once we decide to see life as occurring at the intersection of lineage formation and the (collaborative dimension) of metabolic processes. In this respect, biological entities such as viruses, prions, and plasmids, which are problematic under other frameworks and ontologies for life, should be considered alive when actively collaborating in various metabolic processes. This blurs the distinction between life and non-life and also drops the commitment to an exclusively cellular view of life (ibid., 227–8).

These conceptual and theoretical implications are crucial for our understanding of the living. My goal in this chapter is to clarify these issues by adopting an organizational perspective with respect to the diverse characters of the biological realm. Specifically, I intend to show that a process-based organizational ontology in biology can offer clearer and more explicit alternatives to all these issues, in a way that avoids the problems of pluralism without ignoring or undermining the collaborative nature of the biological realm.

## 3. A Process-Based Organizational Ontology for Biology

### 3.1. Simple self-maintenance

The basic assumption in the study of complex systems is that a certain part of the world (usually, a set of elements and their interrelations) constitutes a system, and that the rest is the environment. As discussed in section 2, promiscuous individualism rejects the possibility of individuating a system in a single, unique way. The identity of the system (the relevant inside–outside dichotomy) would always be a matter of a particular point of view. However, there are systems that draw their own distinctions from the environment; they create their own identity. There should then be a way to explain the formation of this identity so that it may be not entirely dependent on the relativist position of an observer.[3] The adoption of an organizational perspective, which focuses on the current causal relations between the processes constituting a system, is suggested as the basis for specifying the identity of a system in such a way (Bickhard 2000; 2004; Moreno and Barandiaran 2004). Let us elaborate more on this.

Processes exist in relation to other processes, in an organization of processes (Bickhard 2009, 2011a, 2011b; Campbell 2009).[4] Among them, there are some that work together so that they constitute *cohesive systems*, that is, systems that persist and manifest *a form of stability* in the sense of a spatio-temporal integrity—unlike the case of a quantity of gas (Campbell 2009).[5] Some organizations of processes manage to maintain such cohesion in appropriate *far-from-equilibrium* conditions.[6] A well-discussed example is that of a candle flame (Bickhard 2000). The microscopic reactions of combustion generate the flame (a macroscopic pattern), which contributes to the maintenance of the conditions for its own existence by constraining[7] the surroundings (temperature, wax, oxygen) and turning them into appropriate boundary conditions required for its own maintenance. As long as the combustion feeds back to its boundary conditions, the whole organization is self-reinforced and the flame will stably persist. Through its dynamics, the flame manages to maintain its

---

[3] The idea that systems can be individuated completely separately from the situated position of the observer is in principle dismissed. However, there are many causal properties that can serve as a basis for a principled specification of system identity and that are not entirely observer-dependent (Collier 1988; Collier and Hooker 1999; Christensen and Bickhard 2002; Campbell 2009).

[4] Some organizations of processes persist more than others in an environment. It is in the context of a persistent (and relatively robust) organization of processes that collaboration—either in the form of cooperation between processes in the organization or in the form of competition between old and new processes for participating in the organization—could be considered inherent in a process-based metaphysics.

[5] Cohesion is defined as the causal closure of the relations among 'elements' that constitute a 'thing' in a way that keeps it (within a limited range of conditions) from being disrupted by internal and external forces (see Collier 1988, Collier and Hooker 1999, and chapter 4 for a detailed treatment of the concept).

[6] Chapter 7 also discusses far-from-equilibrium systems and their relevance to process views of biology.

[7] Constraints are macroscopic structures that harness microscopic processes by reducing their degrees of freedom, that is, by 'ordering' them (Hooker 2013).

own process of burning. It is in this respect that we can account for the generation of an *identity* as the consequence of a set of far-from-equilibrium processes—which are stably maintained, given certain boundary conditions—and not entirely as a result of an observer's description. A candle flame is a form of self-constitution of the very identity of the system: a cohesive organization of processes that, given adequate initial and boundary conditions, contributes to its own maintenance. Such an organization of processes constitutes a *self-maintaining system*.[8]

At this point, the crucial question is whether the ontology of such organization of processes is sufficient for biology and, as Moreno and Barandiaran (2004) put it, whether self-maintenance is enough for the formation of a genuine in–out dichotomy. As Campbell (2009) points out, a candle flame is a system that contributes to the persistence of the conditions upon which it depends, but the constitutive complexity that enables the candle flame to self-maintain is not internal to the flame itself. There is no doubt that the candle flame is a complex of processes—a complex that actively contributes to its own persistence. However, its persistence is much more dependent on the environmental conditions (the candle and the atmosphere) than on the flame itself, to the extent that the flame cannot do anything to maintain its organization in the face of an abrupt shifting of the conditions of its environment (e.g. running out of wax, suffering a decrease in oxygen availability). As the flame is a cohesive organization of processes that persists far from thermodynamic equilibrium, its openness to the environment is an ontological feature of the system, and not only of our apprehension of it. However, the high degree of explicit and immediate dependence of the flame on its boundary conditions makes this openness so direct that, as pointed out by Bickhard (2011a), there is actually no boundary at which the flame can be isolated, for instance. The flame and its environment are so intertwined that they cannot be distinguished from each other. There is no genuine in–out dichotomy.

### 3.2. Minimal recursive self-maintenance and biological individuality

Nevertheless, one might still suggest that the (minimal) self-maintenance exhibited by the candle flame could be sufficient for defining biological individuality. It has been argued in detail in various studies that this is not the case (Moreno and Ruiz-Mirazo 2009; Mossio and Moreno 2010; Arnellos and Moreno 2012; Moreno and Mossio 2015). The main argument is that functions begin with biology.[9] As I will try briefly to explain, simple self-maintenance is a necessary but not a sufficient condition for grounding functions in the organization of a system. In principle, a self-maintaining system contributes to the existence of its organization. So the reply to the question 'Why does the flame exist?' is: 'Because it contributes (in the way described above) to its maintenance.' In other words, the flame plays a direct causal role in the dynamics responsible for its maintenance. The conditions of existence of

---

[8] As pointed out by Campbell (2009), the ability of a system of processes to be self-maintaining is an emergent causal power of its organization and not of its constitutive processes, which are constantly altered through participation in the organization.

[9] Functions are indispensable to biology. This becomes apparent as soon as one attempts a functional-free description of any biological, or even protobiological, system.

the flame are the norms of its activity, and what the flame does is relevant for its existence. Therefore one can say that self-maintaining systems exhibit a minimal organizational closure (Mossio and Moreno 2010). It is in this sense that Christensen and Bickhard (2002) suggest that any such contribution (e.g. of the flame to its own maintenance) *serves* a function relative to the organization of the processes whose existence it contributes to maintaining. But, despite the intuitive meaning of such an assertion, what is the explanatory (added) value of ascribing functions to the candle flame? In the case of a candle flame, as in all physical dissipative systems, there is a single type of constraints—the macroscopic pattern of organization, in other words the flame—that constrains its surroundings by turning them into appropriate boundary conditions for its maintenance. It doesn't matter how materially complex the system is, or how many constraints we can indicate; what matters is that in simple self-maintenance (i.e. at the level of minimal organizational closure) all constraints are of the same type, thereby producing the same effect.[10] So, if there is no way to distinguish between different contributions to a self-maintaining organization, it makes no sense to ascribe functions to these very contributions.

If we want to speak of functions in a system, *differentiation* (both in terms of the processes of the system and in terms of their specific contribution to its global far-from-equilibrium self-maintaining dynamics) seems to be crucial (Mossio et al. 2009). For this reason, we need to move to higher levels of organizational complexity than the one of minimal organizational closure, which involves only cyclic processes and reactions. But what would the characteristics of such an organization be? As it has been suggested elsewhere (Moreno and Ruiz-Mirazo 2009; Arnellos and Moreno 2012), functional differentiation could emerge in a scenario where a set of different types of constraints come together in a mutually reinforcing manner, thereby establishing a more robust type of self-maintaining dynamics.

An example of such a system would be a prebiotically plausible case of cellular proto-metabolism, where the self-maintenance of the system's organization of processes is a consequence of the interplay between different types of constraints. In this scenario certain processes selectively constrain the low-level microscopic behaviour of different networks of processes in such a way as to enable them to generate more stable macroscopic patterns, which in turn constrain (also selectively) another collection of processes, leading to the production of another macroscopic constraint, and so on—provided that all of them, together, end up depending on one another and the whole organization of processes closes itself recursively.[11] In such an organization of processes, diverse constraints (e.g. the cell membrane and the enzymes) mutually enable their continuous regeneration. By internally synthesizing (at least part of) its own constraints, the system becomes capable of performing a

---

[10] The same (degree of) organizational complexity can also be met in chemical self-maintaining systems such as autocatalytic cycles (Kauffman 2000). The cycle is not maintained only by the external boundary conditions, but also (although in a minimal sense) by catalysts, and the catalysts are maintained by their own action (minimal organizational closure). The point here is that an autocatalytic cycle internalizes many constraints of a single type (catalysts).

[11] In a cell, the membrane and the enzymes constrain the microscopic dynamics of the molecular flows produced by chemical reactions in different ways (see Arnellos and Moreno 2012 for a relevant analysis).

diversified modulation of its own self-maintaining dynamics. And it is in this organizational context that different constraints can be taken to make *distinguishable contributions* to the global self-maintenance of the system, thereby mutually enabling their continuous regeneration and thus serving a function relative to the specific organization of processes (a form of *functional interdependence*). In principle, this is the basic or core type of organizational closure realized by biological systems: a type of robust self-maintenance.

Do systems of processes that exhibit this type of organizational closure achieve a 'genuine' in–out dichotomy? The appearance of cellular systems at large that build their own boundary with the environment introduces the level of organizational complexity necessary for a clear distinction between the inside and the outside of a system. In the organizational framework we have discussed so far, such systems should be considered *biological individuals*.[12] We can distinguish two aspects: (a) the organization of processes of the system and its environment;[13] and (b) the constitutive processes that occur within the system's physical boundary, including those of its construction and maintenance, together with the interactive processes of the system that occur outside of it. The constitutive processes are organized so that they constrain the flow of matter and energy between the environment and the system, and especially the interactive processes, in such a way that certain relations of the system to the environment hold in order for the identity of the system (as an organization of processes) to be both defined and maintained. Therefore, *the ontological requirement for biological individuality (a clear far-from-equilibrium in-out dichotomy) is an organization of processes that achieves closure on the basis of two kinds of process: constitutive and interactive.*

What is of importance for our discussion is whether this type of self-maintenance of a system of processes is sufficient to describe *the organizational complexity of present-day biological systems*. In cases of cellular proto-metabolism, a plausible scenario is that the constraining of interactions was presumably happening in the form of simple homeostatic reactions implemented via feedback relations through the dynamics of the constitutive processes (see e.g. Ruiz-Mirazo and Mavelli 2008). The constitutive processes would not only produce and maintain part of the semi-permeable membrane of these cells, but also modulate it so that it would be able to mediate different relations of the system to the environment. This kind of process stability would require mutual dependence between different constraints capable of compensating against a range of perturbations by virtue of the plasticity of their dynamics and the resulting global constitutive organization. This kind of biological individuals would overcome certain variations. However, the nature of the interactive processes would not be different from the mere physico-chemical reactions happening all the time across the membrane of all cells. As pointed out by Moreno and Mossio (2015), this type of (constitutive) stability would be limited, since it wouldn't allow the exploration of other regimes of closure by the system

---

[12] Self-reproductive capacities are presupposed in the prebiotic scenario we discuss. In case they were not, such protocellular organizations would be what Wilson (1999: 60) describes as a *functional* (biological) *individual*. See section 5 for further discussion.

[13] The internal organization appears significantly more integrated with respect to its environment than a borderless autocatalytic network or a candle flame.

(provided that there are available ones). Such an organization of processes can't actively change its environment in such a way as to use it again for its viability, nor can it leave its environment for another one. For these systems, the environment is just a source of indistinguishable perturbations. Organizations of processes that manage to maintain cohesion within a small range of changes of very local conditions can be said to exhibit *limited recursive self-maintenance*.

## 3.3. Recursive self-maintenance and the organismality of unicellular organizations

All present-day unicellular organizations have the capacity to adapt to their environments to some extent by switching between different regimes of closure (achieved by their organizations of processes) so as to contribute to their self-maintenance, and thereby to their self-reproduction, in response to some changes in the conditions they detect in the environment.[14] A typical, well-discussed example is the bacterium *Escherichia coli* and its capacity for chemotactic interactions (Campbell 1974). When the bacterium is close to a sugar gradient, it will swim up this gradient—instead of the usual tumbling movement within periods of not encountering increasing concentrations. Swimming and tumbling are two different interactions that are both *functional* in the sense that they contribute to the bacterium's self-maintenance in different conditions, and the bacterium can usually *switch* between them accordingly. The effects of these interactions—materialized as changes produced at the system–environment interface (e.g. more sugar)—become inputs for the coordination of interactions, and consequently for the modulation of the maintenance of the whole bacterial organization. This is an organization that manages to maintain its cohesion in substantially different environments. It achieves self-maintenance across a (relatively) wide range of changes of environmental conditions; in other words, it exhibits *(genuine) recursive self-maintenance* (Bickhard 2004; Campbell 2009).

What are the organizational characteristics of such systems of processes? Given that the appropriate physico-chemical conditions are not always immediately available, robust self-constitution (reproduction and maintenance) involves not only the control of metabolic processes (for instance through modulation of the metabolic organization—e.g. the lac operon mechanism) but also the control of the production, maintenance, and modulation of boundary conditions through the regulation of the system's interaction with the environment.[15] We have mentioned the capacity of *E. coli* to switch between different interactions (swimming and tumbling). It would be rational to assume that 'switching' between different sets of self-maintaining processes needs some sort of infrastructure that can make the relevant shifts in relation to the detected environmental input. As a matter of fact, in *E. coli*, the operation of the two-component signal transduction (TCST) subsystem of processes relates the metabolic regime with the two types of motility-based interaction in a way that

---

[14] In this respect, all unicellular organizations are considered *evolutionary (functional) individuals*.
[15] Cellular reproduction is not just controlled through genetic replication. There is a key role played by metabolism and, consequently, by the ongoing interactions with the environment in the cell division cycle.

modifies boundary conditions so that metabolic maintenance is ensured.[16] We can leave aside the physico-chemical details of the processes involved; but the organizational characteristics of the TCST are important. As all types of functions in biological systems, the TCST can be considered a set of constraints that act in the bacterium. But, contrary to the metabolic constraints (e.g. enzymes) operating on basic thermodynamic processes, the TCST operates by coordinating the action of other functional constraints. Specifically, the operation of the TCST results in the coordination between membrane receptors and motor mechanisms. And, although this coordination is generally mediated by metabolic pathways of the bacterium, the whole process is organized so as to operate over the basic metabolic functioning of the bacterium, and with dynamics that exhibit a degree of *decoupling* from those of the basic metabolic organization; otherwise the maintenance of the bacterial organization could be disrupted.[17] So, in the organizational framework we are considering, we could say that switching in *E. coli* is organized on the basis of 'second-order' constraints that operate on constitutive constraints while being decoupled from them. Importantly, as suggested by Bich et al. (2015), in such organizations we already find all that is required for ascribing *regulatory functionality* to the set of processes that control the activation of its various self-maintaining processes according to environmental changes. In this respect, the TCST operates as a *regulatory subsystem that functionally coordinates the bacterium's interaction with the environment.*[18]

Chemotaxis can thus be considered a form of higher-order control of bacterial metabolism. Accordingly, a chemotactic interaction (and its regulation) is functional as long as the selected action (tumbling or swimming) satisfies the endogenously generated *norm* associated with the maintenance of the bacterial organization, namely that the constitutive processes must occur as they do in order for the bacterium itself to exist (Barandiaran and Moreno 2008). The nature of chemotaxis as well as of other motility-based interactions performed by such organizations is different from the nature of the interactions performed by limited, recursively self-maintaining organizations. Chemotaxis is more than a mere physical interaction, as the biological individual's own viability is affected by it. Chemotaxis is a functional action on the environment, in the sense that it is performed outside the system's own boundaries and it actively modifies the conditions of the system–environment relation in such a way that the whole bacterial organization can use the new conditions for its own maintenance and, consequently, for its reproduction. The environment is no longer simply a source of indistinguishable perturbations. On the contrary, the cell itself (its identity) determines what is relevant for its maintenance. In this way the interactive dimension of the organization can be viewed as a function of its constitutive normativity.

---

[16] The TCST mediates the temporal detection of differences in the concentration of environmental nutrients to flagella's motor output in a global way that enables adaptation (Bijlsma and Groisman 2003).

[17] This is characteristic of both metabolism-independent and metabolism-based chemotaxis. For details, see Alexandre 2010.

[18] Overall, *E. coli* chemotaxis is regulated by the TCST as the selective choice of a subset of particular metabolic pathways among the available repertoire (see van Duijn et al. 2006 for a relevant discussion).

From an organizational perspective, the endogenously produced regulation that dynamically coordinates the constitutive with the interactive processes entails a particular form of functional integration that seems to characterize all present-day cells.[19] Extant unicellular entities are organized on the basis of functional coordination, which in turn requires multiple sets of functional interdependence. This form of integration entails that cellular organizations show a functional and reciprocal relationship between interactive processes (and the regulatory subsystem that coordinates them) and the overall constitutive organization that supports this subsystem materially and energetically. This form of integration requires a type of closure and mutual interdependence between constitution and interaction that is realized via the endogenous production of regulation for both self-construction and self-maintenance, through functional interactions with the environment. This form of integration is characteristic of unicellular, (genuinely) recursively self-maintaining organizations; in other words, it is characteristic of a basically autonomous biological organization (Ruiz-Mirazo and Moreno 2004). An ontological consequence of this form of integration is that, ultimately, in such a biological individual *the confluence between metabolism and reproduction is organized in such a way that it is not really possible to separate the individual's 'being' from its 'doing'*. It is in this respect that such a unicellular organization is considered an *organism* (Ruiz-Mirazo et al. 2000; Arnellos and Moreno 2016).

## 4. Collaboration and Multicellular Systems

As argued in section 3, different types of collaboration among processes result in different types of self-maintaining organizations. One of the conceptual challenges of multicellularity is that the organizational variety of the diverse types of cellular associations is hidden under their common capacity for adaptation. All types of multicellular (MC) collaborations appear to exhibit the same form of integration among their unicellular constituents. Indeed, all MC collaborations manage to generate and maintain relatively cohesive forms of integrated organization. This integration—the various functional interactions between the cells—enhances the overall adaptive capacity of MC associations, as they can occupy new niches and increase the possibilities of survival of the constituent units, as well as of the association as a whole. Thus, considering functional coordination for adaptation as the main criterion for organismality,[20] all MC systems—at least from a phenotypic point of view—seem to be integrated as *individual organisms*.

Elsewhere I have argued that this is not a secure position, as it does not enable us to distinguish different characters of multicellularity and their respective organizational basis (Arnellos and Moreno 2016). Below I elaborate on this claim by considering

---

[19] All bacterial chemosensory systems are variations on the chemotaxis TCST (Kirby 2009). Eukaryotes such as paramecia detect nutritious gradients in the same way, but they use detectors at both of their ends, and then compare the inputs.

[20] The common conception is that an organism is an integrated biological entity, spatially separated from others and made out of interdependent parts that are integrated so that they work in coordination with each other for the proper function of the organized whole (Wilson 1999).

examples of bacterial, early eukaryotic, and early eumetazoan multicellularity. Specifically, I will apply the organizational ontology sketched in section 3 to distinguish different schemes of collaboration in multicellular systems and to assess which types of MC organization can be said to exhibit the form of organismal integration found in unicellular organizations. As I will argue, the form of unicellular integration does not manifest itself in all types of MC collaboration and, when it does, there are still important differences with respect to the unicellular case.

## 4.1. A case of collaboration among single-species bacteria

Collaboration among *Myxococcus xanthus* bacteria is intense and results in a collective form of motility, which provides the MC system with the ability to engage in 'wolf-pack'-like hunting and feeding (Berleman and Kirby 2009).[21] Independently of the intensity of collaboration between the myxobacteria, the integration between the motility of the swarm and its constitutive organization is loose. Swarming is a mere result of self-organizing chemotactic movements, and the collective interaction is the emergent net effect of numerous local interactions (Zhang et al. 2012). As argued in Arnellos and Moreno 2015, swarming exhibits no coordination attributable to the whole MC system. On the contrary, each individual cell participates in the swarm by coordinating its local chemotactic interactions with its immediate environment through its own regulatory organization (see Arnellos and Moreno 2015 for details).[22] Actually it is the richness of the environment that maintains the swarm. The cells will stop moving in a swarm when nutrients are exhausted. This is because, at the swarm level, there is no functional differentiation; each bacterium executes the same chemotactic interactions. There is therefore no collaboration on the basis of functional interdependence. Interestingly, also during 'fruiting body' formation, where some cells differentiate into spores and some others retain their rod shape, the whole system is frozen and immotile, waiting to be carried to a nutrient-rich environment by a passing animal. In general, associations of cells (e.g. of myxobacteria or of slime moulds) cannot at the same time be motile and participate in the construction and maintenance of their fruiting bodies.

So we see that, despite the apparent collaboration in single-species biofilms, not only is there no functional coordination at the MC level, but the whole integration is so loose that the MC organization's constitutive identity does not require interactivity,[23] while the organization itself can even be reversed to the stage where the constituents may disperse and exist autonomously as unicellular individuals. This implies that this

---

[21] *M. xanthus* cells form structured single-species biofilms with motility-mediated expansion (formation of tentacle-like packs, cell groups, and synchronized rippling waves of oscillating cells) when other microbial nutrients are available in the environment, and massive spore-filled aggregates that rise upwards from the substratum to form fruiting bodies, mainly when exposed to low or no nutrients.

[22] Within the swarm, individual cells are constantly moving, transiently interacting with one another, and independently reversing their gliding direction (Kaiser and Warrick 2011). Cells alignment and the formation of clusters could even be due to pure mechanical interactions among cells and between the cells and the substrate (Balagam and Igoshin 2015).

[23] The interactive activity is not even compatible with the constitutive one, in the sense that the fruiting body of the biofilm cannot be motile.

type of collaboration between unicellular prokaryotic organizations does not reflect the form of unicellular integration at the MC level.

### 4.2. A case of early eukaryotic collaboration

A well-studied case of relatively simple eukaryotic collaboration is *Volvox carteri*. In its adult stage, this MC alga normally consists (a) of almost 2,000 biflagellate, terminally differentiated somatic cells engaging in phototaxis and (b) of sixteen germ cells, which are non-motile but can grow through photosynthesis and reproduce (Kirk 2005). The way interaction is organized in *V. carteri* is qualitatively different from the one found in biofilms (see Arnellos and Moreno 2015 for details). There is no known direct communication among the somatic cells (Ueki et al. 2010). So, without considering the anatomical characteristics of the spheroid, the whole organization of interaction is of the type of swarms; each somatic cell swims according to its own detection of the local environment. However, because of the morphological and anatomical constraints introduced during development (the spheroid's polarity and asymmetry, combined with the immersion of all cells in the extracellular matrix; a proper orientation of the cells with respect to one another; and the reorientation of the flagella, beating in each cell towards the always heavier posterior), the collaboration in *V. carteri* is tighter and much stricter than in the myxobacteria. This structural arrangement is necessary for the proper execution of phototactic swimming and will stay unaltered during the MC alga's lifetime. It could be considered collaboration on the basis of functional interdependence. The movement of the flagella of each somatic cell requires the movements of the other cells, in the sense that proper phototaxis is the net effect of all cells' movements. One could argue that the existence of structural components spanning the whole MC system that ensure its adaptive interaction is a form of global coordination. Still, although the alga's swimming is adaptive (it manages to stay or move to euphotic conditions), *there is no central functional coordination in the sense of a regulatory system (like the TCST) that would switch the MC system between two different organizational regimes.*[24]

In accordance with the account we have discussed, *V. carteri* exhibits a more integrated organization than biofilms. Contrary to *M. xanthus* biofilms, the alga's constitutive identity is compatible with its interactive dimension, and its MC organization cannot be reversed to the unicellular stage of its constituents because of the absolute germ/soma division of labour. However, compared to the organismal integration of a cell, the integration achieved by this MC alga is weak. Isolated germ cells of *V. carteri* would still grow and divide under euphotic conditions (Koufopanou and Bell 1993). This means that the constitutive identity of the MC alga could be reproduced and maintained even without its interactivity; in other words, unlike unicellular organizations, the alga's relatively weak interdependence between constitution and interaction is not indispensable for the development and

---

[24] There is just one self-maintaining process organization in *V. carteri*: that of swimming. Turning (changing direction) is just the net effect of somatic cells moving their flagella (swimming) differently as a result of detecting different light intensities due to their being placed at opposite sides in the spheroid (see Ueki et al. 2010 for details).

maintenance of its organization. This type of early eukaryotic collaboration results in a recursively self-maintaining organization but it still does not achieve the form of integration its constituents entertain when in their unicellular form.

## 4.3. A case of eumetazoan collaboration

As I have argued, *different types of cellular collaboration result in different forms of integration for MC organizations*. This has implications for the constitutive and interactive characteristics of the cellular collaboration as well as for their interrelations. I have recently suggested that the level of minimal organizational complexity necessary and sufficient for the appearance of a MC form of integration analogous to the one exhibited by unicellular systems is met in eumetazoan organizations (Arnellos and Moreno 2015, 2016). Such organizations can deploy several different interactions with their environment in order to contribute to their self-maintenance.[25] This is possible due to the existence of the nervous system (NS)—a specialized subsystem that exerts a fine-tuned control on the numerous different processes responsible for adaptive behaviour in a given environment. However, such a subsystem is functional only in the context of specific body plans adapted to various interactions. The NS is integrated in a body plan with a set of primitive and differentiated organs, which provide the MC animal with the metabolic and biomechanical requirements for its behaviour. Besides its role as a controller of behaviour via the neuroendocrine system, the NS also regulates the development and maintenance of the metabolic processes of the body by which it is also being developed and maintained.[26]

What is of importance for our discussion is that adaptive behaviour requires unified body movement (Keijzer and Arnellos 2017). And this in turn requires functional coordination with metabolic and even developmental processes. In eumetazoan organizations, each one of the various subsystems of processes operates according to its local norms. Consequently, the control of eumetazoan behaviour cannot be achieved without functional coordination of all its different local regulatory subsystems. What is required is higher-order integration. This is what a regulatory centre provides: it functionally integrates all local norms according to a higher-level normativity (Arnellos and Moreno 2015). This is precisely the role of the NS in eumetazoa. It does not only regulate contractile epitheliomuscular tissues that generate sensorimotor interactions; it also regulates processes of development, growth, and global homeostasis (Jekely et al. 2015).

It is in this respect that such MC organizations exhibit a form of functional integration analogous to that of a unicellular organization. The MC organization's constitutive identity (encompassing both the capacity for self-reproduction and the capacity for self-maintenance) is so strongly entangled with its interactive dimension (the actions in the environment) that one cannot exist or be explained without the other. Nevertheless, in such a MC organization, the organismal form of integration is not realized in a way that is totally symmetrical to the one found in unicellular

---

[25] For instance, a jellyfish can swim fast when escaping predation and slow when eating or migrating.

[26] Neuropeptides are abundant in jellyfish and play an important role in regulating a variety of developmental and physiological processes (Hartenstein 2006).

entities, since, in accordance with the organizational framework I have proposed, unicellular entities are organized on the basis of a different form of functional integration from the one exhibited by a eumetazoan collaboration. Overall, a MC organization exhibiting such a form of integration can be considered *organismal* (Arnellos and Moreno 2016).[27]

## 5. Revisiting the Implications of Promiscuous Individualism and an Excessively Collaborative View of Life

In section 2 I discussed some implications of the thesis of promiscuous individualism and the collaborative view of life that accompanies it. I will now try to critically assess those implications and to suggest specific alternatives based on the analysis in section 3 and the results of its application in section 4.

*5.1. Organisms and other biological individuals*

The first implication we noted was that it leads to a rather blurred definition of organisms and of the relation between organisms and other biological individuals. Let's start with the concept of the organism as a MDCL. This is an idealization of organisms on the basis of a false presupposition that all constituent (and differentiated) cells have the same genome (see Buss 1987). Dupré questions but does not dismiss the MDCL view. For the sake of promiscuous individualism, he states: 'I have not wanted to say that the MDCL is an erroneous conception... The mistake is to think that it involves a discovery of what the organism really is, and must therefore be the right conception for all purposes' (Dupré 2012: 241). From the organizational point of view, Dupré is right. The requirement for a biological organization to qualify as an organism is an indispensable closure between its constitutive and its interactive dimensions. In this respect, a biological organism need not be a MDCL.

For instance, let us take the case of the symbiosis between pea aphids and the bacterium *Buchnera aphidicola* (Moran 2006).[28] First of all, this is an obligate endosymbiosis, where the bacterium resides in a specialized compartment inside the cytoplasm of aphid cells and is transmitted vertically, through maternal eggs, roughly in the same way mitochondria are transmitted from our mothers. Second, this is more or less a case of genetic mosaicism, except that not all cases of mosaicism have lethal consequences when they are not satisfied. Third, it seems there are no aphids that do not have the *B. aphidicola* bacterium. For the aphid, the bacterium is just another cellular part, another component process of the whole aphid organization. It is more in the context of the putative application of the MDCL concept that

---

[27] From this perspective, plants qualify as basically organismal, since their interactive dimension coincides with the constitutive one (see Arnellos and Moreno 2015 and 2016 for more detailed discussions).

[28] The problems symbiosis imposes on aspects of individuality and organismality deserve a separate treatment. Here I aim to provide some preliminary directions and ideas to deal with this phenomenon from the organizational perspective. The implications of symbiosis for the prospects of a process ontology for biology are also considered in chapters 1, 5, 9, and 15.

this particular biological organization is considered symbiotic than on the basis of fact that the collaboration is between a eukaryotic and a prokaryotic individual. In any case, as I have argued, organismality is a property of specific biological organizations, and the organization of the aphid (including the bacterial symbiont) satisfies the pertinent requirements. So Dupré is right to suggest that, in light of the dependence of MC organisms on diverse sets of symbiotic microbes for their successful differentiation as well as for their survival, the MDCL assumption should be questioned.

Keeping in line with promiscuous individualism, Dupré (2012: 88) suggests that there are various ways in which cells (homogeneous or not) collaborate to form integrated wholes. This is not very informative. As we have discussed, all MC collaborations seem to exhibit a degree of integration. Moreover, it is important to note that there are MDCLs that form integrated wholes but nevertheless do not qualify as *bona fide* organisms. For example, *V. Carteri* is a case of a MDCL (it achieves alignment and export of fitness, see Folse and Roughgarden 2010) yet, as I argued in section 4, it does not qualify as an organism (see also Arnellos et al. 2014 for the same argument from a complementary perspective).

Dupré (2012) avoids providing a single definition for an organism. On several occasions he seems to suggest a 'functional' organism concept: 'we might also want to approach the question of what constitutes an organism from a functional perspective: what are the systems of cells that interact with the surrounding environment as organized and generally cooperative wholes?' (ibid., 152; see also 172, 203). And, quite contrary to what promiscuous individualism would prescribe, he seems to be taking organisms 'that interact functionally with their biological and non-biological surroundings' to be more fundamental than 'organisms that are parts of evolutionary lineages' (ibid., 124-5). Dupré also uses a functional individual concept according to which organisms collaborate in symbiotic wholes, forming functioning biological individuals (ibid., 9). In principle this is not wrong. However, it seems too generic to capture the diverse characteristics of the biological world. Although Dupré is clear that we should not equate organisms to biological individuals (ibid., 207), promiscuous individualism does not inform us how and in what particular contexts the notions should be distinguished.[29]

In the organizational view we have discussed, several schemes of collaboration of biological processes result in the constitution of a system's identity, but each constitutive organization is not necessarily integrated in an organismal way. In this respect, all organisms are also biological individuals, but not all biological individuals achieve organismal status. Wilson (1999: 89) suggested that a biological entity is a *functional individual* when it is composed of causally integrated heterogeneous parts in such a way that the entity would (typically) suffer impaired function if some of its parts were removed or damaged. Very few authors would disagree with this definition. And, even if Wilson emphasizes the current causal relation between the parts of a whole,

---

[29] An immediate predicament is that, if organisms are integrated biological wholes (i.e. functional individuals), then in what sense does this differ from the standard view of organisms (organisms are unicellular entities, microbes, and multicellular ones, as Dupré mentions), assuming that we do not reject polygenomic wholes?

this definition could be modified to include the evolutionary dimension of individuality, resulting in what has been described as the *reproducible–functional (evolutionary) individual*. This conception includes the capacity of an organization to reproduce itself and to be maintained in its environment (see also Arnellos and Moreno 2016).

According to the analysis offered here, MC organizations such as the one exhibited by *V. carteri* do not achieve organismal status, but are nevertheless evolutionary individuals. Each cell type needs the other for the whole colony to be maintained and to reproduce its organization. Somatic cells are necessary for the spheroid to be moved into euphotic conditions and germ cells are needed for the reproduction of the colony. An inverse case is the squid–*Vibrio* non-obligate symbiosis (McFall-Ngai 1994). From the organizational point of view, the squid–*Vibrio* organization qualifies as a functional individual on the basis of functional interdependence with respect to the bearer of the trait of bioluminescence. However, since self-reproduction is a capacity only of the squid and of the bacteria and not of the symbiotic organization, the latter should not be considered an evolutionary individual.[30]

## 5.2. The individual and the organismal status of microbial communities

The second implication I outlined pertains to the organismal status of microbial communities. Dupré (2012: 88, 152) begins by suggesting that the general conception of microbes as single-celled organisms should be revised, since most of the time microbes do not function simply as isolated individuals but rather in complex associations, often composed of highly diverse kinds of cells (e.g. the biofilms on the surfaces of our teeth). For O'Malley and Dupré (2007), these microbial communities are self-organizing entities operating as functional units. Therefore, once the MDCL concept is out of the way, the only reason for not regarding microbial communities as MC organisms is that 'the definition of multicellularity is closely based on knowledge of multicellular eukaryotes' (ibid., 176). Let us examine this in more detail.

Dupré and O'Malley suggest that biofilms possess many (but not all) of the characteristics of multicellular organisms (Dupré 2012: 88, 177). However, this does not automatically mean that microbial communities should be regarded as organisms. This is not just because of the adoption of a definition of multicellularity based on MC eukaryotes. From the organizational perspective I have discussed, apart from claiming that early eukaryotic multicellularity is not organismal, I have also argued that the requirements for MC organisms are not satisfied in systems with a lower organizational complexity than that of eumetazoa. So, irrespectively of eukaryotic multicellularity, the possession of some of the characteristics of so-called 'paradigmatic organisms' does not say much about the requirements for organismality. The objection is stronger because, if we take biofilms as self-organizing functional units, then these should qualify as organisms. But in this case, as Werndl (2013) points out, the question is: on what basis is the assumption that biofilms are self-organizing entities that operate as functional units sufficient (or even necessary) for organismality? Self-organization is a very broad term, which can be used to describe the processes in a gas container being externally heated up as well as what happens in

---

[30] See chapter 9 for a complementary discussion of this case study.

our whole body. At any rate, as discussed in section 3, self-organization per se is barely adequate for simple self-maintenance. More seems to be needed to postulate that a self-organizing entity operates as a functional unit.

At this point, O'Malley and Dupré (2007) could say that microbial communities are much more than just individuals that happen to have mingled with one another; although they do not possess the level of physiological integrity that individual organisms do, they nevertheless exhibit a degree of integration that allows them to form organism-like communities (ibid., 177). In this case, I think we should try to examine the forms of such integrations. As far as single-species biofilms formed by *M. xanthus* are concerned, I have argued that not only is constitution completely independent of interaction at the MC level, but that there is not even functional interdependence between the individual bacteria that make up the swarm. Therefore, from an organizational perspective, such microbial communities do not even qualify as functional individuals, let alone as organisms.[31]

## 5.3. The distinction between life and non-life

Does a collaborative view of life blur the distinction between life and non-life? According to our discussion, it can be said that life and non-life are roughly distinguished by the realization of organizational closure on the basis of functional differentiation—that is, on the basis of the mutual dependence (for continuous regeneration) of constraints with *distinguishable contributions* to the global far-from-equilibrium self-maintenance of an organization. Viruses and prions do not exhibit such closure. Therefore, according to the organizational account I have adopted, the individual microbe remains the fundamental ontological unit in microbiology. Let us discuss this in some more detail.

Although it is not quite clear what O'Malley and Dupré (2007: 184) mean when they say that 'the individual microbe is not the fundamental ontological unit in microbiology', promiscuous individualism implies that there is no such uniquely identified unit. In line with their emphasis on collaboration, Dupré and O'Malley (2012b: 220) push even further against the view of cells as the fundamental ontological units in biology when they say that, 'even when single cells are considered in isolation, each cell is a complex of collaborating parts'. However, this is not inconsistent with the fact that it is exactly such complex collaborations that bring about (genuine) recursively self-maintaining organizations. At this point, one could ask why such organizations should be considered alive. Dupré and O'Malley stress the capacity for reproduction and for metabolic sustenance as two fundamental features the intersection of which becomes sufficient for considering an entity as living. But from an organizational perspective, metabolic self-sustainability, reproduction, and their intersection should be interpreted as properties *of* an organization. Extant cells are recursively self-maintaining organizations with exactly such capacities. And it is

---

[31] The case of multispecies biofilms is more complicated and deserves a separate treatment. From an organizational point of view, the ascription of a loose form of functional individuality to multispecies biofilms seems possible, whereas the requirements for evolutionary individuality don't seem to be satisfied. At any rate, the latter is a highly controversial topic even from adaptationist and evolutionary perspectives (see e.g. Ereshefsky and Pedroso 2015; Clarke 2016).

from this perspective that viruses should not be considered alive, as viruses are neither self-reproductive nor metabolically self-sustaining organizations.

Dupré and O'Malley say that such commitment to an exclusively cellular view of life implies a single leap from fully non-living to fully living (ibid., 228). I do not think this is the case. In section 3 I argued that organizations exhibiting limited recursive self-maintenance should be considered biological individuals. However, in a prebiotic scenario, such a type of self-maintenance could be instantiated (a) by non-compartmentalized protocells whose self-maintaining networks are very partially coupled to the vesicles that encapsulate them and that could trigger some protoselective process for stability and persistence (Keller 2007; Budin and Szostak 2011); as well as (b) by protocells with reproductive capacities and with such a coupling between their proto-metabolic network and the components of their membranes that their population could undergo evolution by some kind of (proto-)natural selection (Budin and Szostak 2011). And it is also very likely that the evolutionary selective paths of such individuals would not make it to the organization of existing, fully fledged living beings (see Moreno and Mossio 2015: ch. 5 for a detailed discussion). So, according to the organizational view, collaboration does not blur the distinction between life and non-life. Rather, types of collaboration result in different forms of biological organization that, in the process of originating life, could have been associated with several types of systems, from (hypothetical) protocellular living individuals to fully fledged living unicellular organisms.

## 6. Conclusions

The importance of collaboration as a central characteristic of life has sometimes been neglected due to the dominant neo-Darwinist view that evolution favours competent selfish competitors. There should be no doubt that collaboration is one of the main characteristics of life and its evolution. Dupré and O'Malley (2012b) are absolutely right to suggest that it is very hard to imagine life (both at the intra- and at the intercellular level) that is not collaborative. A processual view of the living realm suggests that evolution proceeds by weaving a collective network of increasingly complex and entangled processes among biological entities (which are themselves constituted by sets of processes) at different phenomenological levels and with different cohesive strengths. This leads to a conceptualization of living systems as being embedded in evolutionary and ecological webs of metabolic and reproductive interactions. However, an excessively collaborative view of life implies that the essence of biological organization lies in complex webs of processes that span across different living systems. As a result, biological entities appear to become blurred in a series of dynamic and diverse collaborations with flexible and unfixed boundaries (Dupré and O'Malley 2012b). Accordingly, promiscuous individualism declares that there are multiple ways to define individuals, and that we can divide the biological world in as many different ways as we see fit. As I have attempted to argue, this leads to an unnecessary pluralism that in many cases also becomes unmanageable. Considering that we should keep our descriptions as concrete and specific as possible, another, monistic approach to the problem is to accept that biological entities and their boundaries should not be taken for granted and that their emergence and importance should be explicated.

This is exactly what I have tried to do in this chapter, from an organizational point of view. Specifically, adopting a process-based organizational ontology, I have suggested that the essential features of unicellular organismality are captured by a self-maintaining organization of processes that is integrated on the basis of a special type of collaboration, realized by regulatory processes that functionally coordinate its constitutive and interactive aspects. I have used this ontology to describe different types of cellular collaborations at the multicellular level, but also to examine which types of MC organization exhibit the form of organismal integration found in unicellular organizations. I have claimed that, although several types of cellular collaborations result in MC biological individuals, the exportation of the core organizational characteristics of unicellular organismality at the MC level require a genuine functional integration (a special type of collaboration realized through a regulatory centre—not just regulatory processes) between the constitutive and the interactive processes of the system. I concluded by arguing that, despite its processual basis, an organizational ontology for biology can provide specific suggestions for several implications raised from the consideration of a hypercollaborative view of life—especially relating to the distinction between life and non-life, between symbionts and host, and between organisms and other biological individuals—without undermining the importance of collaboration in understanding life and its evolution.

Acknowledging the collective dimension of life does not mean ignoring the individual organization of living systems, which has, after all, played a key role in the history of life as the locus of mechanisms, of adaptations, and of selective–evolutionary dynamics. An understanding of the diversity of the biological world requires, I have suggested, an appeal to the organization of individual living systems. I think this is a safe and promising direction, since a theoretically well-founded notion of an individual biological organization is conceptually coextensive with a naturalized account of other fundamental concepts of living systems, such as genetic information, functionality, agency, autonomy, and cognition (Ruiz-Mirazo and Moreno 2012). Moreover, it seems to me that, without such a notion, it would be difficult to make a clear-cut distinction between organisms, parts of organisms, groups of organisms, and other forms of cooperative or 'ecological' networks, and thereby to understand the diversity of the biological realm (Arnellos et al. 2013; Arnellos et al. 2014; Arnellos and Moreno 2016).

## Acknowledgements

I am grateful to Dan Nicholson and John Dupré for organizing this collection of papers in the very hot topic of process philosophy and for inviting me to express my ideas. Many of the ideas presented here have been shaped during my collaboration with Alvaro Moreno. I would also like to thank an anonymous reviewer for critical comments that clarified the structure of the paper. Many thanks to James DiFrisco for highly stimulating discussions during the preparation of the manuscript.

## References

Alexandre, G. (2010). Coupling Metabolism and Chemotaxis-Dependent Behaviours by Energy Taxis Receptors. *Microbiology* 156: 2283–93.

Arnellos, A. (2016). Biological Autonomy: Can a Universal and Gradable Conception be Operationalised? *Biological Theory* 11 (1): 11–24

Arnellos, A. and Moreno, A. (2012). How Functional Differentiation Originated in Prebiotic Evolution. *Ludus Vitalis* 37: 1–23.

Arnellos, A. and Moreno, A. (2015). Multicellular Agency: An Organizational View. *Biology & Philosophy* 30 (3): 333–57.

Arnellos, A. and Moreno, A. (2016). Integrating Constitution and Interaction in the Transition from Unicellular to Multicellular Organisms. In K. Niklas and S. Newman (eds), *The Origins and Consequences of Multicellularity* (pp. 249–75). Cambridge, MA: MIT Press.

Arnellos, A., Moreno, A., and Ruiz-Mirazo, K. (2014). Organizational Requirements for Multicellular Autonomy: Insights from a Comparative Case Study. *Biology & Philosophy* 29 (6): 851–84.

Arnellos, A., Ruiz-Mirazo, K., and Moreno, A. (2013). Autonomy as a Property That Characterises Organisms among Other Multicellular Systems. *Contrastes* 18: 357–72.

Balagam, R. and Igoshin, O. A. (2015). Mechanism for Collective Cell Alignment in *Myxococcus xanthus* Bacteria. *PLoS Computational Biology* 11: e1004474.

Barandiaran, X. and Moreno, A. (2008). Adaptivity: From Metabolism to Behavior. *Adaptive Behavior* 16 (5): 325–44.

Berleman, J. and Kirby, J. (2009). Deciphering the Hunting Strategy of a Bacterial Wolfpack. *FEMS Microbiology Review* 33 (5): 942–57.

Bich, L., Mossio, M., Ruiz-Mirazo, K., and Moreno, A. (2015). Biological Regulation: Controlling the System from Within. *Biology & Philosophy* 31 (2): 237–65.

Bickhard, M. (2000). Autonomy, function and representation. *Communication and Cognition* 17: 111–31.

Bickhard, M. (2004). Process and Emergence: Normative Function and Representation. *Axiomathes* 14 (3): 135–69.

Bickhard, M. (2009). The interactivist model. *Synthese* 166 (3): 547–91.

Bickhard, M. (2011a). Some Consequences (and Enablings) of Process Metaphysics. *Axiomathes* 21 (1): 3–32.

Bickhard, M. (2011b). Systems and Process Metaphysics. In C. Hooker (ed.), *Handbook of Philosophy of Science: Philosophy of Complex Systems*, vol. 10 (pp. 91–104). Amsterdam: Elsevier.

Bijlsma, J. J. and Groisman, E. A. (2003). Making Informed Decisions: Regulatory Interactions between Two-Component Systems. *Trends in Microbiology* 11: 359–66.

Budin, I. and Szostak, J. W. (2011). Physical Effects Underlying the Transition from Primitive to Modern Cell Membranes. *Proceedings of the National Academy of Sciences* 108 (13): 5249–54.

Buss, L. W. (1987). *The Evolution of Individuality*. Princeton: Princeton University Press.

Campbell, D. T. (1974). Evolutionary Epistemology. In P. Schilpp (ed.), *The Philosophy of Karl Popper* (pp. 413–63). LaSalle: Open Court.

Campbell, R. J. (2009). A Process-Based Model for an Interactive Ontology. *Synthese* 166 (3): 453–77.

Christensen, W. D. and Bickhard, M. H. (2002). The Process Dynamics of Normative Function. *Monist* 85 (1): 3–28.

Clarke, E. (2016). Levels of Selection in Biofilms: Multispecies Biofilms Are Not Evolutionary Individuals. *Biology & Philosophy* 31 (2): 191–212.

Collier, J. (1988). Supervenience and Reduction in Biological Hierarchies. In M. Matthen and B. Linsky (eds), *Philosophy and Biology* (pp. 209–34) (Supplementary volume 14 of the *Canadian Journal of Philosophy*). Calgary: University of Calgary Press.

Collier, J. and Hooker, C. A. (1999). Complexly Organised Dynamical Systems. *Open Systems and Information Dynamics* 6: 241–302.

Dupré, J. (2012). *Processes of Life: Essays in the Philosophy of Biology*. Oxford: Oxford University Press.

Dupré, J. and O'Malley, M. A. (2012a). Metagenomics and Biological Ontology, In J. Dupré, (ed.), *Processes of Life: Essays in the Philosophy of Biology* (pp. 188–205). Oxford: Oxford University Press.

Dupré, J. and O'Malley, M. A. (2012b). Varieties of Living Things: Life at the Intersection of Lineage and Metabolism. In J. Dupré (ed.), *Processes of Life: Essays in the Philosophy of Biology* (pp. 206–29). Oxford: Oxford University Press.

Ereshefsky, M. and Pedroso, M. (2015). Rethinking Evolutionary Individuality. *Proceedings of the National Academy of Sciences* 112 (33): 10126–32.

Folse, H. J. III and Roughgarden, J. (2010). What Is an Individual Organism? A Multilevel Selection Perspective. *Quarterly Review of Biology* 85 (4): 447–72.

Fox Keller, E. (2007). The Disappearance of Function from 'Self-Organizing Systems'. In F. Boogerd, F. Bruggeman, J.-H. Hofmeyr, and H. V. Westerhoff (eds), *Systems Biology: Philosophical Foundations* (pp. 303–17). Amsterdam: Elsevier.

Hartenstein, V. (2006). The Neuroendocrine System in Invertebrates: A Developmental and Evolutionary Perspective. *Endocrinology* 190: 555–70.

Hooker, C. (2013). On the Import of Constraints in Complex Dynamical Systems. *Foundations of Science* 18: 757–80.

Jékely, G., Keijzer, F., and Godfrey-Smith, P. (2015). An Option Space for Early Neural Evolution. *Philosophical Transactions of the Royal Society B* 370 (1684): 20150181.

Kaiser, D. and Warrick, H. (2011). *Myxococcus xanthus* Swarms Are Driven by Growth and Regulated by a Pacemaker. *Journal of Bacteriology* 193: 5898–904.

Kauffman, S. (2000). *Investigations*. Oxford: Oxford University Press.

Keijzer, F. and Arnellos, A. (2017). The Animal Sensorimotor Organization: A Challenge for the Environmental Complexity Thesis. *Biology & Philosophy* 32: 421–41.

Kirby, J. R. (2009). Chemotaxis-like Regulatory Systems: Unique Roles in Diverse Bacteria *Annual Review of Microbiology* 63: 45–59.

Kirk, D. L. (2005). A Twelve-Step Program for Evolving Multicellularity and a Division of Labor. *BioEssays* 27: 299–310.

Koufopanou, V. and Bell, G. (1993). Soma and Germ: An Experimental Approach Using Volvox. *Proceedings of the Royal Society in London Series B* 254: 107–13.

McFall-Ngai, M. J. (1994). Animal-Bacterial Interactions in the Early Life History of Marine Invertebrates: The *Euprymna scolopes/Vibrio fischeri* Symbiosis. *American Zoologist* 34: 554–61.

Moran, N. A. (2006). Symbiosis. *Currents in Biology* 16: R866–R871.

Moreno, A. and Barandiaran, X. (2004). A Naturalized Account of the Inside-Outside Dichotomy. *Philosophica* 73: 11–26.

Moreno, A. and Mossio, M. (2015). *Biological Autonomy: A Philosophical and Theoretical Enquiry*. New York: Springer.

Moreno, A. and Ruiz-Mirazo, K. (2009). The Problem of the Emergence of Functional Diversity in Prebiotic Evolution. *Biology & Philosophy* 24 (5): 585–605.

Mossio, M. and Moreno, A. (2010). Organizational Closure in Biological Organisms. *History and Philosophy of the Life Sciences* 32 (2–3): 269–88.

Mossio, M., Saborido, C., and Moreno, A. (2009). An Organizational Account of Biological Functions. *British Journal for the Philosophy of Science* 60: 813–41.

O'Malley, M. A., and Dupré, J. (2007). Size Doesn't Matter: Towards a More Inclusive Philosophy of Biology. *Biology & Philosophy* 22: 155–91.

Rosslenbroich, B. (2014). *On the Origin of Autonomy: A New Look at the Major Transitions in Evolution*. Berlin: Springer.

Ruiz-Mirazo, K., Etxeberria, A., Moreno, A., and Ibañez, J. (2000). Organisms and Their Place in Biology. *Theory in Biosciences* 119: 43–67.

Ruiz-Mirazo, K. and Mavelli, F. (2008). Towards 'Basic Autonomy': Stochastic Simulations of Minimal Lipid–Peptide Cells. *BioSystems* 91 (2): 374–87.

Ruiz-Mirazo, K. and Moreno, A. (2004). Basic Autonomy as a Fundamental Step in the Synthesis of Life. *Artificial Life* 10 (3): 235–59.

Ruiz-Mirazo, K. and Moreno, A. (2012). Autonomy in Evolution: From Minimal to complex Life. *Synthese* 185 (1): 21–52.

Ueki, N., Matsunaga, S., Inouye, I., and Hallmann, A. (2010). How 5000 Independent Rowers Coordinate Their Strokes in Order to Row into the Sunlight: Phototaxis in the Multicellular Green Alga Volvox. *BMC Biology* 8 (103).

Ulanowicz, R. E. (2009). *A Third Window: Natural Life beyond Newton and Darwin*. West Conshohocken: Templeton.

Ulanowicz, R. E. (2013). Process-First Ontology. In B. G. Henning and A. C. Scarfe (eds), *Beyond Mechanism: Putting Life Back into Biology* (pp. 115–31). Lanham: Lexington Books.

Van Duijn, M., Keijzer, F., and Franjen, D. (2006). Principles of Minimal Cognition: Casting Cognition as Sensorimotor Coordination. *Adaptive Behavior* 14 (2): 157–70.

Werdl, C. (2013). Do Microbes Question Standard Thinking in the Philosophy of Biology? Critical Notice of 'Processes of Life: Essays in the Philosophy of Biology' (John Dupré). *Analysis* 73 (2): 380–7.

Wilson, J. (1999). *Biological Individuality: The Identity and Persistence of Living Entities*. Cambridge: Cambridge University Press.

Wilson, R. A. (2012). Review of John Dupré, 'Processes of Life: Essays in the Philosophy of Biology'. *Notre Dame Philosophical Reviews*. http://ndpr.nd.edu/news/processes-of-life-essays-in-the-philosophy-of-biology.

Zhang, Y., Ducret, A., Shaevitz, J., and Mignot, T. (2012). From Individual Cell Motility to Collective Behaviors: Insights from a Prokaryote. *Myxococcus xanthus FEMS. Microbiology Review* 36: 149–64.

# PART IV
# Development and Evolution

# 11

# Developmental Systems Theory as a Process Theory

*Paul Griffiths and Karola Stotz*

## 1. Introduction

Developmental systems theory (DST) builds on a long tradition of ideas about the systems nature of development among biologists and psychologists, predominantly ideas from workers in the field of developmental psychobiology. Elements of DST are derived from C. H. Waddington's views on developmental systems (Waddington 1952) and on the 'epigenotype' (Waddington 1942), from Daniel Lehrman's influential critique of the concept of innate behaviour (Lehrman 1953), from Gilbert Gottlieb's theory of probabilistic epigenesis (Gottlieb 2001), and from Susan Oyama's book *The Ontogeny of Information* (Oyama 1985). Developmental psychobiologists Donald Ford and Richard Lerner integrated many of these ideas into a formal framework in their book *Developmental Systems Theory* (Ford and Lerner 1992).

DST analyses development, heredity, and evolution in a way that avoids dichotomies such as nature versus nurture, genes versus environment, and biology versus culture. In this framework, *development* (ontogeny) is the reconstruction of a life cycle using resources passed on by previous life cycles. DST takes *heredity* to encompass both the stability and the plasticity of biological form, which are complementary aspects of the recurrence and modification, in each generation, of a system of genetic, epigenetic, and exogenetic developmental resources (Stotz and Griffiths 2017). The prime focus of a DST account of *evolution* is the life cycle—the series of events that occurs in each generation of a lineage. The process of evolution is the differential reproduction of variant life cycles. The end of one life cycle and the beginning of the next is marked by the reconstruction of the various mechanisms that allow the life cycle to reproduce itself from relatively simple resources. The replication of genes is simply one aspect of the replication of a life cycle. Many classes of developmental resource are replicated: genes, methylation patterns, membrane templates, cytoplasmic gradients, centrioles, nests, parental care, habitats, and cultures are all at least partly constructed by past generations and interact to construct future generations.

DST attracted the interest of philosophers of biology in the 1990s, mostly in response to the work of Susan Oyama (Godfrey-Smith 2000; Gray 1992; Griffiths and Gray 1994; Moss 1992; Robert et al. 2001). However, while most scientific work in the developmental systems tradition was on behavioural development, including child development, philosophical discussion of DST focused on its implications for 'gene-centred' views of molecular developmental biology and evolutionary biology. In this vein, Kim Sterelny, Michael Dickison, and Kelly Smith (1996) proposed to assimilate developmental systems theory to the replicator/interactor view of evolution put forth by Richard Dawkins (1976) and David Hull (1988). They suggested that the evidence and arguments used to support DST could be accommodated by the concession that there are some non-genetic replicators in addition to the genetic ones. Evolution is the result of competition between members of an *extended* class of replicators.

The 'extended replicator' approach was rejected by developmental systems theorists Griffiths and Gray (1997), who pointed to some paradoxical consequences of trying to describe developmental systems and their evolution in a replicator framework. A developmental system includes a 'developmental niche' that contains the reliably inherited developmental resources needed to reconstruct that developmental system—or to modify it, in the case of phenotypic plasticity. Some of the resources that make up the developmental niche are actively constructed by the parents—for example breast milk, or the incubation mounds built by male brush-turkeys (Goth 2004). Others, such as the flock structure required for normal behavioural development in cowbirds, are constructed through the activity of many conspecifics, and not only by the parents (West and King 2008). But some merely persist, independently of the activities of previous generations in the developmental systems—for example the territories inherited by male scrub jays from their fathers (West et al. 1988: 49). All these resources are potentially part of the evolved developmental system:

> There is a fundamental similarity between building a nest, maintaining one built by an earlier generation, and occupying a habitat in which nests simply occur (for example, as holes in trees). In all three cases, there may be an evolutionary explanation of the interaction of the nest with the rest of the developmental system. (Griffiths and Gray 1994: 291)

Griffiths and Gray used examples of habitat and host imprinting to show that one lineage can outcompete another even when the feature in which the two lineages differ falls into the category of 'merely persistent' resources (ibid., 288–90). They reiterated this point in 1997, in order to rebut the objection that habitat and host associations themselves do not evolve by natural selection; only the behaviours that *produce* habitat and host associations do. This objection fails because populations with exactly the same behavioural mechanism for habitat or host imprinting may differ in their level of fitness as a result of the specific persistent resource with which they (re-)establish a relationship: 'In cases of host imprinting in parasitic insects or cuckoos...much of the rest of the parasite group's evolution may result from the success of lineages with one relationship rather than another' (Griffiths and Gray 1997: 485).

The relationship between a developmental system and a persistent resource can explain aspects of both development and evolutionary success. But the persistent resource itself is not replicated in development, nor does it increase its representation relative to alternatives as a result of the evolutionary success of the system, so the persistent resource cannot be treated as a replicator. What is replicated, and may increase in representation, is the *relationship* between system and resource. Griffiths and Gray (ibid.) argued that treating these relationships as replicators independent of their relata would be a *reductio ad absurdum* of the replicator concept. DST offers a less paradoxical treatment of persistent resources:

> We conceive of an evolving lineage as a series of cycles of a developmental process, where tokens of the cycle are connected by the fact that one cycle is initiated as a causal consequence of one or more previous cycles, and where small changes are introduced into the characteristic cycle as ancestral cycles initiate descendant cycles. The events which make up the developmental process are developmental interactions—events in which something causally impinges on the current state of the organism in such a way as to assist production of evolved developmental outcomes. (Griffiths and Gray 1994: 291)[1]

A process ontology for DST allows it to reconcile two otherwise paradoxical facts: some components of the evolved developmental system persist without reference to the rest of the system, but the presence of these components in the system can be explained by natural selection. These facts cease to seem paradoxical if we focus on how life cycles—processes rather than systems of entities—reproduce themselves and on how variant life cycles reproduce themselves more or less effectively. It is the developmental *process* that replicates itself across the generations, making use of persistent resources as well as of resources created by earlier cycles of that process. Re-establishing or breaking a relationship with a persistent resource, for example by becoming imprinted on a new habitat, is an event that is part of a developmental process.

## 2. Process Biology

Griffiths and Gray argue that the fundamental unit of analysis in DST is a developmental process. This process is better described as a *life cycle*, since it encompasses the entire period between conception and death. Does this mean that DST is a form of process biology? In this chapter we will argue that it does. As a first step, we discuss the inspiration that many early twentieth-century biologists drew from process philosophy, and we ask which if any of their ideas correspond to those found in DST. A direct link between DST and these earlier thinkers is the embryologist, developmental geneticist, and theoretical biologist Conrad Hal Waddington (1905-75), whose ideas have been cited by many advocates of DST.[2] In *The Evolution of Developmental Systems*, Waddington describes his understanding of physiology, development, and evolution as nested processes:

> Biologists have always been forced by their subject matter to take time seriously. But it is only gradually that they have realised to the full the necessity always to consider living things as essentially processes, extended in time, rather than static entities.

---

[1] This proposal has a parallel within the replicator tradition, namely G. C. Williams' view of the organism as a region of space-time structured by evolved information (Williams 1992). By that stage in his career Williams had abandoned the view that objects compete to replicate themselves and embraced the differential replication of information, an approach that lends itself more easily to a process perspective and that seems to accommodate the replication of relationships.

[2] Waddington's process biology is also discussed in chapter 12.

> The day-to-day activities of living things are carried on by processes which occur anything from a fraction of a second to an hour or two... [the organism] is gradually carried along through another series of processes, those of development, each phase of which occupies a time which is fairly long compared with the life-cycle. Each life-cycle is again nearly circular, and gives rise to new animals, the descendant generation... The accumulation of such changes gives rise to the slow process of evolution, which is on a still larger scale than either the physiological or developmental ones. (Waddington 1952: 155)

Waddington was deeply influenced by the process philosophy of Alfred North Whitehead, as were many of his British contemporaries, for instance the biochemist Joseph Needham, the theoretical biologist J. H. Woodger, and the leading Cambridge behavioural biologist W. H. Thorpe (see chapter 1 here and Abir-Am 1987; for Thorpe's views, see Thorpe 1956). Whitehead's influence on biology was equally strong in Australia, where it extended to the Nobel Prize-winning immunologist Frank Macfarlane Burnet (Anderson and Mackay 2014), to the geneticist and ecologist L. Charles Birch (Birch 1965; see also chapter 15), and to the geneticist Wilfrid E. Agar. It was Agar (1882–1951) who introduced Australian biologists to process philosophy, through works such as 'Whitehead's Philosophy of Organism: An Introduction for Biologists' (Agar 1936). Agar sought to convince his fellow biologists that Whiteheadian process philosophy was a good framework for biological research, and so his work is particularly useful when we ask what aspects of this older enthusiasm for process thought are relevant to DST.

There are some aspects of Whitehead's thought that were important to these biologists but have little relevance in the current context. Whitehead rejected the idea that the mental must be explained in physical terms, insisting instead on a monistic 'panexperientialism'. Agar noted that '[t]he conception that the world, including the physical world, is composed of entities which are "drops of experience" or feelings will seem to many people a strange one' (ibid., 18). He hastened to reassure biologists that '[w]e must bear in mind that "feeling" is here used throughout as the purely general term for any kind of acting or being acted upon, in such a way that the make-up of the subject is affected' (ibid., quoting Emmet 1932: 142). Through these aspects of his system, Whitehead offered biologists a way to make room for consciousness in the physical world, as well as a route to reconcile science and religion; and, for some biologists, this was an important source of their attraction to Whitehead (Bowler 2014).

However, as historian Peter Bowler has noted, 'Waddington had no interest in encouraging scientists to revive an interest in religion' (Bowler 2014: 80). His interest in Whitehead's metaphysics was strictly biological. The lessons he derived from Whitehead were that biological phenomena should be explained at the level of the whole system and that the reconstruction of form in development should be explained dynamically, not as a result of the transmission of something that concretely embodies that form. In his autobiography he outlined some specific impacts on his work in embryology and genetics:

> In the late '30s I began developing the Whiteheadian notion that the process of becoming (say) a nerve cell should be regarded as the result of the activities of large numbers of genes, which interact together to form a unified 'concrescence'.... Again a few years earlier it had become

apparent that the 'gene-concrescence' itself undergoes processes of change; at one embryonic period a given concrescence is in a phase of 'competence' and may be switched into one or other of a small number of alternative pathways of further change—but the competence later disappears and if you've missed the bus the switch won't work.... If I had been more consistently Whiteheadian, I would probably have realized that the 'specificity' [the fact that the switch sends the cell down a particular developmental pathway] involved does not need to lie in the switch at all but may be a property of the 'concrescence' and the ways in which it can change. Because of course what I have been calling by the Whiteheadian term 'concrescence' is what I have later called a *chreod* [canalised developmental trajectory].

(Waddington 1975: 9–10)

As we will see in this chapter, what makes DST a process theory is that it seeks to explain developmental outcomes as the result of a dynamic process in which some of the interacting factors are products of earlier stages of the process, rather than as the result of the arrangement of preexisting factors into a static mechanism. Even when factors exist independently of the developmental process, they are drawn into it and made part of a developmental 'system' by the unfolding process, as we have already discussed above. It is the process that defines the system. In these respects, DST is the direct inheritor of Waddington's process biology.

Another reason why biologists were drawn to Whitehead was his *organicism*—the idea that collectives have an enduring identity that cannot be reduced to the continuity of their parts.[3] This was at the heart of Agar's interest in Whitehead:

It is in [Whitehead's] conception of the unity of a nexus that we strike the main idea of theories of organism as usually understood by biologists, namely, the idea that the whole is more than the sum of its parts, and indeed imposes its own character on its parts. As Ritter puts it, the whole acts causally on its parts, as well as being acted on causally by its parts. This is only understandable if we get away from the idea of substance and fix our attention on process. We must not think of the molecule as composed of ultimate particles of matter in motion. But the molecule is a pattern of processes, and each constituent process conforms to its place in the pattern, and resists factors tending to alter it. (Agar 1936: 29; see also Agar 1943)

Agar finds in Whitehead an account of the identity of an organism that is reassuringly Darwinian: 'It is cardinal to Whitehead's philosophy that the subjective aim of an actual entity is not merely at self-realization, but at self-realization as an agent creative of other entities like itself, or at least of the production in other actual entities of feelings like its own' (Agar 1936: 22). Elsewhere Agar identifies 'subjective aim' with final causation, so part of what he is saying here is that the *telos*[4] of an organism is to reproduce itself.

We do not think that the details of Agar's Whiteheadian organicism contain much that is of relevance to DST, but the issue he is addressing—the identity of processes through time and their distinctness from one another—is a vital one for DST. A biological individual is a process that may intersect with other organismic processes, but it has a principle of identity that marks just this series of events out as

---

[3] On organicism in biology, see chapters 1, 7, 12, and 13.
[4] Its end or purpose; in Aristotelian terminology, 'that for the sake of which' the organism exists and acts.

one biological individual. DST thus requires an account of what is known in process philosophy as 'genidentity' or identity as continuity of organization. This principle of identity also determines, by identifying a process, the boundaries of the developmental system—the matrix of resources required for that process to proceed to completion. We return to this issue in section 6, where we will see that the identity of a developmental process and its distinction from other individual processes are indeed given by its *telos*, although not in the exact sense envisaged by Agar.

## 3. Process in the Developmental Systems Tradition

> An animal is, in fact, a developmental system, and it is these systems, not the mere adult forms which we conventionally take as typical of the species, which becomes modified during the course of evolution.
>
> (Waddington 1952: 155)

Waddington's idea of a developmental system and his early attempt to explain development as the result of the dynamic structure of that system are important precursors to DST (Griffiths and Tabery 2013). However, he remained a profoundly gene-centred thinker. The dynamic structure is an 'epigenotype'—the global expression of all the organism's genes—and it explains, through the presence of 'chreods', the resistance of some developmental trajectories to environmental perturbation. Waddington does not embrace the idea that the evolved developmental system actually *includes* aspects of the environment rather than merely being designed to function in an environment and cope with its variations.

For this central theme in DST we must turn to the comparative psychologist Daniel S. Lehrman, perhaps the single most important figure in the development of DST, and indeed of the scientific field of developmental psychobiology:

> Natural selection acts to select genomes that, in a normal developmental environment, will guide development into organisms with the relevant adaptive characteristics. But the path of development from the zygote stage to the phenotypic adult is devious, and includes many developmental processes, including, in some cases, various aspects of experience.
>
> (Lehrman 1970: 36)

Lehrman was not a Whiteheadian, and so it is all the more significant that his efforts to place development at the heart of the study of animal behaviour led him to adopt a process view of the organism:

> The use of 'explanatory' categories such as 'innate' and 'genetically fixed' obscures the necessity of investigating developmental *processes* in order to gain insight into the actual mechanisms of behavior and their interrelations. The problem of development is the problem of the development of new structures and activity patterns from the resolution of the interaction of *existing* structures and patterns, within the organism and its internal environment, and between the organism and its outer environment. At any stage of development, the new features emerge from the interactions within the *current* stage and between the *current* stage and the environment. The interaction out of which the organism develops is *not* one, as is so often said, between heredity and environment. It is between *organism* and environment! And the organism is different at each different stage of its development. (Lehrman 1953: 345)

Lehrman is not merely being pedantic in insisting that organisms, not genes, interact with environments. The impact of a genetic or environmental factor at some point in development depends on how the organism has developed up to that point. It is an organism at some stage of development that interacts with both genes and environment to produce the next stage of development. Development is essentially a dynamic process in which, as Waddington insisted, we need to take account of time, as well as of a list of ingredients.

The historical contingency of individual development was at the heart of what developmental psychobiologist Gilbert Gottlieb, another major source for DST, called the 'developmental psychobiological systems view' (Gottlieb 1970). Gottlieb contrasted what he called 'probabilistic epigenesis' with what he saw as the prevailing view of 'predetermined epigenesis'. The latter concept, he argued, covered up the persistence of preformationist thought in modern biology (Gottlieb 2001; see also Robert 2004). According to Gottlieb, 'the cause of development—what makes development happen—is the relationship of the components, not the components themselves' (Gottlieb 1997: 91). The impact of any causal factor depends on the order in which the system is exposed to that and other factors. This places limits on our ability to predict the results of development from a list of measured factors. Gottlieb's influence lives on in DST's emphasis on contingency.[5]

The idea of a formal developmental systems *theory* is due to Donald Ford and Richard Lerner (1992), who identify two core theses of DST. The first, which they call 'developmental contextualism', is derived from Gottlieb's concept of probabilistic epigenesis. Development proceeds at several levels—for example gene expression, the formation of tissues, and the state of the environment—and the interactions between levels are the prime focus of research, rather than one level being focal and the others being background against which the former unfolds. Developmental contextualism is a modern version of the epigenetic—as opposed to predeterminist—view of development. Ford and Lerner's second core thesis is 'dynamic interactionism', which they contrast with conventional, 'static interactionism'. This reflects Lehrman's distinction between organism–environment and gene–environment interaction mentioned earlier. Ford and Lerner regard interaction as an ongoing process that can transform the interactants themselves. In other words, the parts that interact with one another and with the developmental system are products of the developmental system. Overall, Ford and Lerner present a thoroughly processual view of the developmental system: a view in which we can see the same ideas that we encountered in Waddington and that he claimed to have derived from Whitehead.

This process view of development led to the radical reformulation of the distinction between nature and nurture proposed by Susan Oyama (2002). In the conventional picture of (static) gene–environment interaction, nature and nurture are simply interacting causes. Genes, or genes plus 'epigenes', represent nature. The environment represents nurture. Added together, they cause development. In DST, however, nature and nurture are product and process. Nurture is the interaction between the current state of the organism and the resources available to

---

[5] To the best of our knowledge, Gottlieb was not influenced by Whitehead though he did acknowledge John Dewey as a major source of inspiration (Griffiths and Tabery 2013), and Dewey was acknowledged by Whitehead as an important influence.

it—environmental *and* genetic. The nature of the organism at each stage is simply the state of the organism—including the modified state of its genome and of its developmental niche, both of which have been transformed by earlier processes of nurture. Oyama rejects the very idea that nature exists separately from, and before, nurture. One way she conveys this is by insisting that the developmental information expressed in the organism is not present in the starting point of development, but is itself created by the process of development, through feedback from the current state of the organism to the states of the resources that will influence future development. This is what she means by the 'ontogeny of information'.

In *The Ontogeny of Information* Oyama pioneered the parity argument, or the 'parity thesis', concerning genetic and environmental causes in development (see also Griffiths and Gray 1994; Griffiths and Gray 2005; Griffiths and Knight 1998; Stotz 2006; Stotz and Allen 2012). Oyama relentlessly tracked down failures of parity of reasoning in earlier theorists. The same feature is accorded great significance when a gene exhibits it, only to be ignored when a non-genetic factor exhibits it. When a feature thought to explain the unique importance of genetic causes in development is found to be more widely distributed across developmental causes, it is discarded and another feature is substituted. Griffiths and Gray (1994) argued in this spirit against the idea that genes are the sole or even the main source of information in development. Other ideas associated with 'parity' are that the study of development does not turn on a single distinction between two classes of developmental resources, and that the distinctions useful for understanding development do not all map neatly onto the distinction between genetic and non-genetic. Ulrich Stegmann has argued that, because DST has not identified a single, essential way for genetic and non-genetic resources to be treated with parity of reasoning, the idea of parity is too vague to be useful (Stegmann 2012). It is hard to know what to make of this criticism. Other critics of DST have dismissed parity as the wildly holistic view that no distinctions of any kind can be made among developmental causes: 'Parity arguments then claim that picking out one cause, when in fact there are many, cannot be justified on ontological grounds because, after all, causes are causes' (Waters 2007: 533).[6] Developmental systems theorists have repeatedly rejected this interpretation and provided examples of how one developmental cause can be more significant than another in ways that are consistent with DST (Oyama 2000; Griffiths and Knight 1998; Griffiths and Gray 2005). In recent work, Griffiths and colleagues have constructed a quantitative measure of relative causal contribution and used it to assess the parity thesis in specific cases (Griffiths et al. 2015).

As we have seen, the ideas that Waddington derived from process philosophy can be found in all the major figures who inspired and developed DST. In the next section we discuss in more detail two core ideas that define DST as a distinctive approach to development and we argue that both support the notion that the fundamental unit of analysis for DST is the developmental process or the life cycle.

---

[6] For other examples, see Weber 2006: 607; Rosenberg and McShea 2008: 174; Okasha 2009: 724; Woodward 2011: 249; French 2012: 197. Okasha calls this kind of wild holism 'causal democracy', a term introduced by Philip Kitcher (2001). Kitcher's principle of causal democracy states that biology should not assume that the genes are the most significant causes but should assess the issue empirically, on a case-by-case basis. It is thus very similar to Oyama's parity thesis, and neither is committed to any kind of holism.

## 4. Core Ideas in DST: Epigenesis and Developmental Dynamics

Two ideas recur in all the authors presented above and are at the heart of DST: *epigenesis* and *developmental dynamics* (Griffiths and Tabery 2013). Both of them support the view that the central focus of DST should be on developmental processes rather than on sets of objects (developmental resources).

*4.1. Epigenesis*

The term 'epigenetics' derives from the much older word 'epigenesis'. 'As a continuation of the concept that development unfolds and is not preformed (or ordained), epigenetics is the latest expression of epigenesis' (Hall 2011: 12). It was coined by Waddington as a fusion of 'epigenesis' and 'genetics', to refer to the processes by which genotype gives rise to phenotype and to the study of those processes (Waddington 1942). Waddington suggested that existing knowledge from experimental embryology supported a view of how genes are connected to phenotypes that is broadly in line with the older idea of epigenesis. The interaction of many genes produces an emergent level of organization that Waddington called the 'epigenotype', and development is explained by the dynamics of the developmental system at this level.

Waddington's 'epigenotype' is a global expression of the genetic causal factors that influence development. The effect of changing any one gene depends on how that gene interacts with the rest of the system. The epigenotype as a whole interacts with the environment to determine the phenotype. DST expands this vision to include non-genetic factors that influence development. The epigenotype is replaced by a more inclusive vision of a developmental system, a global expression of all the causal factors that influence development. The developmental system still does not determine a unique phenotype, both because development is a probabilistic process, as Gottlieb emphasized, and because development is plastic by design. The environment provides many requirements for normal genome expression (the 'ontogenetic niche' of West and King 1987), and thus partly constitutes the developmental system; but the environment also determines the specific values of variables in an individual life cycle and thereby selects the particular course that development will take from those available to the system. The genome also plays these two roles, as some variables are determined by genetic individual differences (see Tabery 2009).

In 1958 the biologist David L. Nanney introduced another sense of 'epigenetics': the sense in which it is primarily used in molecular biology today (Haig 2004). Epigenetics in this sense is the study of mechanisms that determine which genome sequences will be expressed in the cell. These mechanisms control cell differentiation and give the cell an identity that is often passed on through mitosis. Writing in the year in which Francis Crick first stated his 'sequence hypothesis' that the order of nucleotides in DNA determines the order of amino acids in a protein and thus encodes the biological specificity of the protein (Crick 1958), Nanney wrote:

On the one hand, the maintenance of a 'library of specificities', both expressed and unexpressed, is accomplished by a template replicating mechanism. On the other hand, auxiliary mechanisms with different principles of operation are involved in determining which specificities are to be expressed in any particular cell.... [T]hey will be referred to as 'genetic systems' and 'epigenetic systems'. (Nanney 1958: 712)

Epigenetics in this narrower, modern sense allows a major role for the environment in development: 'As the past 70 years made abundantly clear, genes do not control development. Genes themselves are controlled in many ways, some by modifications of DNA sequences, some through regulation by the products of other genes and/or by [the intra- or extracellular] context, and others by external and/or environmental factors' (Hall 2011: 12). The regulated expression of the coding regions of the genome depends on mechanisms that differentially activate and select the information in coding sequences, depending on context. Biological information is distributed between the coding regions in the genome and regulatory mechanisms, and the specificity manifested in gene products is the result of a process of 'molecular epigenesis' (Stotz 2006; Griffiths and Stotz 2013).

So the idea of epigenesis is alive and well in contemporary biology. As Waddington argued, developmental outcomes are to be explained at the level of the whole system, and not by single causes that 'encode' or 'instruct' that outcome (for an influential restatement of this view, see Noble 2006). Developmental outcomes are also explained dynamically, as trajectories in a space of possible states of genome expression. The role of epigenetic marks in development is to successively differentiate cells as a result of earlier stages of development, making genome expression in one tissue at a time a function of the history of these cells. The complexity of biological networks makes it plausible that in many cases this process will display emergent dynamics that can only be studied through simulation, a point we expand on in the next subsection.

DST adds to this modern epigenetic vision of development the same element that it added to Waddington's original vision of epigenetics, namely a constructive role for the environment. The networks that regulate gene expression extend outside the cell and outside the organism. Evolution designs developmental processes that draw these wider resources into the developmental system by re-establishing relationships with them. The presence of suitable external resources is in many cases explained by the activity of parents and of conspecifics more generally ('developmental niche construction', see Griffiths and Stotz 2013), and sometimes by the feed-forward effects of earlier stages of the developmental process itself.

## 4.2. Developmental dynamics

The idea that development is a dynamic process is central to DST. Ford and Lerner contrast 'dynamic interaction' with a more conventional conception of interaction associated with analysis of variance techniques, such as those used in behavioural genetics (Ford and Lerner 1992). In this 'static interaction' the values of two variables measured before development, such as shared genes and shared environment, are shown to interact with each other. In contrast, dynamic interaction must be studied as a temporally extended process. For example, in Celia Moore's iconic work on sexual development in male rat pups, male sexual development depends on differential licking of the genital area of male and female pups by the mother. But this response of the mother to male pups depends on differences in their urine, which are the result of earlier processes of sexual differentiation (Moore 1984, 1992). The presence of this environmental influence is a feed-forward from earlier development in the pup itself. The patterns of gene expression that underlie sexual development in the rat arise through interaction with an environment that has been partially structured by an earlier stage of the rat's development. The notion of developmental dynamics embodies one of the basic ideas of process biology, namely that the developmental system is defined, and in part physically produced, by the process of development.

If interaction is a dynamic process, then the temporal dynamics of the interaction may play an independent role in explaining the outcome. This is why many DST advocates have also been attracted to explanations of development that draw on dynamical systems theory (abbreviated here DyST, to avoid confusion). Griffiths and Tabery (2013) argue that there is nothing about the basic idea of dynamic interaction found in DST that *requires* the use of DyST. The example of rat development just given, for example, is a sequential mechanism that can be described without using DyST. But in other cases DyST provides additional explanatory resources (Thelen and Smith 1994; Bechtel and Abrahamsen 2013). DyST exemplifies, even more strongly than the bare notion of developmental dynamics, the idea that developmental outcomes should be explained at the level of the whole system (see chapter 12 for a detailed discussion of DyST).

In this section we have argued that the core principles of DST, epigenesis and developmental dynamics, embody the very same ideas that featured in Waddington's process biology. Developmental outcomes are explained at the systems level, and in identifying the components of a developmental system we start with the developmental process, not the other way around. In the next section we develop these thoughts further by examining how DST has conceptualized the constituents of developmental systems.

## 5. An Ontology for DST: Genomes, Epigenomes, and Developmental Niches

One of the most controversial features of DST is its conceptualization of the developmental system as an organism–environment system. Rather than it being the case that an organism develops in an environment, aspects of the developmental environment are part of the developmental system. As well as talking of developmental systems, advocates of DST have talked of sets, collections, or matrices of developmental resources—and, more recently, of ontogenetic or developmental niches that provide the developmental context for organisms or genomes.

DST has always resisted the belief that there is a single way to divide the inputs to development that will be useful for every scientific question about development (Hinde 1968; Johnston 1987; Oyama 1985). Instead, distinctions should be introduced locally to suit the question at hand. For some purposes, as an alternative to 'organism and environment' or 'genes and environment', the resources that make up a developmental system can be partitioned into three: the genome, the epigenome (chemical modifications of DNA that are transmitted through meoisis), and the developmental niche. Since the fundamental unit of analysis for DST is the complete developmental process or the life cycle, we can think of that process as occurring within, and as feeding forward into the construction of, a developmental system with these three components. Or we can think of the life cycle as consisting of the regulated expression of an epigenetically modified genome through its interaction with a developmental niche.

The genome is a familiar concept, and the epigenome increasingly so. The notion of a developmental niche will be less familiar to many readers. Developmental psychobiologists Meredith West and Andrew King (1987) introduced the term 'ontogenetic niche' to capture the idea that environmental resources form a social and ecological legacy inherited by a developing organism. We have used 'developmental niche' as a synonym for West and King's term (Stotz 2008; Stotz 2010; Griffiths and Stotz 2013). Species-specific phenotypes depend on species-typical environments of development. These are often the result of parental activities, but their construction can also involve other conspecifics, past and present, and, importantly, the offspring itself. The idea of the construction of a developmental niche answers a fundamental question about inheritance: How do parents reliably influence the phenotype of their offspring and promote healthy development? Organisms do not rely on chance to provide their offspring with the resources for normal development: they actively intervene to modify environments to this end. West and King described the ontogenetic niche as an 'information centre' in the sense that it makes the interaction between organism and environment more *specific* than it would otherwise be. The idea of an information centre was initially developed in order to capture the experiences necessary for species-typical learning (Galef and Wigmore 1983). These are the 'aspects of experience' that Lehrman identified as part of the developmental system (Lehrman 1970, 36 and quoted above). However, this idea can be applied to the much broader category of any environmental stimulus that acts as a specific cause of normal development (Griffiths and Stotz 2013).

Dividing the developmental system into genome, epigenome, and developmental niche may be useful in the study of evolution, because it parallels one way of dividing the mechanisms of heredity. It is now fairly conventional to recognize epigenetic heredity mechanisms as a genuine form of heredity alongside genetic inheritance, although arguments about whether these mechanisms have equal *evolutionary* significance continue.[7] But DST proponents, just like other recent theorists

---

[7] It is often asserted that epigenetic change will only affect evolution if the changes themselves persist for more than one generation (e.g. Wilkins 2011). But in conventional quantitative genetics the evolutionary significance of genetics does not result from tracking individual alleles from one generation to the next—quantitative genetics does not do this. Instead, Mendelian assumptions let us work out what phenotypes (and hence fitnesses) will appear in the next generation as a function of the phenotypes in the last generation. Epigenetic and exogenetic inheritance both change this mapping from parental phenotype to offspring phenotype, and therefore affect evolution. Both epigenetic and exogenetic inheritance appear in quantitative genetics as 'parental effects': correlations between parent and offspring phenotypes above and beyond the correlations between parent and offspring genotypes, which are not the result of a shared environment independently influencing both parent and offspring. It has long been understood that one-generation parental effects can substantially alter the dynamics of evolutionary models and can change the state to which a population will evolve as an equilibrium (Lande and Price 1989; Wade 1998). The argument that epigenetic inheritance needs to be stable for several generations to have evolutionary significance appears to be a non sequitur.

(e.g. Jablonka and Lamb 2005), recognize a wider range of heredity mechanisms. It is unfortunate that this wider class of mechanisms is often also referred to as 'epigenetic inheritance', which makes that term ambiguous, as it is used more narrowly to refer only to epigenetic marks inherited through meiosis. In earlier work we have suggested keeping 'epigenetic inheritance' for the narrower class of mechanisms and using West and King's term 'exogenetic inheritance' (West and King 1987: 5) for the broader class of mechanisms. It is this broader class of heredity mechanisms that constructs the developmental niche: 'Organisms construct their life cycles through the interaction of the contents of the fertilized egg, the genome and its narrowly epigenetic surroundings, with a "developmental niche" which is the result of epigenetic inheritance in a wider sense...  "exogenetic inheritance"' (Griffiths and Stotz 2013: 5).

It is worth noting that this broader exogenetic form of inheritance may be more stable than narrow epigenetic inheritance. Some exogenetic inheritance occurs through the induction of epigenetic modifications in the offspring through parental behaviour. This can have long-term, often lifelong effects on the offspring phenotype. In some known cases these offspring phenotypes include the very parental behaviour that induced them, so that the offspring reproduce the effect in the next generation, and so forth (Champagne and Curley 2009). These behaviorally transmitted but epigenetically mediated effects contribute to the long-term stabilization of aspects of the developmental niche, and hence may be more long-lived than meiotic epigenetic inheritance.

It is important not to conflate the developmental niche with the 'niche' of niche construction theory (Odling-Smee et al. 2003). Niche construction theory concerns the influence of past generations on the selective pressures that act on future generations. This activity partially constructs a *selective niche*, the set of parameters that determine the relative fitness of competing types in the population. The *developmental* niche, however, is the set of parameters that must be within certain bounds for an evolved life cycle to occur (or, in more traditional terms, for the organism to develop normally). The two niches will often share many parameters. They are, however, conceptually quite distinct. For example, signals from parent to offspring that induce transgenerational adaptive phenotypic plasticity, as when *Daphnia* signal their offspring to grow additional defences against predators, are a clear example of developmental niche construction: the parent *Daphnia* is structuring the developmental environment of its offspring. But this is no more a case of *selective* niche construction than is the inheritance of an advantageous mutation! The *Daphnia* embryo alters itself to fit the selective environment rather than altering the selective environment.

## 6. DST as a Process Theory of the Organism

One reason why early twentieth-century biologists were drawn to process philosophy was that it offered a 'theory of the organism'—an account of the unity of living systems (see chapter 1). Recent interest in process ontologies for biology has revived interest in the concept of 'genidentity', or identity as continuity of organization (see Guay and Pradeu 2015, as well as chapters 2, 4, 5, and 7 here). Distinct stages are stages of the same entity because one developed from the other, rather than because they share some common properties:

> [Genidentity] says that the identity through time of an entity X is given by the continuous connection of states through which X goes.... In this view, the individual X is never presupposed or given initially, because the starting point is the decision to follow a specific and appropriate *process* P, and the individual X supervenes on this process.... In other words, for the genidentity view, what we single out as an 'individual' is always the by-product of the *activity* that is being followed, not its prior foundation (not a presumed 'thing' that would give its unity to this activity). (Guay and Pradeu 2015: 317–18)

Guay and Pradeu here exemplify themes familiar from our earlier discussion of Waddington and Agar. The persistence of biological form should be explained dynamically, not by the transmission of something that concretely embodies that form. The identity of an individual through time is a dynamic continuity of form. If the fundamental unit of analysis in DST is the developmental process or the life cycle, if heredity in DST is a relation between one life cycle and another, and if natural selection occurs in populations of life cycles (Griffiths and Gray 2001), then DST needs to give an account of the genidentity of these processes. It needs to say where one developmental process ends and the next begins. This problem arises in a dramatic form when organisms have alternating haploid and diploid phases of comparable length (Godfrey-Smith 2015). Is each phase a life cycle, or is a life cycle the combination of a haploid and a diploid phase?

The principle of genidentity of a life cycle also needs to explain how a life cycle can consist of a different series of events in different generations of that cycle. This problem arises in its most dramatic form when a species has a range of substantially different ways to get from conception to death. Some newts, for example, exhibit facultative paedomorphosis, in which individuals respond to differences in their environment either by retaining the morphology of their aquatic, larval stage and becoming reproductively mature in that state or by going through metamorphosis to become a terrestrial 'adult' reproductive form (Denoël, Joly, and Whiteman 2005). Nevertheless, the same issue arises in principle whenever an organism exhibits adaptive phenotypic plasticity, so that successive life cycles in a single lineage do not contain the same developmental events.

DST has often been criticized for replacing the common-sense idea of an individual organism with a novel and nebulous 'system' (Sterelny et al. 1996; Merlin 2010; Pradeu 2010; see also the references in n. 6). This criticism has become increasingly unfair over the past twenty years. It is no overstatement to say that conventional theories of biological individuality are in a state of crisis brought on by new empirical and theoretical developments in biology. These developments include research on evolutionary transitions in individuality, the realization of the extent to which core physiological processes in multicellular organisms are carried out by microbial commensals, the discovery of ever more complex and highly integrated functional associations between microbes themselves, as well as increased attention paid by philosophers and theoretical biologists to the full diversity of life, in all its glorious weirdness! The result has been a wave of new work in philosophy and theoretical biology on the nature of individuality, a literature that shows few signs of reaching a consensus (cf. Calcott and Sterelny 2011; Pradeu 2012; Ereshefsky and Pedroso 2013; Bouchard and Huneman 2013; Guay and Pradeu 2015; see also chapters 9 and 10 here). Statements like the following are not hard to find in the recent literature:

> Individuals can be defined anatomically, embryologically, physiologically, immunologically, genetically, or evolutionarily.... [E]ach stems from the common tenet of genomic individuality: one genome/one organism. As such, all classical conceptions of individuality are called into question by evidence of all-pervading symbiosis. (Gilbert et al. 2012: 325)

It is not very reasonable to complain that DST has a more problematic conception of a biological individual than the traditional organism, when that traditional conception of a population of physiologically integrated cells with a single genotype is itself so widely regarded as problematic. In light of this, the initial response of Griffiths and Gray to this line of criticism continues to be effective. They argued that the idea of an individual organism was in fact quite problematic, and that DST did not need to offer a *more* watertight account of the individuality of developmental processes in order to make itself a viable competitor to conventional accounts of the units of evolution and development (Griffiths and Gray 1994; Griffiths and Gray 2001).

Griffiths and Gray sketched how DST would approach the problem, using much of the same apparatus that biologists were already using to address problems with the traditional conception of an individual organism. They argued that a DST account of the individuality of developmental processes—what we are now calling genidentity— would define individuality in terms of the ability to act as a unit of selection (Griffiths and Gray 1994: 292–8; Griffiths and Gray 2001: 209–14). They drew on accounts of evolutionary individuality from the emerging framework of multilevel selection theory to suggest that 'an individual is a life cycle whose components cannot reconstruct themselves when decoupled from the larger cycle' (Griffiths and Gray 2001: 213) and to recognize that, just like cells, organisms, and superorganisms, life cycles might exist at several different levels of biological organization.

Looking back at theories of the organism in early twentieth-century process biology, we can see a distinct similarity between Griffiths and Gray's ideas about the identity of developmental processes and Agar's intuition that an organismic process is united by its 'subjective aim' or *telos*. A series of developmental events is a single process because those events serve a common evolutionary goal, namely to maximize the representation of cycles descended from them in future generations vis-à-vis the representation of the variant cycles with which they compete. We can draw on conventional evolutionary theory to make this suggestion a little more precise: *an individual life cycle is a token of a life history strategy, and that strategy is its* telos *and its principle of genidentity*.

Life history theory is a powerful and remarkably general framework for addressing many basis questions about organismic design (Stearns 1992; Roff 2002; see also chapter 4). In life history theory, the goal of an organism is to find the optimal way to parcel the resources available to it into offspring. This problem is modelled as the simultaneous optimization of two parameters, the probability of surviving to each age class and the number of offspring produced in each age class, integrated across all age classes. The primary constraint on this optimization problem is the quantity of resources available to the organism. But the problem is also constrained by multiple trade-offs between the two key parameters: an overall trade-off between survival and reproduction; a trade-off between reproduction in the current age class and in later age classes; another between current reproduction and growth, between growth and survival to later age classes, and so forth. Solving this complex optimization problem under different sets of constraints and in different ecological settings leads to the many different life history strategies observed in nature. Since life history theory already conceives of an organism as a series of events (age classes), it is readily applicable to a life cycle that consists of a series of developmental interactions, each one of which moves the life cycle forward.

Life history theory embodies a powerful principle of genidentity, because the evolutionary rationale for the choice of strategy at each life history stage is conditional on what choices have been or will be made at the other stages. A life cycle conceived of as the implementation of a life history strategy is held together by the trade-offs between its stages. If these events were not part of a single life history serving a single, Darwinian *telos*, then they would not trade off against one another in this way. It makes sense for me to accept an elevated risk of cancer in later life in return for *my* increased reproductive success, but not for *your* reproductive success, unless that is discounted by our coefficient of relationship. The life history strategy also defines where one process ends and another begins, namely at the points between which a single set of such trade-offs exists.[8] Life history theory explains how life cycles that do not contain the same developmental events can nevertheless constitute a single lineage of cycles, which succeed in reproducing themselves. Adaptive phenotypic plasticity is part of a life history strategy, and individuals who exhibit different developmental outcomes as a result of this plasticity are individuals who shared the same, plastic strategy.

Introducing a life history perspective makes it clear why it is legitimate for developmental systems theorists to help themselves to whatever is currently the best evolutionary account of biological individuality[9] and to 'processualise' that account. Evolutionary accounts of individuality seek to identify collections of biological material that are evolving as one: they are more or less successful in reproducing themselves as a whole, and that success cannot be reduced to the successes of each part of the whole, or to the success of some larger whole, of which this is a part. Admittedly, many discussions of this problem make it seem a matter of finding which *spatial* parts make a *spatial* whole. But this is an illusion; any such unit will in fact be extended in time and will embody a life history strategy. DST will use this strategy to identify the events that make up a single, processual biological individual.

---

[8] Life history theory is in practice conducted as a branch of population genetics, and we anticipate the objection that the implicit definition of the limits of an individual that we have made use of in this section is in fact derived from genetic identity. But this cannot be the case, as the theory applies perfectly well to asexual organisms where parents and offspring are genetically identical. Moreover, so far as we can see, life history theory could be extended unproblematically to cases in which heredity is epigenetic and exogenetic as well as genetic.

[9] For the current views of one of the present authors, see Bourrat and Griffiths (under review).

## 7. Conclusions

DST has a natural affinity with process views of the organism. The theorists who inspired and created DST all shared the view that development is a dynamic process whose study requires an investigation of its dynamic form as well as of the static constituents on which it draws. In Waddington's case, this conviction was directly inspired by process philosophy. The idea of epigenesis, perhaps the single most important idea in the developmental systems tradition, is fundamentally processual. In development, something new comes into being that is not prefigured in any of the inputs to development. Developmental dynamics, another idea that has been central to all the major contributors to DST, is also essentially processual. The impact of a genetic or environmental factor at some point in development depends on how the organism has developed up to that point. Development is, essentially, a dynamic process and cannot be reduced to a list of ingredients and their interactions. The entities that make up a developmental system—which we can divide for some purposes into a genome, an epigenome, and a developmental niche—are picked out as elements of a single system by the unity of the process to which they contribute, and not vice-versa. That principle of unity—the genidentity of a life cycle—we have argued, is simply its Darwinian *telos*: a life history strategy.

## Acknowledgements

This project/publication was made possible through the support of a grant from the Templeton World Charity Foundation. The opinions expressed in this publication are those of the authors and do not necessarily reflect the views of the Templeton World Charity Foundation. We have benefited greatly from comments by Thomas Pradeu, Warwick Anderson, Peter Fairleigh, and an anonymous referee.

## References

Abir-Am, P. (1987). The Biotheoretical Gathering, Trans-Disciplinary Authority and the Incipient Legitimation of Molecular Biology in the 1930s: New Perspective on the Historical Sociology of Science. *History of Science* 25 (1): 1–70.

Agar, W. E. (1936). Whitehead's Philosophy of Organism: An Introduction for Biologists. *Quarterly Review of Biology* 11 (1): 16–34.

Agar, W. E. (1943). *A Contribution to the Theory of the Living Organism*. Melbourne: Melbourne University Press.

Anderson, W. and Mackay, I. R. (2014). Fashioning the Immunological Self: The Biological Individuality of F. Macfarlane Burnet. *Journal of the History of Biology* 47 (1): 147–75. doi: 10.1007/s10739-013-9352-1.

Bechtel, W. and Abrahamsen, A. A. (2013). Thinking Dynamically about Biological Mechanisms: Networks of Coupled Oscillators. *Foundations of Science* 18 (4): 707–23.

Birch, L. C. (1965). *Nature and God*. London: SCM Press.

Bouchard, F. and Huneman, P. (eds). (2013). *From Groups to Individuals: Evolution and Emerging Individuality*. Cambridge, MA: MIT Press.

Bourrat, P. and Griffiths, P. E. (under review). Multi-Species Individuals.
Bowler, P. J. (2014). *Reconciling Science and Religion: The Debate in Early-Twentieth-Century Britain*. Chicago: University of Chicago Press.
Calcott, B. and Sterelny, K. (eds). (2011). *The Major Transitions in Evolution Revisited*. Vienna Series in Theoretical Biology. Cambridge, MA: MIT Press.
Champagne, F. A. and Curley, J. P. (2009). Epigenetic Mechanisms Mediating the Long-Term Effects of Maternal Care on Development. *Neuroscience and Biobehavioral Reviews* 33 (4): 593–600.
Crick, F. H. C. (1958). On Protein Synthesis. *Symposium of the Society for Experimental Biology* 12: 138–63.
Dawkins, R. (1976). *The Selfish Gene*. Oxford: Oxford University Press.
Denoël, M., Joly, P., and Whiteman, H. H. (2005). Evolutionary Ecology of Facultative Paedomorphosis in Newts and Salamanders. *Biological Reviews of the Cambridge Philosophical Society* 80 (4): 663–71.
Emmet, D. (1932). *Whitehead's Philosophy of Organism*. London: Palgrave Macmillan.
Ereshefsky, M. and Pedroso, M. (2013). Biological Individuality: The Case of Biofilms. *Biology and Philosophy* 28 (2): 331–49.
Ford, D. H. and Lerner, R. M. (1992). *Developmental Systems Theory: An Integrative Approach*. Newbury Park: Sage.
French, S. (2012). The Resilience of Laws. In D. Dieks, W. Gonzalex, S. Hartmann, M. Stoltzner, and M. Weber (eds), *Probabilities, Laws, and Structures* (pp. 187–200). Heidelberg: Springer Science & Business Media.
Galef, B. G. and Wigmore, S. W. (1983). Transfer of Information Concerning Distant Foods: A Laboratory Investigation of the 'Information-Centre' Hypothesis. *Animal Behavior* 31: 748–58.
Gilbert, S. F., Sapp, J., and Tauber, A. I. (2012). A Symbiotic View of Life: We Have Never Been Individuals. *Quarterly Review of Biology* 87 (4): 325–41.
Godfrey-Smith, P. (2000). Explanatory Symmetries, Preformation, and Developmental Systems Theory. *Philosophy of Science* 67: S322–31.
Godfrey-Smith, P. (2015). Individuality and Life Cycles. In T. Pradeu and A. Guay (eds), *Individuals Across the Sciences* (pp. 85–102). New York: Oxford University Press.
Goth, A. (2004). Social Responses without Early Experience: Australian Brush-Turkey Chicks Use Specific Visual Cues to Aggregate with Conspecifics. *Journal of Experimental Biology* 207 (13): 2199–208.
Gottlieb, G. (1970). Conceptions of Prenatal Behavior. In L. R. Aronson, E. Tobach, D. S. Lehrman, and J. S. Rosenblatt (eds), *Development and Evolution of Behavior: Essays in Memory of T. C. Schneirla* (pp. 111–37). San Francisco: W. H. Freeman.
Gottlieb, G. (1997). *Synthesizing Nature-Nurture: Prenatal Roots of Instinctive Behavior*. Hillsdale: Lawrence Erlbaum.
Gottlieb, G. (2001). A Developmental Psychobiological Systems View: Early Formulation and Current Status. In S. Oyama, P. E. Griffiths, and R. D. Gray (eds), *Cycles of Contingency: Developmental Systems and Evolution* (pp. 41–54). Cambridge, MA: MIT Press.
Gray, R. D. (1992). Death of the Gene: Developmental Systems Strike Back. In P. E. Griffiths (ed.), *Trees of Life: Essays in the Philosophy of Biology* (pp. 165–210). Dordrecht: Kluwer.
Griffiths, P. E. and Gray, R. D. (1994). Developmental Systems and Evolutionary Explanation. *Journal of Philosophy* 91 (6): 277–304.
Griffiths, P. E. and Gray, R. D. (1997). Replicator II: Judgment Day. *Biology & Philosophy* 12 (4): 471–92.

Griffiths, P. E. and Gray, R. D. (2001). Darwinism and Developmental Systems. In S. Oyama, P. E. Griffiths, and R. D. Gray (eds), *Cycles of Contingency: Developmental Systems and Evolution* (pp. 195–218). Cambridge, MA: MIT Press.

Griffiths, P. E. and Gray, R. D. (2005). Three Ways to Misunderstand Developmental Systems Theory. *Biology and Philosophy* 20 (2): 417–25.

Griffiths, P. E. and Knight, R. D. (1998). What Is the Developmentalist Challenge? *Philosophy of Science* 65 (2): 253–58.

Griffiths, P. E., Pocheville, A., Calcott, B. Stotz, K., Kim, H., and Knight, R. (2015). Measuring Causal Specificity. *Philosophy of Science* 82 (4): 529–55.

Griffiths, P. E. and Stotz, K. (2013). *Genetics and Philosophy: An Introduction*. New York: Cambridge University Press.

Griffiths, P. E. and Tabery, J. G. (2013). Developmental Systems Theory: What Does It Explain, and How Does It Explain It? In R. M. Lerner and J. B. Benson (eds), *Embodiment and Epigenesis: Theoretical and Methodological Issues in Understanding the Role of Biology within the Relational Developmental System, Part A: Philosophical, Theoretical, and Biological Dimensions* (pp. 65–94). Waltham: Academic Press.

Guay, A. and Pradeu, T. (eds). (2015). *Individuals Across the Sciences*. Oxford: Oxford University Press.

Hall, B. K. (2011). A Brief History of the Term and Concept of Epigenetics. In B. Hallgrimsson and B. K. Hall (eds), *Epigenetics: Linking Genotype and Phenotype in Development and Evolution* (pp. 9–13). Berkeley: University of California Press.

Haig, D. (2004). The (Dual) Origin of Epigenetics. *Cold Spring Harbor Symposia on Quantitative Biology* 69: 67–70.

Hinde, R. A. (1968). Dichotomies in the Study of Development. In J. M. Thoday and A. S. Parkes (eds), *Genetic and Environmental Influences on Behaviour* (pp. 3–14). New York: Plenum.

Hull, D. L. (1988). *Science as a Process: An Evolutionary Account of the Social and Conceptual Development of Science*. Chicago: Chicago University Press.

Jablonka, E. and Lamb, M. J. (2005). *Evolution in Four Dimensions: Genetic, Epigenetic, Behavioral, and Symbolic Variation in the History of Life*. Cambridge, MA: MIT Press.

Johnston, T. D. (1987). The Persistence of Dichotomies in the Study of Behavioural Development. *Developmental Review* 7: 149–82.

Kitcher, P. (2001). Battling the Undead: How (and How Not) to Resist Genetic Determinism. In R. Singh, K. Krimbas, D. Paul, and J. Beatty (eds), *Thinking about Evolution: Historical, Philosophical and Political Perspectives (Festchrifft for Richard Lewontin)* (pp. 396–414). Cambridge: Cambridge University Press.

Lande, R. and Price, T. (1989). Genetic Correlations and Maternal Effect Coefficients Obtained from Offspring-Parent Regression. *Genetics* 122: 915–22.

Lehrman, D. S. (1953). A Critique of Konrad Lorenz's Theory of Instinctive Behavior. *Quarterly Review of Biology* 28: 337–63.

Lehrman, D. S. (1970). Semantic and Conceptual Issues in the Nature-Nurture Problem. In D. S. Lehrman (ed.), *Development and Evolution of Behaviour* (pp. 17–52). San Francisco: W. H. Freeman.

Merlin, F. (2010). On Griffiths and Gray's Concept of Expanded and Diffused Inheritance. *Biological Theory* 5 (3): 206–15.

Moore, C. L. (1984). Maternal Contributions to the Development of Masculine Sexual Behavior in Laboratory Rats. *Developmental Psychobiology* 17: 346–56.

Moore, C. L. (1992). The Role of Maternal Stimulation in the Development of Sexual Behavior and Its Neural Basis. *Annals of the New York Academy of Sciences* 662: 160–77.

Moss, L. (1992). A Kernel of Truth? On the Reality of the Genetic Program. *Philosophy of Science Association Proceedings*, vol. 1: 335–48. East Lansing, MI: Philosophy of Science Association.

Nanney, D. L. (1958). Epigenetic Control Systems. *Procedings of the National Academy of Sciences* 44 (7): 712–17.

Noble, D. (2006). *The Music of Life: Biology Beyod Genes*. Oxford: Oxford University Press.

Odling-Smee, F. J., Laland, K. N. and Feldman, M. W. (2003). *Niche Construction: The Neglected Process in Evolution*. Princeton: Princeton University Press.

Okasha, S. (2009). Causation in Biology. In H. Beebee, P. Menzies, and C. Hitchcock (eds), *The Oxford Handbook of Causation* (pp. 707–25). Oxford: Oxford University Press.

Oyama, S. (1985). *The Ontogeny of Information: Developmental Systems and Evolution*. Cambridge: Cambridge University Press.

Oyama, S. (2000). Causal Democracy and Causal Contributions in Developmental Systems Theory. *Philosophy of Science* 67: S332–S347.

Oyama, S. (2002). The Nurturing of Natures. In A. Grunwald, M. Gutmann, and E. M. Neumann-Held (eds), *On Human Nature: Anthropological, Biological and Philosophical Foundations* (pp. 163–70). New York: Springer.

Pradeu, T. (2010). The Organism in Developmental Systems Theory. *Biological Theory* 5 (3): 216–22.

Pradeu, T. (2012). *The Limits of the Self: Immunology and Biological Identity*, trans. by E. Vitanza. Oxford: Oxford University Press.

Robert, J. S. (2004). *Embryology, Epigenesis and Evolution: Taking Development Seriously*. Cambridge: Cambridge University Press.

Robert, J. S., Hall, B. K., and Olson, W. M. (2001). Bridging the Gap between Developmental Systems Theory and Evolutionary Developmental Biology. *Bioessays* 23: 954–62.

Roff, D. A. (2002). *Life History Evolution*. New York: W. H. Freeman.

Rosenberg, A. and McShea, D. W. (2008). *Philosophy of Biology: A Contemporary Introduction*. New York: Routledge.

Stearns, S. C. (1992). *The Evolution of Life Histories*. Oxford: Oxford University Press.

Stegmann, U. E. (2012). Varieties of Parity. *Biology and Philosophy* 27 (6): 903–18.

Sterelny, K., Dickison, M., and Smith, K. (1996). The Extended Replicator. *Biology & Philosophy* 11 (3): 377–403.

Stotz, K. (2006). Molecular Epigenesis: Distributed Specificity as a Break in the Central Dogma. *History and Philosophy of the Life Sciences* 28: 533–48.

Stotz, K. (2008). The Ingredients for a Postgenomic Synthesis of Nature and Nurture. *Philosophical Psychology* 21: 359–81.

Stotz, K. (2010). Human Nature and Cognitive–Developmental Niche Construction. *Phenomenology and the Cognitive Sciences* 9: 483–501.

Stotz, K. and Allen, C. (2012). From Cell-Surface Receptors to Higher Learning: A Whole World of Experience. In K. Plaisance and T. Reydon (eds), *Philosophy of Behavioural Biology* (pp. 85–123). Boston: Springer.

Stotz, K. and Griffiths, P. E. (2017). Genetic, Epigenetic and Exogenetic Information. In R. Joyce (ed.), *Routledge Handbook of Evolution and Philosophy* (pp. 106–19). London: Routledge.

Tabery, J. G. (2009). Difference Mechanisms: Explaining Variation with Mechanisms. *Biology & Philosophy* 21 (5): 645–64.

Thelen, E. and Smith, L. (1994). *A Dynamic Systems Approach to the Development of Cognition and Action*. Cambridge, MA: MIT Press.

Thorpe, W. H. (1956). *Learning and Instinct in Animals*. Cambridge, MA: Harvard University Press.
Waddington, C. H. (1942). The Epigenotype. *Endeavour* 1: 18–20.
Waddington, C. H. (1952). The Evolution of Developmental Systems. In D. A. Herbert (ed.), *Proceedings of the Twenty-Eighth Meeting of the Australian and New Zealand Association for the Advancement of Science* (pp. 155–9). Brisbane: A. H Tucker, Government Printer.
Waddington, C. H. (1975). *The Evolution of an Evolutionist*. Ithaca, NY: Cornell University Press.
Wade, M. J. (1998). The Evolutionary Genetics of Maternal Effects. In T. A. Mousseau and C. W. Fox (eds), *Maternal Effects as Adaptations* (pp. 5–21). Oxford: Oxford University Press.
Waters, C. K. (2007). Causes That Make a Difference. *Journal of Philosophy* 104 (11): 551–79.
Weber, M. (2006). The Central Dogma as a Thesis of Causal Specificity. *History and Philosophy of the Life Sciences* 28 (4): 595–609.
West, M. J. and King, A. P. (1987). Settling Nature and Nurture into an Ontogenetic Niche. *Developmental Psychobiology* 20: 549–62.
West, M. J. and King, A. P. (2008). Deconstructing Innate Illusions: Reflections on Nature–Nurture–Niche from an Unlikely Source. *Philosophical Psychology* 21: 383–95.
West, M. J., King, A. P., and Arberg, A. A. (1988). The Inheritance of Niches. In E. M. Blass (ed.), *Developmental Psychobiology and Behavioral Ecology*, vol. 9 of *The Handbook of Behavioral Neurobiology* (pp. 41–61). New York: Plenum Press.
Wilkins, A. (2011). Epigenetic Inheritance: Where Does the Field Stand Today? What Do We Still Need to Know? In S. B. Gissis and E. Jablonka (eds), *Transformations of Lamarckism: From Subtle Fluids to Molecular Biology* (pp. 389–93). Cambridge, MA: MIT Press.
Williams, G. C. (1992). *Natural Selection: Domains, Levels and Challenges*. New York: Oxford University Press.
Woodward, J. (2011). Causes, Conditions, and the Pragmatics of Causal Explanation. In G. J. Morgan (ed.), *Philosophy of Science Matters: The Philosophy of Peter Achinstein* (pp. 247–57). Oxford: Oxford University Press.

# 12

# Waddington's Processual Epigenetics and the Debate over Cryptic Variability

*Flavia Fabris*

## 1. Introduction

From the perspective of the emerging framework of the extended evolutionary synthesis (EES), developmental plasticity appears as an epigenetic process rather than as the predetermined outcome of a program encoded in the genome (Pigliucci and Muller 2010). The EES further suggests that organisms play a central role in evolution, though it rejects the causal primacy of natural selection. It argues that organisms act in conjunction with selection in shaping their particular developmental trajectories. Development is thus viewed as an act of 'co-construction' involving the organism and its environment, which together constitute an integrated causal system (see Oyama et al. 2001; Gilbert and Epel 2015).[1]

Back in the 1940s, Conrad Hal Waddington anticipated various features of the EES framework in his pioneering research into the nature of developmental plasticity. Waddington believed that the study of the phenotype should include an account of how the developing organism is able to change in response to genetic and environmental perturbations. This represented a major point of disagreement with the architects of the modern synthesis. While the latter explained the phenotype straightforwardly as a genotypic product, Waddington suggested that phenotypes are temporally extended epigenetic *trajectories*, as opposed to being entities that occur 'one gene at a time' or 'one trait at a time' (Wilkins 2008). Moreover, he hypothesized that development plays a directive role in evolution (Waddington 1942).

---

[1] The idea of conceptualizing the relation between organism and environment as a 'co-construction process' is often associated with Richard Lewontin, who famously stressed that 'organisms fit the world so well because they have constructed it' (Lewontin 1996: 10; see also Lewontin 1983, 2000). An organism, according to this view, is an active agent that is capable of constructing an environment suited to its own ends. The importance of this perspective for evolutionary biology has been recently highlighted by the theory of niche construction (Odling-Smee et al. 2003), which maintains that environments are shaped by the niche-constructing activities of organisms. In this view, the environment is not only deemed to be involved in the selection of genetic variation (as conceived by the modern synthesis), but is also considered to be instrumental in the developmental construction of the organism's phenotype.

According to Waddington, the process of development takes places concurrently at different levels of organization. At the cellular level, for instance, it takes the form of a complex exchange of information between the nucleus and the cytoplasm that contributes to the construction of the phenotype. This act of construction is directed, in the sense that it constrains the trajectory of development through a series of successive bifurcations that lead to a stable phenotypic state.[2] For Waddington, the stability of the phenotype reflects a dynamic balance between robustness and plasticity. That is to say, the phenotype exhibits a tendency to resist internal and external perturbations, thereby buffering the effects of the variability responsible for evolutionary change.

In agreement with the proponents of the modern synthesis, Waddington maintained that genetic variability accumulates over time and forms an evolutionary substrate, which is 'hidden' from the purview of natural selection. This 'hidden variability' constitutes an active potential that explains how organisms adapt to their environment when they are subjected to rapid environmental changes. However, in contrast to proponents of the modern synthesis, Waddington argued that this source of evolutionary change should not be understood as a concrete repository filled with neutral genetic information, randomly created and progressively stored. Rather, this substrate should be construed as an *epigenetic process* that builds up variability in response to perturbations. Importantly, Waddington believed that this hidden variability could help account for the 'inheritance of acquired characters' in a neo-Darwinian way. In particular, he appealed to it to explain the phenomenon of *genetic assimilation*, whereby environmentally induced phenotypic variation becomes constitutively produced (i.e. it loses dependency for its expression on the original environmental trigger and becomes an inherited trait).

Recently, a lively debate has re-emerged concerning the nature of this hidden variability—or 'cryptic variability', to use Waddington's preferred terminology—and its putative role in explicating the genetic assimilation of acquired characters. According to some authors, acquired characters are manifested when a certain stress threshold is passed; and a buffering mechanism of preexisting genetic variation is invoked to account for how this phenomenon occurs (Rutherford and Lindquist 1998; Rutherford 2000; Masel 2013). Other authors, however, have argued that this model fails to explain how cryptic variability causally accounts for the generation of acquired characters, as these can also be produced by *de novo* (as opposed to preexisting) mutations (Specchia et al. 2010).

The processual perspective that is currently resurfacing in the philosophy of biology (see Dupré 2012 and the rest of the chapters in this volume) provides an ideal tool for shedding light on this ongoing scientific debate. This perspective calls for the adoption of a dynamic understanding of living entities, as opposed to the more conventional one afforded by traditional substance ontology.[3] Substances are

---

[2] By 'bifurcations' I mean sudden qualitative changes in the developmental trajectory.

[3] I am aware that 'substance ontology' is a rather broad term for a wide range of positions within metaphysics, and my usage of it is not intended to do justice to all of them. Here I am using it primarily for the purposes of contrasting this position with the processual one I describe in the next section.

typically conceived of as static entities that exist prior to any forms of change or activity. In contrast, process ontology takes change to be fundamental and regards seemingly static entities as transient stabilities of continuous processes. What I will argue in this chapter is that the contemporary debate over cryptic variability reflects different ontological assumptions about the nature of development, and these assumptions result in conflicting conceptualizations of the relationship between variability and inheritance. By examining the ontological commitments of the participants in this debate, I will show that we are better able to make sense of the different ways in which cryptic variability is currently being construed.

The structure of the chapter is as follows. In section 2 I distinguish between substance-ontological and process-ontological frameworks for biology, paying particular attention to the conflicting presuppositions of the modern synthesis on the one hand, and of Waddington's epigenetics on the other. In section 3 I discuss Waddington's epigenetics and its grounding in dynamical systems theory. Then, in section 4, I discuss robustness and plasticity as opposite yet complementary features of development, understood, in Waddingtonian terms, as a homeorhetic (as opposed to a homeostatic) process. After this, in section 5, I discuss the evolutionary implications of Waddington's view of development as a homeorhetic process. Finally, in section 6, I analyse the two sides of the current debate over cryptic variability by examining their respective models of the phenomenon. As my examination will illustrate, the conflict between the assumptions of a substance view and those of a process view reflects the different capacities of these models to make sense of the inheritance of acquired characters.

## 2. Substance versus Process: Two Conflicting Ontologies for Biology

Although organisms are the main targets of biological inquiry, the actual nature of organisms seldom receives any attention. This is because the question 'What is an organism?' is often viewed as an insoluble problem, mostly left to philosophers of biology and biologists with philosophical inclinations. The problem can be tackled from different epistemological perspectives. From a synchronic point of view, we can scrutinize organisms by considering their essential morphological structures at a specific timescale: an organ, a tissue, a cell, and so on. The two fundamental questions to answer, in this case, are (1) how organisms are organized into different hierarchical levels; and (2) how these levels relate to one another. Alternatively, we can conceive of organisms as diachronic entities, looking at how they change over time, in order to come to terms with how these structures are modified during development. In this case, the fundamental questions to be answered concern the persistent nature and the directive character of organisms. It is important to realize, however, that we are likely to arrive at different answers to these questions, depending on the ontological assumptions we begin with.

For example, let us assume that we are in the business of studying the developmental trajectory of a cell, from its initial undifferentiated state to its final differentiated adult state. To explain this transition, we need to look at the cell at each temporal instant. At time $t_1$ the cell is entirely undifferentiated, at time $t_2$ it is at a more differentiated stage than at time $t_1$, and so on, until its process of differentiation ends. According to this picture, an organism's development—as a multicellular lineage—is the temporal succession of all its stages of differentiation, each one exhibiting a specific structure and characteristic properties. Development, in other words, is the sum of all stages of cellular differentiation, in orderly succession. We can think of this as a *substance* view of development.

It could be argued, however, that thinking about development as a succession of discrete, ordered stages does violence to the very notion of development. Without wanting to dispute the heuristic usefulness of this perspective as a means of modelling change, we might remain unconvinced by a view of development that portrays it as a static order composed of atomic temporal elements (i.e. stages). Rather, we might want to say that what development is corresponds to the whole, temporally extended process—the one that denotes the organism's entire life cycle. In this view, development is neither localizable nor decomposable. Any particular developmental stage is a mere abstraction, cut off from the integrated spatio-temporal process. We can think of this as a *process* view of development.

Methodologically, if we adopt a substance view, we will start by examining each stage of development independently and explain the dynamicity of the ontogenic process in terms of the temporal succession of these stages. On the other hand, if we adopt a process view, we will examine the process of development as a temporally extended whole and explain the stability of each stage within the developmental cycle.

The substance and the process views I have just discussed can be said to correspond to two opposing ontologies of the living world (see Dupré 2012). In the biology of the past century, these two ontologies have had a deep impact on the way biologists have conceived of organisms, their development, and their evolution. The modern synthesis seems to have been quite clearly associated with the substance view. The understanding of development shared by most of the architects of the modern synthesis was characterized by a sort of revamped preformationism: the idea that developmental change consists in an 'unfolding' or 'unrolling' of something that is already present and in some way preformed (Oyama 2000; Lewontin 2000). This neo-preformationist conception legitimated the substance view of development as an ordered succession of stages by emphasizing the role of DNA as the instigator of this process. Development in this picture is a deterministic process executed by a *genetic program*, which stores all the necessary instructions for the construction of the organism.[4] From this perspective, development is not all that different from a domino sequence whose initial trigger is provided by the decryption of the genetic text. Today neo-preformationism is assumed in genomic quantitative analyses that attempt to formalize the hereditary material as a static structure bearing a 'code script' (as Schrödinger famously called it)

---

[4] One of the major implications of a substance ontology in biology is *essentialism*, the thesis that 'essential properties' are necessary and sufficient conditions for the existence of things. Within a substance-ontological framework, essential properties are offered as a causal explanation of why a thing persists despite the changes it undergoes (see chapter 1 for a more detailed exposition of this point). The metaphor of the genetic program is a clear example of essentialism, in which DNA sequences are conceived of as the essential properties that determine the developmental outcome of organisms.

for the architecture of the organism. The organism, in this view, is reduced to an epiphenomenon of its genes (Gilbert and Sarkar 2000; Nicholson 2014).

Although preformationism was—and arguably, still is—the predominant framework in the explanation of development (see e.g. Eric Davidson's much publicized work on decoding the regulatory genome of the sea urchin embryo), not too long ago a number of biologists put forward a dynamic view of biology, and of development in particular, grounded in process ontology. Just as the modern synthesis was being forged, a different intellectual movement in biology developed, known as *organicism*, which sought to articulate a non-reductionist and dynamic understanding of organisms inspired by the writings of Alfred North Whitehead. The organicists, who included Waddington, viewed organisms not as organized assemblages of material things but as integrated functional units in which the whole and the parts causally influence each other (see Peterson 2014, Nicholson and Gawne 2015, and chapters 1, 7, 11, and 13 here).

The tacit ontological disagreement between organicists and proponents of the modern synthesis was not confined to the nature of organisms and their development, but also extended to the relation between inheritance and evolution. According to the substance-ontological framework of the modern synthesis, development is construed as a morphological change that has no impact on inheritance or on evolution. The organism is a genetic product; genes are just code scripts, inert entities waiting to be read and transcribed. In contrast to this picture, the organicists, and Waddington in particular, believed that development exerts a direct influence on both inheritance and evolution. Specifically, Waddington suggested that organisms are capable of shaping their own developmental trajectories, thereby actively contributing to their adaptive persistence. He coined the term 'epigenetics' to designate the causal study of molecular processes that sustain organisms through their development. Following his processual inclinations, he articulated a novel account of organisms as dynamical developmental systems, in which genes do not act as scripts but interact with their transcriptional products and their cytoplasmic environment. The genome does not really instruct development (as simplistically assumed by the modern synthesis); it is rather the developmental system as a whole that actively reads and interprets the genome. Having briefly outlined Waddington's processual view of the organism, let us now examine his epigenetic theory in more detail.

## 3. Waddington's Epigenetics in the Context of Dynamical Systems Theory

Waddington sought to provide a firm scientific grounding for his process-ontological views. He found such a foundation in the theoretical framework known as dynamical systems theory (DyST).[5] Contemporary proponents of DyST describe changes in

---

[5] For the purposes of this chapter, I will use the abbreviation DyST to distinguish dynamical systems theory from developmental systems theory (which is often referred to in the philosophy of biology by its acronym, DST). On the relation between DyST and DST, see chapter 11.

a system as transitions between stages (e.g. Slack 2002; Fagan 2012; Ferrel 2012; Huang 2012; Jaeger and Monk 2014). The dynamism of the system resides in the succession of one stage after another and, as the system develops, this succession describes the progression of the system through time. Applied to embryogenesis, DyST enables the system's development to be analysed in terms of a succession of stages. A change in the system is conceptualized as the shift from one stage, with its particular embryological features, to another. We should not be tempted, however, to understand development as something *composed* of these stages, as stages are not ontological constituents of the developmental process. Instead, the stages described by DyST are mere abstractions: mathematical representations of stable sections of the whole process. In general, a system that undergoes such changes is referred to as a 'coupled dynamical system'. This is how Waddington thought of organisms.

From the perspective of DyST, a dynamical system can be described with a set of independent variables that represent numerically the properties that the concrete system manifests. The value of all variables at any given time is referred to as the 'state' of the system (see Van Gelder 1998). Coupled dynamical systems are those whose variables vary in relation to the external parameters that lead the system to shift from one state to another. Parameters, in turn, often depend on the state of the system. For instance, a particular environment modifies and is modified by the particular organism that inhabits it. For this reason, the variables of the dynamical system and the parameters with which it is coupled can be thought to constitute a larger causal system (provided that both jointly account for the changes it undergoes). This larger system as a whole, Waddington claimed, is stabilized by a flux of activities, which flow in and out of it (Waddington 1959).

During development, nucleus and cytoplasm interact by means of feedback loops to selectively stimulate different networks of genes. These networks do not always act in the same sort of way. Rather, they exhibit dynamic behaviours only in specific embryological stages, in which they can be switched (see Waddington 1956, 1961). Waddington called the switching of these networks 'competence'. Each competence is a stage at which the organism may change its developmental path. However, the more the organism develops, the faster it loses its competence to differentiate further. Thus, the developmental space progressively restricts or constrains the possible developmental outcomes—and thereby the possible phenotypes—that the organism might exhibit through time. Waddington called this phenomenon of progressive restriction of competences 'canalization'.

In order to better convey his idea of canalized development, in *The Strategy of the Genes* Waddington (1957) represented the progressive restriction of competences as a multidimensional developmental surface of the egg cell, the famous 'epigenetic landscape' model (see Figure 12.1). The landscape depicts the development of the cell from its undifferentiated, regionalized state to its final stage. The surface is composed of 'chreods', which are formed under the action of selective pressures (see ibid., 29). Chreods, which in the landscape resemble a system of valleys and pits, act as possible developmental pathways, connecting early undifferentiated cytoplasmic states— which in Waddington's picture are represented at the top of the hill—with alternative

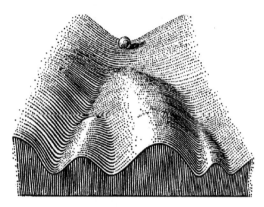

**Figure 12.1** Pictorial representation of (a part) of an epigenetic landscape. From Waddington 1957.

discrete end states.[6] As the ball rolls down the hill, it exhibits a tendency to restrict its developmental potentialities—its competence—over time (Waddington 1940, 1956, 1957, 1961, 1968a; see also Gilbert 2000; Slack 2002; Fagan 2012; Griffiths and Stotz 2013; Fusco et al. 2014).

As we have just noted, the general developmental potentialities of the organism during embryogenesis become restricted over time, leading to stable end states or *steady states*.[7] These resulting states correspond to the final stage of the landscape, when the cell comes to be fully differentiated. It is important to note, however, that competences are also steady states. All of these stages are phases of dynamic equilibria or stable metabolic regimes. Accordingly, phenotypes are best construed not as fixed end products, but as transient (and therefore partial) stabilities of a continuous process (Waddington 1968a).

To illustrate how transient stability underlies a processual view of the phenotype, consider the life cycle of *Drosophila melanogaster*. At an early developmental stage, the drosophila is just a larva. It exhibits certain stable features and characteristics particular to this stage. Later on, a subsequent form of partial stability emerges when the drosophila becomes a pupa. The pupa and the larva differ substantially in their properties and abilities. Nevertheless, both are temporal stages of the same developmental entity (i.e. the drosophila). Thus, the phenotype of the drosophila is not just its end state, that is, the adult stage it reaches in the latter phase of its life cycle. Rather, the phenotype comprises each dynamically stable regime that the organism manifests in the course of its developmental trajectory. If we want to consider the larva and the

---

[6] Waddington coined the term 'chreod' to refer to a canalized trajectory. A modern compound based on the combination of two ancient Greek words, namely the verb χρῆ ('is necessary, must') and the noun ὁδός ('way, road'), this term should be understood as meaning 'obliged pathway' (< *chrē-hodos*; see Waddington 1961, 1968b). In Waddington's own words, it indicates 'a path of change which is determined by the initial conditions of a system and once entered upon cannot be abandoned' (Waddington 1961: 64).

[7] The notion of steady state refers to 'a time-independent state where the system remains constant as a whole and in its (macroscopic) phases, though there is a continuous flow of component materials' (Bertalanffy 1968: 125). The thermodynamic basis of the steady state concept and its relevance to a processual biology are examined in chapter 7.

pupa as being one and the same organism, they have to be understood as stable phases of a plastic, temporally extended developmental entity; that is, as stages at which the developmental entity, though adaptively plastic, manifests specific as stable equilibria. It is by virtue of these equilibria that the organism possesses different competences over time. Let us now see how this dynamic stability is achieved.

## 4. Development as the Homeorhetic Balance between Robustness and Plasticity

I have already mentioned that, even though developmental competences degrade over time, the organism's ability to maintain itself in a dynamic equilibrium and to interact adaptively with its environment remains invariant. While its adaptive capacities diminish as development unfolds, the organism's compensatory ability seems to be 'hardwired' (i.e. it does not deteriorate). In his experimental work Waddington identified a global characteristic—an independent parameter of the system as a whole—that enables it to resist perturbations (be they genetic or environmental), while at the same time allowing developmental resources to be used in different ways during embryogenesis. The phenotype, understood by Waddington in processual terms, manifests two specific (though seemingly contradictory) properties: *robustness* and *plasticity*. Robustness is the ability to display stability in the face of perturbations, and it accounts for how an organism that develops (and is thus subjected to internal and external changes) maintains certain configurations constant for prolonged periods of time. Plasticity, on the other hand, is the capacity to alter these same configurations over time, in other words, to produce different yet coherent somatic states in response to internal and external stimuli.

Waddington proposed to take robustness and plasticity as mutually complementary properties resulting from a single compensatory process, which he referred to as 'homeorhesis'. In analyses of development, it is crucial to distinguish homeorhesis from homeostasis. Waddington appealed to cybernetics to make sense of this distinction. Within a cybernetic framework, homeostasis refers to the tendency of a system to revert to its original configuration and thereby to restore the stability of its internal environment. It thus denotes the persistence through time of a specific static configuration. There are many physiological examples of homeostatic responses. The way in which the human body reacts to sudden changes in temperature is one of them. Despite wide variations in the external temperature, the body constantly maintains its internal temperature within a relatively narrow interval; if the boundaries of this interval are violated, the body faces dramatic consequences that may, in some cases, lead to death. The body's internal temperature, in other words, does not vary as a consequence of changes in the temperature of the external environment. Assuming initial parity, if the latter drops to 12°C, the former does not experience the same decrease. The body here exhibits a homeostatic response: it actively maintains its internal temperature following changes in the external temperature so as to ensure that it does not fall outside a narrowly defined physiological range.

Homeorhesis, like homeostasis, also refers to the regulatory ability of a system to reach a dynamic form of stability by compensating against perturbations within a specific range of responses. The difference is that, while a homeostatic response concerns the maintenance of a single, fixed steady state, a homeorhetic response refers to the *stability of the temporally extended trajectory of the system* (Waddington 1957, 1968b). In the context of development, homeorhesis is what enables the embryo to undergo differentiation in a robust yet plastic way, guaranteeing the normal operation of the physiological processes in the system that otherwise would be disrupted. Homeorhesis is, in a sense, a more general biological property than homeostasis, as it maintains the organism in a stable state over the course of its development by means of a range of specific homeostatic responses.

Having explained how homeorhesis differs from homeostasis, we can now consider its bearing on the epigenetic landscape (Figure 12.1). As noted above, the chreods in this model correspond to the slopes along the valley. Waddington referred to this as the 'chreodic profile': the branching system of temporal trajectories through which an egg cell is robustly canalized. The stability of the entire developmental pathway is explained by the robustness of chreods, which preserves the system in 'continual change along a certain pathway' (Waddington 1977: 105). At the same time, robustness also secures the system's plasticity, that is, its ability to produce different somatic states in response to stimuli. From both a mathematical and a cybernetic point of view, the concept of homeorhesis is intimately related to the concept of chreod. Influenced by his friend, the French topologist René Thom, Waddington came to understand chreods in mathematical terms. Within the framework of DyST, a chreod can be described 'as a multidimensional domain that contains a vector field converging on a time extended attractor' (Waddington 1968a: 526). This move allowed Waddington to offer a more precise description of the contrast between homeostasis and homeorhesis. In Waddington's own words,

the fact that the vector fields converge on to the attractors gives rise to a process of homeorhesis, which can be contrasted with the more conventional idea of homeostasis in which the vector fields converge on to a *static point* which is not time-extended.

(Waddington 1968a: 526; emphasis added)

To summarize, then, Waddington introduced the concept of homeorhesis in order to resolve the apparent contradiction between the robustness of development and the plasticity of phenotypes as developmental products. Homeorhetic processes differ from homeostatic ones, even though both lead to stability. So far, I have discussed homeorhesis in the context of development. Since Waddington's primary intellectual enterprise was to bridge the gap between embryology, genetics, and evolution, I will now discuss how the compensatory behaviour described by homeorhesis is crucial for understanding the inheritance of acquired characters.

## 5. Evolutionary Implications: The Genetic Assimilation of Acquired Characters

In the previous section I have discussed inflexibility (i.e. robustness) and flexibility (i.e. plasticity) as the characteristic features of development. The developmental system is inflexible in the sense that it is canalized in a robust way despite environmental or genetic perturbations. And it is flexible in the sense that its ontogenetic path can be modified through different steady states. The homeorhetic regulatory capacity of ontogeny connects developmental products—the phenotypes—with the underlying epigenetic network. Together, they guarantee robustness in a highly plastic developmental path.

Waddington believed that the genome is continuously modified by the behaviour of the developmental system, that is, by the way in which the system acts and reacts to inducing signals. However, these signals cannot influence the genome directly; in other words, they cannot induce any heritable modification (i.e. variability) by simply exerting a selective pressure. Local inducing signals are 'absorbed' by the system and may result in internal genetic changes, but these local changes do not affect the global epigenetic network; they remain underneath the chreodic profile (Waddington 1957). Although they are causally active, they are prevented from affecting the global ontogenetic path. The regulatory behaviour of the whole epigenetic network guarantees the invariance of the phenotypes, thereby explaining why the phenotype is not constantly reshaped by external stimuli. According to Waddington, this regulatory behaviour is what grounds the evolutionary capacity for the genetic assimilation of acquired characters (Waddington 1953, 1956, 1975).

Waddington's understanding of development and of its regulatory capacities marked an important departure from the modern synthesis, as it provided an alternative explanation of variability and its evolutionary role. The architects of the modern synthesis were deeply suspicious about the possibility of acquiring variation by means of simple interactions with the environment. Recognizing the possibility of this acquisition seemed to imply collapsing the canonical Weissmanian distinction between germ plasm and soma. Inheritance and evolution were considered to be relevant only to the former. Conflating the two lines was tantamount to dismissing the causal priority of the DNA over all other cytoplasmatic elements. And, in the context of molecular biology, it also meant violating the so-called central dogma (see Jablonka and Lamb 2005).

The modern synthesis, as a framework that combines Darwinism and Mendelism, links phenotypic variability to genetic variability, which exists independently of the environmental context. The ability of lineages to undergo evolutionary change is taken to derive from their tendency to produce and preserve genetic variability. In this view, what matters for evolution is the heredity of a static, genetic substratum. This substratum resides in the nuclear chromosomes and is not significantly affected by its surroundings.

The idea of hidden genetic variability dates back to the forging of the modern synthesis, when population geneticists provided explanations of the evolution of organisms in terms of changes in the distribution of genetic alleles in populations. Theodosius Dobzhansky, in particular, postulated the existence of a pool of allelic variants capable of bringing about beneficial effects to phenotypes in unusual circumstances. More specifically, he claimed that the adaptive plasticity of organisms should be explained by an underlying store of concealed genetic variability, which underlies the organism's capacity to adapt to rapid environmental changes. The causal effects of random hidden mutations, he argued, fuels evolution by producing beneficial phenotypes under new circumstances (see Paaby and Rockman 2014; Ledon-Retting et al. 2014).

From a modern synthesis perspective, the phenotypic manifestation of hidden variability is explained by the crossing of a threshold in a polygenic system. Since polygenic inheritance is related to the small additive effect of many alleles, the effect of each individual allele is too small to be noticed by natural selection (Mather 1941, 1943). This variation accumulates over time without being manifested; as long as it stays under a certain threshold, it remains phenotypically hidden. However, when organisms are in unusual or stressful environmental circumstances, this concealed variation can be phenotypically manifested, and therefore selected. In this view, the capacity to inherit phenotypic plasticity resides in a static, preexisting, adaptive substratum.

Waddington's proposal is fundamentally different. The proponents of the modern synthesis, in order to explain adaptive plasticity, postulated what present-day scholars refer to as an 'evolutionary substratum' (Paaby and Rockman 2014). Waddington, by contrast, claimed that this so-called 'substrate' is rather something that is actively maintained and modified by the developmental system as a whole. Following Dobzanhsky (1951), he agreed that hidden variability explains how organisms adapt to their environment when subjected to rapid environmental changes. But, in contrast to Dobzhansky, he stressed that the crypticity of these mutations must be understood in terms of the canalization of development, and hence in terms of the *epigenotype* (i.e. the whole dynamic developmental network that connects the genotype to the phenotype; see Waddington 1942).[8] What this implies is that this hidden variability is an effect of the controlled action of the organism on its own development. The more the organism is able to organize and shape such variability, the more it is able to develop selectable adaptive capacities. In this way, hidden variability should be understood, according to Waddington, by appealing to the homeorhetic dynamics of the canalization of development, instead of by resorting to the progressive accumulation of discrete genetic variation.

Genetic assimilation, as we have already noted, is the process by which particular phenotypic answers to environmental stimuli can be incorporated into the genotype through a process of selection. Waddington called these phenotypes 'heritable acquired characters', because they can be manifested again in the offspring even in the absence of the original environmental stimulus (Waddington 1975). He found evidence of genetic assimilation in drosophila in experiments he conducted in the 1950s. By using a heat-shock treatment to induce the *crossveinless* phenocopy and, in another experiment, by using ether to induce *bithorax*, Waddington showed how somatic mutations could become heritable.[9] He demonstrated that these characters, if selected for a certain number of generations in the presence of the same stress, could be assimilated in the germline. Interpreting these results in Darwinian (as opposed to Lamarckian) terms, Waddington appealed to the modern synthesis notion of hidden variability. However, he referred to it as *cryptic* variability, as it

---

[8] On the concept of the 'epigenotype', see Gilbert 2012 and Jablonka and Lamm 2012.

[9] The term 'phenocopy' refers to 'the appearance of a phenotype which mimics that produced by some recognized mutant allele' (Goldschmidt 1935, quoted in Waddington 1975: 77). Waddington adopted this term in his experiments on the genetic assimilation of acquired characters (see Waddington 1953) to describe the mutation that results in broken posterior *crossveins* in the wings of drosophila.

seemed to be concealed beneath the robustness of the developmental paths. Waddington's idea was that, when developmental trajectories diverge from their ordinary path, processes that buffer variability act so as to guarantee the control of their alterations (Waddington 1977). However, if an environmental stress is strong enough to overcome this robustness, an alternative adaptive path can become available through the expression of genetic variants. These variants can then be selected and become heritable through the process of genetic assimilation.

Overall, by applying the concept of homeorhesis in the context of evolution, Waddington was able to offer a novel Darwinian interpretation of the old Lamarckian problem of the inheritance of acquired characters. In the next section, I shall discuss how Waddington's hypothesis is being interpreted by contemporary authors.

## 6. Assessing Two Contemporary Models of the Canalization of Development

In current discussions, authors seldom distinguish between the concept of *hidden* variability, postulated by the architects of the modern synthesis, and Waddington's own preferred notion of *cryptic* variability. This is the case in spite of the fact that most of them ostensibly lean towards an interpretation of adaptive plasticity along Waddingtonian lines. Despite decades of empirical research, the nature of this variability is still hotly debated (see e.g. Rutherford and Lindquist 1998; Queitsch et al. 2002; Specchia et al. 2010). As Waddington's original model involved a buffering mechanism that conceals variability, several contemporary authors have attempted to investigate its molecular makeup. In this final section, I will examine the two predominant models for the buffering process of the canalization of development originally hypothesized by Waddington. Ultimately, I will argue that only one of the two is consistent with Waddington's homeorhetic processual view, which conceives of variability as an epigenetic phenomenon.

### 6.1. How the two models differ in their interpretation of cryptic variability

In contemporary genetics and molecular biology, there are two major models of canalization, both involving the function of the chaperone heat-shock protein Hsp90—a protein that responds naturally to environmental changes. Although these models are able to account for the same experimental data regarding the genetic assimilation of acquired characters, they differ radically in their respective assumptions concerning variability. While one conceives the phenotypic variation as the manifestation of preexisting hidden genetic information (Rutherford and Lindquist 1998), the other explains acquired characters in terms of *de novo* mutations (Specchia et al. 2010). Let us now examine both models and compare their respective ontological assumptions regarding the nature of variability and its putative role in the inheritance of acquired characters.

In 1998 the geneticists Rutherford and Lindquist observed that, in flies and plants, a reduced activity of Hsp90 was correlated with the induction of a wide spectrum of phenotypic variants (Rutherford and Lindquist 1998). These variants play an evolutionary role: under selection they can be assimilated and passed to subsequent generations, even after the function of Hsp90 is restored. This experimental finding provided the impetus for new investigations of Waddington's theory of genetic assimilation, taking Hsp90 as the molecular buffering mechanism responsible for the phenomenon. In both flies and plants, when the activity of Hsp90 is reduced via silencing from mutations or treatment with inhibitors, a wide spectrum of phenotypic variants is induced (Rutherford and Lindquist 1998; Queitsch et al. 2002). It was thus suggested that Hsp90 acts as a sort of 'capacitor' of morphological evolution, that is, as an on/off switching mechanism that affects 'the visibility of a particular set of conditionally neutral variants' (Masel 2013: 1). Hsp90 buffers preexisting accumulated genetic variation and, when it is inhibited, it induces the release of this variation.

The second model, formulated by Specchia and colleagues, suggests a totally different understanding of the canalization of development (Specchia et al. 2010). According to these researchers, the buffering, storage, and release of preexisting genetic variation does not represent a general evolutionary mechanism for the genetic assimilation of acquired characters. Instead, they hypothesize that Hsp90 regulates silencing mechanisms mediated by Piwi-interacting RNAs (piRNAs); a class of germline-specific small interfering RNAs (siRNAs) known to play a role in maintaining repetitive sequences and transposons in a repressed state (Piacentini et al. 2014). When Hsp90 is altered and the respective products are absent, transposable elements move into the germline. Consequently, a wide range of phenotypic variants can potentially be induced (Piacentini et al. 2014). Heritable phenotypes are thus explained by *de novo* mutations correlated with the insertion of transposons (Elgart et al. 2015; Paaby and Rockman 2014; Sato and Siomi 2010). This model results in a rather different interpretation of Waddington's theory. The phenomenon of genetic assimilation is not explained by an actualization of a hidden inner genetic variability, but as a co-selection process between transposable elements and the germline (Piacentini et al. 2014).

Specchia and colleagues are not alone in advocating this novel conception of variation and its role in adaptive plasticity. Decades earlier, Barbara McClintock argued that genomes are dynamic entities that do not react in a programmed fashion, but rather constantly reorganize their resources (McClintock 1984). She suggested that the activation of transposons by stress reshapes the genome, leading to the formation of new species through the creation of new adaptive resources. Specchia and colleagues have confirmed this evolutionary role, and have suggested that genomes exhibit an adaptive plasticity that enables organisms to reshape their developmental paths. According to this view, when an organism is subjected to an environmental stress, processes of silencing transposons—which usually keep them in a repressed state—can be disrupted, inducing the mobilization of transposons that become active. This process is thus deemed to be responsible for the creation of new variability, which can serve as a potential source of adaptive evolution (see Piacentini et al. 2014 for a more detailed discussion of these models).

Overall, according to the first model, proposed by Rutherford and Lindquist, cryptic variability is grounded in the progressive accumulation of genetic information.

The buffering mechanism contributes to create a storage of nucleotidic information that is gradually accumulated during the organism's development. According to the second model, proposed by Specchia and colleagues, cryptic variability is a *process* that leads to modifications at the level of the whole epigenotype. Development is construed here as the interplay between the flexibility and the inflexibility of the genome and of its products. While in the first model the focus is on how genetic variability is maintained and conserved, in the second it is more on how it is actually produced. By resisting the assumption of a preexisting repository of genetic mutations as the main source of variability, the second model provides a more accurate interpretation of Waddington's conception of cryptic variability. Indeed, for Waddington, crypticity is a property of the developing organism *as a whole*, rather than one localized in specific preexisting genetic mutations.

## 6.2. Do both models capture the homeorhetic nature of canalization?

In evolutionary biology, plasticity is often described as a 'conservative or *homeostatic* factor in evolution that prevents, rather than promotes, change' (West-Eberhard 2003: 8, emphasis added). More generally, evolutionary biologists have tended to construe evolution in terms of a frequency-dependent equilibrium theory and to describe evolutionary causes (such as mutation, selection, and drift) as departures from this equilibrium. This is exemplified by the Hardy-Weinberg law of equilibrium, which was traditionally adopted by population geneticists to explain the evolutionary tendency of populations to resist modification 'unless perturbed by definite force or chance events' (ibid.). This orthodox viewpoint contrasts with Waddington's homeorhetic conception of developmental stability, which he regarded as the active, dynamic process that buffers modifications despite constant environmental and genetic perturbations. The concept of cryptic variability is, for Waddington, an expression of this developmental capacity. In the remainder of this section, I shall assess whether the contemporary models we have just discussed conform to the requirements of Waddington's concept of canalization: more specifically, whether they capture its homeorhetic (rather than homeostatic) nature.

Let us start with the first model. According to Rutherford and Lindquist, Hsp90 is responsible for the buffering of canalization in development. In normal conditions, Hsp90 conceals hidden genetic mutations and buffers the system against internal and external perturbations. However, when its function is compromised, these variants become manifested, leading to a wide range of abnormal phenotypes in both flies and plants (Rutherford and Lindquist 1998; Queitsch et al. 2002). The data from this model are represented as a spike threshold (see Sato and Siomi 2010: 2). In these graphs, variations accumulate quantitatively in peaks or spikes. These spikes represent genetic variations that tend to accumulate over time during the organism's development. According to this model, once the spike is high enough to pass the threshold, the information accumulated is manifested. When the organism is highly stressed, these thresholds become lowered and the variation depicted in the spikes begins to be uncovered. As a consequence, preexisting variations become manifested in the system. The manifestation of variants is thus a responsive phenomenon, represented by a threshold that moves up and down in response to environmental stimuli. There is therefore no room in this model for any compensatory process of the system as a whole vis-à-vis its own variability. Canalization is just an equilibrium between a pool of stored variants and the environmental stimuli. As a result, this model fails to capture Waddington's homeorhetic conception of canalization.

In the second model, the buffering is not performed by any particular molecular component. Here the threshold is not described in homeostatic terms, with respect to a specific parameter (as in the postulated buffering role of Hsp90 in the first model). Instead, it is represented by a bundle of different parameters that contribute to describe the phenomenon. Consequently, there is no need to postulate a specific threshold responsible for the manifestation of hidden variants, as variability itself does not derive from a storage of preexisting genetic information. In this model there is no causal priority of the genetic variants with respect to the manifestation of acquired characters. Consequently, this model can be said to capture Waddington's homeorhetic conception of canalization.

## 7. Conclusions

In this chapter I have examined Waddington's epigenetics and his processual account of cryptic variability, which is based on his notion of homeorhetic stability. I have shown that Waddington offered an account of development that represented a novel alternative to more traditional preformationist interpretations, which in many ways have prevailed to the present day. I have discussed how Waddington's epigenetics, cashed out in terms of DyST, brought together his process-ontological inclinations (deriving from his adoption of Whitehead's antireductionist metaphysics) and his cybernetic understanding of development. His proposal, homeorhetic stability, represents an important and, as it turns out, still valuable way of understanding (a) phenotypes as developmental products, and (b) development itself as a dynamic balance between robustness and plasticity. Moreover, Waddington's concept of homeorhesis is able to account for the phenomenon of genetic assimilation, and thus enables us to bridge the gap between evolutionary and developmental explanations. Today, Waddington's epigenetics is the standard theoretical reference point for the molecular explanation of developmental canalization. I have argued, however, that not all contemporary models satisfy the requirements that the dynamic nature of homeorhesis imposes on the explanation of the genetic assimilation of acquired characters. More generally, I submit that whether any such model ultimately succeeds in contributing to the emerging conceptual framework of the EES will depend on its capacity to capture the original insights of Waddington's processual theory of epigenetics.

## Acknowledgements

I would like to thank Dan Nicholson and John Dupré for the opportunity to participate in this project. I'm also grateful for the comments and advice I received from two anonymous referees, Dan Nicholson, Eva Jablonka, Marion Lamb, Paul Griffiths, Jan Baedke, Alan Love, Staffan Muller-Wille, Rani Lill Anjum, Stephen Mumford, Andrea Raimondi, Davide Serpico, Anne Sophie Meincke, and the Exeter ProBio Group.

# References

Bertalanffy, L. von. (1968). *General System Theory: Foundations, Development, Applications.* New York: George Braziller.

Dobzhansky, T. (1951). *Genetics and the Origin of Species.* New York: Columbia University Press.

Dupré, J. (2012). *Processes of Life: Essays in the Philosophy of Biology.* Oxford: Oxford University Press.

Elgart, M., Snir, O., and Soen, Y. (2015). Stress-Mediated Tuning of Developmental Robustness and Plasticity in Flies. *Biochimica et Biophysica Acta* 1849 (4): 462–6.

Fagan, M. B. (2012). Waddington Redux: Models and Explanation in Stem Cell and Systems Biology. *Biology & Philosophy* 27 (2): 179–213.

Ferrell, J. E. Jr. (2012). Bistability, Bifurcations, and Waddington's Epigenetic Landscape. *Current Biology* 22 (11): R458–R466.

Fusco, G., Carrer, R., and Serrelli, E. (2014). The Landscape Metaphor in Development. In A. Minelli and T. Pradeu (eds), *Toward a Theory of Development* (pp. 114–28). Oxford: Oxford University Press.

Gilbert, S. F. (2000). Diachronic Biology Meets Evo-Devo: C. H. Waddington's Approach to Evolutionary Developmental Biology. *American Zoologist* 40: 729–37.

Gilbert, S. F. (2012). Commentary: 'The Epigenotype' by C. H. Waddington. *International Journal of Epidemiology* 41 (1): 20–3.

Gilbert, S. F. and Epel, D. (2015). *Ecological Developmental Biology: The Environmental Regulation of Development, Health, and Evolution.* Sunderland: Sinauer Associates.

Gilbert, S. F. and Sarkar, S. (2000). Embracing Complexity: Organicism for the 21st Century. *Developmental Dynamics* 219 (1): 1–9.

Goldschmidt, R. (1935). Gen und Ausseneigenschaft: I. Zeitschrift für induktive Abstammungs- und Vererbungslehre 69: 38–69.

Griffiths, P. and Stotz, K. (2013). *Genetics and Philosophy: An Introduction.* Cambridge: Cambridge University Press.

Huang, S. (2012). The Molecular and Mathematical Basis of Waddington's Epigenetic Landscape: A Framework for Post-Darwinian Biology? *BioEssays* 34 (2): 149–57.

Jablonka, E. and Ehud Lamm, M. J. (2012). Commentary: The epigenotype—A Dynamic Network View of Development. *International Journal of Epidemiology* 41 (1): 16–20.

Jablonka, E. and Lamb, M. J. (2005). *Evolution in Four Dimensions: Genetic, Epigenetic, Behavioral, and Symbolic Variation in the History of Life.* Cambridge, MA: MIT Press.

Jaeger, J. and Monk, N. (2014). Bioattractors: Dynamical Systems Theory and the Evolution of Regulatory Processes. *Journal of Physiology* 592: 2267–81.

Ledon-Rettig, C. C., Pfennig, D. V., Chunco A. J., and Dworkin, I. (2014). Cryptic Genetic Variation in Natural Populations: A Predictive Framework. *Integrative and Comparative Biology* 54 (5): 783–93.

Lewontin, R. C. (1983). Gene, Organism, and Environment. In D. S. Bendall (eds), *Evolution from Molecules to Men* (pp. 273–85). Cambridge: Cambridge University Press.

Lewontin, R. C. (1996). Biology as Engineering. In J. Collado-Vides, B. Magasanik, and T. F. Smith (eds), *Integrative Approaches to Molecular Biology* (pp. 1–10). Cambridge, MA: MIT Press.

Lewontin, R. C. (2000). *The Triple Helix: Organism, Environment, and Evolution.* Cambridge, MA: Harvard University Press.

Masel, J. (2013). Q and A: Evolutionary Capacitance. *BMC Biology* 11: 103.

Mather, K. (1941). Variation and Selection in Polygenic Characters. *Journal of Genetics* 41: 159–93.

Mather, K. (1943). Polygenic Balance in the Canalization of Development. *Nature* 151: 68–71.
McClintock, B. (1984). The Significance of Responses of the Genome to Challenge. *Science* 226: 792–801.
Nicholson, D. J. (2014). The Machine Conception of the Organism in Development and Evolution: A Critical Analysis. *Studies in History and Philosophy of Biology and Biomedical Sciences* 48: 162–74.
Nicholson, D. J. and Gawne, R. (2015). Neither Logical Empiricism nor Vitalism, but Organicism: What the Philosophy of Biology Was. *History and Philosophy of the Life Sciences* 37 (4): 345–81.
Odling-Smee, F. J., Laland, K. N., and Feldman, M. W. (2003). *Niche Construction: The Neglected Process in Evolution*. Princeton: Princeton University Press.
Oyama, S. (2000). *The Ontogeny of Information. Developmental Systems and Evolution*. Durham, NC: Duke University Press.
Oyama, S., Griffiths, P. E., and Gray, R. D. (2001). *Cycles of Contingency: Developmental Systems and Evolution*. Cambridge, MA: MIT Press.
Paaby, A. B. and Rockman, M. V. (2014). Cryptic Genetic Variation, Evolution's Hidden Substrate. *Nature Reviews Genetics* 15: 247–58.
Peterson, E. L. (2014). The Conquest of Vitalism or the Eclipse of Organicism? The 1930s Cambridge Organizer Project and the Social Network of Mid-Twentieth-Century Biology. *British Journal for the History of Science* 47 (2): 281–304.
Piacentini, L., Fanti, L., Specchia, V., Bozzetti, M. P., Berloco, M., Palumbo. G., and Pimpinelli, S. (2014). Transposons, Environmental Changes, and Heritable Induced Phenotypic Variability. *Chromosoma* 123: 345–54.
Pigliucci, M. and Muller, G. B. (2010). *Evolution: The Extended Synthesis*. Cambridge, MA: MIT Press.
Queitsch, C., Sangster, T. A., and Lindquist, S. (2002). Hsp90 as a Capacitor of Phenotypic Variation. *Nature* 417: 618–24.
Rutherford, S. L. (2000). From Genotype to Phenotype: Buffering Mechanisms and the Storage of Genetic Information. *BioEssays* 22: 1095–5.
Rutherford, S. L. and Lindquist, S. (1998). Hsp90 as a Capacitor for Morphological Evolution. *Nature* 396: 336–42.
Sato, K. and Siomi, H. (2010). Is Canalization More Than Just a Beautiful Idea? *Genome Biology* 11 (3): 109.
Slack, J. M. W. (2002). Conrad Hal Waddington: The Last Renaissance Biologist? *Nature Reviews Genetics* 3: 889–95.
Specchia,V., Piacentini, L., Tritto, P., Fanti, L., D'Alessandro, R., Palumbo, G., Pimpinelli, S., and Bozzetti, M. P. (2010). Hsp90 Prevents Phenotypic Variation by Suppressing the Mutagenic Activity of Transposons. *Nature* 463: 662–5.
Van Gelder, T. (1998). The Dynamical Hypothesis in Cognitive Science. *Behavioral and Brain Sciences* 21 (5): 615–28.
Waddington, C. H. (1940). *Organisers and Genes*. Cambridge: Cambridge University Press.
Waddington, C. H. (1942). The Epigenotype. *International Journal Epidemiology* 41: 10–13.
Waddington, C. H. (1953). Genetic Assimilation of an Acquired Character. *Evolution* 7: 118–26.
Waddington, C. H. (1956). *Principles of Embryology*. London: George Allen & Unwin.
Waddington, C. H. (1957). *The Strategy of the Genes: A Discussion of Some Aspects of Theoretical Biology*. London: George Allen & Unwin.
Waddington, C. H. (1959). *Biological Organization: Cellular and Sub-Cellular*. London: Pergamon Press.
Waddington, C. H. (1961). *The Nature of Life*. London: The Scientific Book Club.

Waddington, C. H. (1968a). Towards a Theoretical Biology. *Nature* 218: 525–7.
Waddington, C. H. (1968b). *Towards a Theoretical Biology*, vol. 1: *Prolegomena*. Edinburgh: Edinburgh University Press.
Waddington, C. H. (1975). *The Evolution of an Evolutionist*. Ithaca, NY: Cornell University Press.
Waddington, C. H. (1977). *Tools for Thought*. St Albans: Paladin.
West-Eberhard, D. M. J. (2003). *Developmental Plasticity and Evolution*. Oxford: Oxford University Press.
Wilkins, A. S. (2008). Waddington's Unfinished Critique of Neo-Darwinian Genetics: Then and Now. *Biological Theory* 3 (3): 224–32.

# 13
# Capturing Processes
## The Interplay of Modelling Strategies and Conceptual Understanding in Developmental Biology

*Laura Nuño de la Rosa*

## 1. Introduction

Historically, embryology (and developmental biology later on) has concerned itself with the process of development from egg or seed to adult. Following processes is a—if not *the*—characteristic activity of science, and visual representations play a major role in this endeavour (Griesemer 2007). 'Developmental series' are the main illustrations of ontogeny, and, since they emerged in the late nineteenth century, they have shaped the conceptualization, comparison, and explanation of developmental processes. As any other representation of natural processes, developmental series involve a form of abstraction or idealization wherein some features are selected while others are ignored, depending on the epistemic goals of the inquiry (Love 2010a; see also Griesemer 2007). Developmental series have undergone deep transformations in virtue of the different research goals they have served over the years (Hopwood 2005, 2007), but there are three main aspects of development that have been repeatedly abstracted away during their construction. First, in developmental series normal stages are meant to represent 'normal' development (i.e. the developmental pattern that is common to most members of a species) and thus explicitly exclude individual embryonic variation. Second, in representing ontogeny as a linear and temporally delimited sequence that covers a certain period of the life history of an organism, developmental series delimit the temporal boundaries of development, marking a beginning and an end of development. And, third, developmental series represent ontogeny as a sequence of successive forms or structures rather than as a continuous process.

These three dimensions of idealization have served different epistemic goals in different historical periods. In the comparative framework of pre-evolutionary morphology, individual variation needed to be abstracted away in the establishment of homologies, and the representation of development as a sequence of successive stages enabled embryologists to compare the structures characteristic of each stage and to trace them back to their embryonic precursors. With the advent of evolutionary biology, the identification of homological relationships (reinterpreted as evidence of common descent) remained the main goal of morphology (Brigandt 2003). In Haeckel's theory of recapitulation, ontogenetic stages were seen as a record of evolutionary patterns, and comparative embryology was devoted to uncover the parallelisms between series of ontogenetic and phylogenetic patterns. At the end of the nineteenth century, experimental embryology radically transformed the epistemic goals of embryology. Embryologists abandoned the description and comparison of developmental patterns subordinated to the study of evolution, and reoriented their efforts towards the experimental study of the causal processes responsible for the generation of form (Maienschein 1991). In this new disciplinary context, developmental series became a tool for standardizing the experimental work carried out by researchers in different laboratories (Hopwood 2005, 2007).

The multifaceted idealization of development embodied by developmental series has therefore been instrumental to the development of embryology as a discipline. However, abstraction practices can also constrain our ability to recognize and study certain phenomena (Love 2010a), particularly when the methods of representation are conflated with the phenomenon itself. For well over a century, biologists have challenged the three aforementioned dimensions of idealization involved in the construction of developmental series. Probably the most recurrent concern has been with the first of these, that is, with the abstraction of inter- and intraspecific developmental variation. The exclusion of interspecific variation in Haeckel's comparative plates led to overestimating homological relationships and underrating the role of heterochrony as a mechanism of evolutionary change (de Beer 1958; Richardson 1995). As for intraspecific variation, Alan Love has investigated how the practice of developmental staging has led to a neglect of the phenomenon of developmental plasticity in contemporary evo-devo (Love 2010a). With regard to the second form of idealization, there is a long tradition in developmental biology, from Joseph H. Woodger to current evo-devo biologists, that has opposed the linear sequencing of development (Woodger 1929; Minelli 2003; Bonner 2015). Instead of viewing development programmatically, as a teleological phenomenon where the egg is the first stage of a process that leads to the creation of a mature organism, these authors take the life cycle to be the primary research subject of developmental biology, the egg and the adult no longer being the beginning and the end of a linear causal process, but rather temporal parts of a life history where change takes place at different speeds (Nuño de la Rosa 2010).[1] Finally, with regard to the third dimension of idealization outlined above, many authors have warned that the characterization of development as a sequence of disconnected morphological stages prevents the recognition of the profoundly dynamic nature of developmental processes. In this chapter I will focus on this particular idealization of developmental series.

The process approach to development has deep roots in the history of embryology. Historians of biology have recently shown that the principles of development worked out by the founders of modern embryology went far beyond the mere temporalization of ontogeny. For example, Karl Ernst von Baer aimed not simply to explain

---

[1] On this conception of development, see also chapter 11.

temporal changes, but to inscribe the generation of new organisms into a continuous process (Vienne 2015). In this respect, the notion of 'rhythm' played a major role in his explanation of morphogenesis (Wellmann 2015). After a long period of supremacy of the morphological approach to development, in the 1950s, organicist biologists, deeply influenced by Whitehead's process metaphysics, revolted against the anatomist's timeless concept of the organism.[2] Adopting a radically dynamic perspective on the living organization, they defined the organism as a spatio-temporal process and understood organic form 'as a cross-section through a spatio-temporal flow of events' (Bertalanffy, quoted in Rieppel 2006: 531). Opposing the view of development as a series of discrete patterns, Conrad H. Waddington became the main advocate of a new 'diachronic biology', which understood organisms as developmental systems that undergo an endless process of becoming. To be able to account for this essentially dynamic character of the living organization, Waddington needed to introduce a whole range of new dynamic terms such as 'chreod', 'canalization', and 'homeorhesis' (Waddington 1957), together with new visual representations of development such as his influential epigenetic landscape. After suffering a long period of exclusion from mainstream biology, Waddington's process approach to development and evolution has been recovered and taken up by current developmental and evolutionary biologists (Gilbert 2000; Jamniczky et al. 2010).[3]

Nevertheless, the process approach to development has not resulted in a transformation of the major representations of development—at least not until very recently. While the dynamic understanding of development has had an impact on the formal explanatory models of development and on the graphical illustrations used to represent such models (Baedke 2013; Fusco et al. 2014), developmental series of normal stages have remained the main visual representations of ontogeny. Time-lapse microscopy was introduced in developmental biology all the way back in the early twentieth century, and Waddington himself declared, already in 1962, that films of development were necessary for counteracting the deanimating effects of the microscope (Landecker 2006: 126). However, not until the first decade of the twenty-first century have the new 'digital embryos' (Keller et al. 2008), built from in vivo microscopy, started to replace the static series of normal stages as the standard representations of development. The recent convergence of microscopy, molecular, and computer technologies in live imaging is at present prompting a shift in our perception of development and in the theories we use to conceptualize it. Taking into account the radical interweaving of technological and conceptual advances in the history of embryology (Hopwood 1999), this chapter looks at how new techniques for reconstructing developmental processes are contributing to a processual understanding of development.

I proceed as follows. First, I investigate how time-lapse imaging has brought with it a radical dynamization, not only of the descriptive models of development, but also of the theories of development themselves (section 2). Next, I explore the role played

---

[2] Organicist conceptions of the organism are examined in chapters 1 and 7.
[3] It has been argued that Waddington's commitment to Whitehead's process metaphysics was one of the reasons for his lack of impact on the evolutionary biology of his time (Peterson 2011). For detailed discussions of Waddington's processual views, see chapters 11 and 12.

by imaging technologies in the return of organicism to developmental biology, and I argue that the reduction of the methodological trade-off between spatial and temporal resolution rendered by 4D imaging has also served to shorten the theoretical distance between processual and structural approaches to development (section 3). Finally, I focus on how the revolution in computational imaging and visualizing techniques is opening up new ways of explaining (not only describing) developmental processes (section 4).

## 2. In Vivo Imaging and the Four-Dimensional Conceptualization of Life

A fundamental aspect of the descriptive modelling of developmental processes is the construction of 'embryological time' (Griesemer 2000, 2002). One of the main motivations for building staging systems has been the desire to ascertain the age of an embryo. Individual organisms develop at different speeds, and therefore the chronological age of an embryo (typically defined as the number of hours or days after fertilization) is not an accurate indicator of its structural age. This lack of correlation between chronological and structural age makes developmental stages the very markers of embryological time. In other words, it is the qualitative morphological features, not the chronological age of an embryo, that indicate the phase of development (i.e. the developmental stage) an embryo belongs to. In the staging systems based on morphological criteria, embryological timing is reduced to temporal 'ordering', that is, to the coordination of sequential events in time, which tends to assume an underlying causal relationship between the successive states (Webb and Oates 2015).

However, classical staging systems based on morphological criteria are not without problems (Boehm et al. 2011). First, embryonic parts may develop at different speeds, so that the same embryo may simultaneously belong to different developmental stages, depending on the organ that one takes as a reference. Second, some developmental events occur in shorter time frames than those captured by time-point microscopy and are therefore excluded from the characterization of normal developmental stages. Finally, there are also dynamic traits of embryos that cover longer periods of development but cannot be captured by time-point microscopy. This is the case of those types of biological timing, such as interval timing and rhythms, that cannot be reduced to a mere sequence of events (Webb and Oates 2015). Interval timing is a process with a well-defined duration between two events. Here the key feature is not so much the sequence of successive states as the kinetics of the process. Intracellular developmental timers, for example, control when vertebrate precursor cells stop dividing and start differentiating. Rhythms are continuous sequences of repetitive events with regular periods. A classical example is the sequential formation of body segments characteristic of most animal phyla.

A continuous description of the developmental state of an embryo over time is the obvious way to overcome the limitations of time-point microscopy. In this regard, the introduction of live imaging constitutes a radical revolution in developmental biology. While the sequential representations of normal stages involve thousands of individuals, each of them fixed at different moments of its life, in vivo imaging makes it possible to witness the development of one and the same organism (Kelty and Landecker 2004). More importantly, in vivo imaging renders a much more accurate temporal resolution of development than in vitro microscopy. One of the main advantages of in vivo imaging over time-point microscopy is that it allows us to capture developmental processes over time through the use of time-lapse imaging. While traditional microscopy acquires images at distinct time points (e.g. daily), in time-lapse microscopy living embryos are cultured on an imaging device that captures images almost continuously, at much shorter intervals (e.g. a minute). The resulting film (a series of film frames) is then projected at a much higher speed (e.g. sixteen frames per second; see Wong et al. 2013). The ability to manipulate the time of observation through projection (minutes) compared to the time of the experiment (hours or days) turns time-lapse imaging into an instrument for the investigation of biological time (Landecker 2006). The biologist and cinematographer Jean Comandon was particularly aware of the theoretical potentialities of microcinematography as an instrument of research. According to Comandon, just as microscopes had opened up the spatial dimension of investigation, the film camera enhanced the *temporal* dimension of perception, allowing us to see well-known phenomena in a new way or to discover previously imperceptible processes (Landecker 2005, 2006, 2009).[4] Ever since the introduction of time-lapse imaging, the development of methods for acquiring, analysing, and understanding images in order to generate numerical information has been the main technological breakthrough in enhancing the manipulability of time. With the so-called 'computer vision', filming allows the observer not only to see events that are not visible in static images, but also to subsequently deal with time as a measurable variable in experiments (Stramer and Dunn 2015).

The origins of time-lapse imaging go back to the invention of the cinematograph in the late nineteenth century and were intimately intertwined with the study of life, particularly of morphogenesis. In fact, one of the first time-lapse films (made in 1907 by the Swiss biologist Julius Ries) was a two-minute film of the process of sea urchin fertilization and development (Landecker 2009). A century later, making movies of cells, tissues, and embryos has become a familiar practice in the laboratory, to the point that it can be said without exaggeration that 'most cell biologists these days are also cinematographers' (Stramer and Dunn 2015: 9).[5] The ability to track in real time the dynamic processes that take place at the cellular and tissue levels has revolutionized developmental biology over the last two decades. Since the middle of the last

---

[4] Curiously enough, Henri Bergson—one of the most passionate advocates of a process metaphysics—did not show any enthusiasm for the film camera as an enhancer of temporal perception (Totaro 2001). On the contrary, in his *Creative Evolution*, he argued that the cinematographer was an instrument analogous to the human intellect, insofar as it acted as a mechanism to spatialize time (Bergson 1911). According to Bergson, the intellect is by nature a spatializing mechanism that can only acquire knowledge by expressing movement, the essence of reality, in static and discontinuous terms. In the same way, the camera breaks down real movement into a series of still frames and then re-creates (through projection) an illusion of movement. The cinema, just like our intellect, is incapable of capturing what Bergson calls 'duration', a process where past, present, and future overlap.

[5] For a review of the current state of the art in time-lapse microscopy imaging, see Meijering et al. 2008.

decade, examples of continuous live recording of single organs have increased greatly, and the visualization in real time of the early development of whole embryos is revolutionizing the field of human-assisted reproduction, where non-invasive methods are imperative (Wong et al. 2013). For instance, using time-lapse videography, Connie Wong and co-workers have demonstrated that two morphologically identical eight-cell human embryos that would have been classified under the same developmental stage in a time-point analysis were actually products of different developmental processes (Wong et al. 2010).

Nonetheless, in toto imaging, or the dynamic imaging of whole embryos over the entire course of development, remains a major challenge (Keller et al. 2008). There are still many species, especially mammals, whose development can be studied only by interrupting it at static time points; and representing ontogeny as a continuous process is not a trivial task for bioinformatics (Davidson and Baldock 2001). As a consequence, most embryo atlases still present a stage-by-stage view of development. Nonetheless, in the last few years, new methods have appeared that reconstruct the continuity of development in embryos whose growth cannot be recorded in vivo (Wong et al. 2015) and light sheet-based fluorescent microscopy has been applied to analyse cellular dynamics in the early development of two model organisms: zebrafish (Keller et al. 2008) and drosophila embryos (Tomer et al. 2012). In both cases, the application of automated image analysis provides 'digital embryos', that is, 'comprehensive databases of cell positions, divisions, and migratory tracks' of the early development of entire embryos (Keller et al. 2008: 1065). Although the dynamic imaging of whole embryos over the entire course of embryogenesis is still an unrealized project, in toto imaging promises to replace the static and discontinuous views of ontogeny rendered by developmental series.

How have these new dynamic modes of representing ontogeny impacted our understanding of development? On the one hand, the introduction of time-lapse imaging in biological research has rendered visible a whole new realm of processual phenomena that were too slow to reach the threshold of human perception and that escaped static means of representation such as histology, photography, or drawing (Landecker 2006). But the adoption of time-lapse imaging does not only allow us to see new phenomena hidden by the static representations of development generated by in vitro microscopy. A much more telling indicator of how the introduction of new dynamic modes of representation has influenced *theories* of development is how they have allowed embryologists to see well-known phenomena in a radically new light. Hannah Landecker has shown how, when the first films of biological processes were projected in the early twentieth century, the experience of watching living processes on screen enabled the perception of familiar phenomena as dynamic entities rather than as fixed structures. This shift in perception was seen as the result of a kind of 'reanimation' of the still images of biological phenomena rendered by static means of representation, and the resulting films were interpreted as a manifestation of the (processual) essence of life (Landecker 2005). Also in botanical research, the first time-lapse images of plant growth were seen as providing evidence of the vitality of plants (Gaycken 2012). Importantly, we should not interpret the visualization of processes as a mere illustration of dynamic theories of life. Rather, these new images involve a 'reanimation' of the actual theories themselves

(Landecker 2012). Since the introduction of in vivo imaging, embryologists have endorsed the realistic character of embryo films in place of the static representations of ontogeny rendered by developmental series. The representation of development as a series of still images of embryos at different discrete developmental stages is now seen as an artificial representation of what is actually a continuous process. Embryo films have put into motion (and have therefore reanimated) the series of sections, photographs, and drawings of dead, fixed embryos that had previously been the only perceptual evidence of ontogeny.

In her research on the evolution of biological theories in the field of cell biology, Landecker has argued that the introduction of fluorescent imaging has been the main technological breakthrough in the development of a new dynamic perception, and hence conception, of life. The visualization over time of molecular structures through fluorescence first allowed us to 'watch the genetic code running', but has ended up challenging the gene-centred view of biology altogether:

[P]rocesses which were thought to be programmed...—particularly those unfolding in organismal development—are shown by live-cell imaging to arise out of a messier, looser set of molecular relations and interactions...With live-cell imaging and a host of other developments in protein sciences, it seems that the cell composed of functional structures is dissolving into molecular entities that constantly but always changeably constitute structures. It is not so much that the structures begin to move, but movements—for example in the assembly and self-organization of the cytoskeleton—begin to constitute structure. (2012: 393–4)

The dynamic visualization of subcellular processes has radically changed contemporary theories of development, bringing with it new process-based theories of 'the inner life of the cell', to use the title of the famous Harvard animation of the workings of a blood cell. According to Landecker, these new theories of life endorse a new 'molecular vitalism' (Kirschner et al. 2000) where the gene has lost its causal supremacy with regard to the other molecules of the cell (RNA, proteins, or calcium ions), and explanations are sought in terms of macromolecular self-organization (Landecker 2012). While this might certainly be the case for cell biology, I do not think that the major theoretical issue that is at stake in the new dynamic images of development is 'the molecular foundation of life' (ibid., 393). As I will argue in the next section, biological disciplines, notably developmental biology, deal with different levels of organization, and the new modes of capturing biological processes are playing a major role in characterizing and understanding the dynamic nature of biological hierarchy, particularly at the cell and tissue levels.

## 3. Resolution, Contextuality, and the Return of Organicism

Organicism flourished between the First and the Second World War as a materialist but non-reductionist alternative to the dichotomous explanations of life given by mechanicists and vitalists (Nicholson and Gawne 2015). The organism was conceived of as an integrated whole whose parts, essentially related to one another, cannot be understood in isolation. Furthermore, organicists conceded a central role to the irreducible hierarchical nature of biological organization: the principles that govern the behaviour of the parts at a higher level cannot be deduced from principles that apply to lower levels of the hierarchy. After the rise of molecular biology in the 1950s and the ascendancy of the modern synthesis view of evolution, organicist philosophy of biology was expelled from mainstream biology. However, since the early 1980s, an increasing dissatisfaction with adaptationism and genetic reductionism has led to a revival of organicism in several fields of biology (Gilbert and Sarkar 2000).

In developmental biology, the two main arguments against the reductionist and deterministic view of development as programmed in the genes echo the major theoretical tenets of early organicism. First, developmental biologists have emphasized the importance of the cellular, tissue, and organismal context in understanding the role of genes in development (Laubichler and Wagner 2001). Moreover, when we aim to explain global patterning, local specification alone is not enough to explain the generation of the functionally coordinated structures that make up an organism. Taking the context of the organism as a whole is essential to identifying the mechanisms responsible for orchestrating the time and place of local factors (Winslow et al. 2007). This reference to the various contextual levels where gene action needs to be situated leads to the other major theoretical argument against genetic reductionism. The ontogeny of an organism is conceived of as 'a hierarchy of developmental processes at different levels of organization' (Hall 2003: 226). The properties at one level of complexity (e.g. cells or tissues) cannot be ascribed directly to their component parts (e.g. genes), because they emerge through interactions among the parts at different levels of organization. A cell interacts with its neighbours and with the extracellular medium, cells aggregate in germ layers and tissues, tissues interact in organogenesis, organs interact with the rest of the body, and the organism itself interacts with the surrounding environment. In this explicitly hierarchical view, presuming that the molecular level is the most fundamental ontological level in biology is an unwarranted metaphysical assumption (Laubichler and Wagner 2001; see also chapter 1 here). Rather, different rules are appropriate for each level of the irreducible hierarchy of the living organization (Gilbert and Sarkar 2000). In short, development cannot be reduced to the mechanisms of gene activity, and a new *holistic* and *multilevel* approach to developmental processes is needed (Salazar-Ciudad and Jernvall 2004).

This multilevel and systemic view of development has been articulated before, quite independently of the recent advancements undergone by imaging technologies. Still, the descriptive modelling of development seems essential if we are to capture both the holistic and the hierarchical dimensions involved in the development of complex multicellular organisms. For one thing, imaging captures embryological data in their full living context. Moreover, if we aim to understand how different regimes of causality operate at each scale of organization in the developing embryo, the first essential methodological step is to accurately characterize these organizational levels. Depending on which biological phenomenon we aim to explain, we will pay attention to one or another level of organization, and therefore to one or another developmental mechanism. In this context, modelling should be understood as 'the art of... choosing an appropriate level of abstraction' (Wolkenhauer and Ullah 2007: 164; see also Brigandt 2015). This approach to modelling, widely endorsed in the philosophical discourse on explanatory models, is equally valid for descriptive models. Depending on the organizational scale one is interested in studying, different microscopy methods will be chosen for reconstructing development at the anatomical, histological, or cellular level.

As I noted at the start of this chapter, any mode of representation involves a form of idealization wherein some features are selected and others are ignored, in accordance with our descriptive and explanatory needs. In particular, the dynamic dimension of development has traditionally been abstracted away in the static representations of development rendered by developmental series. However, the difference between still modes of representation (histology, drawing, and photography) and dynamic ones (film) cannot be reduced to a conflict between true and false representations (Landecker 2012). Both static and dynamic modes of imaging represent the same developmental phenomena, and therefore the split between structural and processual approaches to development should be analysed (as Griesemer has argued for the research styles of genetics and embryology) as 'a matter of what is represented in the foreground versus the background of attention' (Griesemer 2007: 382). Importantly, the underlying ontological commitments of biologists differ according to whether the biologists in question hold pattern-based or process-based philosophical views: the former mainly consider timeless structures, whereas the latter perceive reality as consisting of systems in a permanent state of flux (Nuño de la Rosa and Etxeberria 2012). It can be argued that the theoretical distance between process and structure reflects the methodological trade-off between spatial and temporal resolution: time-lapse imaging allows higher temporal resolution, but images with higher spatial resolution take longer to collect (Brainerd and Hale 2006).

Still, the methodological trade-off between spatial and temporal resolution has been steadily weakening in the last decades. At the beginning of the twentieth century, one of the main limitations of applying time-lapse imaging to the study of development was precisely its lack of resolution. As living cells are translucent, the only way to see cells under a traditional light microscope was to stain and therefore to kill them. Embryologists were faced with the dilemma of having to decide between representing high-resolution but dead structures or visualizing live yet structureless processes. The structural and the processual modes of description seemed to be, as Nils Bohr (1933) famously claimed for the mechanistic and the finalistic understandings of living systems, complementary. That is to say, to obtain a full understanding of biological phenomena, the static and the dynamic modes of description were mutually exclusive, yet equally necessary. Therefore, choosing a dynamic medium of representation of development over a static one could not be interpreted as adopting a more 'realistic' perspective. Rather, opting for a processual mode of description of development was also a kind of idealization, insofar as cell resolution was sacrificed for the sake of the dynamic visualization of development. And the opposite was also the case: the ascendancy of histological methods in the late nineteenth century was crucial to the success of a structure-based (as opposed to a process-based) approach to life (Landecker 2009). However, over the course of the twentieth century there have been three major advances in microscopy that have challenged the presumed inevitability of a trade-off between spatial and temporal resolution, allowing researchers to combine in vivo high-resolution microscopy and time-lapse imaging. These are the introduction of the phase-contrast microscope in the 1930s, the invention of the laser scanning confocal microscope in 1986, and the cloning of the green fluorescent protein (GFP) of jellyfish in 1992 (Stramer and Dunn 2015: 12).

Having said this, time-lapse microscopy has been incapable, until very recently, of rendering 3D representations of development. Before the introduction of non-destructive 3D imaging, the only way of reconstructing the three-dimensionality of embryos was to physically slice the embryo into hundreds of sections and reconstruct the 3D morphology from these sections. At first, the series of sections, mounted on glass slides, was used to generate physical 3D models. With the digitalization of photography, sections were photographed and the digital images were virtually stacked together. In both cases, looking at processes was incompatible with appreciating the three-dimensionality of development. Embryos still needed to be frozen at different moments of their development in order for the researcher to 'look' inside them.

The major revolution in 3D imaging has come at the beginning of the present century, with the development of various non-destructive imaging modalities. These new microscopic techniques, such as optical, ultrasound, microcomputed tomography, as well as magnetic resonance imaging, allow us to obtain stacks of digital images of optical or virtual sections through a specimen. 3D visualization and the analysis of such massive data sets are performed with the aid of reconstruction software (Weninger et al. 2004). Thus, thanks to 'virtual histology' (Sharpe 2008), embryos can now not only be imaged in their natural environment, but also be manipulated without being destroyed. These new imaging modalities can be used to generate images from complex living embryos throughout embryonic development (Gregg and Butcher 2012), leading to time-lapse imaging in three dimensions, or 4D microscopy. Thus, whereas when choosing a microscopic technique there is always a compromise to make between spatial resolution and temporal sampling (Luengo-Oroz et al. 2011), 4D imaging finally makes it possible to acquire data that have high spatial context and which are longitudinal over time.

The ability of time-lapse imaging to characterize development at different levels of organization makes the resulting images the locus of integration of the different explanatory approaches in developmental biology. Thus Khaled Khairy and Philipp Keller (2011: 488) set 'the system-level understanding of developing organisms' as the long-term goal of a descriptive model of embryogenesis that encompasses different levels of biological organization. According to Sean Megason and Scott Fraser (2007), this should also be the goal of developmental bioinformatics, namely the creation of a complete database with information on developmental processes at different levels of organization, including gene expression patterns, cell lineages, and cellular dynamics throughout the duration of development. This descriptive model 'may allow the extraction of a fundamental set of mechanistic rules in a normalized morphogenetic scaffold and thus pave the way for a developmental computer model with truly predictive power' (Khairy and Keller 2011: 488). Understanding how these new descriptive modelling strategies contribute to contemporary explanations of development will be the subject of the last section of this chapter.

## 4. Reconstructing and Explaining Developmental Processes

Philosophers of science have traditionally dismissed visual representations of natural phenomena either as data summaries or as purely illustrative tools of scientific theories. According to this view, descriptive models only become a legitimate part of scientific research when they are subordinated to explanatory models. Against this backdrop, recent studies of representation in scientific practice are deeply concerned with the role of visual representations in scientific explanation, from diagrams (Perini 2005) to 3D models (Chadarevian and Hopwood 2004). In particular, since the pioneering work of Nick Hopwood on the early history of representational practices in embryology (Hopwood 1999, 2005, 2007), the last few years have witnessed an increasing interest in the role of descriptive modelling strategies in the explanatory models of developmental biology and evo-devo (Griesemer 2002; Love 2010b; Fusco et al. 2014).

In the writings of the early cinema theorists and film-makers such as Sergey Eisenstein, Jean Epstein, or Béla Balázs, there was a recurring metaphorical connection between seeing life under a microscope and seeing it through a camera. In these writings, the movie shot was compared to the cell, the montage to the organism formed by cell division, and film-making to embryogenesis. Landecker (2005) has shown how the recurrence of this analogy illustrates the deep connection that tied cinema and biological research in the early twentieth century. I believe that these kinds of comparisons also point to another dimension of representation that reveals the connection between describing and explaining. In the same way in which watching a film is, in reality, an illusion that hides the fact that the spectator is actually viewing a succession of still pictures, not to mention all the editing that goes into making a movie, looking at development under a microscope or watching development on a screen is also artificial in other respects. Dynamic images of development are highly contrived representative or phenomenal models of development. Paradoxically, the most realistic images of development are also the most crafted reconstructions of it.

In the predigital age, reconstructions of development were derived from microscopical observations that relied on the visual inspection and manual analysis of drawings or photographs. Analyses of dynamic processes were particularly arduous. For instance, the only way to analyse cell movements was to print a series of individual frames at regular intervals and manually quantify the displacement of individual cells. Since the digitization of the video time-lapse in the mid-1990s and the increasing number of megapixels per image (Gordon 2009), computational tools for automated image processing and data analysis (from cell segmentation to cell-tracking algorithms) have become indispensable for achieving the reconstruction of developmental processes in space and time (Khairy and Keller 2011; Luengo-Oroz et al. 2011; Rittscher 2010). A major technological breakthrough of the digital revolution that has enabled the widespread diffusion of dynamic images of development has been the possibility to upload videos to online publishing platforms. In recent years, time-lapse movies both of microscopical recordings and of reconstructions have become more and more common in developmental biology (see e.g. Keller et al. 2008).

The quantification of observation introduced by computer images has inverted the received view that developmental biologists had commonly assumed about descriptive modelling. While representative models of development have been traditionally assimilated to a qualitative understanding of ontogeny,[6] the main advantage of live imaging with regard to explanation is 'quantitation' (Pantazis and Supatto 2014). Indeed, the past two decades have seen a phenomenal increase in the number of tools for capturing quantitative data from living embryos, including new microscopy methods, the use of fluorescent molecules to probe gene function, and, above all, image analysis software. The combination of these new tools has turned microscopy into a quantitative methodology able to measure in great detail the spatio-temporal dynamics of developmental processes at the molecular, cellular, and tissue level (Oates et al. 2009). Some authors even claim that the quantitation of imaging is transforming developmental biology as a whole into 'a new interdisciplinary field where biologists' verbal descriptions are turned into more quantitative and formal descriptions amenable to automated quantitative analysis and comparison' (Luengo-Oroz et al. 2011: 630).

However, while the role of imaging in providing comprehensive and accurate descriptions of developmental processes is well appreciated, its role in the explanation of morphogenesis deserves more recognition. In a recent paper, Fengzhu Xiong and Sean Megason caution about the underestimation of the role of imaging in generating and testing models of development: 'imaging is often dismissively considered as "descriptive" at best whereas perturbation based approaches are automatically considered more "mechanistic"' (Xiong and Megason 2015: 632). Even among biologists who are explicitly sympathetic to a theoretical process-based approach to biological phenomena, the descriptive modelling of the dynamics of development, made possible by contemporary live imaging technology, is regarded as a minor contribution by comparison to explanatory models. For example, after recognizing the 'spectacular recent advances in live imaging technology', Johannes Jaeger and Nick Monk caution that 'it is important to keep in mind that they [the movies of developmental processes made possible by these new technologies] *only provide descriptions—not explanations*—of the phenomenally complex and orchestrated dynamic organization of cells and developing organisms' (Jaeger and Monk 2015: 1067; emphasis added).

By contrast, advocates of imaging techniques in developmental biology have adopted an explanatory approach to development that opposes the traditional explanatory approach of developmental genetics (e.g. Lippincott-Schwartz 2011; Xiong and Megason 2015). Developmental genetics conceives of development as a spatio-temporal sequence of gene expressions, a linear approach to causality that allows developmental genetics to explain ontogeny. This *interventionist approach* begins with an induced mutation and a discrete phenotypic consequence, and then establishes a perturbation-to-consequence chain of events in which the mutant gene is taken to be the cause of the developmental process under study (von Dassow and Munro 1999). These causal sequences, visually represented

---

[6] In particular, when they were introduced in cell biology in the early twentieth century, dynamic modes of imaging were criticized as unscientific on account of their qualitative nature (Landecker 2009).

by gene regulatory networks, are further completed with other genes that intervene in the generation of a particular organ. The interventionist mode of explanation assumed in developmental genetics seems to fit well with contemporary mechanistic accounts of how phenomenal and explanatory models relate in scientific research. Carl Craver (2006: 356) characterizes purely phenomenal models as those representations 'that scientists construct as more or less abstract descriptions of a real system'. By contrast, explanatory models are meant to account not merely for how the system behaves, but also for 'how it will behave under a variety of interventions' (ibid., 358). The explanatory models of developmental genetics abstract away genes and gene interactions on the basis of the effects of their manipulation. The details recorded in the phenomenal model (e.g. the mechanical properties of the cells, the geometry of tissues) are assumed to be irrelevant for explanation, insofar as they are seen as somehow encoded by the causal (genetic) factors identified in the explanatory model.

The major handicap of the interventionist approach is that, while it allows identification of some of the causal factors (the genes and their products) that participate in a given developmental process, it cannot unravel the dynamics leading to the final outcomes of development (Xiong and Megason 2015). In other words, the interventionist approach cannot articulate the genotype–phenotype map (von Dassow and Munro 1999). Many developmental biologists hold an alternative view, which does not reduce development to a problem of gene expression. In what can be called the *morphogenetic view* of development, genes do not make structures in an autonomous way; instead they confer certain properties to cells, which self-organize in the construction of organs and structures in accordance with physico-chemical laws (Alberch 1991). The properties resulting from the interactions at the molecular, cellular, and tissue levels (e.g. the physical properties of biological materials, the self-organizing capacities of cell aggregates, the geometry of tissues) are not codified in the genome but emerge from the dynamics of developmental systems (Oster et al. 1988).

The morphogenetic view of development implies a very different approach to the non-trivial relation between merely phenomenal and genuinely explanatory models in developmental biology. As recognized by Wilhelm His—one of the founders of experimental embryology and the inventor of the microtome—organismal form is not a self-evident problem awaiting mechanical explanation. Rather, if embryologists aim to understand morphogenesis, they should actively engage in reconstructing the embryo by reproducing the causal relationships they want to understand (Hopwood 1999). Whereas in His' case the 3D reconstructions of development were intimately tied to his topological explanations of morphogenesis, contemporary developmental biologists converge in claiming that the current level of technological progress associated with processual phenomenal models allows a switch from the perturbation approach to 'explanation in terms of dynamical formulation and behavior' (von Dassow and Munro 1999: 310).

In this view, the explanations of developmental genetics are seen as qualitative in nature, whereas the progress in quantitative imaging of dynamic phenotypes is considered as paving the way towards the quantitative explanation of development (Oates et al. 2009). Thus, in a recent review, Jennifer Lippincott-Schwartz identifies two approaches to development. She describes the molecular explanation of development as instantiating a 'structure approach' to the developing organism, insofar as it attempts to obtain the blueprint of normal development from knowledge of the epigenomic state of the organism. In contrast, those explanations of development that aim to characterize the cells' relationships with their cellular and tissue environment during development are seen as instantiating a 'process approach', insofar as they need to characterize cell behaviours throughout the entire duration of development. The 'imaging approach' appears as the bridge between the two explanatory approaches, owing to the ability of imaging to provide quantitative descriptions of spatio-temporal relationships among genes and the associated cell and tissue outputs (Lippincott-Schwartz 2011).

It might be argued that the phenomenal models of development provided by current live imaging techniques are closer to explanations in the morphogenetic approach than to explanations in the interventionist approach. In the intreventionist approach of developmental genetics, phenomenal models of development are 'detached' from the explanatory models, being mere illustrations of the effect of genes in development. In contrast, in the morphogenetic approach, phenomenal models act as a kind of scaffolding for the explanatory models of developmental processes. Xiong and Megason (2015) highlight two ways in which this interaction between imaging and modelling (or between phenomenal and explanatory models, in my terminology) takes place. First, a detailed observation of the phenomena to be explained allows formulation and testing of explanatory models. Thus cell migratory tracks, cell division patterns, and lineage trees can be tested against biophysical models of cell behaviour and cell mechanics. Here quantitation plays an essential role, since mathematical modelling requires precise numbers to use the data for modelling. In this way, explanatory models of a given developmental process can only become testable once the process is translated into quantitative descriptions. Second, since the morphogenetic approach aims to provide quantitative explanations of the dynamics of development, modelling demands imaging to have the highest possible temporal amplitude. It is in the context of specific, testable models that perturbation can become a truly powerful tool for explanation.

An illustrative example of how imaging can play a crucial role in formulating morphogenetic explanations of development is the unravelling of the physical forces that intervene in the epiboly process in zebrafish embryos (Xiong and Megason 2015). In zebrafish epiboly, the enveloping cell layer (EVL) surface epithelium at the animal pole spreads over the yolk cell, dragging deep cells along via adhesion. According to the prevalent model, the force-generating mechanism driving the spreading is the contraction exercised by a contracting ring of actomyosin on the frontier of the EVL. However, imaging of epiboly at high resolution at the marching frontier has allowed us to identify friction-resistant actomyosin flows as an equally important force-generating mechanism driving the spreading of the EVL.

# 5. Conclusions

A processual view of biological entities might be said to be congenial to embryologists. Throughout the history of embryology, philosophers and naturalists have denounced the artificiality of the static, purely morphological characterization of development, blaming the anatomists' bent for forms frozen in time. However, the intractability and speed of developmental processes progressively led to an epistemological abandonment of processes in favour of discretizing ontogenies in arrays of patterns. Very recently, however, new microscopic, video, and bioinformatic techniques for visualizing and analysing developmental processes are finally making it possible to bring processes back from the noumenal darkness, so that they may be treated as proper objects of research in their own right. Since the early years of the new millennium, films of 'digital embryos' built from in vivo microscopy not only have started to replace the static series of normal stages as the standard representations of development, but also are contributing to a processual understanding of development.

In this chapter I have shown that the introduction of time-lapse imaging in developmental biology has opened the door to a whole new realm of processual phenomena and has cast well-known developmental processes under a radically new dynamic light. I have also discussed how imaging technologies are important actors in the current return of organicism in developmental biology, being essential tools for capturing both the holistic and the hierarchical dimensions of morphogenesis. Moreover, I have examined how, in overcoming the methodological trade-off between spatial and temporal resolution, 4D imaging has simultaneously shortened the theoretical distance between processual and structural approaches to development. Finally, I have argued that the quantitative information encoded in contemporary images of development transcends the traditional role associated with visual models as mere illustrations of causal theories. In fact, in morphogenetic approaches to development, the construction of phenomenal models appears to take on a crucial role in the process of formulating and testing explanatory models.

## Acknowledgements

This work was funded by a Juan de la Cierva fellowship (FJCI-2014-22685) from the Spanish Ministry of Economy, Industry and Competitiveness, at the University of the Basque Country. I would like to thank Dan Nicholson, John Dupré, and three anonymous reviewers for their suggestions and valuable comments on earlier versions of this chapter.

## References

Alberch, P. (1991). From Genes to Phenotype: Dynamical Systems and Evolvability. *Genetica* 84 (1): 5–11.

Baedke, J. (2013). The Epigenetic Landscape in the Course of Time: Conrad Hal Waddington's Methodological Impact on the Life Sciences. *Studies in History and Philosophy of Biological and Biomedical Sciences* 44 (4): 756–73.

Bergson, H. (1911). *Creative Evolution*. New York: Henry Holt and Company.

Boehm, B., Rautschka, M., Quintana, L., Raspopovic, J., Jan, Z., and Sharpe, J. (2011). A Landmark-Free Morphometric Staging System for the Mouse Limb Bud. *Development* 138 (6): 1227–34.

Bohr, N. (1933). Light and life. *Nature* 133: 421–3 and 457–9.

Bonner, J. T. (2015). *Life Cycles: Reflections of an Evolutionary Biologist*. Princeton: Princeton University Press.
Brainered, E. L. and Hale, M. E. (2006). In vivo and Functional Imaging in Developmental Physiology. In S. Warburton and W. Burggren (eds), *Comparative Developmental Physiology* (pp. 21–40). New York: Oxford University Press.
Brigandt, I. (2003). Homology in Comparative, Molecular, and Evolutionary Developmental Biology: The Radiation of a Concept. *Journal of Experimental Zoology* 299 (1): 9–17.
Brigandt, I. (2015). Evolutionary Developmental Biology and the Limits of Philosophical Accounts of Mechanistic Explanation. In P.-A. Braillard and C. Malaterre (eds), *Explanation in Biology* (pp. 135–73). Berlin: Springer.
Chadarevian, S. and Hopwood, N. (2004). *Models: The Third Dimension of Science*. Stanford: Stanford University Press.
Craver, C. F. (2006). When Mechanistic Models Explain. *Synthese* 153: 355–76.
Davidson, D. and Baldock, R. (2001). Bioinformatics beyond Sequence: Mapping Gene Function in the Embryo. *Nature Reviews Genetics* 2 (6): 409–17.
De Beer, G. (1958). *Embryos and Ancestors*. Oxford: Oxford University Press.
Fusco, G., Carrer, R., and Serrelli, E. (2014). The Landscape Metaphor in Development. In A. Minelli and T. Pradeu (eds), *Towards a Theory of Development* (pp. 114–28). Oxford: Oxford University Press.
Gaycken, O. (2012). The Secret Life of Plants: Visualizing Vegetative Movement, 1880–1903. *Early Popular Visual Culture* 10 (1): 51–69.
Gilbert, S. (2000). Diachronic Biology Meets Evo-Devo: C. H. Waddington's Approach to Evolutionary Developmental Biology. *American Zoologist* 40: 729–37.
Gilbert, S. and Sarkar, S. (2000). Embracing Complexity: Organicism for the 21st Century. *Developmental Dynamics* 219 (1): 1–9.
Gordon, R. (2009). Google Embryo for Building Quantitative Understanding of an Embryo as It Builds Itself. II: Progress toward an Embryo Surface Microscope. *Biological Theory* 4 (4): 396–412.
Gregg, C. L. and Butcher, J. T. (2012). Quantitative in vivo Imaging of Embryonic Development: Opportunities and Challenges. *Differentiation; Research in Biological Diversity* 84 (1): 149–62.
Griesemer, J. R. (2000). Reproduction and the Reduction of Genetics. In P. J. Beurton, R. Falk, and H.-J. Rheinberger (eds), *The Concept of the Gene in Development and Evolution: Historical and Epistemological Perspectives* (pp. 240–85). New York: Cambridge University Press.
Griesemer, J. R. (2002). Time: Temporality and Attention in Iconographies of the Living. In H. Schmidgen (ed.), *Experimental Arcades: The Materiality of Time Relations in Life Sciences, Art, and Technology (1830–1930)* (pp. 45–57). Berlin: Max Plank Institut für Wissenschaftsgeschichte.
Griesemer, J. R. (2007). Tracking Organic Processes: Representations and Research Styles in Classical Embryology and Genetics. In M. D. Laubichler and J. Maienhein (eds), *From Embryology to Evo-Devo* (pp. 375–433). Cambridge, MA: MIT Press.
Hall, B. (2003). Unlocking the Black Box between Genotype and Phenotype: Cell Condensations as Morphogenetic (modular) Units. *Biology & Philosophy* 18 (2): 219–24.
Hopwood, N. (1999). 'Giving Body' to Embryos: Modeling, Mechanism, and the Microtome in Late Nineteenth-Century Anatomy. *Isis* 90 (3): 462–96.
Hopwood, N. (2005). Visual Standards and Disciplinary Change: Normal Plates, Tables and Stages in Embryology. *History of Science* 43: 239–303.

Hopwood, N. (2007). A History of Normal Plates, Tables and Stages in Vertebrate Embryology. *The International Journal of Developmental Biology* 51 (1): 1–26.
Jaeger, J. and Monk, N. (2015). Everything Flows: A Process Perspective on Life. *EMBO Reports* 16 (9): 1064–7.
Jamniczky, H. A., Boughner, J. C., Rolian, C., Gonzalez, P. N., Powell, C. D., Schmidt, E. J., et al. (2010). Rediscovering Waddington in the Post-Genomic Age. *Bioessays* 32 (7): 553–58.
Keller, P. J., Schmidt, A. D., Wittbrodt, J., and Stelzer, E. H. K. (2008). Reconstruction of Zebrafish Early Embryonic Development by Scanned Light Sheet Microscopy. *Science* 322 (5904): 1065–9.
Kelty, C. and Landecker, H. (2004). A Theory of Animation: Cells, L-Systems, and Film. *Grey Room* 17: 30–63.
Khairy, K. and Keller, P. J. (2011). Reconstructing Embryonic Development. *Genesis* 49 (7): 488–513.
Kirschner, M., Gerhart, J., and Mitchison, T. (2000). Molecular 'Vitalism'. *Cell* 100 (1): 79–88.
Landecker, H. (2005). Cellular Features: Microcinematography and Film Theory. *Critical Inquiry* 31 (4): 903–37.
Landecker, H. (2006). Microcinematography and the History of Science and Film. *Isis* 97 (1): 121–32.
Landecker, H. (2009). Seeing Things: From Microcinematography to Live Cell Imaging. *Nature Methods* 6 (10): 707–9.
Landecker, H. (2012). The Life of Movement: From Microcinematography to Live-Cell Imaging. *Journal of Visual Culture* 11 (3): 378–99.
Laubichler, M. D. and Wagner, G. P. (2001). How Molecular Is Molecular Developmental Biology? A Reply to Alex Rosenberg's Reductionism Redux: Computing the Embryo. *Biology & Philosophy* 16 (1): 53–68.
Lippincott-Schwartz, J. (2011). Bridging Structure and Process in Developmental Biology through New Imaging Technologies. *Developmental Cell* 21 (1): 5–10.
Love, A. C. (2010a). Idealization in Evolutionary Developmental Investigation: A Tension between Phenotypic Plasticity and Normal Stages. *Philosophical Transactions of the Royal Society of London. Series B, Biological Sciences* 365 (1540): 679–90.
Love, A. C. (2010b). Rethinking the Structure of Evolutionary Theory for an Extended Synthesis. In M. Pigliucci and G. B. Müller (eds), *Evolution: The Extended Synthesis* (pp. 403–41). Cambridge, MA: MIT Press.
Luengo-Oroz, M. A., Ledesma-Carbayo, M. J., Peyriéras, N., and Santos, A. (2011). Image Analysis for Understanding Embryo Development: A Bridge from Microscopy to Biological Insights. *Current Opinion in Genetics and Development* 21 (5): 630–7.
Maienschein, J. (1991). Epistemic Styles in German and American Embryology. *Science in Context* 4 (2): 407–27.
Megason, S. G. and Fraser, S. E. (2007). Imaging in Systems Biology. *Cell* 130 (5): 784–95.
Meijering, E., Smal, I., Dzyubachyk, O., and Olivo-Marin, J. C. (2008). Time-Lapse Imaging. In Q. Wu, F. A. Merchant, and K. R. Castleman (eds), *Microscope Image Processing* (pp. 401–40). Amsterdam: Elsevier Academic Press.
Minelli, A. (2003). *The Development of Animal Form: Ontogeny, Morphology, and Evolution.* Cambridge: Cambridge University Press.
Nicholson, D. J. and Gawne, R. (2015). Neither Logical Empiricism nor Vitalism, but Organicism: What the Philosophy of Biology Was. *History and Philosophy of the Life Sciences* 37 (4): 345–81.
Nuño de la Rosa, L. (2010). Becoming Organisms: The Organisation of Development and the Development of Organisation. *History and Philosophy of the Life Sciences* 32 (2–3): 289–315.

Nuño de la Rosa, L. and Etxeberria, A. (2012). Pattern and Process in Evo-Devo: Descriptions and Explanations. In H. de Regt, S. Hartmann, and S. Okasa (eds), *EPSA Philosophy of Science: Amsterdam 2009*, vol. 1 (pp. 263–74). Amsterdam: Springer.

Oates, A. C., Gorfinkiel, N., Gonzalez-Gaitan, M., and Heisenberg, C.-P. (2009). Quantitative Approaches in Developmental Biology. *Nature Reviews Genetics* 10 (8): 517–30.

Oster, G. F., Shubin, N., Murray, J., and Alberch, P. (1988). Evolution and Morphogenetic Rules: The Shape of the Vertebrate Limb in Ontogeny and Phylogeny. *Evolution* 42 (5): 862–84.

Pantazis, P. and Supatto, W. (2014). Advances in Whole-Embryo Imaging: A Quantitative Transition Is Underway. *Nature Reviews Molecular Cell Biology* 15 (5): 327–39.

Perini, L. (2005). Explanation in Two Dimensions: Diagrams and Biological Explanation. *Biology & Philosophy* 20 (2): 257–69.

Peterson, E. L. (2011). The Excluded Philosophy of Evo-Devo? Revisiting C. H. Waddington's Failed Attempt to Embed Alfred North Whitehead's 'Organicism' in Evolutionary Biology. *History and Philosophy of the Life Sciences* 33 (3): 301–20.

Richardson, M. K. (1995). Heterochrony and the Phylotypic Period. *Developmental Biology* 172 (2): 412–21.

Rieppel, O. (2006). 'Type' in Morphology and Phylogeny. *Journal of Morphology* 267 (5): 528–35.

Rittscher, J. (2010). Characterization of Biological Processes through Automated Image Analysis. *Annual Review of Biomedical Engineering* 12 (1): 315–44.

Salazar-Ciudad, I. and Jernvall, J. (2004). How Different Types of Pattern Formation Mechanisms Affect the Evolution of Form and Development. *Evolution and Development* 6 (1): 6–16.

Sharpe, J. (2008). Optical Projection Tomography. In C. W. Sensen and B. Hallgrimsson (eds), *Advanced Imaging in Biology and Medicine: Technology, Software Environments, Applications* (pp. 199–224). Berlin: Springer.

Stramer, B. M. and Dunn, G. A. (2015). Cells on Film: The Past and Future of Cinemicroscopy. *Journal of Cell Science* 128 (1): 9–13.

Tomer, R., Khairy, K., Amat, F., and Keller, P. (2012). Quantitative High-Speed Imaging of Entire Developing Embryos with Simultaneous Multiview Light-Sheet Microscopy. *Nature Methods* 9 (7): 755–63.

Totaro, D. (2001). Time, Bergson, and the Cinematographic Mechanism: Henri Bergson on the Philosophical Properties of Cinema. *Offscreen* 5 (1). http://offscreen.com/view/bergson1.

Vienne, F. (2015). Seeking the Constant in What Is Transient: Karl Ernst von Baer's Vision of Organic Formation. *History and Philosophy of the Life Sciences* 37 (1): 34–49.

Von Dassow, G. and Munro, E. (1999). Modularity in Animal Development and Evolution: Elements of a Conceptual Framework for EvoDevo. *Journal of Experimental Zoology* 285 (4): 307–25.

Waddington, C. H. (1957). *The Strategy of the Genes*. London: Allen & Unwin.

Webb, A. B. and Oates, A. C. (2015). Timing by Rhythms: Daily Clocks and Developmental Rulers. *Development, Growth and Differentiation* 58 (1): 43–58.

Wellmann, J. (2015). Folding into Being: Early Embryology and the Epistemology of Rhythm. *History and Philosophy of the Life Sciences* 37 (1): 17–33.

Weninger, W. J., Tassy, O., Darras, S., Geyer, S. H., and Thieffry, D. (2004). From Experimental Imaging Techniques to Virtual Embryology. *History and Philosophy of the Life Sciences* 26 (3–4): 355–75.

Winslow, B. B., Takimoto-Kimura, R., and Burke, A. (2007). Global Patterning of the Vertebrate Mesoderm. *Developmental Dynamics* 236 (9): 2371–81.

Wolkenhauer, O. and Ullah, M. (2007). All Models Are Wrong. In F. Boogerd, F. J. Bruggeman, J.-H. Hofmeyr, and H. V. Westerhoff (eds), *Systems Biology: Philosophical Foundations* (pp. 163–79). Amsterdam: Elsevier Academic Press.

Wong, C. C., Chen, A. A., Behr, B., and Shen, S. (2013). Time-Lapse Microscopy and Image Analysis in Basic and Clinical Embryo Development Research. *Reproductive BioMedicine Online* 26 (2): 120-9.

Wong, M. D., Eede, M. C. van, Spring, S., Jevtic, S., Boughner, J., Lerch, J., and Henkelman, R. M. (2015). 4D Atlas of the Mouse Embryo for Precise Morphological Staging. *Development* 142 (20): 3583-91.

Wong, C. C., Loewke, K. E., Bossert, N. L., Behr, B., De Jonge, C. J., Baer, T. M., and Reijo Pera, R. A. (2010). Non-Invasive Imaging of Human Embryos before Embryonic Genome Activation Predicts Development to the Blastocyst Stage. *Nature Biotechnology* 28 (10): 1115-21.

Woodger, J. H. (1929). *Biological Principles: A Critical Study*. London: Kegan Paul.

Xiong, F. and Megason, S. G. (2015). Abstracting the Principles of Development Using Imaging and Modeling. *Integrative Biology: Quantitative Biosciences from Nano to Macro* 7 (6): 633-42.

# 14

# Intersecting Processes Are Necessary Explanantia for Evolutionary Biology, but Challenge Retrodiction

*Eric Bapteste and Gemma Anderson*

## 1. Introduction

Ecosystems change, species transform, humans develop and age, tomatoes rot, biofilms grow, genes mutate and recombine: processes are everywhere in biology. Responsible for both stasis and change, they affect and effect multiple levels of biological organization. Essentially, every biologist is engaged in the description of processes. The study of biological evolution is itself largely a study of processes.

Frequently, processes constitute natural explanantia in evolutionary biology (and this is probably true of other historical sciences as well). 'Divergence from a common ancestor' is a well-known explanans for biodiversity. Processes are often described by simple names or phrases such as 'speciation', 'eukaryogenesis', 'endosymbiosis', 'descent with modification', and so on. Yet one cannot help but be struck by the actual complexity of the causal interactions that such terms refer to, if one looks beneath the simple terminology. While metaphysical inquiries into the nature of biological processes may be out of the reach of standard scientific discourse, epistemic approaches to such processes can nevertheless be attempted. In particular, processes can be captured by patterns upon which a scientific discourse can be built.

For example, in phylogenetics, a tree-like (or branching) pattern usually captures a tree-like process of evolution—that is, a series of divergences from a last common ancestor—as changes accumulate within a lineage. Generations of evolutionary biologists have searched for the actual succession of these branching patterns in order to illustrate the split from an ancestral form into novel lineages. This has led to attempts to reconstruct the universal tree of life. Phylogenetics has unquestionably become a central practice in evolutionary biology, and the tree-like pattern a gold standard in investigations of evolutionary processes.

At the same time, however, biologists also describe living things as organized assemblies of more or less interdependent components, abstracted as intertwined and interconnected regulatory, metabolic, protein–protein interaction, genetic, and developmental networks (Alon 2006; Wilkins 2007; Yafremava et al. 2013). Thus, focusing on the evolution of the relationships between organismal components offers another strategy to explain organismal evolution. This perspective, inspired by the science of evolving networks, requires treatments that are complementary to phylogenetic analyses, as the latter are not designed to model a plurality of intersecting processes operating at distinct timescales on interdependent components.

In this chapter we argue that intersecting processes are an important, underestimated explanans for the evolution of biodiversity, but that their due consideration will challenge retrodiction. In section 2 we review how phylogenetics has popularized the study of descent with modification by using a branching (tree-like) pattern, commonly used for retrodiction. However, since the same phylogeny is often compatible with very different processes, tree-based approaches only provide a partial and ambiguous account of evolutionary history. Trees leave out essential intersecting processes that are part of the explanation of evolutionary history. Moreover, the integration of these latter processes into evolutionary studies requires other tools, methods, and patterns than those currently explored in evolutionary trees.

In section 3 we argue that even phylogenetic networks, such as Doolittle's famous web of life, are incomplete representations of the intersecting processes that affect organismal and viral lineages. Current phylogenetic network approaches still underestimate intersecting processes that affect more than one class of closely phylogenetically related agents (e.g. cells, or viruses, or plasmids). To deal with this, we introduce an expanded and updated version of Doolittle's web of life. In our version, gene externalization—the process by which a copy of a gene is placed in an unrelated genome host structure—receives a specific representation. This process implies that genes get disseminated across genetic exchange communities. It thus becomes impossible to infer past history by tracing such a complex network to a single root node. This multiplicity of connections between genomes encourages instead community-level analyses for retrodiction. However, beyond the development of new network patterns, alternative exploratory strategies are still needed to account for additional intersecting processes.

In section 4 we argue that this goal can be achieved by studying the evolution of biological organizations, abstracted as interaction networks that feature elements that can be of different types, of different origins, and from distinct levels. Crucially, the evolution of such biological organizations can be causally decoupled from speciation events. Thus, the branching pattern used to make sense of evolution will not naturally describe or explain changes in these networks, which are sustained by intersecting processes. By contrast, we briefly illustrate how network patterns have been used in studies of molecular evolution. These examples show how evolutionary studies benefit from topological explanations, as there is an important link between the phenotypes that evolutionary biologists seek to explain and the evolution of the organization of the networks that cause these phenotypes. This practice pertains to the network sciences, and more precisely to the study of evolving networks.

In section 5 we complete our epistemic diagnosis that intersecting processes are missing from the explanatory toolbox of evolutionary biology by discussing a fundamental challenge raised by these processes. Since processes evolve and since they are our explanantia, then the latter also evolve. The merging of processes, creating new forms of organization and novel processes, is a common feature of evolutionary transitions, and is therefore a major challenge for retrodiction. These challenges notwithstanding, we predict an increased use of evolving and phylogenetic networks in evolutionary biology. As a result, we propose that the use of additional network patterns to explain the evolution of biological phenomena from the molecular to the organismal level could be achieved by developing a (yet to be introduced) network-based typology of evolutionary processes.

## 2. An Increasingly Appreciated Issue: The Underdetermination of Phylogenetic Trees

Phylogenetics has popularized the study of descent with modification by using a branching (tree-like) pattern (Felsenstein 2004). Tree reconstruction can be achieved for entities from different levels of biological organization, from molecules to organisms. Thus there are gene trees, protein trees, genome trees, and species trees flourishing in the literature. A common use of these phylogenetic patterns is retrodiction, that is, making 'predictions' about the past. For example, the relative order of species divergence is often inferred from gene trees, because branchings in single-copy gene trees are explained as speciation events. However, a gene tree topology is not always so easy to interpret in processual terms (i.e. as speciations; see Lapointe et al. 2010). The reason for this is that the same gene tree is often compatible with very different processes. Consequently, a tree topology provides neither a sufficient nor a complete explanation of the complexity of biological evolution. Rather than offering unequivocal evidence about the actual course of evolutionary events, gene trees provide inconclusive evidence. They require extra-phylogenetic assumptions about the likelihood of various types of evolutionary processes to be interpreted, such as, for example, the likelihood of speciation event versus the likelihood of multiple gene transfers (Ku et al. 2015a; Ku et al. 2015b).

This situation is nicely treated in the work of Ku and colleagues (Ku et al. 2015b), which is concerned with defining the correct interpretation of the observed, yet puzzling phylogenetic positions of eukaryotic sequences in gene trees, which mix sequences from prokaryotes (archaea and bacteria) with eukaryotic sequences (Figure 14.1). Since archaea, bacteria, and eukaryotes belong to three distinct domains, had their genes evolved separately within each of these lineages, it is extremely improbable that eukaryotic sequences would be mixed with prokaryotic sequences in the gene trees. It thus becomes necessary to invoke introgressive processes (i.e. processes in which the genetic material of a particular evolutionary unit propagates into different host structures and is replicated within those host structures) or reconstruction artefacts designed to make sense of these complex topologies. Some introgressive processes must be added to speciation to explain why some eukaryotic genes look like prokaryotic genes. But that recognition raises another question: what introgressive process is the relevant explanans for these puzzling phylogenetic patterns?

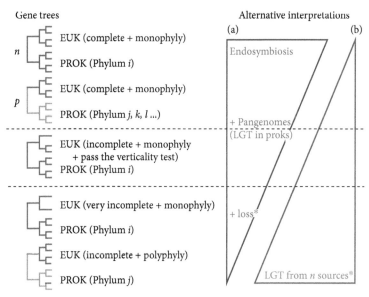

**Figure 14.1** Alternative interpretations of the same gene tree, and sets of gene trees. Phylogeneticists work with gene trees represented on the left (the branches are coloured in red for eukaryotic sequences and in blue for prokaryotic sequences). Gene trees are organized (from top to bottom) by increasing interpretative complexity, according to the messiness and patchiness of the eukaryotic sequences and their sources. Each topology can be interpreted by appealing to different processes, indicated on the right. Interpretations are not mutually exclusive, but their likelihoods differ in accordance with the likelihood of the processes that are deemed to be the explanantia. The width of the each triangle indicates the relative likelihood of the interpretation of the gene trees on the left. The * stresses that, to favour (a), a process must explain massive independent gene losses in eukaryotes; whereas, to favour (b), lateral gene transfer (LGT) by vectors must appear likely, and hence processes of eukaryote-to-eukaryote or prokaryotes-to-eukaryotes gene transfers must be identified.

The answer to that question is hardly ever in the gene tree. Methods with limited flexibility in the patterns they can display, such as phylogenetic tree reconstruction, cannot be used uncritically to infer these processes (Bapteste et al. 2012). As Ku et al. (2015b) put it: '[t]here are at least two competing alternatives to account for prokaryotic genes in eukaryotes—gradual LGT accrual versus episodic gene transfer from organelles'. One process is slow and can involve multiple gene donors (Keeling and Palmer 2008), the other is fast and involves very few donors.[1] This example highlights a fundamental reason for the failure of the branching pattern to serve as a universal explanans in evolutionary biology: trees leave out essential features of evolution in the real world, namely intersecting processes (Bapteste et al. 2009). However, the example also shows that the inclusion of such processes is not straightforward.

---

[1] Furthermore, these donors may themselves already harbour phylogenetically composite genomes, due to LGT between prokaryotes in their own ancestral lineages.

## 3. Intersecting Processes Are Also Absent from Phylogenetic Networks

Phylogenetic tree reconstruction methods are not the only phylogenetic approaches facing a need to better integrate intersecting processes. Phylogenetic networks, representing lateral gene transfer (LGT) and hybridization by lateral edges between diverging branches, also stand to be improved in this respect.

The observation that intersecting processes affect more than one class of closely phylogenetically related agents (e.g. cells, or viruses, or plasmids) is still underappreciated. The famous diagram by Ford Doolittle represents LGTs and endosymbioses across the web of life by a multi-rooted network, in which the three cellular domains progressively emerge from complex ancestral populations (Doolittle 1999). This diagram is often intended to illustrate in an extreme way how much complexity intersecting processes could produce if introgressive processes were massive and widespread (Huson et al. 2010). But in many respects this heuristic drawing is not in the least bit extreme. It does not feature mobile genetic elements (MGEs), which are key players in the introgressive processes that intersect with vertical descent (Halary et al. 2010). Large quantities of gene sharing occur between MGEs and cells, between cells, and between MGEs (Bapteste et al. 2012; Bapteste 2014; Halary et al. 2010; Jachiet et al. 2014; Yutin et al. 2013). Moreover, Doolittle's drawing does not feature the relationships between hosts and microbiotas known to produce holobionts, which presumably span across the eukaryotic web of life. It does not show, more generally, that all animals, plants, fungi, and many protists are associated with prokaryotes in ways that affect their development and evolution (Brucker and Bordenstein 2012; Gilbert et al. 2012). Therefore an expanded (updated) version of Doolittle's web of life is now required. Figure 14.2 offers a first step in this direction.[2]

Figure 14.2 stresses the intersection of unrelated, evolving lineages: organisms and MGEs. It highlights the genetic intertwining between viral and cellular genomes (Bapteste and Burian 2010; Bapteste et al. 2012; Filee et al. 2006; Filee 2014; Raoult 2009; Villarreal and Witzany 2010), in addition to classic horizontal gene transfer. The red lines highlight an additional, overlooked intersecting process, which we call 'gene externalization'. During gene externalization, a copy of a gene is placed in the genome of an unrelated host. As a result, genetic material from a particular genome or from a particular lineage disseminates, leaving copies in numerous other genomes or lineages beyond its original lineage. By this process, copies of genes feature outside their genome of origin, constituting a pool of externalized genes. Such gene externalization is not the typical LGT event found in textbooks, nor is it adequately represented in phylogenetic networks by an arrow between a cellular donor and a cellular recipient. Such an exemplar LGT requires two steps of gene externalization. A gene transfer between two cells mediated by a MGE presupposes movement from the bacterial genome to a MGE, and also from that MGE to the next bacterial genome.

---

[2] One of us is currently developing similarity network methods, searching for communities and for simple patterns in gene-sharing networks: see Bapteste et al. 2012 and Alvarez-Ponce et al. 2013. These methods operate under the expanded version of Doolittle's drawing, enhancing the study of intersecting processes in scientific analyses, with promising results.

**Figure 14.2** Adaptation of Doolittle's diagram (Anderson, G., watercolour on paper, 2015). The famous diagram by Ford Doolittle, slightly updated to represent transfers between archaeal and bacterial lineages, is delimited by a black outline that represents cellular lineages. It shows the vertical and lateral processes of evolution at the origin of cellular diversity. Purple and yellow lines (corresponding to the vertical and lateral evolution of viral and plasmid lineages, respectively) complement and expand this classic drawing of the web of life. Red lines indicate gene externalization events between cellular organisms and mobile genetic elements.

Gene externalization can thus be seen as one half of such a schematic LGT. This difference between gene externalization and LGT means that rules (usually regarding functional biases) discovered about LGT may differ from rules about gene externalization.[3,4] Invisible in phylogenetic networks, gene externalization requires its own specific representation.

Importantly, these considerations regarding the intersection between vertical descent, LGT, and gene externalization suggest an alternative strategy for retrodiction. Rather than assuming that homologous genes coalesce (i.e. trace back to a single ancestral copy in one common ancestral genome; see Dagan and Martin 2007), they suggest an opposite, complementary perspective. Genes disseminate and genomes dissipate across genetic exchange communities (Skippington and Ragan 2011), and retrodiction thus requires community-level analyses of molecular evolution (Bapteste 2014; Corel et al. 2016). This might offer a valuable opportunity. Gene externalization through MGEs may be as old as viruses. Consequently, extant viral genomes may have preserved some early informative imprints about the history of

---

[3] In particular, gene externalization may be random and occur at a high rate, which would not be visible from LGT analyses, if the host cell receiving a transferred gene selects against the residency of some of the externalized genes (for example, informational genes may be more externalized than transferred).

[4] Preliminary analyses suggest that gene externalization is a general property of life: cellular genomes, especially prokaryotic ones, decoalesce; as do viral genomes. They also suggest that this process is likely to have been ongoing for a very long time. This proposition is backed up by the scientific literature on virus evolution by gene accretion (Hatfull 2008; Hendrix et al. 1999; Filee et al. 2007) and on DNA dissemination via gene transfer agents (McDaniel et al. 2010), as well as by the debates regarding the respective origins of viruses and cells (Villareal and Witzany 2010; Filee et al. 2003; Claverie and Ogata 2009).

life inherited from past viral genomes (Filee et al. 2003). The focus on genome decoalescence therefore invites evolutionary biologists to dig into an unusual record of molecular evolution—unusual, yet potentially the largest such record: the copies of externalized genes present in MGEs. Overall, a comprehensive investigation of intersecting processes is likely to require tools, methods, and patterns that complement phylogenetic approaches. We will now show what alternative explanatory strategies are actually being explored in evolutionary studies.

## 4. The Need to Investigate Reticulate Intersecting Processes in Evolutionary Studies

Intersecting processes such as merging (Bapteste et al. 2012; Méheust et al. 2015), autocatalytic cycles (Eigen 1971; Eigen and Schuster 1977, 1981), feedback loops (Milo et al. 2002), and fast-forward loops (Alon 2006) are increasingly being included in evolutionary theory. The recognition of the key explanatory role played by these processes—captured epistemically by merging patterns, cycles, and motifs in interaction networks—is being paralleled by the realization that biological systems are organizations. Specifically, biological systems can be modelled as structured and dynamic sets of interacting components. Accordingly, molecules and organisms are increasingly considered (i) to be evolving as parts of interaction networks, and/or (ii) to be themselves composed of intersecting processes represented by intertwined networks that describe their intimate organizations (Braillard 2008).

The theoretical importance of the concept of organization in biology has long been recognized (Maturana and Varela 1980; Alon 2006; Wilkins 2007; Nicholson 2012). Crucially, the evolution of biological organizations can be causally decoupled from speciation events. More precisely, interacting organismal components can be of different ages, some having appeared earlier than others. As such, organisms are 'temporal mosaics', as is often illustrated by atavisms. Moreover, components may have different persistence spans—more stable components introduce structural and functional continuity in organismal evolution. Interestingly, well-adjusted biological organizations may persist over longer evolutionary periods than that of the given species from which they initially derive. For example, certain reptilian jaw-joint bones evolved into two of the middle-ear bones of different mammalian lineages. Importantly, interactions, and eventually interdependencies, between components can change over time. Preexisting components can get recycled and used to fulfil novel functions, as illustrated in co-options and evolutionary tinkering events. However, the standard epistemic branching patterns used to make sense of lineage divergences from a common ancestor will not naturally describe or explain changes in these networks. As a result, including interaction networks in comparative analyses has the potential to enhance our explanations of biodiversity. Typically, when intersecting processes are considered, one realizes that another form of selection is at play in the history of life. Selection—not to mention neutral processes—does not act only between organisms, but also within organisms (Bouchard 2014; see Figure 14.3). Relationships between components of interaction networks constrain the heritability and the variation of these components.

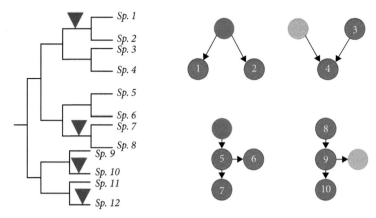

**Figure 14.3** Two complementary explanatory frameworks. The phylogenetic framework on the left panel provides a standard form of evolutionary explanation. Here properties are mapped against the species phylogeny. The purple triangle represents a recurring trait, produced by convergence or parallelism, for example the loss of a flagellum. The distribution of this character, which is at odds with the species phylogeny, suggests multiple independent occurrences during evolution, and as such is not explained by the phylogenetic tree *sensu stricto*. The network framework on the right panel provides an alternative form of evolutionary explanation. Here the nodes represent agents, for example genes, and the arrows represent interactions between them. Red-coloured nodes represent agents which, should they be lost (for example due to the fact that many mutations can make a given gene disfunctional), have a high likelihood of affecting downstream agents owing to the structure of the network, and this can result in convergent evolution. The loss of a feature in a species can be explained by the topology of the interaction network, and its repetition, by the commonality of this topology. By contrast, green nodes could be lost without major effects on the rest of the interaction network.

Therefore, it is also compelling to study the evolution of biological organizations (composed of elements that can be of different types, of different origins, and at distinct levels) as the result of intersecting processes. We can illustrate this move towards a more inclusive study of evolution—one that includes intersecting processes—by considering two emblematic examples at the molecular level.

### 4.1. Explaining the evolution of translation with a hypercycle

In the 1970s Manfred Eigen relied on intersecting processes to offer a historical explanation of the evolution of translation (Eigen and Schuster 1977; Eigen et al. 1991). He sought to understand how the instructions of nucleic acids became translatable into proteins during the early history of life.[5] Specifically, Eigen tried to solve the following conundrum: present-day biosynthesis relies upon two different types of biological entities: enzymes and nucleic acids acting in a complex network of reactions and defining a composite macromolecular system. In that regard, the origin

---

[5] Eigen did not tackle this major question of biology with a tree, showing the divergence of DNA or protein, but with evocative sketches and very complex systems of differential equations, which describe the rate of reproduction of the various nucleic and proteic elements of the system in order to predict its behaviour.

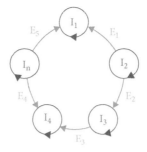

**Figure 14.4** The hypercycle (Anderson, G., from Eigen 1977). A catalytic hypercycle consists of self-instructive units Ii with two-fold catalytic functions. As autocatalysts or—more generally—as catalytic cycles, the intermediates Ii are able to instruct their own reproduction and, in addition, provide catalytic support for the reproduction of the subsequent enzymatic intermediate. The figure depicts the cyclic hierarchy. Eigen illustrated the stabilizing property of the hypercycle as a whole in his simulations. He split a sentence into four words and considered that each of the words was a distinct quasi-species. The four words were thus self-replicating with similar fitness. As these four quasi-species were in competition, only one quasi-species, hence one word, would get fixed as the simulation progresses. When these words are in functional linkage, producing a chain, all the benefit from the linkage goes to the last word, and only this last word is reproduced. (It is the famous last word!) When the words are organized in a hypercycle, the entire sentence gets stabilized. This is why hypercycles have such a remarkable potential role in evolution.

of biosynthesis is a type of chicken–egg problem, as in this system nucleic acids are causally responsible for enzymes and enzymes are causally responsible for nucleic acids. It is thus very difficult to imagine which came first, especially since this process, biosynthesis, evolved several billion years ago. Eigen used two major intersecting processes, autocatalytic loops and self-instructive cycles, to explain the evolution of nucleoproteic biosynthesis from a chaotic mix of molecules.

Eigen introduced as explanans an interaction network organized as a 'hypercycle' (Figure 14.4), which integrates the different processual behaviors of both kinds of molecules: autocatalysis for enzymes, self-instructions for nucleic acids. He insisted that his solution brought about a novel type of process, with fundamentally new evolutionary properties:

It is the object of this paper to show, first that the breakthrough in molecular evolution must have been brought about by an integration of several self-reproducing units to a cooperative system and, second, that a mechanism capable of such an integration can be provided only by the class of hypercycles.   (Eigen and Schuster 1977: 00)

In the hypercycle, as soon as self-instructive nucleic acids produce some enzyme involved in the replication of the nucleic acids, any advantageous mutation in the nucleic acids that would produce a more effective enzyme will be preserved in the system. When enzymes get improved, they become more efficient at copying nucleic acids; thus the error copy rate in nucleic acids decreases and longer nucleic acids can evolve, which enhances the information content of the system. Hypercycles can not only preserve their original information content but also enlarge it and stabilize it.

The patterns of cycles (autocatalytic loops and self-instructive loops) and of hypercycles were thus mobilized to construct a theory about nothing less than the evolution of translation in early life.

### 4.2. Explaining the evolution of biological functions by network analyses

Molecular biologists are increasingly using networks such as protein–protein interaction networks and regulatory networks, as well as the patterns within them (Alon 2006; Milo et al. 2002), to make sense of the evolution of specific biological functions.[6]

They believe that investigating the evolution of a biological function requires more than the phylogenetic study of a gene or protein family, in other words, more than the mapping of substitutions on a gene or a protein tree. The reason for this belief is that most of the time the old equations 'one gene, one protein' and 'one protein, one function' simply do not hold. The way functions evolve and are selected depends on intersecting processes. Causal interactions between molecules belonging to distinct gene and protein lineages sustain functional modules. Relations of interdependency and mutual selective pressures are common (Figure 14.5).

Thus, a complete understanding of the origin of a function—and, by extension, of the evolution of a function—requires knowledge about the topology of the protein-protein interaction network involved in the folding of that enzyme. When this

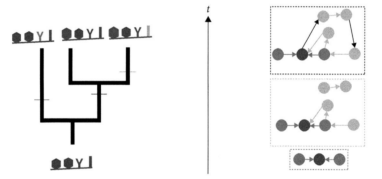

Inherent properties of a sequence
gradual change of isolated components

Relational properties of a sequence
evolution of components relationships

**Figure 14.5** The ins and outs of phylogenetic modelling. The figure is polarized from past (bottom) to present (top). The focal enzyme is represented in purple. The left panel shows how the evolution of a function would be described in a classic framework of molecular phylogeny. The function of the purple protein evolves as substitutions accumulate in the coding genes, as represented by a protein tree. The right panel shows how the evolution of a function would be described from an evolving network perspective. Each node is a protein; an arrow represents a causal interaction between two proteins.

---

[6] Indeed, systems biologists expect biological networks to feature at least a recurring small set of basic building blocks called 'network motifs', which are practically defined as patterns of interconnections occurring in complex networks at numbers that are significantly higher than those in randomized networks (Milo et al. 2002; Alon 2006).

enzyme requires other 'molecular robots' to perform its tasks, it also requires knowledge of the protein–protein interaction network to which the folded focal enzyme contributes (Alon 2006).[7] Since an enzyme function seems to be in part determined by the relational properties of this enzyme, the evolution of a protein function must be described by retracing the evolution—the topological changes—of a network that approximates such intersecting processes.[8]

Overall, evolutionary studies benefit from topological explanations because there is an important link between the phenotypes that evolutionary biologists seek to explain and the evolution of the organization of the networks that caused these phenotypes. This line of thought pertains to network sciences and, more precisely, to the study of evolving networks.

## 5. Processes, and Hence Explanantia, Evolve

While our epistemic diagnosis indicates that evolutionary biology should afford greater attention to intersecting processes, these pose a fundamental challenge, which we shall consider in this section. The merging of processes, creating new organizations and novel processes, is a common feature of evolutionary transitions. Therefore, accounting for the robust merging of preexisting processes is critical for the explanation of evolution. We have already encountered this issue when discussing Eigen's theory on the origin of translation. Eigen described hypercycles as responsible for a major breakthrough in molecular evolution (Eigen and Schuster 1977). Life as we know it could not have been explained before translation evolved. However, translation was not always an active biological process on Earth. The evolution of translation was a major transition, which completely transformed biological evolution (Szathmáry 2015). What this means is that, when dramatically new processes are introduced on Earth, so are novel types of explanantia for biodiversity.[9] If Eigen is right, this was the case when autocatalytic cycles and self-replicative loops became integrated in the form of hypercycles.

Importantly, this is not the only proposed instance of a major transition coupled to the emergence of a new biological process achieved through merging (Szathmáry 2015). Although we probably do not know the exact details through which meiosis—that is, sexual reproduction—evolved, William Martin, a major evolutionary biologist, must be credited with stressing the importance of the evolution of this fundamental process during the history of life (Ku et al. 2015a). Meiosis is a process that distinguishes eukaryotes (which perform it) from prokaryotes. Meiosis is fundamental

---

[7] By contrast, the information contained in the fragment of DNA that codes for this particular enzyme is not sufficient to produce a ready-made operational intracellular robot. Consequently, a phylogenetic analysis of an enzyme offers an underdetermined argument for the evolution of that enzyme's function, and should thus be revisited with an approach that incorporates intersecting processes.

[8] Interestingly, similar views have been expressed by prominent evo-devo scientists such as Duboule and Wilkins (1998). According to these authors, networks explain how internal constraints lead to restrictions in the production of evolutionary novelties.

[9] For example, the 'big bang' in the creation of composite genes has been proposed to have been paralleled by a 'big bang' in the evolution of interaction networks during the early history of life on Earth (see Wang and Caetano-Anolles 2009).

because it generates genetic variability, and consequently it fuels evolution with novelties upon which natural selection can later act. As a result, the tempo and modes of evolution are tremendously different in meiotic (sexual) and ameiotic organisms: the latter present pangenomes, while the former do not (Ku et al. 2015a). Causal explanations of biodiversity must change (i.e. they cannot rely on the same sets of processes) before and after the evolution of meiosis, because this process was not always active.

Why does it matter that meiosis probably evolved from intersecting processes? Meiosis seems to be as old as eukaryotes, and this major domain of life is almost certainly the result of the merging of (at least) two prokaryotic lineages: one ancestral archaeon and one ancestral bacterium, whose remnants can be found in the form of the mitochondrion within the cell of most present-day eukaryotes (McInerney et al. 2014; Williams et al. 2013). Thus, the process responsible for the life cycle of an early bacterium and the process responsible for the life cycle of an early archaeon merged and integrated with each other and, over time, that fusion produced a novel process, responsible for the life cycle of a novel composite organization: the eukaryotic cell. In these new life forms (from which all animals, plants, algae, amoebae, fungi, and other protists derive), meiosis contributed the necessary genetic variations in a fundamentally new way, thus preventing the extinction of eukaryotes that was expected as a result of a Muller's ratchet (had eukaryotes evolved clonally).

Of course, this kind of merging event, however major in its evolutionary scope, is impossible to model with a tree-like pattern. It can, however, be represented with a network (see Figure 14.5). What we want to stress here is the crucial epistemic consequence of the merging of intersecting processes during eukaryogenesis. By introducing meiosis among living things, this merging scenario challenges the simple retrodiction practice that is often associated with phylogeny and its tree-like pattern. To illustrate this, imagine that there were omniscient biologists who lived at the time before meiosis emerged on Earth, and who were perfectly aware of all the biological processes of their time. The predictions of these biologists concerning future biodiversity would in all likelihood have been wrong, because a genuinely new process as fundamental as the mode of generation of variation—the invention of sex—profoundly changed the course of organismal evolution in completely unexpected and almost unimaginable ways. Obviously, these primeval biologists did not exist, but present-day biologists face a comparable challenge. In order to draw inferences about the past, they use all their understanding of processes known today. They rely on uniformitarianism, the notion that yesterday's evolutionary processes were the same as today's processes. That notion, however, cannot be applied uniformly across the history of life because, before eukaryotes evolved, the modes of generation of genetic variation were very different from what they are now.

New processes make it difficult to predict and retrodict life's evolution. It may even be the case that new processes add to or eliminate preexisting processes. Moreover, since the processes changed over time, the patterns used to infer evolution should also change. Using similar patterns as proxies for evolutionary processes before and after the evolution of meiosis could be very misleading, since modes and regimes of molecular variation changed profoundly around that time. The evolution of meiosis modified the very process of speciation. Thus, even those who are tempted to relate all life forms by placing them on a single tree of life should not be naïve about the processual interpretation of a so-called universal branching pattern.

A branch leading to some prokaryotic taxon captures very different processes from a branch leading to some eukaryotic taxon in a tree of life (Bapteste and Dupré 2013). Phylogeny is underdetermined: a diversity of processes is hidden behind a unity of patterns. Arguably, this underdermination comes at the expense of biological knowledge. When one looks at an evolutionary tree with a uniformitarian mind, one is not so strongly compelled to identify transitions; rather, one is more likely to be (overly) inclined to see what looks like a *bona fide* gradual change.

This kind of issue (i.e. conflating processes due to the use of overly simple, apparently unifying representations) may explain why other major mergings of lineages (Nelson-Sathi et al. 2012; Nelson-Sathi et al. 2015) have remained unnoticed with trees of prokaryotes (Abby et al. 2012). Mergings of lineages were recently proposed to have affected—and possibly produced—all main archaeal groups. This is especially clear in the case of Haloarchaea, whose genomes are loaded with introgressed genes of bacterial origin. These findings about archaea are particularly exciting, since they propose that the merging of metabolic genes of bacterial origin would be a common, recurring theme in the history of life, and this hints at the possibility that even intersecting processes follow some rules (Méheust et al. 2015). Even if these proposals of additional major transitions (in the case of haloarchaea, from an anoxic to an oxic lifestyle) remain controversial, debates about the bacterial content of archaeal genomes are sufficient to demonstrate that intersecting processes such as LGTs, which lead to the introgression of genes into genomes, practically challenge retrodiction (López-García et al. 2015; Nelson-Sathi et al. 2012; Nelson-Sathi et al. 2015).

## 6. Conclusions: Towards a Typology of Processes

In this chapter we have discussed the diversity of processes with explanatory value for evolutionary biology that go beyond, and yet are complementary to, the classic notion of 'divergence from a last common ancestor'. We have argued that some of these evolutionary processes may be more appropriately described by patterns of representation that are different from the traditional branching ones. Our considerations are intended, not to belittle the importance of phylogenetic reconstructions, but to stress the need for a further integration of network concepts into evolutionary analyses (Wilkins 2007) in order to account for intersecting processes. While the main objects of study in phylogenetics are lineages, the main objects of study in systems biology are organizations. The evolution of the networks that make up the living world is a central theme of systems biology, but it also lies at the forefront of evolutionary research, as inferring the evolution of these organizations is a complementary way to understand the evolution of life on Earth. We believe that a synthesis between phylogenetic studies and analyses of interaction networks would be highly fruitful, since evolution depends on changes in organizations as well as on the divergence and merging of lineages.

Moreover, an increased use of networks in evolutionary biology could be coupled to the development of a yet to be introduced typology of processes designed to analyse the (big) processual picture of life. In reaching this end, evolutionary studies may benefit from identifying simple patterns in evolving networks and in phylogenetic networks. This enterprise could help with translating the principles of systems biology and network theory into an abstract, unifying language for theoretical biology in order to improve our understanding of the (big) processual picture of life. Typically, when systems biology seeks to identify common guiding principles in interaction networks that represent all sorts of processes—for example from transcriptional regulation to protein-protein interaction, or cellular communication (Alon 2006)—it tries to decompose large graphs into smaller meaningful motifs and modules. Using networks as a new level of abstraction to describe biological reality beyond a list of individual entities, systems biology searches for new regularities in biology (Braillard 2008). These regularities, found in the network topology, are expected to provide a universal 'alphabet' of interaction networks and to reveal a 'periodic table' for functional regulatory circuits (Kitano 2002). These small motifs, the components of networks, are detectable and are already proving to be useful in scientific inquiry. Of course, such simple patterns are abstractions, in the sense that they can be understood as smaller motifs composing a larger network of interconnected processes.[10] Yet graphlets of interaction nonetheless constitute an underappreciated regularity in biological systems.

We wish to propose the same sort of enterprise in order to identify some regularities at an even higher level of abstraction: the networks of processes, using a process typology. If successful, the most important payoff from such a strategy would be the detection of universal trends in processual networks and the possibility of identifying a simple 'alphabet' of processes. A richer and more explicit set of analytical patterns, approximating intersecting processes, could help evolutionary biologists make better sense of the stunning diversity of evolutionary phenomena, such as early transitions in the evolution of life, the genetic sharing involved in microbial social life, or new joint physiologies, organs, and modes of reproduction involved in evolutionary transitions and in adaptations.

At a more general level, the development of a typology of processes would constitute a genuine attempt toward unification within the (logically pluralistic) biological sciences. To paraphrase Bertalanffy (1968: 48), a unitary conception of the biological world may be based, not upon the possibly futile and certainly far-fetched hope to finally reduce all levels of reality to the level of molecules, but rather on the potential isomorphy of processes in different biological fields.

## Acknowledgements

We thank Dan Nicholson, John Dupré, and three anonymous reviewers for comments that greatly improved the manuscript. We also thank Alexander Jaffe for critical revisions of the text. EB thanks the ERC grant FP7/2007-2013 Grant Agreement # 615274 and ANR-13-BSH3-0007 Explabio.

## References

Abby, S. S., Tannier, E., Gouy, M., and Daubin, V. (2012). Lateral Gene Transfer as a Support for the Tree of Life. *Proceedings of the National Academy of Sciences* 109 (13): 4962–7.

---

[10] Moreover, however recurrent, such network motives are not necessarily adaptive. They could be the 'spandrels' of network complexity; a by-product of network building rules (Sole and Valverde 2006).

Alon, U. (2006). *An Introduction to Systems Biology: Design Principles of Biological Circuits.* Boca Raton: CRC Press.

Alvarez-Ponce, D., Lopez, P., Bapteste, E., and McInerney, J. O. (2013). Gene Similarity Networks Provide Tools for Understanding Eukaryote Origins and Evolution. *Proceedings of the National Academy of Sciences* 110 (17): E1594–603.

Bapteste, E. (2014). The Origins of Microbial Adaptations: How Introgressive Descent, Egalitarian Evolutionary Transitions and Expanded Kin Selection Shape the Network of Life. *Frontiers in Microbiology* 5 (83).

Bapteste, E. and Burian, R. M. (2010). On the Need for Integrative Phylogenomics, and Some Steps toward Its Creation. *Biology and Philosophy* 25 (4): 711–36.

Bapteste, E. and Dupré, J. (2013). Towards a Processual Microbial Ontology. *Biology and Philosophy* 28 (2): 379–404.

Bapteste, E., et al. (2009). Prokaryotic Evolution and the Tree of Life Are Two Different Things. *Biology Direct* 4, 34.

Bapteste, E., et al. (2012). Evolutionary Analyses of Non-Genealogical Bonds Produced by Introgressive Descent. *Proceedings of the National Academy of Sciences* 109 (45): 18266–72.

Bouchard, F. (2014). L'évolution par sélection naturelle. In T. Hoquet and F. Merlin (eds), *Précis de philosophie de la biologie* (pp. 251–61). Paris: Vuibert.

Braillard, P.-A. (2008). *Enjeux philosophiques de la biologie des systèmes.* PhD dissertation, University of Brussels.

Brucker, R. M. and Bordenstein, S. R. (2012). Speciation by Symbiosis. *Trends in Ecology & Evolution* 27 (8): 443–51.

Claverie, J. M. and Ogata, H. (2009). Ten Good Reasons Not to Exclude Viruses from the Evolutionary Picture. *Nature Reviews Microbiology* 7: 306–11.

Corel, E., Lopez, P., Méheust, R., and Bapteste, E. (2016). Network-Thinking: Graphs to Analyze Microbial Complexity and Evolution. *Trends in Microbiology* 24 (3): 224–37.

Dagan, T. and Martin, W. (2007). Ancestral Genome Sizes Specify the Minimum Rate of Lateral Gene Transfer during Prokaryote Evolution. *Proceedings of the National Academy of Sciences* 104 (3): 870–5.

Doolittle, W. F. (1999). Phylogenetic Classification and the Universal Tree. *Science* 284 (5423): 2124–9.

Duboule, D. and Wilkins, A. S. (1998). The Evolution of 'Bricolage'. *Trends in Genetics* 14 (2): 54–9.

Eigen, M. (1971). Selforganization of Matter and the Evolution of Biological Macromolecules. *Naturwissenschaften* 58 (10): 465–523.

Eigen, M. and Schuster, P. (1977). The hypercycle. A Principle of Natural Self-Organization. Part A: Emergence of the Hypercycle. *Naturwissenschaften* 64 (11): 541–65.

Eigen, M. and Schuster, P. (1981). Comments on 'Growth of a Hypercycle' by King (1981). *Biosystems* 13 (4): 235.

Eigen, M., et al. (1991). The Hypercycle: Coupling of RNA and Protein Biosynthesis in the Infection Cycle of an RNA Bacteriophage. *Biochemistry* 30 (46): 11005–18.

Felsenstein, J. (2004). *Inferring Phylogenies.* Seattle: University of Washington.

Filee, J. (2014). Multiple Occurrences of Giant Virus Core Genes Acquired by Eukaryotic Genomes: The Visible Part of the Iceberg? *Virology* 466–7: 53–9.

Filee, J., Forterre, P., and Laurent, J. (2003). The Role Played by Viruses in the Evolution of Their Hosts: A View Based on Informational Protein Phylogenies. *Research in Microbiology* 154 (4): 237–43.

Filee, J., et al. (2006). A Selective Barrier to Horizontal Gene Transfer in the T4-Type Bacteriophages That Has Preserved a Core Genome with the Viral Replication and Structural Genes. *Molecular Biology and Evolution* 23 (9): 1688–96.

Filee, J., Siguier, P., and Chandler, M. (2007). I Am What I Eat and I Eat What I Am: Acquisition of Bacterial Genes by Giant Viruses. *Trends in Genetics* 23 (1): 10–15.

Gilbert, S. F., Sapp, J., and Tauber, A. I. (2012). A symbiotic view of life: we have never been individuals. *Quarterly Review of Biology* 87 (4): 325–41.

Halary, S., et al. (2010). Network Analyses Structure Genetic Diversity in Independent Genetic Worlds. *Proceedings of the National Academy of Sciences* 107 (1): 127–32.

Hatfull, G. F. (2008). Bacteriophage Genomics. *Current Opinion in Microbiology* 11 (5): 447–53.

Hendrix, R. W., et al. (1999). Evolutionary Relationships among Diverse Bacteriophages and Prophages: All the World's a Phage. *Proceedings of the National Academy of Sciences* 96 (5): 2192–7.

Huson, D., Rupp, R., and Scornavacca, C. (2010). *Phylogenetic Networks*. Cambridge: Cambridge University Press.

Jachiet, P. A., et al. (2014). Extensive Gene Remodeling in the Viral World: New Evidence for Nongradual Evolution in the Mobilome Network. *Genome Biology and Evolution* 6 (9): 2195–205.

Keeling, P. J. and Palmer, J. D. (2008). Horizontal gene transfer in eukaryotic evolution. *Nature Reviews Genetics* 9 (8): 605–18.

Kitano, H. (2002). Computational Systems Biology. *Nature* 420 (6912): 206–10.

Ku, C., et al. (2015a). Endosymbiotic Gene Transfer from Prokaryotic Pangenomes: Inherited Chimerism in Eukaryotes. *Proceedings of the National Academy of Sciences* 112 (33): 10139–46.

Ku, C., et al. (2015b). Endosymbiotic Origin and Differential Loss of Eukaryotic Genes. *Nature* 524 (7566): 427–32.

Lapointe, F. J., et al. (2010). Clanistics: A Multi-Level Perspective for Harvesting Unrooted Gene Trees. *Trends in Microbiology* 18 (8): 341–7.

López-García, P., et al. (2015). Bacterial Gene Import and Mesophilic Adaptation in Archaea. *Nature Reviews in Microbiology* 13 (7): 447–56.

Maturana, H. R. and Varela, F. J. (1980). *Autopoiesis and Cognition: The Realization of the Living*. Reidel: Boston.

McDaniel, L. D., et al. (2010). High Frequency of Horizontal Gene Transfer in the Oceans. *Science* 330 (6000): 50.

McInerney, J. O., O'Connell, M. J., and Pisani, D. (2014). The Hybrid Nature of the Eukaryota and a Consilient View of Life on Earth. *Nature Reviews Microbiology* 12 (6): 449–55.

Méheust, R., Lopez, P., and Bapteste, E. (2015). Metabolic Bacterial Genes and the Construction of High-Level Composite Lineages of Life. *Trends in Ecology & Evolution* 30 (3): 127–9.

Milo, R., et al. (2002). Network Motifs: Simple Building Blocks of Complex Networks. *Science* 298 (5594): 824–7.

Nelson-Sathi, S., et al. (2012). Acquisition of 1,000 Eubacterial Genes Physiologically Transformed a Methanogen at the Origin of Haloarchaea. *Proceedings of the National Academy of Sciences* 109 (50): 20537–42.

Nelson-Sathi, S., et al. (2015). Origins of Major Archaeal Clades Correspond to Gene Acquisitions from Bacteria. *Nature* 517 (7532): 77–80.

Nicholson, D. J. (2012). The Concept of Mechanism in Biology. *Studies in History and Philosophy of Biological and Biomedical Sciences* 43 (1): 152–63.

Raoult, D. (2009). There Is No Such Thing as a Tree of Life (and of course Viruses Are Out!). *Nature Reviews Microbiology* 7 (8).

Skippington, E. and Ragan, M. A. (2011). Lateral Genetic Transfer and the Construction of genetic exchange communities. *FEMS Microbiology Reviews* 35 (5): 707–35.

Sole, R. V. and Valverde, S. (2006). Are Network Motifs the Spandrels of Cellular Complexity? *Trends in Ecology & Evolution* 21 (8): 419–22.

Szathmáry, E. (2015). Toward Major Evolutionary Transitions Theory 2.0. *Proceedings of the National Academy of Sciences* 112 (33): 10104–11.

Villarreal, L. P. and Witzany, G. (2010). Viruses Are Essential Agents within the Roots and Stem of the Tree of Life. *Journal of Theoretical Biology* 262 (4): 698–710.

von Bertalanffy, L. (1968). The Meaning of General System Theory. In L. von Bertalanffy (ed.), *General System Theory: Foundations, Development, Applications* (pp. 30–53). New York: George Braziller.

Wang, M. and Caetano-Anolles, G. (2009). The Evolutionary Mechanics of Domain Organization in Proteomes and the Rise of Modularity in the Protein World. *Structure* 17 (1): 66–78.

Wilkins, A. S. (2007). Between 'Design' and 'Bricolage': Genetic Networks, Levels of Selection, and Adaptive Evolution. *Proceedings of the National Academy of Sciences* 104 (Suppl. 1): 8590–6.

Williams, T. A., et al. (2013). An Archaeal Origin of Eukaryotes Supports Only Two Primary Domains of Life. *Nature* 504 (7479): 231–6.

Yafremava, L. S., et al. (2013). A General Framework of Persistence Strategies for Biological Systems Helps Explain Domains of Life. *Frontiers in Genetics* 4 (16).

Yutin, N., Raoult, D., and Koonin, E. V. (2013). Virophages, Polintons, and Transpovirons: A Complex Evolutionary Network of Diverse Selfish Genetic Elements with Different Reproduction Strategies. *Virology Journal* 10 (158).

PART V

Implications and Applications

# 15
# A Process Ontology for Macromolecular Biology

*Stephan Guttinger*

## 1. Introduction

Philosophers arguing for the need to adopt a process ontology often turn to the natural sciences to find support for their position. For instance, process philosophers have appealed to quantum physics because it 'puts money in the process philosopher's bank account' (Rescher 1996: 97), given its focus on fields and entangled states rather than on distinct entities. Some process philosophers have turned instead to the biological sciences, using in particular examples at the organism or ecosystem level (see Birch and Cobb 1981; Dupré 2012; Henning 2013; and the other chapters in the present volume). A class of entities that has been almost completely ignored in these discussions, however, is that of macromolecules.[1] This is somewhat surprising, as macromolecules are arguably the entities that biologists have devoted most attention to in the last sixty to eighty years.

The dominant focus on examples at the quantum level and at the organism level raises several problems for current process frameworks. One is the issue of scope: it is not clear how much the existing accounts have to say about the macromolecular level. Due to their size—typical examples of macromolecules are proteins or DNA molecules—it is usually assumed that quantum effects (and hence quantum theory) can be safely ignored when dealing with macromolecules. So, even if there are good reasons to take quantum phenomena to be fundamentally processual, this does not mean that such arguments automatically apply to macromolecules. Moreover, although macromolecules constitute living systems, they are not themselves living, and consequently we cannot assume that what is said about the processual nature of organisms is automatically applicable to macromolecules.

And it is not just the scope of the existing accounts that is a potential issue for the process philosopher. A problem is also that the natural sciences usually represent molecules as well-defined, distinct entities that have inherent properties owing to their material constitution (i.e. their atomic microstructure). According to chemical

---

[1] A rare exception is Ross Stein (2004, 2006), whose work I examine in section 5.

wisdom, molecules—including biochemicals such as DNA or proteins—look much more like substances than like processes.

What we are offered, then, by philosophers and by scientists alike, gives us little reason to assume that macromolecules are processual in nature. The goal of this chapter is to challenge this view and to argue (a) that macromolecules are fundamentally relational entities; and (b) that this relational nature of macromolecules is of a kind that only a process ontology can account for.

As the basis for my discussion, I will use the ecological worldview formulated thirty-five years ago by Charles Birch and John B. Cobb (1981). This ecological model represents an attempt to formulate a universally applicable process view of the world and states that everything, from atoms to populations, is an ecosystem and has to be treated as such. This means, according to Birch and Cobb, that all entities are fundamentally relational in character, which in turn means that only a process ontology can make sense of them. In section 2 I will introduce the ecological model and its assumptions in greater detail, highlighting in particular the importance of what Birch and Cobb call 'internal relations'. This concept is used by them to characterize the specific type of relations we encounter in ecosystems (at all levels). This special form of relations is also the reason why ecosystems need to be understood as fundamentally processual systems. In section 3 I will identify a key limitation of this account, namely its inability to provide an understanding of the capacities of molecules that builds on internal relations. I will show that this problem affects the macromolecular realm in particular and opens the door for a substance view of these crucial entities. The goal of the rest of the chapter will be to formulate a revised ecological account that can overcome this problem. To do so I will look at two case studies. In section 4 I will discuss the case of symbiotic systems, which show how we can develop a fully relational understanding of the capacities of biological entities. Key to this analysis will be the notion of 'integrated capacities', a term I will introduce to refer to capacities that depend on internal relations. In section 5 I will look at protein biology (and enzyme catalysis in particular) to show that this relational understanding of capacities also applies at the macromolecular level. With this revised understanding of capacities it will then be possible to formulate an extended ecological model that covers all entities, including macromolecules.

## 2. The Ecological Model of the World

The ecological worldview proposed by Birch and Cobb in their book *The Liberation of Life: From the Cell to the Community* is an attempt to formulate a general, process-based account of the world (Birch and Cobb 1981). The model is based on two key claims: (1) everything—from atoms to organisms to populations—is an ecosystem, as opposed to some sort of machine or mechanism (ibid., 89);[2] and (2) an ecosystem model of the world goes hand in hand with a process ontology, since ecosystems are

---

[2] Birch and Cobb use the terms 'machine' and 'mechanism' interchangeably. Also, when talking about something's being an ecosystem, Birch and Cobb not only mean that everything has to be treated as if it were an ecosystem but that all entities *are* ecosystems.

fundamentally relational in nature.³ In what follows I will look at these two claims in more detail.

## 2.1. Ecosystems, machines, and the environment

The reason for Birch and Cobb's (1981) first claim is mostly empirical: they argue that, if we look at what the sciences are telling us about the world, we see that a machine view simply does not fit in with how the world works (or is assumed to work).⁴ In the machine/mechanistic framework, entities are seen as distinct beings disconnected from each other, much like separated boxes. What allows each entity to behave the way it does is its internal structure, not the relations it has to its environment. Apart from providing some essential 'enabling' factors (such as a source of power) the environment has no significant bearing on the functioning of the machine parts or on the machine itself.

The ecological model paints a very different picture of the world, as it does not treat it as a set of disconnected boxes. Rather, entities are seen as interconnected complexes that behave the way they do because of the relations they have to other entities and/or processes in their environment (ibid., 83). In this view, the environment always needs to be taken into account when analysing the workings of entities, as it is (part of) what determines how entities behave.

Birch and Cobb claim that the ecological view is actually what the sciences present us with and what practising scientists have long adopted (even those who call themselves mechanists).⁵ Especially in studies of animal behaviour, according to Birch and Cobb, researchers have always been forced to acknowledge the importance of context, as these studies have invariably shown that the behaviour of animals is fundamentally altered by changes in the environment of the animal (ibid., 80).

## 2.2. Internal versus external relations

Adopting a mechanistic view does not mean, of course, that relations are neglected or disposed of. We can easily see this when we look at the importance of factoring in context sensitivity and context dependence when explaining or predicting the behaviour of machines or machine parts. Context dependence means that external relations ('inputs') are required for the correct functioning of the parts or of the machine. A cogwheel, for instance, turns because of the relations it has to other components of the machine it is a part of; what the cogwheel does depends on the relations it has to other entities or processes in its environment. This might seem like a trivial point,

---

³ Birch and Cobb consistently talk of an 'event ontology' rather than a 'process ontology' (in fact the word 'process' does not even appear in the index of their book). However, this does not mean that the term 'process ontology' is out of place to describe their position. Not only are the two widely seen as process philosophers (or, maybe better, as a 'process biologist' in Birch's case), but they also state that their ecological model is in principle the same as Whitehead's philosophy of organism (i.e. his process ontology; see Birch and Cobb 1981: 8).

⁴ For an extended discussion of the problems with the machine view in relation to the project of developing a process ontology for biology, see chapter 7.

⁵ Stein (2004: 10) makes a similar claim about the situation in the chemical sciences.

but it is important as it shows that factoring in relations is at the very core of the machine view.

Context sensitivity also has a role to play in a machine view, as even a simple part such as a steel rod is a context-sensitive entity that will, for instance, expand or contract in response to changes in the temperature of its environment. Such a change in length can have a significant effect on the functioning of a machine, which means that context sensitivity becomes a crucial aspect that needs to be accounted for when analysing or predicting the workings of a machine.

The reason why context sensitivity is compatible with a machine view is that it can be treated as an intrinsic feature of an entity. A steel rod might expand or contract depending on changes in its environment, but the *way* it responds to such changes is usually seen as an intrinsic feature or capacity of the particular type of steel the rod is made out of. The (potential) behaviour of the rod is seen as being determined by its material composition and structure rather than by its relations to other processes or entities. Other rods composed of different materials will display their own 'characteristic' capacity to expand or contract, as their microstructure is different. When we assume such context-insensitive context sensitivity, we are still operating within the disconnected boxes model that Birch and Cobb identify as one of the hallmarks of the machine view. Simply developing a relational account that acknowledges the importance of relations is therefore not enough to get the machine view into trouble.

Birch and Cobb are well aware of this, which is why they introduce a distinction between *internal* and *external* relations to deal with it. With this distinction they want to emphasize that what matters is not whether there *are* relations, but what role they play in the system of interest. In a mechanical system, Birch and Cobb maintain, relations are merely external, which means that the nature of an entity is not affected by the relations it has with other things or processes. The cogwheel or the steel rod are not affected in their nature by their (external) relations or by the change (turning, expanding, contracting) they undergo. The way they react to changes in their context is set by their material constitution.

In an ecological system, what a thing is depends on the relations it has; this means that there are no 'merely external' relations. The relations always have a constitutive role to play and hence are to be treated as internal:

The ecological model proposes that... the constituent elements of the structure [i.e. an entity of interest] operate in patterns of interconnectedness which are not mechanical. Each element behaves as it does because of the relations it has to other elements in the whole, and these relations are not well understood in terms of the laws of mechanics. The true character of these relations is discussed in the following... as 'internal' relations. (Birch and Cobb 1981: 83)

## 2.3. Substance versus process

The contrast between the two different types of relations also brings us to Birch and Cobb's second claim, namely that there needs to be a switch from a substance to a process ontology if we are to switch from a machine to an ecological model. According to Birch and Cobb, a substance is exactly the type of entity that a machine view of the world presupposes, given that a substance is something that is not affected in its nature by the relations it has. Relations are always external to a substance.[6] But, since a machine/mechanism view is not supported by what the natural sciences are telling us about the world, the substance ontology that underlies it has to be replaced by something that fits the ecosystem structure of the world.

The replacement Birch and Cobb have in mind is what they call an event ontology (see n. 3). According to this account, the behaviour of a system has to be explained by reference to events and not underlying substances. Importantly, events are 'constituted by their interconnectedness with other events' (Birch and Cobb 1981: 88), which means that relations are constitutive of (and, in the authors' terminology, internal to) events. This also means that there is no event that simply preexists and then relates to other things; in abstraction from its relation to the environment, the event itself is nothing (ibid., 87).

The switch from a substance to an event ontology also includes a reversal of the explanatory aims: whereas a substance ontologist aims to explain events that happen in the world in terms of substances, someone adopting an ecological worldview or an event ontology will aim to explain how persisting objects come about through an interconnection between events (ibid., 86). Importantly, the focus on events does not mean that there is no place for stable entities in an ecological model of the world. Events are simply seen as the *primary* elements of the world and stable entities are treated as 'enduring patterns among changing events' (ibid.: 95).

What is interesting about the account Birch and Cobb develop (and where, I think, it goes beyond the work of Whitehead it builds on) is the notion of 'ecological system' that it brings into the discussion. This concept gives us a different and powerful framework within which to think about the specific relational—and hence processual—character of entities at different levels of organization. The ecological model, in principle at least, offers a way of formulating a unified process framework that might apply to organisms and quantum systems as well as to the entities in between: macromolecules. But, as I will discuss in the next section, this unification is not without its difficulties.

## 3. Problems with the Ecological Model

It is clear from what we have discussed so far that the ecological account is meant to apply to all levels of entities, from single atoms to populations of organisms. However, beyond some general statements about atoms and molecules being ecosystems, Birch and Cobb offer little evidence to substantiate their radical claims about the nature of molecules. The problem is not that they do not discuss examples taken at the molecular level. It is rather that the examples they give do not seem to support their claims, as they do not establish how molecules are defined by internal rather than external relations.

We see this problem when we consider their discussion of DNA and its workings. As in the case of animal behaviour mentioned earlier, Birch and Cobb point out that empirical work on DNA has shown that its behaviour is highly context-sensitive and

---

[6] Birch and Cobb call a machine/mechanistic model a 'natural expression of substance-thinking' (Birch and Cobb 1981: 85). A similar claim is advanced in chapter 1 of this volume.

context-dependent; the way DNA behaves is always affected by, and depends on, the particular environment it finds itself in.

The problem is that their discussion of DNA behaviour does not give us any reason to go beyond the type of context sensitivity that depends merely on external relations, which we encounter in the case of machines. How close their description of DNA is to the machine view becomes particularly clear when they say: 'The DNA in the nucleus of the fertilised egg contains all the instructions necessary to make all the different proteins and all the different sorts of structures in all the different sorts of cells in the body... But not all the instructions are needed by every cell' (Birch and Cobb 1981: 81). Applying this view of DNA to the example of gene regulation in bacteria they go on to state that 'in their normal life in our intestines... bacteria must be ready to change their enzymes quickly to suit the sort of sugar we send down to them. They are selecting just one from the several that their DNA allows them to produce' (ibid., 81).

What stands out here is that the DNA molecule is portrayed by Birch and Cobb as having an intrinsic capacity to code for different products (proteins and other 'structures'). This capacity does not come about through relations but is given by the structure of the DNA, which means that it is contained within the DNA itself. It is only the *realization* of these capacities that depends on the interaction of DNA with specific enabling factors in its environment; what part of the DNA becomes activated in what way and at what time depends on contextual elements such as transcription factors or methylation patterns. But the set of options is restricted and, most importantly, defined by the intrinsic properties of DNA. The relations with the environment do not change what the DNA is and what capacities it has.

Such a view clearly still adheres to the machine view that Birch and Cobb want to avoid, as it operates with a picture of DNA as an entity for which relations are merely external; the context is what selects externally from a potential that is defined by the structure of the DNA itself. The more radical claims Birch and Cobb make about the fundamentally relational nature of molecules are therefore not supported by their own descriptions of the functioning of DNA.

The DNA example shows that the ecological account potentially faces a serious challenge when it comes to macromolecules, as its current formulation does not offer a fully relational understanding of their capacities. By describing macromolecules such as DNA as entities that ultimately don't depend on internal relations, Birch and Cobb leave the door open for a machine interpretation of this crucial class of entities. The problem is that the ecological model in its current formulation is missing a fully relational understanding of capacities. In the next section I will address this issue and claim that recent research on symbiotic systems can show us a way to such a relational account. To characterize what is special about this alternative view of capacities, I will introduce a distinction between 'component' and 'integrated' capacities.

## 4. Symbiosis and the Importance of Integrated Capacities

In recent decades, research in biology and ecology has greatly advanced our knowledge of how complex organisms and ecosystems persist and reproduce themselves.

This research has also shown that symbiotic life forms are the rule rather than the exception in the biological realm.[7]

Symbiotic systems are systems in which (in the simplest case) two organisms form a tight relationship with each other from which either one side or, in the case of mutualism, both sides benefit. Often the organisms involved depend on the symbiotic relationship for their survival, in which case scientists talk of obligate symbionts. In this section I will discuss the case of termite colonies in order to show how symbiotic systems help us gain a more relational understanding of capacities. In section 5 I will then show how this alternative understanding of capacities can also be applied to macromolecules.

## 4.1. Termites and their capacity to survive and reproduce

My discussion of symbiosis will focus on the work of J. Scott Turner, who studies termites of the genus *Macrotermes*, which live in southern Africa.[8] These termites form large structures, within which they cultivate a specific type of fungus, which can digest the cellulose in wood or grass—the termites' main source of food.

A prominent feature of these structures is the tall mounds that mark the location of a termite colony. The mounds can grow up to thirty feet in height and contain an intricate internal structure consisting of a central 'chimney' connected to a network of passages and thin-walled tunnels. Interestingly, these mounds do not serve as housing for the termites, which live in a spherical nest below the mound. Turner's (2000) work suggests that the mounds rather serve as something like a lung, helping to maintain specific atmospheric conditions inside the colony.

These specific atmospheric conditions are required to guarantee the survival of the colony. The reason for this requirement lies in the demands of the fungi that form part of the colony: to work at the optimal rate, they require sufficient oxygen and the right temperature and humidity. And it is here that the mound and its complex network of chimneys and tunnels come into play: the mound, Turner proposes, harvests air currents on the surface and channels it into the mound, allowing a tightly regulated turnover of the air within the termite colony and thereby creating (part of) the conditions needed for its functioning. Turner found that the oxygen concentration within the nest is kept at 17 per cent (which is 2 per cent lower than the atmospheric concentration of oxygen) and humidity is kept at 70 per cent (as compared to an average of 20 per cent on the outside).

The termites constantly rework the structure of the mound, to make sure that these conditions are maintained. This makes the mound, as Turner puts it, a process rather than a static object. Interestingly, even though the termites are a key force that shapes the mound and the nest, Turner found that this shaping takes place also because of the activity of the fungus. This is because the growth of the fungus can cause it to break through the surface of the mound, thereby creating leaks that the termites then fill in and repair (Turner 2005). This 'dance of agency', to use Andrew Pickering's term (Pickering 1995), turns out to be a key force in the shaping of the mound and

---

[7] The philosophical significance of symbiosis is also examined in chapters 1, 5, 9, and 10.

[8] For an overview, see Turner 2002, but also Henning 2013, which uses the termite case to argue for the need for a process ontology. I discuss Henning's work in section 4.2.

the nest: the formative power here is a meshwork of activities rather than the activity of just one particular entity (e.g. the termites).

The termites' case nicely illustrates that (as in other cases of symbiosis) all the elements of the system have to work together to bring about a colony that can survive and reproduce. As Turner puts it, '[t]he termite colony—insects, fungus, mound, and nest—becomes like any other body that is composed of functionally different parts working in concert and is ultimately capable of reproducing itself. Taken as a whole, the colony is an extended organism' (Turner 2002: 66).[9]

## 4.2. Process and individuality

The termites' case illustrates an important and more general challenge, which has gained significant attention in philosophy of biology in recent years: as even the soil-based mound has to be seen as an integral part of the superorganism, our traditional understanding of what a biological individual is becomes seriously undermined by such life forms (Clarke 2010; Gilbert et al. 2012; Bouchard and Huneman 2013; Ereshefsky and Pedroso 2013; Guay and Pradeu 2015). The problem of biological individuality is brought up not only by the termites' case but by all forms of symbiosis and has also been used to argue for the need to adopt a process ontology (see Dupré 2012, Henning 2013, and—again—chapters 1, 5, 9, and 10 here).

Brian Henning in particular argues that through Turner's work on termites we come to realize that 'a single termite is unintelligible apart from the collective organism of which it is a member' (Henning 2013: 240). He then adds:

> Individuals normally have clearly defined boundaries, a membrane that demarcates where they begin and end. Here we find that, as a single superorganism, the termite colony is extended in space and time, without clearly defined boundaries or a skin to define where the environment stops and the superorganism begins. Normally we would say that a single insect crawling on the ground is a proper individual. However, Turner's research shows that a single termite is no more an individual than a single cell in a petri dish solution. (Ibid., 241)

Henning claims that this blurring of boundaries challenges the substance-based metaphysics our traditional notions of individuality rely on; in other words a process ontology (à la Whitehead, in this case) is required to make sense of the ontological status of what Henning calls 'collective individuals'.

In his discussion Henning emphasizes the absence of sharp boundaries between the superorganism and the environment. But it is not immediately obvious in what sense boundaries become blurred here, as he is clearly able to talk about the individual termites, which he identifies as distinct, well-defined entities and which Turner tracks and studies without any (conceptual) problems. Turner is also clear about what distinct entities belong to the colony (mound, termites, fungus, etc.). So certainly not all dividing lines we are used to drawing around entities are undermined by symbiosis. But what does Henning mean, then, when he claims that the single

---

[9] For more on the heteronomy of organisms and other biological entities, see Dupré and O'Malley 2009.

termite is 'unintelligible' to us apart from the larger system and that we need a process ontology in order to be able to account for what makes a 'proper' individual?

### 4.3. Distributed capacities

What I think is crucial here—and this is something that Henning (2013) does not directly emphasize—is the strong focus on capacities. This observation is not limited to Henning's paper but applies more generally to discussions of symbiotic systems. If we look at the previous example, we see that what is not intelligible in the case of the isolated termite is how some of its capacities come about. The discussion shows that the boundaries that are being brought down by the termite case are not the physical boundaries around the different entities, but the boundaries we draw around the autonomous being, the thing that 'does' something. The significance of symbiosis examples in general is in the realization they give us that the entities we used to refer to as single living things (for instance termites moving around, eating, and digesting food) turn out to be less autonomous than we might have thought them to be: in order to function the way they do, they need the other elements of the symbiotic system. This is the sense in which they are not 'proper' individuals anymore.

What is shifting in our understanding of the termite (or other organisms, for that matter) is the *attribution of the power to do things*.[10] We are moving from considering only one thing as the carrier of a certain power to considering a network of entities as the legitimate carrier of that power: it is not just the termite that shapes the mound and has the power to digest, reproduce and survive, but a system of interconnected entities that has to be treated as the centre or origin of these capacities or powers. What changes is our understanding of where to place capacities we normally ascribe to entities demarcated through clearly distinct boundaries (e.g. a membrane or some sort of skin).

### 4.4. The ecology of powers

The symbiosis example shows that collaboration is at the core of what defines living systems: different entities have to work together to achieve a particular goal, such as survival or reproduction (Dupré and O'Malley 2009). But why should the importance of working together mean that we have to abandon a substance ontology and adopt a process view instead? After all, the idea of collaboration is, by itself, fully compatible with a substance-based understanding of the world. Quite trivially, the capacity of a machine—for example, the capacity of a car to turn left or right—usually comes about because several parts of that machine (with their individual properties and capacities) work together. In a machine view, it is assumed that each of the parts involved has its own set of subcapacities, which contribute to the overall system capacity, and that the parts have these subcapacities because of their specific composition (and not because of their relations to the other parts of the machine).

If symbiosis could be reduced to such an understanding of collaboration, then the termite example would not force us to move away from a machine view. And it is

---

[10] In what follows I will use the terms 'power' and 'capacity' interchangeably. Note that my use of the term 'power' does not necessarily correspond to the more technical meaning this concept often has in metaphysics (see e.g. chapter 3 in this volume).

tempting to treat the termite system in this way, for instance when considering its capacity to digest grass or wood. At first sight this capacity seems to be the outcome of two distinct parts working together, namely the termite that delivers wood or grass to the nest and the fungus cells that digest the cellulose contained in this material, breaking it into single sugar molecules (which in turn serve as food for the termites).

But the working together of the different parts in symbiosis is not the same as the working together of the parts in a machine, as we cannot simply take the termite as a moving (and mowing) device that collects grass or wood and the fungus as a digestive apparatus, then plug the two together, each coming with its own independent and preexisting set of capacities. What is crucial about the powers we ascribe to the termite colony and its parts is rather the constant interaction between (the activities of) the entities of the larger system. All these activities and interactions are interconnected and interdependent in such a way that, if they were to stop, then the key properties of the system and of what we describe as its parts would disappear.

This interdependence becomes clear if we consider how the different parts come to have the subcapacities we usually ascribe to them. If we, for instance, take a closer look at the ability of fungus cells to digest cellulose, we quickly learn that the actual degradation of cellulose molecules is performed by enzymes that are produced by the fungus cells (see Baldrian and Valaskova 2008 for an overview). We also know, from basic cell biology, that the fungus cells need specific subcellular compartments (e.g. the so-called 'endoplasmic reticulum') to function properly. But, to be able to form the different intracellular structures and have them work as they should, the fungus cell needs particular conditions, including a specific temperature or oxygen concentration. Importantly, as the work of Turner shows (section 4.1), these parameters are not simply inherent properties of the nest or the soil but themselves the results of ongoing processes such as the movement of air through the tunnels that are created and constantly maintained by the termites, or the digesting activity of the fungus cells. The capacities we find in complex biological systems are therefore not simply preexisting features of its parts that merely need to be activated by (external) relations but features that only come about through what Birch and Cobb call internal relations. To find terminology that allows us to describe this specific form of relational capacities, I will turn now to the work of William Bechtel and Robert Richardson.

## 4.5. Component versus integrated capacities

In their book *Discovering Complexity*, Bechtel and Richardson develop a framework that can be used to analyse different types of complex systems (Bechtel and Richardson 2010). This framework, I think, can also be helpful for the present discussion about complex capacities. Following the work of William Wimsatt and also Herbert Simon, Bechtel and Richardson distinguish between aggregative systems and composite systems, the latter being further subdivided into *component* and *integrated* systems. Aggregative systems are those in which (a) each component has an intrinsic function or behaviour; and (b) the organization of the components does not affect their behaviour. As a consequence, the overall behaviour of the system is a function of component behaviour and does not depend on the organization of the parts (ibid., 25).

In composite systems the organization of the system (and hence the environment of the different parts) starts to play a role in the behaviour of the system and its parts. In a *component* system the behaviour of each component is still determined by its intrinsic features, but the organization of the system affects the behaviour of the whole and its parts (this is basically what Birch and Cobb call a machine or a mechanism). In an *integrated* system the behaviour of each component is no longer intrinsically determined, as the organization of the whole becomes a key determining factor of each component's capacity to act (this corresponds to what Birch and Cobb call an ecosystem).

This distinction between component and integrated systems can also be useful for the discussion about capacities and the different forms of collaboration we encounter in the machine and ecosystem cases. In a substance ontology/machine view, the capacities of the individual parts are treated as inherent in the self-contained, distinct entities that compose the system. These capacities are intrinsic and simply need an external 'trigger' or stimulus and a 'nurturing' or 'enabling' environment. When working together, they might respond to inputs from the other elements of the system but their way of reacting—their repertoire of possible behaviours—is intrinsically determined. These are what I will call 'component capacities'. In the case of symbiotic systems, the property of having a power or capacity is not some intrinsic feature of preexisting entities, but a relational feature that comes about within the system of interest through the intersection of different processes (or, in the terminology of Birch and Cobb, events). The context becomes a constitutive factor for these capacities, which is why I will refer to them as 'integrated capacities'.

## 4.6. Integrated capacities at all levels?

As we discussed in section 2, the Achilles heel of the ecological account is that it does not offer a fully relational understanding of the behaviour of macromolecules. Interestingly, the symbiosis example offers an understanding of capacities as relational properties and can therefore provide an important extension of the ecological view. The question is whether this relational understanding of capacities can be extended to the macromolecular level and therefore lead us to a process framework that applies to all levels of entities.

At first sight such an extension seems problematic, as macromolecules don't seem to be prone to individuality issues, like symbiotic systems. There are no 'super-molecules' for which the boundary between entity and environment becomes blurred, as it was the case for the termite colony; the ascription of capacities simply does not seem to be an issue at the molecular level. This is illustrated not only by the case of DNA, discussed in section 3, but also by other macromolecules, for instance proteins. In the molecular life sciences proteins are usually treated as 'molecular machines'. The capacities of these machines do not depend on relations, but on being the right kind of entity, that is, a molecule with a particular structure (a three-dimensional fold, in the case of proteins) and composition (the amino acid sequence of a protein). The capacities of proteins are therefore treated as component, and not as integrated, capacities. However, as I will show in the next section, this picture of macromolecules as carriers of component capacities quickly falls apart if we look closer at how these entities actually work. Capacities at the macromolecular level ultimately turn out to be as integrated as those ascribed to symbiotic systems, at least if we follow how scientists themselves try to make sense of the powers of proteins.

## 5. Proteins, Structure, and Capacities

Proteins are linear polymers of amino acids, linked to one another via peptide bonds (which is why they are often called 'polypeptides'). Proteins are often portrayed as the 'doers' or the 'workhorses' of the cell, a characterization that is mostly reserved for enzymes—the proteins that have the ability to catalyse chemical reactions. An example of such workhorses are the kinases, a class of enzymes that have the ability to mediate phosphorylation, that is, the transfer of a phosphate group from a donor to a target molecule (usually another protein). That the kinases are presented as the carriers of specific intrinsic powers can be seen in any review or research article that talks about them. If we pick a random example of a paper that has the word 'kinase' in its title (here I chose Ubersax and Ferrell 2007), we quickly find claims such as these: 'Despite sharing a common fold, kinases bind to and phosphorylate different protein substrates'; 'Non-systematic studies and biochemical lore suggest that kinases vary greatly in the number of these sites that they phosphorylate.'[11] Clearly kinases are presented in these passages as the carriers of the capacity or power to phosphorylate target molecules. Importantly, this capacity is not treated as a relational feature of a larger system but as an inherent (and defining) property of the kinase itself. This means that it is treated as a component rather than as an integrated capacity.

### 5.1. From structure to power?

But how do enzymes like kinases obtain their power? Enzymes are catalysts that function by lowering the activation energy of a chemical reaction. According to the transition-state theory of chemical reactions, each chemical reaction has to go through a transition state, which represents the maximum energy point along the reaction pathway. The reason for this is that a high-energy intermediate between the substrate and the product of the reaction is being formed in the transition state. The energy required to reach this intermediate is called the activation energy of the reaction.

The role of the catalyst is to lower the energy required to reach the transition state, in other words, to lower the activation energy of the reaction. By doing so, the catalyst allows the reaction to take place at a higher rate at a given temperature, since more substrate molecules in the mixture will have the required energy to overcome the activation barrier.

How do enzymes achieve this feat? Explanations of how proteins function are regularly given using what some refer to as the sequence–structure–function (SSF) paradigm (see e.g. Wright and Dyson 1999 or Redfern et al. 2008). The SSF has been central to protein biology, roughly, for the past hundred years; it postulates that the

---

[11] We also find similar statements in textbooks, for instance the key textbook *Molecular Biology of the Cell*, where the authors state: 'The protein kinases that phosphorylate proteins in eukaryotic cells belong to a very large family of enzymes, which share a catalytic (kinase) sequence of about 290 amino acids" (Alberts et al. 2008: 176).

function of a protein is determined by its three-dimensional structure, which in turn is determined by its unique amino acid composition and sequence (i.e. by its atomic composition and organization). The interactions between the atoms of these different amino acids allow the polypeptide to take on a specific three-dimensional conformation. The interactions between amino acid residues (both backbone and side chain atoms) can range from hydrophobic interactions to electrostatic or covalent bonds. According to the SSF paradigm, it is the composition and the sequence of the protein that determines the structure it can obtain. This structure then determines what the protein can do, that is, what specific powers it has. The SSF paradigm therefore presents proteins as distinct entities that possess specific capacities due to their inherent properties—a picture that is perfectly in line with a substance-based understanding of molecules.

What is special about the structure of enzymes is that they possess a so-called 'active site' in which the side chains of specific amino acid residues are spatially arranged in a particular way. This specific arrangement of the amino acids forms a chemical environment that allows the enzyme to function as a catalyst, as it enables it to bind to the high-energy intermediate and thereby stabilize it (or so the SSF paradigm would imply). This stabilization indicates that the energy required to reach the transition state of the reaction is lowered, which means that the rate at which the reaction proceeds at a given temperature is increased owing to the presence of the enzyme.

This explanation of enzyme function using the SSF paradigm shows us an important qualification of the picture of the enzyme as a 'doer' that we encountered earlier and that is so prevalent in the ways scientists talk and write about molecules: if we look at the story of enzyme action in more detail, we see that what the enzyme does is to provide a particular chemical environment, a surface that allows different molecules to interact in specific ways (the active site). These interactions ultimately are what allows the system as a whole to undergo a transformation along a different reaction pathway, which means a path with a lower activation energy. Enzyme catalysis is a much more collaborative enterprise than one might think, judging by how scientists usually talk about enzymes. And the picture of the enzyme as some sort of machine or workhorse that actively phosphorylates a target is clearly a highly metaphorical way of portraying what is going on.

But, as we have seen earlier, the mere fact that collaboration is important does not mean that we have a system that operates with integrated rather than component capacities. If we want to argue that a machine view could not account for the functioning of enzymes, we need to show that, also in this case, we have integrated capacities at work. To do so I will turn to the work of the chemist and process philosopher Ross Stein and to his discussion of recent developments in enzyme research.

*5.2. More than just collaboration*

Stein's work is interesting for our current discussion because he follows Birch and Cobb in claiming that molecules (and enzymes in particular) are ecosystems. To support this claim, Stein considers new models of enzyme function developed in enzymology. What is special about these models is that they no longer portray three-dimensional structure as some sort of stiff scaffold, but rather as a dynamic feature of the protein. A protein, according to these models, is better represented by an ensemble of similar but different structures rather than by one fixed structure. And, importantly, the protein is assumed to constantly cycle through the conformations that form its ensemble.[12]

What makes this new view of enzymes particularly interesting is that this activity of cycling through different conformations is taken to be crucial for the functioning of the enzymes. The idea is that enzymes have the power to stabilize the high-energy transition state of a specific reaction *because* of the dynamics of their structure. And the enzyme can only cycle through its different states because it is coupled to the environment, that is, to the thermal motion of the bulk water that surrounds it. It is this complex interaction with the surrounding water that ultimately shapes the way the enzyme behaves and that gives rise to the power to catalyse a chemical reaction.

This provides a crucial reinterpretation of the way power is attributed to an enzyme: the three-dimensional structure, the only element deemed relevant in the SSF paradigm, is no longer enough to bring about function. What matters for the power of the enzyme to catalyse a reaction is rather the constant change in its polypeptide. Importantly, this change only comes about because of the enzyme's interaction with the surrounding water.

To Stein, this is the key change, as the two 'parts'—the enzyme and the surrounding bulk of water—are now treated as a unit (Stein 2004: 15). And it is out of this unit that the capacity to catalyse a chemical reaction arises. As he puts it:

> In the end, we will not be able to locate the origins of the catalytic power of an enzyme in a certain 3-dimensional arrangement of active site residues nor in a certain fold of the protein; rather, enzymatic catalysis will have to be analyzed as structurally specific substrates bound to an active site of definite chemical potential embedded in a dynamic protein matrix that is in thermal exchange with the aqueous environment of bulk solvent. This holistic description of enzyme catalysis can be solidly grounded in the metaphysical foundation of Strong Chemical Processism. (Ibid., 15)

Stein's discussion of recent work in enzymology illustrates how the structure of the protein is no longer seen by scientists as the factor that brings about the capacity of an enzyme to catalyse a chemical reaction. It is also no longer the case that the environment is simply treated as the provider of an energy input that then activates the capacity of the enzyme. The interaction between water and enzyme (neither of which is now demoted to the role of mere external environment) is what brings about the capacity to catalyse the reaction. The boundaries between the enzyme and its environment therefore become blurred, much as in the case of the symbiotic system discussed in section 4. And what we end up with—once we focus on this question of where the capacity resides and how it comes about—is a picture of integrated rather than component capacities of proteins.

---

[12] This view of proteins is in line with a more general development in protein biology that acknowledges the fundamentally dynamic nature of proteins. In particular, the discovery of what is now called 'intrinsically disordered proteins' (IDPs) has provoked a change in our understanding of proteins (for a general overview, see Dunker et al. 2001).

## 5.3. Towards a general process account for macromolecular biology

Stein's approach is of course not without its problems, especially in light of its explicit goal to provide a general process framework that applies to all molecules (and other entities). One issue is that the model Stein discusses might not apply to all enzymes, as there could well be some that do not depend on the environment in the same way his chosen examples do. Furthermore, even if the model applied to all enzymes, it is not clear in what sense proteins that are not enzymes could be equated with ecosystems, as they might have very different modes of functioning.

There are at least two ways to answer such worries about the scope of Stein's account. First, it is crucial, I think, to put his discussion of enzymes into the context of current developments in protein biology more generally, in particular the discovery of intrinsically disordered proteins mentioned earlier (see n. 12). The IDP case convincingly shows that the dynamic nature of the polypeptide has a crucial role to play in the functioning of many more proteins than just enzymes.[13] The prevalence and importance of IDPs for the functioning of the cell undermines the strict link between structure and function that the SSF paradigm postulates, which has important consequences for our understanding of the nature of proteins more generally.

But there is also a second way of making Stein's discussion broader; and this is by questioning the first part of the SSF paradigm, that is, the link between the sequence (microstructure) of a protein and its three-dimensional structure. Stein focuses his discussion of enzymes on the second part of the paradigm, the link between the structure and function of proteins. But the question of how proteins obtain their powers also comes up, of course, when we consider the first part of the SSF paradigm, which states that the sequence of the protein is what defines the structure(s) it can adopt. As in the case of catalytic power, the capacity to adopt a particular fold is treated as an intrinsic feature of the protein (if we follow the SSF paradigm) that only depends on (external) relations.

Interestingly, once we dig deeper into the question of how a protein can adopt a particular three-dimensional conformation (or an ensemble of conformations), we are immediately led to talk about forces. As briefly mentioned in section 5.1, different physical forces are at work when the crucial interactions of a particular conformation of a molecule are formed. One such force that is crucial for the fold of a protein (but also for the double-helix structure of DNA) is the hydrophobic force, a sort of repulsion from water felt by hydrophobic (i.e. apolar) molecules.[14] All structured proteins have a hydrophobic core in which apolar amino acids are 'buried', that is, kept away from the protein's aqueous environment. The formation of this hydrophobic core is an important step in the folding process and is also what to a large extent explains the relative stability of folded proteins. A similar process is at work in DNA, where the stability of the double helix depends not only on hydrogen bonds

---

[13] It is estimated that 30–50 per cent of all proteins are IDPs (Dunker et al. 2001). Note that most of the known IDPs are not enzymes.

[14] In the case of proteins, these are mostly the apolar side chains of specific amino acids such as alanine or leucine. In the case of DNA, these are apolar parts of the nucleotides that form the DNA.

formed between matching base pairs but also on hydrophobic interactions between the stacked nucleotides (Yakovchuk et al. 2006).[15]

The hydrophobic force is interesting for our current discussion because it is not something a single molecule simply possesses, given its intrinsic properties. It is rather a phenomenon that comes about through the interaction of a larger system of (polar and apolar) molecules. It is also not just the mere existence of polar and apolar entities that gives rise to the hydrophobic force. The force only comes about in a context of constant interaction and repulsion; it is a force born out of becoming, and not out of simple being. The structure that the protein (or any molecule, for that matter) adopts is therefore the outcome of a complex process, which takes place within a larger dynamic system. Within this system it is not clearly defined what should be seen as 'internal' and what as 'external', since the boundaries between the entity of interest and its environment are blurred. The capacity to adopt a particular fold is therefore not something that the protein simply possesses and that is then triggered or activated by some external input from the environment but it is, like the catalytic power Stein discusses, or like the capacities of a termite colony, an integrated capacity that emerges from within an integrated whole.

## 6. Conclusions

The goal of this chapter was to show that macromolecules such as proteins or DNA molecules are ecosystems in the sense of Birch and Cobb, in other words, that they are relational entities that can only be fully accounted for by a process ontology. As I argued in section 3, the original ecological model suffers from the problem that it does not offer a fully relational understanding of the capacities of macromolecules. I used the example of termite colonies to develop a relational understanding of capacities ('integrated capacities') and I then argued that such integrated capacities are also what defines the behaviour of molecular systems, be that catalytic activity or the activity of folding into a particular three-dimensional structure. With this fully relational understanding of macromolecules it becomes clear that they, too, are ecosystems, as Birch and Cobb postulated. Being ecosystems means that they are fundamentally processual entities, which in turn means that the ecological account can now fill the gap between the quantum and the organism level and offer a general process framework that also includes macromolecules.

## Acknowledgements

I would like to thank John Dupré, Anne-Sophie Meincke, Dan Nicholson, Nicolas Wüthrich, and the participants of the Biological Interest Group (BIG) meeting at the Egenis centre for

---

[15] DNA is in general a much more dynamic entity than is often assumed. The depiction of DNA as some sort of stable double helix that can at most wind itself around some histones ignores such interesting features as 'DNA breathing', the constant opening and closing of the double helix that is—again—crucial for its proper functioning (see von Hippel et al. 2013; Fei and Ha 2013).

very helpful comments on an earlier version of this chapter. Research for this work was funded by the European Research Council under the European Union's Seventh Framework Program (FP7/2007-2013)/ERC grant agreement n° 324186

## References

Alberts, B., Johnson, A., Lewis, J., Raff, M., Roberts, K. and Walter, P. (2008). *Molecular Biology of the Cell*, 5th edn. New York: Garland Science.
Baldrian, P. and Valášková, V. (2008). Degradation of Cellulose by Basidiomycetous Fungi. *FEMS Microbiology Reviews* 32 (3): 501–21.
Bechtel, W. and Richardson, R. C. (2010). *Discovering Complexity: Decomposition and Localization as Strategies in Scientific Research*. Cambridge, MA: MIT Press.
Birch, C. and Cobb, J. B. (1981). *The Liberation of Life: From the Cell to the Community*. Cambridge: Cambridge University Press.
Bouchard, F. and Huneman, P. (eds). (2013). *From Groups to Individuals: Evolution and Emerging Individuality*. Cambridge, MA: MIT Press.
Clarke, E. (2010). The Problem of Biological Individuality. *Biological Theory* 5 (4): 312–25.
Dunker, A. K., Lawson, J. D., Brown, C. J., Williams, R. M., Romero, P., Oh, J. S., et al. (2001). Intrinsically Disordered Protein. *Journal of Molecular Graphics and Modelling* 19 (1): 26–59.
Dupré, J. (2012). *Processes of Life: Essays in the Philosophy of Biology*. Oxford: Oxford University Press.
Dupré, J. and O'Malley, M. (2009). Varieties of Living Things: Life at the Intersection of Lineage and Metabolism. *Philosophy & Theory in Biology* 1:e003.
Ereshefsky, M. and Pedroso, M. (2013). Biological Individuality: The Case of Biofilms. *Biology & Philosophy* 28 (2): 331–49.
Fei, J. and Ha, T. (2013). Watching DNA Breath One Molecule at a Time. *Proceedings of the National Academy of Sciences* 110 (43): 17173–4.
Gilbert, S. F., Sapp, J., and Tauber, A. I. (2012). A Symbiotic View of Life: We Have Never Been Individuals. *Quarterly Review of Biology* 87 (4): 325–41.
Guay, A. and Pradeu, T. (eds). (2015). *Individuals across the Sciences*. Oxford: Oxford University Press.
Henning, B. G. (2013). Of Termites and Men. In B. G. Henning and A. C. Scarfe (eds), *Beyond Mechanism: Putting Life Back into Biology* (pp. 233–48). Plymouth: Lexington Books.
Pickering, A. (1995). *The Mangle of Practice*. Chicago: University of Chicago Press.
Redfern, O., Dessailly, B., and Orengo, C. (2008). Exploring the Structure and Function Paradigm. *Current Opinion in Structural Biology* 18 (3): 394–402.
Rescher, N. (1996). *Process Metaphysics: An Introduction to Process Philosophy*. New York: SUNY Press.
Stein, R. L. (2004). Towards a Process Philosophy of Chemistry. *HYLE: International Journal for Philosophy of Chemistry* 10 (1): 5–22.
Stein, R. L. (2006). A Process Theory of Enzyme Catalytic Power: The Interplay of Science and Metaphysics. *Foundations of Chemistry* 8 (1): 3–29.
Turner, J. S. (2000). Architecture and Morphogenesis in the Mound of *Macrotermes michaelseni* in Northern Namibia. *Cimbebasia* 16: 143–75.
Turner, J. S. (2002). A Superorganism's Fuzzy Boundaries. *Natural History* 111 (6): 63–7.
Turner, J. S. (2005). Extended Physiology of an Insect-Built Structure. *American Entomologist* 51 (1): 36–8.
Ubersax, J. A. and Ferrell, J. E. Jr. (2007). Mechanisms of Specificity in Protein Phosphorylation. *Nature Reviews Molecular Cell Biology* 8: 530–41.

Von Hippel, P. H., Johnson, N. P., and Marcus, A. H. (2013). Fifty Years of DNA 'Breathing': Reflections on Old and New Approaches. *Biopolymers* 99 (12): 923–54.

Wright, P. E. and Dyson, H. J. (1999). Intrinsically Unstructured Proteins: Re-Assessing the Protein Structure–Function Paradigm. *Journal of Molecular Biology* 293 (2): 321–31.

Yakovchuk, P., Protozanova, E., and Frank-Kamenetskii, M. D. (2006). Base-Stacking and Base-Pairing Contributions into Thermal Stability of the DNA Double Helix. *Nucleic Acids Research* 34 (2): 564–74.

# 16
# A Processual Perspective on Cancer

*Marta Bertolaso and John Dupré*

## 1. Introduction

Life can be characterized as a hierarchy of processes. It is dynamic at all levels (molecules, cells, tissues, organisms, lineages). This chapter thus starts from the premise that living systems should be understood not within a mereological framework (as things) but within a processual one (as processes).[1] For present purposes the most crucial corollary of this ontological observation is the following: whereas for a thing the default is persistence, for a process it is persistence that requires explanation. A table may sit in the attic for decades, undergoing little change that could not immediately be reversed with a duster. If you accidentally shut your cat in the attic, it will soon be a dead cat and, before too long, the skeleton of a cat. Without food and water, it is no longer able to perform the activities that sustain it. To put the point rather crudely, for things, what needs explanation is, typically, changes that they undergo; for processes, it is rather, or at least equally, their continued existence that calls for an explanation. For living systems in particular, this point is well captured by the familiar observation that these are systems far from thermodynamic equilibrium. To keep them there requires work, and this work comprises the processes that sustain the system.

The topic of this chapter is cancer, and the relevance of the above remarks to this topic should be obvious. It is generally supposed that the key to understanding cancer is to find out what causes it. What is it that interacts with a healthy individual organism to initiate the eventually fatal neoplastic process? From a processual perspective, on the other hand, one might rather ask, why do organisms generally *not* develop neoplasms? In slightly more detail, the persistence of an organism requires an exquisite balance between cell division, cell differentiation, and cell death. Many departures from this balance constitute cancers. Given how much activity and appropriate higher-level context are required to maintain the dynamic

---

[1] If and to what extent this assumption, and the following discussion, can be fruitfully extended to a metaphysics of life in general and to a revision of the classical notion of substance is beyond the scope of this chapter. One of the authors, however, is committed to a more general extension of this sort (Dupré 2012 and chapter 1 of this volume, co-authored with Daniel Nicholson), while the other has shown why a mereological framework is not adequate to account for cancer and how a processual perspective also makes sense of scientific practice (Bertolaso 2016).

stabilization of a multicellular organism, it is remarkable that these conditions can be sustained at all over long periods of time. Cancer, which in this chapter we regard as the progressive destabilization of the coupling of cell division and differentiation required to sustain the organism's life cycle, is a phenomenon very much to be expected.

As Denis Noble has stressed, the core issue in accounting for the dynamic stability of living systems is *modelling across scales*. '[I]n order to unravel the complexity of biological processes we need to model in an integrative way at all levels: gene, protein, pathways, subcellular, cellular, tissue, organ, system' (Noble 2002: 1). Given that the stable persistence of an organism requires such interactions between multiple systems at multiple organizational levels, it is no surprise that cancer has proved to have many causes at many levels. If there is anything general to be said about the phenomenon of cancer, then, it is more likely to be about what *prevents* it—that is, about what enables the healthy stabilization of cell populations—than about what causes it. Seeing health as depending on the highly regulated intertwining of different organismic processes implies, in turn, the necessity of adopting a processual view in defining and explaining carcinogenesis.

This dynamic, process-centred perspective has various other important implications for our general approach to biological systems. Focusing on the causes of stability draws attention to the relations of a system to the environment to which it must respond appropriately if it is to persist. More generally, we suggest that the stabilization of a biological system will always, or almost always, depend not merely on the properties of its parts, but also on its position within a larger system. Attention to this point resonates with various very active areas of contemporary research: epigenetics tells us how the effects of the environment on the organism can reach right into the nucleus, affecting gene expression in ways that are presumably often adaptive (Goldberg et al. 2007). Niche construction theory explores the ways in which much of the behaviour of an organism is directed towards creating and maintaining features of the environment that are favourable to the survival and reproduction of the organism (Odling-Smee et al. 2003). And so on.

An equally important context is what we might think of as the wider temporal environment, the reproductive lineage of which a particular organism is part. To exist at all, an organism must of course be part of such a lineage. And a stable lineage must be one that produces individuals that contribute to the maintenance of the lineage, namely by reproducing. Reproduction must result in the production of new organisms that have the capacities to maintain the (lineage) process of which they are part; and it is now appreciated that there are a variety of processes by which this transmission of adaptive features is effected, for example the niche construction just mentioned.[2]

Scientists are increasingly aware that the methodologies required to model across scales contrast with the traditional, reductive approach, which is grounded on a simplistic mereological view of the natural world. The peculiar complexity of biological systems readily accounts for the difficulties we experience in scientific practice in separating the dynamics at any given level of organization from the coupled dynamics of all other levels, including that of the environment within which the

---

[2] For a detailed processual account of organismic lineages and their evolution, see Dupré 2017.

system is embedded. These issues have been explicitly addressed by one of us in various places (Dupré 1993, 2010; O'Malley and Dupré 2005; Powell and Dupré 2009) in terms of the ineliminability of context dependency and top-down causation from biological explanation, and by the other more explicitly in studies on the evolution of explanatory models of cancer (Bertolaso 2009; 2016). From an ontological point of view, the inadequacy of a simple, hierarchical view of living systems in which explanatory relations are exclusively from parts to wholes, culminating in the absurdity of considering every feature of biological form as somehow encoded in a single molecule, is argued to be a central problem, which can be disposed of by adopting a more adequate process ontology for living beings (Dupré 2012). Growing insights into the importance of epigenetic changes and the evolutionary and developmental co-dependencies between different organisms further reinforce this processual view of the biological world.

This chapter will discuss the biology of cancer from the processual perspective just sketched, which will be further elaborated upon in the next section. It is acknowledged that cancer is a process that disrupts the functional coupling of organismic lifetime processes, notably cell proliferation and differentiation. We shall argue that the proliferation–differentiation coupling (in a normal individual) or uncoupling (in a pathological one) as well as the time-context dependence of the neoplastic process are best understood within a relational process-ontological framework that emphasizes the interdependence of these and other key processes (section 3). This perspective stresses the mutual determination of, and regulation between, different processes involved in the development and physiological maintenance of the organism. Such relations also hold between the various organizational levels of a multicellular organism, so we shall need to consider from this processual perspective the balance between autonomy (intrinsically determined behaviour) and connectedness (contextually determined behaviour) of cells or tissues belonging to a multicellular organism. We shall finally suggest that these various dimensions, both spatial and temporal, might be integrated through the concept of *morphogenetic fields*, a notion commonly used by scientists but still almost unexplored by philosophers (section 4). This concept refers to 'large-scale systems of physical properties that have been proposed to store patterning information during embryogenesis, regenerative repair, and cancer suppression that ultimately controls anatomy' (Levin 2012: 243).

## 2. Cancer as a Process

A process view of cancer in the scientific literature is generally motivated by the causal complexity of the disease, the explanatory problems of a reductionist genetic view of carcinogenesis, and the limits of molecular treatments and target therapies that derive from tumour latency and its dependency on time and context (Harris 2005; Sonnenschein and Soto 1999; Sporn 1991, 2006). Cancer is a paradigmatically complex disease. It appears to involve causes of many different kinds—for example environmental, tissue-level, genetic, and epigenetic causes—and it is well known that there are many different varieties of cancer. There is, however, one general characterization that applies to all forms of cancer: it is a failure of the processes that regulate the production and destruction of cells. A functional metazoan requires an exquisite balance between the various different cell types that make up its organs and systems. This balance is maintained by processes of cell division, cell differentiation, and cell death. We noted earlier that, in a sense, what most fundamentally requires explanation is not so much the presence as the absence of cancer, because the regulation and alignment of these processes is a remarkable achievement, and it is no surprise that it should be liable to fail in many different ways and for many different reasons. The proper balance of cell types is not something that is achieved once and then maintained by inertia; its maintenance requires a continuous and dynamic set of activities. Moreover, this functional balance is itself something that changes over the life cycle of an organism, and cancer can be seen as a failure to follow the 'correct' developmental trajectory. It has been described as development gone awry (Soto el al. 2007), as blocked ontogenesis (Potter 1978), and as a disease of the developmental or morphogenetic process (Biava 2002; Bizzarri and Cucina 2014).

Historically, cancer was deemed to be a disease that was due as much to the environment as to endogenous factors. However, this pathology was not subject to careful scientific investigation until the end of the nineteenth and the beginning of the twentieth century, when Rudolf Virchow (1821–1902) explored its relationship to both genetic and cellular factors. Subsequent development of new biochemical tools, advances in microscopy, and more sophisticated analyses of environmental data led to more thorough studies of cancer at various levels. Several hypotheses on the origin of cancer began to emerge, ranging from the 'biological theory' (Rous 1910) to the 'chemical theory' (Potter 1964; Doll and Hill 1956; Colditz et al. 2006; Parkin 2004), which envisaged an alteration of the cells' physiological balance caused by toxic substances in the environment, and to the popular 'viral theory' (Duesberg 1980; Klein 2002; Burmeister 2001). In the 1950s and 1960s, the 'genetic theory' (Knudson 1971) began to dominate oncological thinking and, as a result, decreasing attention was paid to the relevance of the organismic environment and of the immune system. This theory was originally supported by the evidence, first described by Boveri (2007 [1914], 1929), of the highly disorganized character of chromatin in cancer cells. Following the identification of DNA as the molecular basis of genetic inheritance, this disorganization was reinterpreted in genetic terms, resulting in what is today commonly referred to as the 'somatic mutation theory' (see section 3).[3]

Throughout the twentieth century, the aetiology of cancer remained at the forefront of epidemiological research, and the relevance of environmental, epigenetic, immune system, and microenvironmental (e.g. tissue architecture) factors in carcinogenesis was never completely overlooked. Indeed, a number of historical studies showed conclusively that both environmental and organismic factors play a causal role in carcinogenesis, and their effects can be surprisingly precise. A telling example is the wave of cases of cancer diagnosed in female survivors of the atomic bombs of the Second World War (Tokunaga et al. 1979): tumours arose only after a period of time and in an almost synchronous manner for many of those who had been exposed to atomic bomb radiation.

---

[3] For a broader and more comprehensive historical review of theories of carcinogenesis, see Loeb and Harris 2008; Bizzarri et al. 2008; Bertolaso 2009, 2016; and Weinberg 2006.

It has long been recognized that cancer is an inherently complex phenomenon. This complexity manifests itself in the diversity of causal factors and in the timescales on which these factors operate. This should not be all that surprising, given the multiplicity of processes at different timescales that are involved in stabilizing cell division and differentiation. As we have already mentioned, long-range signals provide positional information to regulate organism-wide systemic properties like anatomical polarity, size control, and the multilevel functional integration of cell behaviour. When these various modes of stabilization fail in sufficient number, cancer ensues. This is why cancer cannot be considered a discrete event; still less can a tumour be regarded as a static thing. Rather, it is a process, or a set of processes, which extends over considerable periods of time and generally involves various levels of organization in the body.

The causal complexity of cancer is not simply due to a multiplicity of causes, among which researchers can pragmatically select in their investigations. Multiple alterations at different levels (e.g. genes, proteins, membrane structures, tissues) are all required for a cancer to develop. The eventual outcome is an aberrant proliferation of cells, with the progressive disruption of functional integration, and thus of biological order, at different scales, from genes to cells and tissues. The morphological order that usually characterizes tissues and organs is lost. Although cancer is usually diagnosed on the basis of a morphological disruption of tissue organization, it typically requires the loss of dynamic stability at other levels of organization as well (Bertolaso 2013, 2016). The diversity of the phenomenon of cancer becomes evident when we consider the *temporal dynamics* of the neoplastic process. The course of the disease varies between different patients and generalizations seem hard to find, not for lack of experimental data but because the variables relevant to the timing of the neoplastic process vary widely from case to case.

The appearance of clinical symptoms of a tumour is preceded by a variable period of time called the *latency period*, which in many human tumours can last for years. Throughout this period of latency, cancer already exists as an aggregation of cells whose normal process of proliferation–differentiation has been somehow compromised, but is not yet clinically identifiable. A number of experimental studies have suggested that the onset of cancer may be more directly dependent on the alteration of morphogenetic dynamics over time than on specific genetic mutations. Compelling evidence for the requirement of a properly ordered timing of inputs for the development of neoplastic phenomena dates back to studies conducted in the 1970s on the carcinogenic effects of chemicals applied to the skin of mice (Boutwell 1978). It was found that these animals develop skin cancer if repeatedly exposed to potentially mutagenic chemical carcinogens (such as benzopyrene or dimethylbenzoanthracene). More importantly, this research provided evidence that both the *sequence* and the *frequency* of these exposures are relevant to the onset of cancer.

By the 1980s it became apparent that carcinogenesis could be characterized as a disruption of the organism's regulatory processes (Lotem and Sachs 1974; Potter 1978; Soto and Sonnenschein 1985). The loss of certain cellular functions began to be seen in correlation not so much with various molecular causes of tumour formation as with the decoupling of regulatory mechanisms. This led to a shift in focus from the causal relevance of molecular mechanisms and the effects of their alteration to the coordination of crucial regulatory processes, specifically those pertaining to cell proliferation, differentiation, and apoptosis. This change of emphasis has had important theoretical ramifications. The development of cancer had long been seen as processual in the narrow sense of constituting a linear sequence of events, such as a cumulative series of genetic 'knockouts'. However, recent models of carcinogenesis construe it in a more thoroughly processual vein, as a progressive change in the *rhythm* that orchestrates development and synchronizes the coupled regulatory processes that are part and parcel of healthy physiological function. The distinguishing characteristic of tumours remains the heterogeneity of cell behaviour, which results in an accumulation of cells with aberrant phenotypes within the tumour (Hanahan and Weinberg 2011; Dalerba et al. 2007; Ailles and Weissman 2007), but this is no longer viewed as constituting the final stage of a linear process. Instead, it is increasingly regarded as merely one of the various levels at which the breakdown of order occurs during carcinogenesis.

The disruption of the organism's normal physiological rhythm is manifested in the disturbance of the balance between cell proliferation and differentiation on the one hand, and apoptosis on the other. In addition, there are disruptions in the regulation of gene expression and in epigenetic modifications at a lower level of organization, as well as a breakdown of tissue organization at a higher level of organization. It is misleading to attribute a determinate causal order to the relations between these failures. Rather, as the orchestrating rhythm of the overall system degenerates, order begins to break down concurrently at multiple levels. Because the regulation of each level is achieved by means of cyclical processes acting at both higher and lower levels, multiple causal relations can be identified in this multilevel breakdown of organization. It is a mistake to privilege any particular level over others. It is the hierarchy of cyclical processes and their mutual regulation that generates the remarkable dynamic stability of the multicellular organism. When this stability deteriorates, it does so simultaneously at these interlinked levels.

Accumulating evidence is converging in support of the characterization of cancer as a progressive disorganization of a variety of organizational levels. As an organism's self-maintaining organization is fuelled by the continuous input of energy and information, compromise of the channels through which these are supplied constitutes the most systemic root of cancer. Recent research suggests that synchronized physiological rhythms play a key role in mediating the flow of energy and information at different levels (Plankar et al. 2012). Cancer, rather than being a genetic disease, is more fundamentally 'characterised by a global impairment of energy and information flow through the system, as manifested in genomic, transcriptomic and proteomic dysregulation' (ibid., 21). Moreover, as biological organization does not only pertain to topological order, but also has dynamic manifestations, such as synchronization, cancer is also rooted in the breakdown of the rhythmic coordination of different levels that facilitates these flows.[4]

---

[4] Synchronization is defined as a natural tendency of biological systems to adjust their internal rhythms to a collective operational regime due to their mutual interactions, and it has become one of the main areas of research in non-linear science. Synchronous behaviour of coupled elements enables a powerful response of a system to external stimuli, efficient coordination between different systems (e.g. temporal

## 3. The Relational Ontology of Levels

A historical–epistemological view of the evolution of the interpretive models of carcinogenesis (see Bertolaso 2013, 2016) highlights how models at different levels of organization (from genes to tissues) converged towards analogous morphogenetic and morphostatic principles in their effort to account for the distinctive phenomenology of the neoplastic process we have described in section 2. Such principles do not have their basis in the intrinsic features of molecules, but rather refer to the systemic properties and functions of cells, tissues, and organs, which are processual in nature. In explaining the stability of a multicellular organism and its destabilization following the onset of cancer, it is the dynamic relationships among parts at different levels of the organizational hierarchy that are causally more relevant than the intrinsic properties of those parts. Describing tumours as a progressive disruption of the hierarchical organization, with predictable consequences at the organ level as well as at the level of cells and genes, requires a specific, non-reductionist explanatory approach. The *relational ontology* of levels (Bertolaso 2016, 2017) clarifies the epistemological status and ontological foundations of the explanatory categories needed to account for the interactions of different levels of organismic organization, as well as clarify how their reciprocal dependency works within a processual philosophy. It submits that there is a hierarchy of levels in the functional organization of a metazoan; and, rather than assume a privileged causal level in the explanatory account of diseases such as cancer, it calls for a pluralistic account, able to fully accommodate the temporal coordination between distinct levels of organization.

Let us illustrate the fruitfulness of this perspective with an example. Despite the popularity of genetic accounts of cancer, the majority of mutations actually seem to be the result, rather than the cause, of carcinogenesis (Baker 2014). This does not fit a standard, bottom-up, mechanistic account of carcinogenesis. It is, on the other hand, consistent with an account in which morphogenetic and morphostatic constraints play the most important role in the maintenance of the developmental process—a process actively stabilized at many levels by various robust sub-processes. It can be hypothesized that these stabilization processes are realized in multiple attractor states at various levels of organization (Ingber 2008; Huang and Ingber 2000; Huang 2011). Against this background, microenvironmental factors mediated by epigenetic mechanisms acquire a major causal relevance in the onset of cancer (see also Capp 2006). In this way, cancer can be said to involve a deviation from the range of normal functional states of an organism's developmental trajectory.

This relational perspective also casts light on the ongoing debate between two main positions that compete in explaining carcinogenesis: somatic mutation theory (SMT) and tissue organization field theory (TOFT). The former considers the neoplastic process to be triggered by genetic mutations that lead to alterations in

---

compartmentalization), and information encoding; it also maximizes energy and information flow throughout the system, thereby increasing the organizing potential of biological processes. There is no gene for the rhythm of a pacemaker, or coherent brain oscillations, which represent a dynamic physical basis of cognition. In the same way, there is no gene or genetic pathway for cancer, because cancer is, like any physiological rhythm, a dynamically emergent process (Plankar et al. 2012; Winfreee 1967; Haus and Smolensky 2006).

cellular behaviour, while the latter locates the origin of cancer in the deterioration of tissue organization (Sonnenschein and Soto 1999; Soto and Sonnenschein 2005). Broadly speaking, SMT models adopt a reductive approach to biological explanation: higher-level phenomena are explained in terms of the properties and interactions of their parts, and changes at higher levels are traced to changes at the lower level. Those who reject this reductionistic viewpoint typically accept so-called downward causation, that is, casual explanation of the behaviour of parts in terms of autonomous causal powers of the whole (Noble 2008). As we have discussed, the processual view presented here takes carcinogenesis to involve the disruption of processes at multiple levels of organization, which mutually regulate and stabilize each other and which, collectively, constitute the lifetime process of ontogenic development.

We thus reject the reductionist presuppositions often associated with SMT. At the same time, however, we do not endorse an exclusive holistic focus on tissue organization of the kind that proponents of TOFT seem to support.[5] Instead, we propose a *pluralistic* explanatory approach. Because cancer is a multilevel phenomenon, we believe that what is needed is the deployment of a variety of models which address using different methodologies and over different timescales the effects of the same overall process on various biological entities (genes, cells, organs, etc.).

Overall, SMT is incapable of providing a satisfactory explanation of the characteristics of tumour cells, as well as of the neoplastic process as a whole. In principle, SMT is not incompatible with an explanation of cancer that construes it as an aberrant process of development, or as the disruption of the homeostatic mechanisms governing the normal proliferation of the cells (Hahn and Weinberg 2002a, 2002b). However, it does not possess the explanatory resources needed to account for the multiple levels at which neoplastic processes act and interact. As we have argued, a tumour is formed when a metazoan cell undergoes aberrant changes in the coupling of processes that normally regulate organic growth.

Defining carcinogenesis in terms of mistakes in the proliferation of cells, or as genetically programmed apoptosis, is to construe it as an autonomous cellular process, depending upon the intrinsic (mainly genetic) properties of cells (Weinberg 2006; Hanahan and Weinberg 2000, 2011). But, even if this understanding is not deemed to imply a form of genetic determinism, its exclusive focus on intrinsically driven cellular processes is problematic. Processes such as cell proliferation and differentiation are not self-sufficient, autonomous determinants of development and growth. Development is a multifactorial process, dependent on both external and internal factors (as emphasized for many years now by developmental systems theorists; see e.g. Griffiths and Gray 1994 and chapter 11 in this volume). These cellular processes are both causes of and caused by the developmental processes of which they are part.

The processes of cell division and differentiation that lie at the basis of healthy morphogenesis and are pathologically disrupted in carcinogenesis are not autonomous and internally generated, but are inextricably dependent on the wider context

---

[5] For a discussion of TOFT's merits in opening up a new view regarding the importance of a field theory of development, and on the epistemological implications of adopting this perspective, see Bertolaso 2016.

in which they act. For instance, a stem cell, which is usually considered the paradigmatic source of biological differentiation, is highly dependent upon its spatiotemporal position in the organism (Fagan 2013). Once extracted from its 'biological context', the cell loses its pluripotency and undergoes a process of proliferation and differentiation, like any other somatic cell. In other words, it loses the asymmetry that characterizes the divisions of stem cells. This point illustrates the inadequacy of a further reductionistic perspective, termed 'biological atomism' (see Nicholson 2010), which attempts to derive all organismic activity from the activity of fundamental biological units, in this case cells.

The shortcomings of this atomistic approach highlight the fact that distinguishing tumour cells from normal cells, or tumour cells among themselves, is far from trivial, since not all of the tumour cells have the same proliferative and invasive capacity. The aim of these distinctions is to characterize cells in terms of their activities, as processes. And it is generally supposed that this can be done by characterizing their internal structure, just as one describes the essential features of an object. However, as is so often the case in biological systems, in which the behaviour of an entity invariably depends on its role in a wider context, it appears highly doubtful that this reduction of process to object can be accomplished. The shift in perspective from object to process, from internal structure (the nature of stem cells) to external environment (the 'stemness' of a cell under specified circumstances), draws attention to the more transitory, for example epigenetic, effects of the environment—effects that are often described in terms of networks (Giuliani et al. 2014).

## 4. Morphogenetic Fields

The claims we have advanced in the previous section reinforce the idea that in the onset of the neoplastic phenotype the balance between cell proliferation and cell differentiation is disrupted. As we have seen, molecular elements alone do not offer a satisfactory or sufficient causal explanation of cancer. Something further is needed that can illuminate the structuring of developmental patterns and the coupling of the relevant processes at a variety of spatial and temporal scales. In this context, we believe that it may prove helpful to appeal to an explanatory notion that is commonly used by scientists but which remains virtually unexplored by philosophers.

In the early decades of the twentieth century, organicist theoretical biologists Joseph Needham (1936) and Conrad Hal Waddington (1935) already speculated that cancers represented an escape from the control of a *morphogenetic field*. With the advent of molecular biology, this perspective was abandoned. Only when 'morphogen' gradients became visualized in the 1990s did developmental biology resuscitate this old concept. Morphogens are diffusible substances that contribute to orienting the cell differentiation process and thus to determining the pattern of tissue development.[6]

---

[6] For a historical review of the origin and use of the morphogenetic field concept, see Soto and Sonnenschein 2006. For a contemporary methodological overview of its use in biological modelling, see Levin 2012. For a discussion of its explanatory import, see Bertolaso 2016. For a deeper analysis of the relationship between the neoplastic process and embryogenesis, see Bizzarri et al. 2011.

In more recent literature, this concept has re-emerged in various authors who refer to 'fields of cancerization': these are groups of cells from which specific morphological structures develop in response to biophysical and biochemical cues, generally mediated by epigenetic changes. In cancer, these epigenetic changes are aberrant (Ushijima 2007). Such aberrant epigenetic fields of cancerization promote carcinogenesis (a) by inducing epithelial cell growth, angiogenesis, degradation, and remodelling of the extracellular matrix and basal lamina; and (b) through paracrine signalling that induces epithelial cells to secrete further tumour-promoting factors. More generally, it appears that developmental, regenerative, and cancer biology requires a focus on 'the spatially distributed nature of instructive patterning signals' (Levin 2012: 244). The multilevel phenomenology of cancer involves organizational dynamics that are progressively compromised in the transition to a neoplastic process. The concept of a morphogenetic field aims to make sense of such organizational dynamics.

From this emerging field perspective, cancer is primarily characterized by an unspecific progressive disorganization, which can result from the impairment of the coherent dynamics at some specific level. The multiple interactions both within and between levels that constitute the impaired functional field make it difficult to identify a unique causal factor; the impairment may be induced by multiple, causally interconnected disruptions of the dynamics of the system. For example, damage to oxidative respiration and prolonged dependence on glycolysis may induce structural abnormalities in the mitochondrial inner membrane; it may also severely disturb cell homeostasis, and it will generally lead to genomic instability (e.g. Plankar et al. 2012). Epigenetic instability, which may be caused by stochastic perturbations, can result in a range of genomic changes (e.g. variation in gene expression, chromosomal instability, activation of mobile genetic elements, etc.), any of which may eventually translate into instability of the genome itself. In fact, it has been known for some time that environmental stress can induce genomic rearrangements (Ingber 2008; Levin 2012). One last important factor that should be mentioned in the context of the integrated nature of the functional field is the cytoskeleton (Nelson and Bissel 2006), which integrates many signalling pathways, influences gene expression, coordinates membrane receptors and ionic flows, and localizes many cytosolic enzymes and signalling molecules, ultimately forming an integrated system throughout the tissues and organs (Plankar et al. 2012).

Overall, the concept of field integrates various kinds of phenomena, providing a theoretical principle that can account for different aspects of carcinogenesis from both biological and physical perspectives. Specifically, the term 'field' as it is used in the literature denotes 'a construct that encapsulates key properties of instructive growth and patterning control' (Levin 2012: 243). This is probably why this term is so often used to describe features of the biology of cancer.

One important reason for shifting the perspective from the genetic to the morphogenetic field view is the apparent reversibility of carcinogenesis. The neoplastic phenotype of tumour cells can revert to a normal healthy phenotype when located in a healthy microenvironment (Mintz and Illmensee 1975; Hochedlinger et al. 2004; Kenny and Bissell 2003; Lotem and Sachs 2002; Erdo et al. 2003; Soto and Sonnenchein 2005). Clearly, this is hard to reconcile with the idea that there is some intrinsic property of a cell—the cancer stem cell—that inexorably leads it to express the neoplastic phenotype. The breakdown in spatio-temporal organization of the morphogenetic field—in which cues from both lower (molecules, cells) and higher or functional (tissue, endocrine and nervous stimuli, dietary factors) levels converge—disrupts the coherence between the cell and the overall organismic functional dynamics, thereby contributing to the onset of cancer. According to this perspective, the dynamic switch towards different cell fates is ultimately under the control of a morphogenetic field. Indeed, studies in which cancer cells have been cultured in specific morphogenetic fields (3D, embryonic, or maternal) or by modifying supra-molecular control factors (as the overall tissue stiffness or the endocrine stimulus) show how a strong, 'normal' morphogenetic field successfully induces the reversion of the tumour phenotype (see Bizzarri et al. 2012). On the whole, it is difficult to reconcile this heterogeneity with the hypothesis that there is some unitary unidirectional process that leads from a set of conditions internal to the cell to the full-blown phenomenology of cancer at a hierarchy of levels, without assuming a processual perspective on cancer itself.

## 5. Conclusions

There are various respects in which adopting a process-centred perspective can provide greater clarity in our thinking about cancer. The first step is to move from a simple contrast between a thing, the organism, and something that happens to it, the pathological process. From the point of view we are advocating, the organism itself is a process, specifically a developmental process. Development is not something that happens contingently to the organism; it is a core and structuring activity without which the organism could not be the kind of process it is. At a fine scale, the central constituents of the ontogenetic process are the divisions, differentiations, and deaths of cells. Cancer is a disruption of this ontogenetic process. Given the complexity and multiplicity of the systems that are involved in the reliable regulation of the ontogenetic process, it is no surprise that there are ways in which this can go wrong, or that the longer the system persists, the more likely it is that the regulation of the underlying, ordered production and destruction of cell types will fail.

There are various ontological and epistemological implications of this processual viewpoint on cancer. First and foremost, there are serious limits to the possibility of explaining carcinogenesis in terms of the linear, bottom-up causal sequences that have been dominant in much of recent biology. A process ontology should qualify the way we think scientifically about biological entities at all levels. Neither the organism nor the organ whose pathology we are trying to explain, nor the structural elements (cells, genes, etc.) in terms of which we do so, are static and univocally defined entities. The extent to which they may seem to be static entities is a consequence of their status as processes dynamically stabilized over various, often very short, time-scales. This stabilization is accomplished both by the interactions of parts and by the relations that constitute the parts as parts of a larger system. The proper sense in which 'the whole is greater than the sum of the parts' is that the parts are, to some degree, constituted as the kinds of entities they are by their relation to the whole.

This network of relations in which stabilities emerge at multiple levels and are maintained by the simultaneous activities of entities at further multiple levels is a long way from the linear mechanistic causality in terms of which we are more accustomed to think. Nevertheless, it is what we must come to grips with if we are to understand the processes that maintain the stable features and the stable cycles of living organisms, as well as the ways in which these features and cycles can (too frequently) become disrupted. Some projects under the general rubric of 'systems biology' do appear to be heading in this direction.

In case this characterization of the complexity of biological processes in general and of cancer in particular is discouraging, we end with a more optimistic thought. In exploring regularities at various levels of organization, from patterns of evolutionary adaptation to patterns of gene expression, it seems that outcomes are frequently more robust than the processes that produce them. It is plausible that this is a necessary requirement in order for systems to exhibit both high complexity and high stability; which is to say, a requirement for life. One way of expressing this requirement is in terms of the concept of an 'attractor state', a state that the system has a tendency to reach one way or another. If the state of a properly functioning cell is an attractor state, then presumably cells are capable of reaching many such states, as demonstrated by the processes of cell differentiation. Cancer could then be seen as providing a new, dysfunctional attractor state or, perhaps better, as breaking down the processes that lead the cell to the contextually correct state (as in the extreme case of the teratoma, in which different parts of the tumour mimic bits of different organs).[7]

We describe this as an optimistic thought because, unlike a linear path that has been lost, a scenario in which there may be no attainable way of rejoining it, there is no reason why the locally healthy attractor state would be lost just because the cell has found its way to another. This thought leads to another. It is common to think of medicine as engaged in trying to restabilize a system that some injury or insult has destabilized. But in the case of cancer, at least, the aim, paradoxically perhaps, should be to destabilize rather than stabilize the system. An attractor state, after all, is by definition something stable. States in its vicinity will revert to it; small perturbations will not lead the system away from an attractor. If cancer itself is a stabilized state, a pathological attractor state, then it may be that a large shock is needed to destabilize the system, in the hope that the healthy state will be regained. Perhaps this is a general explanation of why the standard treatments for cancer, the familiar repertoire of slash, poison, and burn, are sometimes successful.

This last thought is of course speculative, but some empirical studies seem to point in this direction (e.g. Cirkel et al. 2014). What we hope it illustrates is that ontological reflections on the nature of cancer, which in turn will require ontological reflections on the nature of living systems more generally, can profoundly affect how we think of the medical problems of oncology. If we are right that living systems are dynamically stabilized processes, this will have profound consequences on how we attempt to keep that process running for an optimal period in a desired condition.

---

[7] See Huang 2011 for a more detailed attempt to characterize cancer along these lines.

## Acknowledgements

We would like to thank Dan Nicholson, Marco Buzzoni, Alfredo Marcos, Mariano Bizzarri, and two referees for invaluable comments on earlier drafts. JD received funding for this research from the European Research Council under the European Union's Seventh Framework Programme (FP7/2007-2013)/ERC Grant Agreement 324186.

## References

Ailles, L. E. and Weissman, I. L. (2007). Cancer Stem Cells in Solid Tumours. *Current Opinion in Biotechnology* 18: 460–6.
Baker, S. G. (2014). A Cancer Theory Kerfuffle Can Lead to New Lines of Research. *Journal of the National Cancer Institute* 107 (2).
Bertolaso, M. (2009). Towards an Integrated View of the Neoplastic Phenomena in Cancer Research. *History and Philosophy of the Life Sciences* 31: 79–98.
Bertolaso, M. (2013). *How Science Works: Choosing Levels of Explanation in Biological Sciences*. Rome: Aracne.
Bertolaso, M. (2016). *Philosophy of Cancer: A Dynamic and Relational View*. Dordrecht: Springer.
Bertolaso, M. (2017). A System Approach to Cancer: From Things to Relations. In S. Green (ed.), *Philosophy of Systems Biology: Perspectives from Scientists and Philosophers* (pp. 37–47). Dordrecht: Springer.
Biava, P. M. (2002). *Complessità e biologia: Il cancro come patologia della comunicazione*. Milan: Mondadori.
Bizzarri, M. and Cucina, A. (2014). Tumor and the Microenvironment: A Chance to Reframe the Paradigm of Carcinogenesis? *BioMedical Research International*.
Bizzarri, M., Cucina, A., Biava, P. M.. Proietti, S., D'Anselmi, F., Dinicola, S., et al. (2011). Embryonic Morphogenetic Field Induces Phenotypic Reversion in Cancer Cells. *Current Pharmaceutical Biotechnology* 12: 243–53.
Bizzarri, M., Cucina, A., and D'Anselmi, F. (2008). Beyond the Oncogene Paradigm: Understanding Complexity in Cancerogenesis. *Acta Biotheoretica* 56: 173–96.
Bizzarri, M., Dinicola, S., and Manetti, C. (2012). Systems Biology Approach to Metabolomics in Cancer Studies. In A. Azmi (ed.), *Systems Biology in Cancer Research and Drug Discovery* (pp. 3–37). Dordrecht: Springer.
Bizzarri, M. and Giuliani, A. (2011). Representing Cancer Cell Trajectories in a Phase-Space Diagram: Switching Cellular States by Biological Phase Transitions. In M. Dehmer., F. Emmert-Streib, A. Graber, and A. Salvador (eds), *Applied Statistics for Network Biology: Methods in Systems Biology* (pp. 377–403). Weinheim: Wiley.
Boutwell, R. K. (1978). Biochemical Mechanism of Tumour Promotion. In T. J. Slaga, A. Sivak, and R. K. Boutwell (eds), *Mechanisms of Tumour Promotion and Carcinogenesis* (pp. 49–58). New York: Raven Press.
Boveri, T. (1929). *The Origin of Malignant Tumors*. Baltimore: Williams & Wilkins.
Boveri, T. (2007 [1914]). *Concerning the Origin of Malignant Tumours*, trans. H. Harris. Plainview: Cold Spring Harbor Press.
Burmeister, T. (2001). Oncogenic Retroviruses in Animals and Humans. *Reviews in Medical Virology* 11: 369–80.
Capp, J. P. (2006). Cancer Cell Undifferentiation: A Matter of Expression Rather Than Mutations? *Bioessays* 28: 102.

Cirkel, G. A., Gadellaa-van Hooijdonk, C. G., Koudijs, M. J., Willems, S. M., and Voest, E. E. (2014). Tumor Heterogeneity and Personalized Cancer Medicine: Are We Being Outnumbered?. *Future Oncology* 10: 417–28.

Colditz, G. A., Sellers, T. A., and Trapido, E. (2006). Epidemiology: Identifying the Causes and Preventability of Cancer? *Nature Reviews Cancer* 6: 75–83.

Dalerba, P., Cho, R. W., and Clarke, M. F. (2007). Cancer Stem Cells: Models and Concepts. *Annual Reviews of Medicine* 58: 267–84.

Doll, R. and Hill, A. B. (1956). Lung Cancer and Other Causes of Death in Relation to Smoking: A Second Report on the Mortality of British Doctors. *British Medical Journal* 233: 1071–6.

Duesberg, P. H. (1980). Transforming Genes of Retroviruses. *Cold Spring Harbor Symposium on Quantitative Biology* 44: 13–29.

Dupré, J. (1993). *The Disorder of Things: Metaphysical Foundations of the Disunity of Science.* Cambridge, MA: Harvard University Press.

Dupré, J. (2010). It Is Not Possible to Reduce Biological Explanations to Explanations in Chemistry and/or Physics. In J. Ayala and R. Arp (eds), *Contemporary Debates in Philosophy of Biology* (pp. 32–47). Oxford: Wiley Blackwell.

Dupré, J. (2012). *Processes of Life: Essays in the Philosophy of Biology.* Oxford: Oxford University Press.

Dupré, J. (2017). The Metaphysics of Evolution. *Internet Focus.* Published online, 18 August. http://rsfs.royalsocietypublishing.org/content/7/5/20160148.

Eisenhauer, E. A., Therasse, P., Bogaerts, J., Schwartz, L. H., Sargent, D., Ford, R., et al. (2009). New Response Evaluation Criteria in Solid Tumours: Revised RECIST Guideline (version 1.1). *European Journal of Cancer* 45: 228–47.

Erdo, F., Buhrle, C., Blunk, J., Hoehn, M., Xia, Y., Fleischmann, B., et al. (2003). Host-Dependent Tumorigenesis of Embryonic Stem Cell Transplantation in Experimental Stroke. *Journal of Cerebral Blood Flow and Metabolism* 23: 780–5.

Fagan, M. B. (2013). The Stem Cell Uncertainty Principle. *Philosophy of Science* 80: 945–57.

Giuliani, A., Filippi, S., and Bertolaso, M. (2014). Why Network Approach Can Promote a New Way of Thinking in Biology: Hosted by Xiaogang Wu, Hans Westerhoff, Pierre De Meyts, Hiroaki Kitano. *Frontiers in Genetics* 83 (5): 1–5.

Goldberg, A. D., Allis, C. D., and Bernstein, E. (2007). Epigenetics: A Landscape Takes Shape. *Cell* 128: 635–8.

Griffiths, P. E. and Gray, R. D. (1994). Developmental Systems and Evolutionary Explanation. *Journal of Philosophy* 151: 277–304.

Hahn, W. C. and Weinberg, R. A. (2002a). Modelling the Molecular Circuitry of Cancer. *Nature Reviews Cancer* 2: 331–42.

Hahn, W. C. and Weinberg, R. A. (2002b). Rules for Making Human Tumor Cells. *New England Journal of Medicine* 347: 1593–604.

Hanahan, D. and Weinberg, R. A. (2000). The Hallmarks of Cancer. *Cell* 100: 57–70.

Hanahan, D. and Weinberg, R. A. (2011). Hallmarks of Cancer: The Next Generation. *Cell* 144: 646–74.

Harris, H. (2005). A Long View of Fashions in Cancer Research. *Bioessays* 27: 833–8.

Haus, E. and Smolensky, M. (2006). Biological Clocks and Shift Work: Circadian Dysregulation and Potential Long-Term Effects. *Cancer Causes Control* 17: 489–500.

Hochedlinger, K., Blelloch, R., Brennan, C., Yamada, Y., Kim, M., Chin, L., and Jaenisch, R. (2004). Reprogramming of a Melanoma Genome by Nuclear Transplantation. *Genes and Development* 18: 1875–85.

Huang, S. (2011). On the Intrinsic Inevitability of Cancer: From Foetal to Fatal Attraction. *Seminars in Cancer Biology* 21: 183–99.

Huang, S. and Ingber, D. E. (2000). Shape-Dependent Control of Cell Growth, Differentiation, and Apoptosis: Switching between Attractors in Cell Regulatory Networks. *Experimental Cell Research* 261: 91–103.

Ingber, D. E. (2008). Can Cancer be Reversed by Engineering the Tumor Microenvironment? *Seminars Cancer Biology* 18: 356–64.

Kenny, P. A. and Bissell, M. J. (2003). Tumor Reversion: Correction of Malignant Behaviour by Microenvironmental Cues. *International Journal of Cancer* 107: 688–95.

Kim, J. (1999). Making Sense of Emergence. *Philosophical Studies* 95: 3–36.

Klein, G. (2002). Perspectives in Studies of Human Tumor Viruses. *Frontiers in Bioscience* 7: 268–74.

Knudson, A. G. (1971). Mutation and Cancer: Statistical Study of Retinoblastoma. *Proceedings of the National Academy of Sciences* 68: 820–3.

Levin, M. (2012). Morphogenetic Fields in Embryogenesis, Regeneration, and Cancer: Non-Local Control of Complex Patterning. *Biosystems* 109: 243–6.

Loeb, L. A. and Harris, C. C. (2008). Advances in Chemical Carcinogenesis: A Historical Review and Prospective. *Cancer Research* 68: 6863–72.

Lotem, J. and Sachs, L. (1974). Different Blocks in the Differentiation of Myeloid Leukemic Cells. *Proceedings of the National Academy of Sciences* 71: 3507–11.

Lotem, J. and Sachs, L. (2002). Epigenetics Wins Over Genetics: Induction of Differentiation in Tumor Cells. *Seminars in Cancer Biology* 12: 339–46.

Mintz, B. and Illmensee, K. (1975). Normal Genetically Mosaic Mice Produced from Malignant Teratocarcinoma Cells. *Proceedings of the National Academy of Sciences* 72: 3585–89.

Needham, J. (1936). New Advances in the Chemistry and Biology of Organized Growth. *Proceedings of the Royal Society of Medicine* 29: 1577–626.

Nelson, C. M. and Bissell, M. J. (2006). Of Extracellular Matrix, Scaffolds, and Signaling: Tissue Architecture Regulates Development, Homeostasis, and Cancer. *Annual Review of Cell and Developmental Biology* 22: 287–309.

Nicholson, D. J. (2010). Biological Atomism and Cell Theory. *Studies in History and Philosophy of Biological and Biomedical Sciences* 41: 202–11.

Noble, D. (2002). Chair's Introduction. In G. Bock and J. A. Goodie (eds), *'In Silico' Simulation of Biological Processes: Novartis Foundation Symposium 247* (pp. 1–3). Chichester: John Wiley & Sons, Ltd.

Noble, D. (2008). *The Music of Life: Biology Beyond Genes*. Oxford: Oxford University Press.

Odling-Smee, F. J., Laland, K. N., and Feldman, M. W. (2003). *Niche Construction: The Neglected Process in Evolution*. Princeton: Princeton University Press.

O'Malley, M. A. and Dupré, J. (2005). Fundamental Issues in Systems Biology. *BioEssays* 27: 1270–6.

Parkin, D. M. (2004). International Variation. *Oncogene* 23: 6329–40.

Plankar, M., Del Giudice, E., Tedeschi, A., and Jerman, I. (2012). The Role of Coherence in a Systems View of Cancer Development. *Theoretical Biology Forum* 105: 15–46.

Potter, V. R. (1964). Biochemical Perspective in Cancer Research. *Cancer Research* 24: 1085–98.

Potter, V. R. (1978). Phenotypic Diversity in Experimental Hepatomas: The Concept of Partially Blocked Ontogeny: The 10th Walter Hubert Lecture. *British Journal of Cancer* 38: 1–23.

Powell, A. and Dupré, J. (2009). From Molecules to Systems: The Importance of Looking Both Ways. *Studies in History and Philosophy of Biological and Biomedical Sciences* 4: 54–64.

Rous, P. (1910). A Transmissible Avian Neoplasm (Sarcoma of the Common Fowl). *Journal of Experimental Medicine* 12: 696–705.

Sonnenschein, C. and Soto, A. M. (1999). *The Society of Cells: Cancer and Control of Cell Proliferation.* New York: Springer.

Soto, A. M., Maffini, M. V., and Sonnenschein, C. (2007). Neoplasia as Development Gone Awry: The Role of Endocrine Disruptors. *International Journal of Andrology* 30: 1–5.

Soto, A. M. and Sonnenschein, C. (1985). The Role of Estrogens on the Proliferation of Human Breast Tumor Cells (MCF-7). *Journal of Steroid Biochemistry* 23: 87–94.

Soto, A. M. and Sonnenschein, C. (2005). Emergentism as a Default: Cancer as a Problem of Tissue Organization. *Journal of Biosciences* 30 (1): 103–18.

Sporn, M. B. (1991). Carcinogenesis and Cancer: Different Perspectives on the Same Disease. *Cancer Research* 51: 6215–18.

Sporn, M. B. (2006). Dichotomies in Cancer Research: Some Suggestions for a New Synthesis. *National Clinical Practice Oncology* 3: 364–73.

Tokunaga, M., Norman, J. E., Asano, M., Tokuoka, S., Ezaki, H., Nishimori, I., and Tsuji, Y. (1979). Malignant Breast Tumors among Atomic Bomb Survivors, Hiroshima and Nagasaki, 1950-74. *Journal of the National Cancer Institute* 62: 1347–59.

Ushijima, T. (2007). Epigenetic Field for Cancerization. *Journal of Biochemistry and Molecular Biology* 40: 142–50.

Waddington, C. H. (1935). Cancer and the Theory of Organizers. *Nature* 135: 606–8.

Weinberg, R. A. (2006). *The Biology of Cancer.* London: Garland Science.

Winfree, A. T. (1967). Biological Rhythms and the Behavior of Populations of Coupled Oscillators. *Journal of Theoretical Biology* 16: 15–42.

# 17
# Measuring the World
## Olfaction as a Process Model of Perception

*Ann-Sophie Barwich*

## 1. Introduction: Why Things Stink

What is the first thing you do when you open a box of milk, especially if it has stayed a few days longer in the fridge and may have gone off? You take a whiff. Although popular opinion sticks to the idea that the human sense of smell is declining and unimportant, this is a blatant misconception.[1] Your nose actually is the most accurate and sensitive chemosensor on earth. It detects the slightest changes in the chemical composition of your environment, and it does so with striking precision. A difference in one atom of two otherwise perfectly similar molecules can cause your perception of their odour quality to vary entirely. For instance, take nonanoic acid ($CH_3(CH_2)_7COOH$; Figure 17.1a), which smells of cheese. If you add only one carbon atom, you get decanoic acid ($CH_3(CH_2)_8COOH$; Figure 17.1b), which you will perceive distinctly different, as smelling rancid!

While your olfactory system is mind-bogglingly precise in its capacity to detect the slightest changes in chemical variation, it is also incredibly flexible in its processing. Think about the wide range of responses to smells: certain odours evoke an immediate and almost universal dis-liking (e.g., the smell of cadaverine, $NH_2(CH_2)_5NH_2$, is not something you will consider pleasant), but many other odours tend to carry individually variable associations—variable depending on their familiarity and on memories of previous encounters.

This dual character, the flexibility of perceptual interpretation in parallel with the precision of its molecular detection mechanism, makes olfaction an excellent model system for renewing philosophical attention to perception. Perceptual analysis has traditionally concentrated on visual perception. In recent years, the neglect of what is

---

[1] The most popular and persistent opinions about our sense of smell are that it is declining, that it is evolutionary unimportant to humans, and that other animals such as dogs are much better at smelling. Nothing could be further from the truth. Neither is the sense of smell declining nor is the human sense of smell considerably worse than the olfactory abilities of dogs. For a popular science account of smell and for debunking such myths, see Gilbert 2008; Shepherd 2012; and Barwich 2016; for a scientific review of the importance of the olfactory system in molecular biology and neuroscience, see Firestein 2001; Shepherd 2004; and Barwich 2015b.

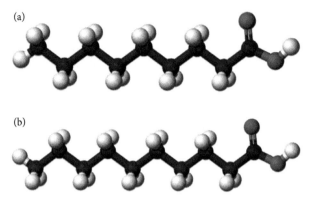

**Figure 17.1** Ball-and-stick models of nonanoic acid (A; top), which smells of cheese, and decanoic acid (B; bottom), which smells rancid (Jynto and Mills 2010a, 2010b). While they differ only in one carbon atom, this difference is responsible for their distinctly different smells.

often referred to as 'the other senses' has started to become a matter of correction, though.[2] Nonetheless, olfaction remains the most neglected sense among these laudable developments.

Smell has long constituted the problem child for philosophers of perception because of its apparent lack of representational capacities. Odours are experienced, but in what way does this constitute representational content? This question has been traditionally addressed by debating the representational nature of odours as corresponding to objects (see the essays in the thematic issue of Keller and Young 2014). Such approach centres on the nature of the stimulus as defining perceptual content. In light of recent developments in cognitive neuroscience, I propose an alternative to that view in this chapter.

Perception here is analysed as a *process*. The thrust of my argument is, in brief, that we need to abandon a stimulus-centred point of view where we think of smells as stable percepts that are computationally linked to external objects such as odorous molecules. There are no stable and intrinsic links between chemicals or input sources and our perceptions, such as of odour qualities. Denying that input sources are the primary element in perceptual analysis does not lead to a denial of their causal and functional significance, however. Once this proposition is clear, a very large part of the philosophical motivation to oppose the perceptual model advanced here should vanish. Instead, we must consider flexible and contextual aspects of the process to understand what it is that we perceive from odorous molecules through our sense of smell.

Smells, the argument proceeds, are not so much about objects and stable object perception as about changes in the chemical composition of the environment and flexibility in terms of its contextual evaluation. In the course of percept formation, sensory input is filtered and structured by different anticipatory processes. What we perceive is highly dependent on a signal's combination with other sensory cues, previous experiences, and expectations of what options a signal affords.

---

[2] For reviews, see audition (O'Callaghan 2014), touch (Fulkerson 2015), taste and flavour (Smith 2012; Spence 2013).

The informational content of smell must not be analysed as perceptual instances in terms of classes of 'odour objects' (e.g. rose), but with respect to 'odour situations' where input cues are integrated in terms of their temporal and contextual associations with other external sensory cues, internal hidden states (experience and memory-based, or internally inferred), expectations or predictions, and feedback processes of error correction. A number of processes can cause certain odour qualities to become more prominent in a percept, allowing for semantic associations with previously encountered smells. Other processes facilitate the variability of semantic associations in smell perception. In order to understand the informational content and identify the perceptual dimensions in olfaction we must model odours after the processes that facilitate signal pattern separation and completion.

In this chapter I elaborate on the scientific foundations and philosophical implications of this idea. That said, the perspective on perception advanced in what follows is not meant to carve out olfaction as necessarily different from the visual or auditory systems. Rather, it is intended to refine our perspective on the variable factors that determine perceptual content. Olfaction in this context bears interest, as it seems to possess a less intuitive perceptual structure than vision and, as a result, a less deceptively straightforward relation between sensory input and perceptual content.

The starting point of this chapter is to engage with the received view in philosophical studies that considers the distal stimulus as the central element for the analysis of perceptual content (Lycan 2000; Batty 2010a, 2912b, 2013; Keller and Young 2014; Keller 2016). Having outlined what constitutes the general challenge here, namely the inadequacy of talk about odour objects, I turn to current scientific studies on the neural basis of olfaction. These studies highlight the non-linearity of stimulus processing and demonstrate the impact of top-down mechanisms in olfaction, and I analyse these experimental developments in the context of an alternative framework as emerging in cognitive neuroscience. The central proposal here is to model the brain in terms of two complementary and simultaneous processes as the integrated proximal stimulus: perceptual bias as anticipation and bias correction as revision. I conclude with a disambiguation of the different meanings of anticipatory processes that regulate perception; and I present perception as a process that measures changing signal ratios in the environment and is shaped by expectancy effects in perceptual content formation.

## 2. The Received View: The Input Determines the Perceptual Experience

In some ways the philosophical analysis of perception used to suffer from the same problems as certain parts of theoretical physics: concepts originate purely from theory, and there often is no way to see how foolproof the grounds for the relevant theoretical convictions really are without guided experimental manipulation. Contemporary philosophy of perception has experienced a great deal of change and challenges in parallel with the rise of cognitive neuroscience over the past decade, however. The essential tension surrounds the double understanding and analysis of perception (1) as a representation of external objects (distal stimulus) and (2) as a result of the neural processes generating stimulus patterns (proximal stimulus).

Take the common-sense idea that our perceptions are shaped by what we perceive: we consider our perceptions to be representations of objects and their features in the world. Philosophers of perception have been careful not to confound perceptual representation with neural representation, and instead have focused on the distal stimulus input as the measure by which we must judge the content of our perceptions. What are the grounds for this view, and what reason is there to reconsider the relation between perceptual and neural representation?

Let's start with the traditional philosophical notion of a 'percept'. Although there seems to be no formal definition of what a percept is, it is commonly used to refer to the perceptual experience that results from the act of perceiving. Our percepts are considered to be *about* things in the world, and understanding this aboutness or intentionality of perception is one of the major occupations in philosophical discourse (Peacocke 2008). For example, my perception of the cup of coffee in front of me is going to tell me something about it, such as its colour, shape, and size. But how shall we model and analyse the content of our perceptual attention to the world?

The philosophical literature has produced numerous arguments on this topic (for a review, see Crane and French 2016). Large parts of the debate concern whether such perceptual experiences are truthful or *accurate* representations of the things we perceive in the world (Akins 1996). Central to this inquiry is the distinction between perceptual appearance and reality. What unites the bulk of philosophical arguments on this topic is a concern about the source of perception and its elemental primacy for perceptual analysis. The shared hypothesis about the directionality of the perceptual process is clear: *the input structures the perceptual content.* What does that mean? And does this apply to olfaction?

The common-sense idea that perception is about objects originates from our dominantly 'visuocentric' theories. It has led some philosophers of mind to the question of what might constitute 'odour objects'. Four suggestions are offered in the literature: (i) smells represent ordinary objects (like roses, wine, or Brussels sprouts); (ii) smells represent clouds of odorous molecules; (iii) smells represent chemical features of molecules; or (iv) smells may be purely subjective phenomenological experiences or sensations that do not present us with propositions specifying particular objects in the world (for different positions, see Lycan 2000; Batty 2010a, 2010b; Keller and Young 2014). Analysis here centres on the assessment of perceptual 'object failure', meaning 'the failure of an experience to present objects accurately' (or to present any objects at all; see Batty 2010b: 10).[3]

A lot of arguments in this debate concern the effect of the visual presence of a source object on olfactory experience (particularly in the work of Lycan and Batty). Notably, this effect has been characterized by the olfactory physiologist Hans Henning as early as in 1916. Henning drew a conceptual distinction between 'the true odor [*Gegebenheitsgeruch*], which is obtained by the observer who is smelling with closed eyes and is ignorant of the nature of the scent, and the object smell [*Gegenstandsgeruch*], which (like color) is projected upon the objects from which it is

---

[3] Alternatively, a discussion about the question of whether we can perceive absences in olfaction as being objectless can be found in Roberts 2015.

known to come and apt to be distorted by associative supplementing' (Henning in Gamble's translation, in Gamble 1921: 292). Eleanor Gamble's translation of the German *Gegebenheitsgeruch* as 'the true odor' is misleading, however, as the literal meaning is 'the situation odour'. For Henning, such perceptual effects presented important methodological factors for psychophysical measurement, not a measure of the 'truthfulness' of odour objects.

There are several philosophical difficulties involved in defining olfactory objecthood (for an extensive analysis, see Keller 2016). Some of the arguments about the nature of odour objects, namely for their being (ii) clouds of molecules or (iii) particular features of molecules, fail to distinguish between the stimulus as the *cause* of perception and the perceptual object as the *content* of perception. This view also runs into scientific problems. To date, there are no known structure–odour relationship rules (i.e. regularities linking specific chemical features and the smell of a molecule), and the causal features of odorants (i.e. the odorous molecules) are dependent on receptor behaviour, not vice versa (Barwich 2015a, 2015b; Poivet et al. 2016). Moreover, and as I explain in the next section, smell is not only determined by molecule–receptor interactions, but is also significantly dependent on higher-level brain processes.

Arguments for (i) (i.e. a semantic understanding of odour objects as ordinary objects) run into trouble as well. Suffice it to say that some smells, such as artistic perfumes, do not necessarily have associations with ordinary objects. Even ordinary objects give off hundreds of odorants, and each one is not only different from the others but also distributed from its source at a different temporal scale from theirs (this is also the basic principle in the composition of alcohol-based perfumes).

A layered account for odour objects as being a combination of semantic (= (i)) and causal (= (ii), (iii)) objecthood does not present an intuitive or clear criterion for an odour object either. 'Layered' means, according to Lycan (2014), that odour perceptions can be veridical in two independent ways: first, on a lower-level account of representation in terms of its causal objects (i.e. I perceive, correctly, the presence of a cloud of molecules as the causal object); and, second, on a higher-level account of representation in terms of its semantic associations (i.e. I perceive, correctly, a cloud of molecules as a rose). Such a model of differentiating the truth values of (i) and (ii), (iii) remains far too uninformative and further runs into trouble once we consider the different variables regarding the distal stimulus as well as its associated semantic content. For example, the attempt to link the smell of ordinary objects such as roses to particular (clouds of) molecules (or their features) is

[a]n innocent approach when we know that the scent of a rose comprises hundreds of different molecules and that none of them smells like a rose. So far I have not found 'the' rose molecule, but I have discovered that the smells of flowers have a biologically dictated cycle, and that their composition can vary significantly without them losing their identity.

(Ellena 2012: Cabris, Thursday 22 July 2010)[4]

---

[4] To be sure, the basic rose smell is more likely composed of dozens, not hundreds, of molecules. The main point of this statement remains valid, however.

A dominant focus on veridical object representation in perceptual analysis further falls short of several key aspects of olfactory experience. First, it ignores the purpose of smelling: 'Stimulus representation isn't the primary business of olfaction. Rather, its job is solving a problem of valuation, rapidly encoding the biological salience of a stimulus and priming our multisensory representation of it to contextually appropriate action' (Castro and Seeley 2014: 1). As other philosophers and scientists have pointed out (Burge 2010; Keller 2016), the biological function of perception is prior to representational accuracy in an evolutionary reading of sensory systems. Perceptions here are primarily understood to facilitate the achievement of *organismal goals* such as the four Fs (fighting, fleeing, feeding, and courtship). While the truthful representation of the world can coincide with the achievement of these biological functions, it need not.

Second, object-centred representational analysis remains indeterminate and misleading with respect to the perceptual dimensions and the structure of olfactory experience.[5] What is the structure of odour perceptions? To be sure, olfactory information is spatially and temporarily structured in the environment. In humans and other animals it can be used for navigation and active exploration (Porter et al. 2007), and we recognize temporal patterns of changes in the olfactory environment on shorter and longer scales, such as circadian and annual fluctuations of smells (Keller 2016). That said, we must be careful not to equate the external structure of a signal with the structure of our perceptual experience.

Human olfaction is generally characterized as being temporal but as lacking spatial dimensions in its perceptual content. It is temporal in a phenomenological sense, as smells appear to be perceived *now*, and they act as an indicator of the presence of something. Olfaction is also temporal, in the sense that we perceive important changes in the chemical constitution of our environment. As we are constantly surrounded by hundreds of airborne molecules, our olfactory system is tuned to this situation by quickly adapting to stable ratios of odorants, so that neural populations fire more actively when novel stimuli are encountered. In comparison, spatial structure in perception is characterized as exhibiting perceptual relations in terms of position, orientation, or directness (Keller 2016). As odours do not exhibit such spatial structuring and 'we do far less of that sort of objectification' in olfaction, this has led some philosophers to believe that 'smell, in humans, is informationally very poor' (Lycan 2000: 277) and lacks 'articulate individuation' (Lycan 2000: 282).

Such a judgement conveys a blatant misunderstanding of what olfaction is *for*. Information is an ambiguous and multifaceted notion, especially with respect to organisms and their sensory systems. I ask you instead, *how many* different smells can you perceive? Scientifically speaking, olfactory quality space is multiscaled and consists of hundreds or thousands of different odours (though the precise number and the usefulness of counting are matters of debate; see Bushdid et al. 2014; Meister 2015; and Magnasco et al. 2015). Furthermore, why do you consider an odour to be pleasant or unpleasant (and when or for how long)? It is rather

---

[5] I have argued in more detail elsewhere why I consider object-centred representational analyses of smells to be ill informed with regard to categories of sensory measurement (Barwich 2014).

curious how much the hedonic tone of odours seems to escape philosophical ideas about perception; one might blame this on the heritage of the Enlightenment's mirthless philosophy of the senses (Classen et al. 1994). Likewise, how much does the context of your encounter with a stimulus and its combination with other sensory cues shape its perceptual content? The most obvious example of the informational richness and context-sensitivity in olfactory perception is the complexity of flavours (Shepherd 2012, Smith 2012, Spence 2013). Your perception of food and beverage flavour is dominated by your sense of smell, more specifically retro-nasal (or mouth-breathing) olfaction. Humans have developed highly sophisticated discriminatory abilities when it comes to flavours.

Overall, such a differentiated account of perceptual information invites us to rethink our standard approach to perception. Regarding the inadequacy of talk about odour objects, other philosophers have suggested adopting suggestion (iv) and simply rejecting an object-representational account, viewing smells as subjective phenomenological experiences or as 'feels' that are somewhat 'free-floating' or 'objectless' (Batty 2010a, 2010b). It remains unclear what precise understanding of odours is gained through this proposal, however. Detaching philosophical analysis of smell from objects and seeing perceptions as mere sensations does not account for the purpose of odour perception as a measure of chemical changes in your environment. It does not explain how we should understand the role of the stimulus as an informational signal for a specific sensory system. Thus this chapter advocates that the structure of the perceptual image must be modelled after the processes it serves. But what are these processes? And how can we think about the informational dimensions of signals in terms of sensory systems and their regulatory principles?

## 3. The Neural Basis of Olfaction and the Idea of Forecasting in Perception

Input-centred modelling of the senses has not been restricted to philosophical debate. Its equivalent in neuroscience is the view that the organization of a sensory system such as the visual or the auditory one is shaped primarily by the incoming signal. Basically, this expresses the idea that external stimuli are recognized by our sensory systems and translated into internal representations by means of topographically organized brain activation patterns that further facilitate behavioural responses to certain stimuli. The resulting input–output model of perception has been crucial for successful developments in visual research, especially throughout the 1980s and 1990s (Marr 2010 [1982]), as well as influencing research on other sensory systems such as olfaction (Davis and Eichenbaum 1992).

Meanwhile this standard bottom-up version of computationalism has been challenged and modified.[6] Over the past decade, sensory and computational neuroscience

---

[6] The implicitly unidirectional and monocausal input to output interpretation of sensory processing has elicited various criticisms and suggested alternatives over the years, especially in philosophy. Most prominently, theories of action, enactivism, and embodiment have argued against the differentiation between perception, body, and the environment (for an extensive review, see Hurley 2001). Overall, these theories view the body as a condition and constraint for forming percepts so that we are able to

has provided much more advanced models and analysis of higher-level brain processing. While we should not equate the analysis of the sensory processes with the phenomenological character of its perceptual products per se, the neural pathways are the basis on which we must build and correct our perceptual theories. What is more, by identifying and analysing current questions in contemporary neuroscience, we gain a much more detailed and informative picture of what kinds of questions we must ask in order to reconsider some philosophical approaches to perception. Olfaction, again, presents us with a salient case for this.

As a rough sketch (see Figure 17.2), the olfactory pathway is structured as follows (for a review, see Firestein 2001). Odorants are first detected by receptors situated on the sensory nerves in the nasal epithelium. All olfactory sensory neurons expressing one particular receptor gene (encoding a receptor type) are then collected in spherical

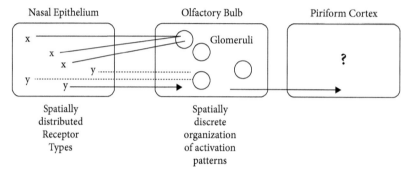

Figure 17.2 Sketch of information flow in the olfactory pathway. Odorants activate receptors situated on the sensory nerves in the nasal epithelium. Receptors are randomly spatially distributed across the mucous. All sensory nerves expressing one receptor gene (coding for a type) are collected in neural spherical structures called glomeruli. As a result, odors here are represented as spatially discrete activation patterns. The synaptic organization of the piriform cortex remains an unresolved issue to date.

interact with our environment through sensory experiences. Or, more briefly put, the state of your body affects the state of your perception and, in turn, of your cognition. On this account, behavioural or motor output and sensory input are coupled and analysed in relation to each other (Varela et al. 1991; Hurley 1998; O'Regan and Noë 2001; and Noë 2004). I have decided to exclude analysis of embodiment ideas in this chapter, for reasons not only of length and focus, but also of appropriateness: strong theories of embodiment that view perception as being somewhat in the body and *out of our heads* (Noë 2009) simply fail to resonate with numerous clinical cases, for example where certain disorders in the right-brain hemisphere can cause feelings of disembodiment in patients (see ch. 3 in Sacks 1998). However, I agree with weaker theories of embodiment that emphasize sensorimotor aspects of sensory systems as influencing perception (e.g. the effects of sniffing patterns on odour perception). One can be of a divided opinion as to whether sensorimotor effects require such an extensive theoretical treatment as in the case of the embodiment movement. In fact some psychologists have objected that the 'basic principles from embodiment theory are either unacceptably vague (e.g., the premise that perception is influenced by the body) or they offer nothing new (e.g., cognition evolved to optimize survival, emotions affect cognition, perception–action couplings are important)' (Goldinger et al. 2016: 959). I remain agnostic on this issue in this chapter.

neural structures (so-called glomeruli) in the olfactory bulb (at the frontal lobe of your brain). At the bulb level, a neat activation pattern shows up (Vassar et al. 1994; Mombaerts et al. 1996). This pattern represents the range of receptors that are activated by certain chemical features of the distal stimulus. So you will get a different activation pattern for a musk molecule from the one you will get for a citrus molecule. Now, as with the visual system, this topographic organization of the bulb was expected to be maintained throughout further processing stages, and the expectation was to find a corresponding topography in the olfactory cortex (Axel 2005). It turns out that this may not be the case.

Olfactory scientists have struggled to find any such topographic organization over the past ten years (Stettler and Axel 2009; Mori et al. 2006). Their efforts have largely concentrated on the so-called piriform cortex, which constitutes the largest part of the olfactory cortex. It was long assumed to be the centre of odour object formation. This means that the piriform was considered to be the domain in the brain where olfactory signals are combined into a unified odour percept. While the piriform cortex does not present us with stable input maps like the bulb, it has been shown to respond to different sorts of organizational regulation, however.

On the one hand, there are findings that suggest that the piriform cortex can get trained into forming more or less temporally stable patterns through innate as well as through learned behaviour (associated with smells). This strategy is pursued for instance in Richard Axel's lab. Taking full advantage of the experimental possibilities offered by novel techniques such as optogenetics, Axel's team traces olfactory signalling from the bulb to the piriform cortex via the amygdala as a sort of 'relay' station (Root et al. 2014). The amygdala is part of the limbic system and deeply involved in processes of memory formation, decision-making, and affective responses.

On the other hand, an alternative model is to 'reverse engineer' and ask what the signal is *for* (i.e. trace its efferent connections) rather than ask where it comes from (i.e. trace its afferent connections). This strategy is employed by Stuart Firestein's lab. Firestein's team was looking at projections from two higher-level domains in the orbitofrontal cortex (the agranular insula and the lateral orbitofrontal cortex) back to the piriform cortex. And indeed, the team found two distinct neural populations with a largely non-overlapping topographic organization (Chen et al. 2014).

To what extent these findings will converge in a unified model of olfactory processing is an empirical question, and it presents an exciting prospect for further research in olfaction. There may or may not be a central domain of synthesizing or unifying olfactory experience. What these approaches have in common thus far is a shared focus on behaviour and learning as fundamental to the formation of odour objects. The amygdala and the orbitofrontal cortex in particular are domains notable for their involvement and centrality in decision-making processes and sensory integration (Shadlen and Kiani 2013; Castro and Seeley 2014).

For philosophical studies of perception, these are interesting experimental developments. They highlight the non-linearity of stimulus activation and representation and demonstrate the impact of top-down neural processing in olfaction. While research on smell lacks a general theory of its subsystems (integrating studies of receptor, bulb, and cortex activity), these experimental inquiries resonate with a general tendency in cognitive neuroscience that has started to pursue an alternative framework for modelling perception and cognition. The growing trend is to think of behavioural systems in computational terms.

While there is no real consensus about theories of the brain to date, there is convergence on what aspects a genuinely alternative conceptual framework for neural processing must build on: bias and revision. 'Bias' refers to the formation of anticipations and preferences through previous experiences, and revisions are processes where these biases are continuously corrected.

Consider a great example of perceptual biases introduced by top-down processing: the role of expectations in flavour perception. Here we encounter numerous phenomena where the perception of colour or texture in foods and beverages affects our judgement about the perceived gustatory qualities of these foods and beverages. In one study, test subjects were given two beverages of the same chemical composition, one being of a brighter colour than the other. Subjects perceived the brighter beverage as sweeter and more intense (Bayarri et al. 2001). Another study appears to ridicule sommeliers and wine tasters who were given wines to test and describe. The subjects in this study were presented with red wines that, unbeknownst to them, were in fact just white wines laced with red food colour. The tasters proceeded to attribute traditional red wine properties to these white wines (Hodgson 2008).

What perceptual puzzles such as these suggest is that seemingly higher-level processes should not be taken as separable modules in the cognitive architecture. They are an integral part of our basic perceptual processing instead. As has become clear by now, perception is not exclusively or even primarily determined by input. Most notably, this is where data from the neural pathway and psychophysical studies of perception converge. What we perceive with the help of our sensory systems is multilayered and multiprocessual: perception is dependent on a signal's combination with other sensory cues, previous experiences, and expectations about the kinds of options this signal affords. These different processual layers are constitutive of the perceptual architecture and the selective biases in percept formation. But how can we model and analyse such seemingly bidirectional causal character of information flow in sensory processing?

Over the past fifty years, a number of neuroscientists have suggested models of neural networks that build on these two processes, anticipation and revision, as complementary mechanisms. In these models, your brain works like a neurocomputer that copes with the plethora of sensory information by predicting stimulus regularities through previous experiences. These sensory regularities provide perceptual templates by which your brain continuously generates an internal virtual model or a simulation of the environment (Friston 2010; Graziano 2013).

To generate such a model, the brain operates by two complementary and simultaneous mechanisms of top-down and bottom-up processing. 'Top-down processing' refers to the information flow from the higher cortical areas to the lower sensory domains. This top-down mechanism makes predictions about the environment on the basis of prior experience of stimulus regularities, and its activity results in so-called 'forward models'. By comparison, bottom-up processing describes the information flow of stimulus input from lower sensory areas to higher-level brain domains. Most crucially, the function of the incoming input from this bottom-up mechanism is defined as an error correction of the forward model. What precisely such top-down processes are and what constitutes the content of predictions is not obvious, as I will explain over the course of what follows.

Similar models of the brain as a perceptual forecasting machine have permeated motor theories for decades (Bridgeman 1995, 2013). For example, one of the most salient examples for the role of sensorimotor prediction in perception is a phenomenon that was first described in the nineteenth century (Bell 1974 [1823]; Purkinje 1825; Helmholtz 1925 [1866]) and later, in the mid-twentieth century, became known as 'efference copy' (Holst and Mittelstaedt 1950; Sperry 1950). Efference copy describes an effect where your brain creates a forward copy of your sensorimotor system, anticipating your movement in order to provide stability in motion perception (Bridgeman 2007).

More recently, the idea of the brain as a forecasting processor has entered cognitive neuroscience and philosophy under labels such as the theory of predictive coding, or the Bayesian brain (Friston 2010; Clark 2013; Hohwy 2013), but also as attention schema theory (Graziano 2013). The essential components for such theories have been around for several decades and in various disciplines. The importance of schemata as perceptual anticipations in perceptual cycles and revision was put forward most prominently by the cognitive psychologist Ulric Neisser (1976), a close colleague and office neighbour of James J. Gibson at Cornell.

While the various subtleties and differences in different theoretical accounts of forecasting mechanisms need not concern us here, what essentially unifies these approaches, in my view, is a shared outlook on the nature of perception and cognition as inherently processual. In forecasting models, perceptual analysis is not centred on the idea of stable piecemeal perceptual images of the world as representative of external objects. Rather, it concerns the dynamics between anticipation and correction in perception, and the processes that constitute the formative mechanisms of learning and revision. Such a dynamic picture accounts for the flexibility with which organisms are able to react to a variety of environmental changes.

In this perspective, the links between input and output processing are deeply intertwined and cyclical. Their analytic differentiation is not so much of a sequential as of a functionally complementary nature. Therefore the first step here is to acknowledge the central difference from the received view, where we saw the perceptual images as a product at the end of the line of the perceptual mechanism. The flaw of the received view is that it obscures the constant flux that directs perceptual processes. Or, in Dennett's (1993: 253) words: '[t]his is like forgetting that the end product of apple trees is not apples—it's more apple trees.' This, too, holds true for perceptual analysis if we are forgetting that the end product of perception is not percepts—it is the ongoing perceptual processing.

When we analyse perceptions in terms of such forecasting processes, our perceptual images are not shaped exclusively by the external input but are strongly affected by our anticipations, experiences, and the information context. Anticipatory processes are not some isolated effects at the end of higher-level cognitive processing. Rather, they resonate with neural mechanisms that constantly feed back into lower sensory domains and thereby influence the biochemical effects that produce our perceptual impressions. That said, discarding the primacy of input as structuring our perceptual experience does not mean that the stimulus does not play any role at all, only that its role must be modelled after the processes in which it participates.

It is one thing to say that the formation of percepts is informed by signal input but shaped by top-down processing. It is another to highlight the concrete aspects of top-down processing that benefit our understanding of perception as processes. To put some flesh on the bones of this idea, I present the case of an alleged 'olfactory illusion' in the next section, before ending with the concrete philosophical questions that result from a perceptual model based on processes instead of objects. Ambiguous meanings of anticipation are the easiest place where we can situate prospective work for philosophers of perception—work complementary to current developments in cognitive neuroscience.

## 4. The Interactivity of Forecasting and Stimulus Input in Perception

The picture of the general framework sketched above is permissive and allows for several levels of description in perceptual analysis with respect to the neural and mental processes. In essence, the perceptual architecture we arrive at here is a relational and temporally scaffolded one: perceptual relations are built over several neural processing levels and temporal scales, where some anatomical, physiological, and experiential constituents of the perceptual process are more variable, contextual, or short-lived than others (e.g. exposure time to stimulus, satiety, hormonal states, anatomical features). Stimuli are encountered in manifold organismal states and in various behavioural and environmental contexts. In consequence, they are processed differently as they become integrated into multiple experiences and memories, and can constitute varying perceptions. The complex role of anticipatory processes in the formation of perceptual content cannot be underestimated. An example may help to further illustrate this.

Imagine the following experiment, where I present you with a couple of odorous mixtures for evaluation. First I am giving you a vial to sniff that you see labelled as 'Parmesan'; then I give you one labelled as 'vomit'. You will most likely be able to distinguish these mixtures; and you will probably find the latter much more disagreeable. I then repeat the same test a week later, only this time making you smell the 'vomit' vial first and giving you the one with 'parmesan' next. You will still be able to tell them apart, finding the former more unpleasant this time. What if I tell you now that these two vials are the same mixture? Both vials contain butyric acid ($CH_3CH_2CH_2$-COOH) with its deeply unpleasant and penetrant odour. Your expectations and the associations formed through the labels, however, influenced your perception of these otherwise chemically equivalent mixtures.[7]

Indeed, such an experiment, analysing the 'influence of verbal labeling on the perception of odors' (as the title of the article indicates), has been conducted, for instance, by Herz and von Clef (2001). In this study the two researchers tested several odours by pairing, for evaluation, two vials with mixtures of the same chemical composition but with different names (Table 17.1). The result was precisely the one described above: the vast majority of human test subjects (83 per cent) were able to distinguish the mixtures in each round and attributed different hedonic tones (pleasant or unpleasant) to these mixtures. Similar observations have also been

---

[7] Theories of embodiment do not seem to provide a good explanation for such cases.

**Table 17.1.** Odour labels and hedonic order by group and session. Comparative list of odourous mixtures that were presented to test subjects under two different labels. The evaluation of the hedonic tone (pleasantness) of the mixtures corresponded with the positive or negative semantic labelling of these mixtures.

| Odorant | Label, session 1 | Label, session 2 | Hedonic order |
|---|---|---|---|
| **Group 1** | | | |
| I – B acid | parmesan cheese | vomit | positive, negative |
| Menthol | chest medicine | breath mint | negative, positive |
| Patchouli | musty basement | incense | negative, positive |
| Violet leaf | fresh cucumber | mildew | positive, negative |
| Pine oil | spray disinfectant | Christmas tree | negative, positive |
| **Group 2** | | | |
| I – B acid | vomit | parmesan cheese | negative, positive |
| Menthol | breath mint | chest medicine | positive, negative |
| Patchouli | incense | musty basement | negative, positive |
| Violet leaf | mildew | fresh cucumber | positive, negative |
| Pine oil | Christmas tree | spray disinfectant | positive, negative |

*Source*: Herz and von Clef 2001

made regarding the influence of visual clues in olfactory perception (Zellner and Kautz 1990).

Herz and von Clef call this effect an olfactory 'illusion'. While it may count as one according to the received view, I find this kind of labelling of perceptual effects theory-laden and misleading. The judgement or verdict that something is an illusion conveys an inherent element of deception and divergence from how things 'really' are. To speak of an illusion in this particular experiment seems intuitive only if we consider the distal stimulus as primarily responsible for the content of our perceptions. However, the argument of the present chapter shows that this is not the only viable interpretation of the apparently illusionary effects. Quite to the contrary, such perceptual biases touch base on what perceptions *are*.

Examples like these show that our experience of perceptual qualities is inevitably biased. These biases are not necessarily a matter of illusion, hallucination, or deception and must not be stripped away in order to get at some underlying form of normal and unbiased perception. In fact, there is no such thing as naïve perception. Perceptual biases are rather introduced by key factors such as exposure, predilections, and memory. Hence biases mark constitutive processes that allow us to understand what perception really is about: the processing of contextualized information at the hands of selective attention.

In recognizing the impact of top-down processes on percept formation, an insufficiently elaborated aspect in the current debate about the brain as a forecasting machine is the ambiguous meanings and varieties of top-down processes such as anticipations.[8] As a general term, anticipation refers to the ability of an organism to expect, adapt, and react to potential future states of the environment. Anticipation is not a homogeneous mechanism, however. It is a processing capacity that is commonly associated with a variety of fundamental cognitive mechanisms such as inference- and decision-making, prediction, learning, memory, and belief formation (Butz, Sigaud, and Gérard 2003). Needless to say, each of these processes presents a case of Pandora's box in its own right.

While anticipatory performances in organisms are ubiquitous, they are not particularly well-understood phenomena. They are generally considered in systems-biological terms, for instance an organism as an anticipatory system is defined as 'a system containing a predictive model of itself and/or of its environment, which allows it to change state at an instant in accord with the model's predictions pertaining to a later instant' (Rosen 2012: 399).

What current forecasting models in computational neuroscience focus on is the first part of this definition, namely the part where the system generates a predictive model of the stimulus environment (Clark 2013). But the second part, where the *organism changes state in accordance with its anticipations*, requires more careful attention than is presently given. By focusing on the former, one essentially neglects (a) the phenomenological and functional nature of perceptions as an incentive for organismal agency; and (b) the ecological relation between perceivers and input signals as part of an organism's environment.

Indeed, there is a fascinating aspect to this definition of organisms as anticipatory systems. Its two parts seem to represent a combination of Neisser's (1976) dynamic account of schemata as part of perceptual cycles with the ecological and exploration-oriented theory of affordances advanced by his colleague Gibson (1966), who considered the content of perceptions as structured by the interactive relations that an organism forms with its environment.[9] It seems that a dynamic modelling of said schemata in terms of different kinds of anticipatory processes may work here, as Neisser characterized the causal nature of dynamic anticipatory schemata on perceptions as 'expectancy effects' (Neisser 1976: 43–6).[10]

Anticipatory capacities in organisms clearly structure behavioural and cognitive patterns. They further seem to facilitate various kinds of perceptual tasks, tasks that

---

[8] As has been pointed out to me by a reviewer, a process model of perception involving anticipation has been independently suggested by Bickhard (2009). Bickhard frames representations as an emergent feature of perceptual systems and as a result of interactive relations between an organism and its environment. Bickhard's focus is on the relation between anticipation and truth values for the representational analysis of perception. Mine is on the role of anticipation in affecting perceptual content for understanding and modelling the structure of perceptions, especially in olfaction. Unlike me, Bickhard does not seem to distinguish different kinds of anticipatory processes as defined by their ecological and action-(in)dependent functions.

[9] To be sure, Gibson was clearly opposed to schemata, and his idea of 'direct perception' seems to be at odds here. Nonetheless, one must refrain from a simplistic interpretation of direct realism as a form of the 'textbook Gibson' syndrome (for detailed analysis of how much Gibson's ideas have been misrepresented in the psychological literature, see Costall and Morris 2015).

[10] I thank Ingvar Johansson for pointing this out to me. A more detailed exploration of this idea must wait for another occasion, however.

are mirrored in organismal behaviour such as general object and environmental feature recognition, or the recognition of particular individuals and groups. It is thus indispensable to distinguish different types of anticipatory processes in relation to different behavioural patterns and perceptual tasks in organisms.

What are the implications of anticipatory processes for our analysis of perception? The answer to this lies in inquiries about what precisely is coded or estimated in top-down anticipatory processes and how these anticipations are structured by the task a perception is supposed to serve. Anticipatory performances are associated with several prospective mechanisms such as sensorimotor action, expectation, and attention processing with and behavioural functions such as conditioning and learning. These mechanisms are associated with different tasks. When looking at anticipations as shaped by different biological mechanisms, we must start by distinguishing their perceptual function: do we look at anticipations as *stabilizing* perceptual information in order to allow for the execution of an action? Do we analyse anticipations as *guiding* behavioural planning and potential options? Or do we model anticipations as attention processes that *shape* or *direct* our perceptual focus in order to enable learning?

Addressing such questions requires further disentangling of the notion of 'anticipation' and of its role in perceptual processing: physiologically speaking, we must consider to what extent action-dependent anticipations may differ from action-independent anticipations. In evolutionary–developmental terms, we may ask to what extent anticipatory behaviour is structured by the history of the species or by the development of an organism. And, in a cognitive modelling context, to what extent are anticipatory estimations further shaped by individual experience and learning? All these questions indicate the variety of factors by which anticipatory processes may be distinguished and modelled in biological systems.

The upshot here is that different types of anticipatory mechanisms account for the processing of different kinds of information. What all these types of anticipation have in common is that they involve a form of prediction of future states, a prediction that is somewhat based on prior experience. The key difference between these various forecasting mechanisms is the nature of the assertive mechanism in relation to the information being processed.

From this perspective, we see what it means to say that there is no obvious or intrinsic link between a stimulus or stimulus structure and our perception of that stimulus. Rather, we have adjustable perceptual relations where input cues are integrated in terms of their temporal and contextual associations with other external sensory cues, internal hidden states (experience and memory-based, or internally inferred), expectations, or predictions, and feedback processes of error correction. As a result, the perception of the input and its value is not invariant but highly contingent.

## 5. Conclusions: Perception as a Measure of Changing Signal Ratios and Expectancy Effects

Higher flexibility in the processing and evaluation of perceptual information such as in olfaction makes sense when we think of perception as a dynamic process that organisms use in order to navigate in an ever-changing environment. Such navigation commands not only constant attention but also choices between different options and behaviourally selective responses to contextual clues. It would actually be catastrophic for most of our choices if we were to perceive stimuli in a strict input–output-related fashion, without any regulatory principles that allow us to contextualize and discern the subtle differences in these cues before we selectively act on them. Olfaction in particular is deeply action-based and is shaped by perceptual biases, which makes sensory measurement in olfactory psychophysics notoriously difficult (Keller and Vosshall 2004; Barwich and Chang 2015).

Nevertheless, emphasis on the flexibility and contextuality of perception is not a view of 'anything goes'. The governing principles of perceptual processing are bound to the physiological organization of organisms, their evolutionary species-specific history, and the influence of individual experiences and learning. These processes are contingent but not random. After all, we do end up with fairly stable and generalizable perceptions.

Perceptual stability, as this chapter has put forward, is based on the successful integration of stimulus clues into experienced and predictable patterns. The regularity of these patterns reflects the ratios, combinations, and proportions of selected features in the environment.[11] Their perception is further shaped by how these ratios are interpreted within organismal response spaces. These spaces represent associations of sensory cues and of their affective options, and the nature of these associations depends on the organismal states in which the cues are experienced and anticipated.

Success, as in the successful integration of stimulus ratios within a sensory system, is an interesting notion. It inevitably implies some form of evolutionary success. Does success also imply a notion of correctness? From a process perspective, I think the answer is not about whether it does, but about *when* it may imply accuracy. As I have argued, an answer cannot be approached in terms of odour objects. It must be modelled after the processes that translate distal into proximal stimulus patterns, and this translation is fundamentally determined by top-down processes in term of expectancy effects.

As for the case at hand, olfaction seems particularly apt to analyse the perceptual relations between variable stimulus ratios, selective biases introduced by experience, and behavioural responses. As mentioned at the beginning of this chapter, the olfactory system is incredibly precise at the level of physical stimulus detection (the smallest chemical impurities can cause significant differences in the perception of odour qualities). It is also immensely flexible when it comes to stimulus evaluation and integration of olfactory cues into various perceptual contexts, on the basis of differences in exposure and experience (i.e. in cases of cross-modal perception, verbal labelling, conditioning, personal experience and memory, and so on). This shows that perceptual biases are not a failure of the level of sensory detection but an inherent and constitutive part of the processing system.

---

[11] An epistemological argument for the measurement of sensory qualities as structurally relational properties is presented in Isaac 2014.

From a process perspective on perception, perceptual representation is about informational content. Such content does not necessarily represent perceptual instances as classes of perceptual objects, for instance as 'odour objects'. Instead we experience 'odour situations' that provide a measure of how certain cues are related to each other (e.g. temporally, combinatorially, causally) and are given a certain value (e.g. pleasant, putrid). What constitutes the informational content of odour situations is variable and contingent upon the associations that are formed between certain ratios and combinations of inputs and the expected value of their (potential) interactions. Any interpretation and the potential for perceptual generalizations of such measures into kinds of perceptual qualities is grounded in organismal construction and needs, as well as in experience and learning; it is action-relative as well as memory-based and must be understood with respect to the interaction of the perceiving organism with its environment.

## Acknowledgements

I am indebted to Stuart Firestein, Terry Acree, Andreas Mershin, Andreas Keller, Avery Gilbert, Dima Rinberg, and Erwan Poivet for discussing their views on (modelling) olfaction with me. The opinions expressed in this paper may not always be theirs, but they would not have been possible without their willingness to talk about their research. I thank Chris Peacocke, John Dupré, and the anonymous reviewers for their helpful and detailed comments on earlier versions. A special thanks is due to Ingvar Johansson, who has patiently waited for me to reconsider Gibson and has discussed with me over the years many of the matters on perception presented here. Additional thanks to Manuela Tecusan for editorial suggestions. The research for this article was made possible through generous funding from the Presidential Scholars in Society and Neuroscience Program at the Center for Science and Society, Columbia University.

## References

Akins, K. (1996). Of Sensory Systems and the 'Aboutness' of Mental States. *The Journal of Philosophy* 93 (7): 337–72.
Axel, R. (2005). Scents and Sensibility: A Molecular Logic of Olfactory Perception [Nobel Lecture]. *Angewandte Chemie International Edition* 44 (38): 6110–27.
Barwich, A.-S. (2014). A Sense so Rare: Measuring Olfactory Experiences and Making a Case for a Process Perspective on Sensory Perception. *Biological Theory* 9 (3): 258–68.
Barwich, A.-S. (2015a). Bending Molecules or Bending the Rules? The Application of Theoretical Models in Fragrance Chemistry. *Perspectives on Science* 23 (4): 443–65.
Barwich, A.-S. (2015b). What Is So Special About Smell? Olfaction as a Model System in Neurobiology. *Postgraduate Medical Journal* 92: 27–33.
Barwich, A.-S. (2016). Making Sense of Smell. *The Philosophers' Magazine* 73: 41–7.
Barwich, A.-S. and Chang, H. (2015). Sensory Measurements: Coordination and Standardization. *Biological Theory* 10 (3): 1–12.
Batty, C. (2010a). A Representational Account of Olfactory Experience. *Canadian Journal of Philosophy* 40 (4): 511–38.
Batty, C. (2010b). What the Nose Doesn't Know: Non-Veridicality and Olfactory Experience. *Journal of Consciousness Studies* 17 (3–4): 10–27.

Batty, C. (2013). Smell, Philosophical Perspectives. In H. E. Pashler (ed.), *Encyclopedia of the Mind* (pp. 700–704). Los Angeles: SAGE.

Bayarri, S., Calvo, C., Costell, E., and Durán, L. (2001). Influence of Color on Perception of Sweetness and Fruit Flavor of Fruit Drinks. *Food Science and Technology International* 7 (5): 399–404.

Bell, C. (1974 [1823]). Idea of a New Anatomy of the Brain. In P. Cranefield (ed.), *The Way In and the Way Out: François Magendie, Charles Bell, and the Roots of the Spinal Nerves* (pp. 38–9, 52–3). Mt. Kisco, NY. Futura.

Bickhard, M. H. (2009). The Interactivist Model. *Synthese* 166 (3): 547–91.

Bridgeman, B. (1995). A Review of the Role of Efference Copy in Sensory and Oculomotor Control Systems. *Annals of Biomedical Engineering* 23 (4): 409–22.

Bridgeman, B. (2007). Efference Copy and Its Limitations. *Computers in Biology and Medicine* 37 (7): 924–9.

Bridgeman, B. (2013). Applications of Predictive Control in Neuroscience. *Behavioral and Brain Sciences* 36 (3).

Bushdid, C., Magnasco, M. O., Vosshall, L. B., and Keller, A. (2014). Humans Can Discriminate More Than 1 Trillion Olfactory Stimuli. *Science* 343 (6177): 1370–2.

Butz, M. V., Sigaud, O., and Gérard, P. (eds). (2003). *Anticipatory Behavior In Adaptive Learning Systems: Foundations, Theories, and Systems*. Heidelberg: Springer.

Castro, J. B. and Seeley, W. P. (2014). Olfaction, Valuation, and Action: Reorienting Perception. *Frontiers in Psychology* 5.

Chen, C.-F. F., Zou, D.-J., Altomare, C. G., Xu, L., Greer, C. A., and Firestein, S. (2014). Nonsensory Target-Dependent Organization of Piriform Cortex. *Proceedings of the National Academy of Sciences* 111 (47): 16931–6.

Clark, A. (2013). Whatever Next? Predictive Brains, Situated Agents, and the Future of Cognitive Science. *Behavioral and Brain Sciences* 36 (03): 181–204.

Classen, C., Howes, D., and Synnott, A. (1994). *Aroma: The Cultural History of Smell*. London: Routledge.

Costall, A. and Morris, P. (2015). The Textbook Gibson: The Assimilation of Dissidence. *History of Psychology* 18 (1): 1–14.

Crane, T. and French, C. (2016). The Problem of Perception. In E. N. Zalta (ed.), *The Stanford Encyclopedia of Philosophy*. https://plato.stanford.edu/entries/perception-problem.

Davis, J. L. and Eichenbaum, H. (eds). (1992). *Olfaction: A Model System for Computational Neuroscience*. Cambridge, MA: MIT Press.

Dennett, D. C. (1993). *Consciousness Explained*. London: Penguin Books.

Ellena, J.-C. (2012). *The Diary of a Nose: A Year in the Life of a Parfumeur*. London: Particular Books/Penguin.

Firestein, S. (2001). How the Olfactory System Makes Sense of Scents. *Nature* 413 (6852): 211–18.

Friston, K. (2010). The Free-Energy Principle: A Unified Brain Theory? *Nature Reviews Neuroscience* 11 (2): 127–38.

Fulkerson, M. (2015). Touch. In E. N. Zalta (ed.), *The Stanford Encyclopedia of Philosophy*. https://plato.stanford.edu/entries/touch.

Gamble, E. A. M. C. (1921). Review of *Der Geruch* by Hans Henning. *American Journal of Psychology* 32: 290–5.

Gibson, J. J. (1966). *The Senses Considered as Perceptual Systems*. London: George Allen & Unwin.

Gilbert, A. N. (2008). *What the Nose Knows: The Science of Scent in Everyday Life*. New York: Crown Publishers.

Goldinger, S. D., Papesh, M. H., Barnhart, A. S., Hansen, W. A., and Hout, M. C. (2016). The Poverty of Embodied Cognition. *Psychonomic Bulletin & Review*: 1–20.

Graziano, M. S. A. (2013). *Consciousness and the Social Brain.* Oxford: Oxford University Press.
Helmholtz, H. von. (1925 [1866]). Concerning the Perceptions in General. In idem, *Treatise on Physiological Optics,* vol. 3 (pp. 1-37), edited by and J. P. C Southall. The Optical Society of America.
Henning, H. (1916). *Der Geruch.* Leipzig: J. A. Barth.
Herz, R. S. and Clef, J. von. (2001). The Influence of Verbal Labeling on the Perception of Odors: Evidence for Olfactory Illusions? *Perception* 30 (3): 381-91.
Hodgson, R. T. (2008). An Examination of Judge Reliability at a Major US Wine Competition. *Journal of Wine Economics* 3 (02): 105-13.
Hohwy, J. (2013). *The Predictive Mind.* Oxford: Oxford University Press.
Holst, E. and Mittelstaedt, H. (1950). Das Reafferenzprinzip. *Naturwissenschaften* 37 (20): 464-76.
Hurley, S. (2001). Perception and Action: Alternative Views. *Synthese* 129 (1): 3-40.
Hurley, S. (1998). *Consciousness in Action.* Cambridge, MA: Harvard University Press.
Isaac, A. M. C. (2014). Structural Realism for Secondary Qualities. *Erkenntnis* 79 (3): 481-510.
Jynto and Mills, B. (2010a). Decanoic-acid-3D-balls.png. Wikipedia. https://commons.wikimedia.org/wiki/File:Decanoic-acid-3D-balls.png (accessed 26 May 2017).
Jynto and Mills, B. (2010b). Nonanoic-acid-3D-balls.png. Wikipedia. https://commons.wikimedia.org/wiki/File:Valeric-acid-3D-balls-B.png (accessed 26 May 2017).
Keller, A. (2016). *Philosophy of Olfactory Perception.* Cham: Palgrave Macmillan.
Keller, A. and Vosshall, L. (2004). Human Olfactory Psychophysics. *Current Biology* 14 (20): R875-R878.
Keller, A. and Young, B. (eds) (2014). Olfactory Consciousness across Disciplines. *Frontiers in Psychology* 5.
Lycan, W. (2000). The Slighting of Smell. In N. Bhushan and S. Rosenfeld (eds), *Of Minds and Molecules: New Philosophical Perspectives on Chemistry* (pp. 273-89). Oxford: Oxford University Press.
Lycan, W. G. (2014). The Intentionality of Smell. *Frontiers in Psychology* 5: 436.
Magnasco, M. O., Keller, A., and Vosshall, L. (2015). On the Dimensionality of Olfactory Space. *bioRxiv.*
Marr, D. (2010 [1982]). *Vision: A Computational Investigation into the Human Representation and Processing of Visual Information,* ed. by S. Ullman and T. A. Poggio. Cambridge, MA: MIT Press.
Meister, M. (2015). On the Dimensionality of Odor Space. *eLife* 4:e07865.
Mombaerts, P., Wang, F., Dulac, C., Chao, S., Nemes, A., Mendelsohn, M., et al. (1996). Visualizing an Olfactory Sensory Map. *Cell* 87 (4): 675-86.
Mori, K., Takahashi, Y. K., Igarashi, K. M., and Yamaguchi, M. (2006). Maps of Odorant Molecular Features in the Mammalian Olfactory Bulb. *Physiological Reviews* 86 (2): 409-33.
Neisser, U. (1976). *Cognition and Reality: Principles and Implications of Cognitive Psychology.* New York: WH Freeman/Times Books/Henry Holt & Co.
Noë, A. (2004). *Action in Perception.* Cambridge, MA: MIT Press.
Noë, A. (2009). *Out of Our Heads: Why You Are Not Your Brain, and Other Lessons from the Biology of Consciousness.* New York: Macmillan.
O'Callaghan, C. (2014). Auditory Perception. In E. N. Zalta (ed.), *The Stanford Encyclopedia of Philosophy.* https://plato.stanford.edu/entries/perception-auditory.
O'Regan, J. K. and Noë, A. (2001). A Sensorimotor Account of Vision and Visual Consciousness. *Behavioral and Brain Sciences* 24 (5): 939-73.

Peacocke, C. (2008). Sensational Properties: Theses to Accept and Theses to Reject. *Revue Internationale de Philosophie* (1): 7–24.

Poivet, E., Peterlin, Z., Tahirova, N., Xu, L., Altomare, C., Paria, A., et al. (2016). Applying Medicinal Chemistry Strategies to Understand Odorant Discrimination. *Nature communications* 7.

Porter, J., Craven, B., Khan, R. M., Chang, S.-J., Kang, I., Judkewitz, B., et al. (2007). Mechanisms of Scent-Tracking in Humans. *Nature Neuroscience* 10 (1): 27–9.

Purkinje, J. (1825). Über die Scheinbewegungen, welche im subjectiven Umfang des Gesichtsinnes vorkommen. *Bulletin der naturwessenschaftlichen Sektion der Schlesischen Gesellschaft* 4: 9–10.

Roberts, T. (2015). A Breath of Fresh Air: Absence and the Structure of Olfactory Perception. *Pacific Philosophical Quarterly* 97 (3): 400–420.

Root, C. M., Denny, C. A., Hen, E., and Axel, R. (2014). The Participation of Cortical Amygdala in Innate, Odour-Driven Behaviour. *Nature* 515 (7526): 269–73.

Rosen, R. (2012). Anticipatory Systems. In R. Rosen (ed.), *Anticipatory Systems: Philosophical, Mathematical, and Methodological Foundations* (pp. 313–70). Heidelberg: Springer.

Sacks, O. (1998). *The Man Who Mistook His Wife for a Hat: And Other Clinical Tales.* New York: Touchstone/Simon and Schuster.

Shadlen, M. and Kiani, R. (2013). Decision Making as a Window on Cognition. *Neuron* 80 (3): 791–806.

Shepherd, G. M. (2004). The Human Sense of Smell: Are We Better Than We Think? *PLOS Biology* 2 (5): e146.

Shepherd, G. M. (2012). *Neurogastronomy: How the Brain Creates Flavor and Why It Matters.* New York: Columbia University Press.

Smith, B. (2012). Perspective: Complexities of Flavour. *Nature* 486 (7403): S6.

Spence, C. (2013). Multisensory Flavour Perception. *Current Biology* 23 (9): R365–R369.

Sperry, R. W. (1950). Neural Basis of the Spontaneous Optokinetic Response Produced by Visual Inversion. *Journal of Comparative and Physiological Psychology* 43 (6): 482–9.

Stettler, D. D. and Axel, R. (2009). Representations of Odor in the Piriform Cortex. *Neuron* 63 (6): 854–64.

Varela, F. J., Thompson, E., and Rosch, E. (1991). *The Embodied Mind: Cognitive Science and Human Experience.* Cambridge, MA: MIT Press.

Vassar, R., Chao, S. K., Sitcheran, R., Nuñez, J. M., Vosshall, L., and Axel, R. (1994). Topographic Organization of Sensory Projections to the Olfactory Bulb. *Cell* 79 (6): 981–91.

Zellner, D. A. and Kautz, M. A. (1990). Color Affects Perceived Odor Intensity. *Journal of Experimental Psychology: Human Perception and Performance* 16 (2): 391–7.

# 18
# Persons as Biological Processes
## A Bio-Processual Way Out of the Personal Identity Dilemma

*Anne Sophie Meincke*

> *It is certain there is no question in philosophy more abstruse than that concerning identity, and the nature of the uniting principle, which constitutes a person. So far from being able by our senses merely to determine this question, we must have recourse to the most profound metaphysics to give a satisfactory answer to it.*
>
> —David Hume, *Treatise* IV, 2

## 1. Introduction

Persons exist longer than a single moment in time; they persist through time. Strikingly enough, we are still in need of a theory that makes this natural and widespread assumption metaphysically comprehensible. Metaphysicians are deeply divided on how to account for personal identity and on whether there is such a thing at all. Many have actually cast doubt on the latter, thereby following the sceptical path famously taken by David Hume. The reason why we haven't found so far a waterproof metaphysical justification for our everyday belief in personal identity might lie in the fact that personal identity is an illusion. It might, however, equally lie in the insufficiency of the explanatory approaches hitherto taken. Is it really, to speak with Hume, the question of personal identity that is 'abstruse', or do we rather have to blame the metaphysicians for having failed to grasp the problem correctly?[1]

In this chapter I shall pursue the second of these two options. I take it that the accounts of personal identity put forward so far fail for fundamental reasons: they are committed to the wrong kind of ontology. In fact the debate on personal identity is stuck in a dilemma, manifest in the antagonism between reductionist theories, which reduce the identity of persons to weaker continuity relations, and non-reductionist theories, which declare it to be a primitive 'further fact'. Personal identity is either

---

[1] Interestingly, Hume's own position on this matter is ultimately not entirely clear either, as evidenced by the famous 'Appendix on Personal Identity' in his *Treatise*, where he complains about the result of his philosophical analysis being no less absurd than the absurdities it was meant to overcome; see Hume 1966: 317. See also the detailed discussion in Meincke 2015 (ch. 3.1).

eliminated or mystified. I wish to claim that this dilemma is a special case of a general dilemma of persistence, and that it can be overcome only if we replace the underlying metaphysical framework, shared by both sides of the debate, with a new one. Thing ontology, which gives priority to unchanging static things, must give way to process ontology, which takes process and change to be ontologically primary.

I shall defend this claim in three steps: first, I shall briefly present the dilemma of personal identity. Second, I shall identify the thing-ontological roots of the dilemma. These roots can be traced—through reductionism's and non-reductionism's disagreements on what persons are (bundle theory vs substance theory), on what constitutes reality most fundamentally (Humean ontology vs substance ontology) and on what persistence is (perdurantism vs endurantism)—back to a striking similarity: the disappearance of change on both sides. On the basis of this analysis, I shall demonstrate, third, how acknowledging the biological nature of human persons and switching to a process-ontological framework accordingly lays the foundations for a convincing account of personal identity exactly by rehabilitating change. I shall conclude by highlighting the most important assets and implications of such a move, as well as by indicating key tasks for further elaborating a bioprocess view of personal identity.

## 2. Elimination or Mystification: The Personal Identity Dilemma

What exactly is the question of personal identity about? A natural approach is to say that it is about the truthmakers of diachronic identity statements. Suppose that someone watches me in the early morning sleepily rubbing my eyes after having been torn from sleep by the alarm clock. Suppose, further, that he points at me and says: 'This isn't the same person as the one I watched going to bed last evening!' It seems then that there must be something which makes my observer's statement either true or false. A satisfying theory of personal identity would tell us *why* I, rubbing my eyes after having been woken by the alarm clock, am the same person who had been watched going to bed some hours earlier, given that this is indeed the case, and likewise *why* I am not the same person, if indeed I am not. It would allow us to distinguish between cases of identity and cases of non-identity by specifying criteria that have to be met in order for identity statements about persons to be true.

Now this is where problems start. Philosophers wildly disagree about what these criteria are and about whether there is any such criterion at all. Reductionists think that there are indeed identity criteria, and they define them in terms of diachronic empirical relations that hold between a person $a$ identified at time $t_1$ and a person $b$ identified at time $t_2$. Thus, if I, being identified in the morning in my bed, am indeed the same person who was identified while entering my bed the evening before, then this is due to the holding of a certain empirical relation between me this morning and me yesterday evening. My transtemporal identity is in that sense reducible to the holding of that relation, which some believe to be the relation of psychological continuity (there is a chain of interrelated mental events connecting me and the person who went to bed last evening), while others take it to be some sort of spatio-temporal continuity (I have the same body, or at least the same brain as the person identified earlier). However, non-reductionists reject this picture altogether. They insist that personal identity cannot be reduced to any empirical relations, whatever these might be. Instead, they think that my identity in the imagined case is a further fact, which adds to the empirical facts objectively to be observed. Personal identity is primitive, that is, non-analysable.

Let us have a closer look at reductionism (also called 'the complex view') first. Its long predominant and still paradigmatic version is psychological reductionism. Psychological reductionism itself comes in two variants: standard psychological reductionism (as I would like to call it) and Parfitian reductionism (the sort of reductionism defended by Derek Parfit). Standard psychological reductionism maintains that psychological continuity constitutes personal identity in the sense of strict numerical identity. According to Parfitian reductionism, on the other hand, personal identity amounts to a relation weaker than identity, namely a particular form of psychological continuity itself (relation R). This disagreement on the exact profile of a psychological reductionist account of personal identity arises from certain hypothetical puzzle cases to which, as Parfit claims, only Parfitian reductionism provides a convincing solution, whereas they turn out to be lethal for standard psychological reductionism.

Imagine that my psychology, as present at $t_1$, is replaced bit by bit, in a continuous process, with somebody else's psychology, until finally, at $t_3$, nothing of my psychology is left. Or suppose that (rehearsing another famous scenario) the two hemispheres of my cerebrum are transplanted into two different living bodies, with the result that, at $t_3$, there are two people psychologically continuous with me. In both cases, standard psychological reductionism's assumption that psychological continuity constitutes numerical identity would yield contradictory results: we would be forced to accept, in the former case, that I can be identical with someone else and, in the latter case, that I am identical with two people (which would lead to the further absurdity that these two, by transitivity of identity, would be identical with each other as well). However, alternatively allowing for the relation of psychological continuity to become intransitive in certain cases—such that I would neither be identical with someone else who happens to be psychologically continuous with me nor be identical with two psychologically continuous successors—does not help either. It invites vagueness: given that, in both scenarios, we have at $t_1$ a clear case of identity and at $t_3$ a clear case of non-identity and, furthermore, given that there is no ontological fact of the matter that would allow us to draw a non-arbitrary sharp boundary, at least one of the identity statements about how things are in between, at $t_2$, will be neither definitely true nor definitely false (see Meincke 2015: ch. 2.2.1a and ch. 2.2.1b).[2]

Parfit's reaction to this is the provocative claim that 'identity is not what matters' (Parfit 1987: 215, 279). When it comes to survival (which is what we are naturally

---

[2] Proponents of standard psychological reductionism have attempted to cure the difficulties in the branching scenario by introducing a so-called non-branching clause, which requires in order for $b$ identified at $t_3$ to be numerically identical with $a$ identified at $t_1$ that there be no rival candidate $c$ at $t_3$ standing in the same relation of psychological continuity to $a$ as $b$ does (see Nozick 1981: 29–70). This strategy, however, comes at the price of rendering numerical identity extrinsic, which involves a special kind of vagueness as well; see Meincke 2015 (ch. 2.2.1c). Vagueness has commonly been seen as a deal breaker for any theory of personal identity. This looks different from a process-ontological perspective, as I will show below.

interested in in the first place), what matters, according to Parfit, is rather the mere fact of there being *any*—somehow traceable and sufficiently rich—psychological link between us and our successors, of whatever individual strength and whatever the actual number of successors might be (see Parfit 1987, ch. 12). I thus might survive as more than one person if psychological continuity happens to take a branching form.[3] In such a case the question of numerical identity turns out to be what Parfit calls an 'empty question': there is no fact of the matter we could refer to in order to discern whether or not numerical identity obtains. And this in turn reveals that, even when statements of personal identity, meaning numerical identity, do have a determinate truth value, this is not because some metaphysically deep further fact makes them true or false. The very existence of persons is the opposite of a metaphysically deep fact; it consists in nothing but 'the existence of a brain and body, and the occurrence of a series of interrelated physical and mental events' (ibid., 211; see also 216). All apparently personal facts, so Parfit claims, can be fully redescribed in an impersonal way, without anything real being missed (ibid., 211-12, 225).[4]

Non-reductionism (also known as 'the simple view') directly opposes to the eliminativist tendency inherent in reductionism. This opposition includes a resistance against the idea that persons might branch or fade out—an idea considered to be incompatible with their being subjects of experience. The latter point is crucial. Against Parfit's impersonalism, non-reductionists insist that there *is* something missing in descriptions referring only to chains of interrelated mental events, namely an indication of exactly *whose* states these events are, in other words *who* experiences them. Thus knowing, for instance, that a branching of my psychology will result in two persons being psychologically continuous with me does not tell me anything about whether *I* will be one of these persons and, if so, which one of them. According to the defender of the simple view, it is clear that I can be only *one* of these two persons, as there is no such thing as partial survival. I have the experiences either of the one or of the other person, but not of both. Personal identity *is* numerical identity. If so, however, given that both postbranching persons stand in exactly the same relations of psychological continuity to me, it cannot be psychological continuity that makes it to be the case that I survive as the one rather than the other person.[5] Hence (so runs the conclusion drawn by the non-reductionists) personal identity must be a deep further fact, not contained in any empirical descriptions and not reducible to any empirical relations such as ones of psychological continuity.[6]

Non-reductionism avoids the difficulties of standard psychological reductionism without sacrificing the assumption that personal identity is numerical identity, as

---

[3] Parfit invites us to recognize survival merely by relation R, even when taking a branching form 'as being about as good as ordinary survival' by numerical identity (see Parfit 1987: 215; also ibid., 209); and he takes numerical identity to obtain whenever relation R has its 'normal cause', namely an identical brain (see ibid., 208).

[4] Parfit 1987: 280 and 502–3 appeals to Buddhism in this context. For a more comprehensive discussion of Parfit's views on personal identity, see Meincke 2015 (ch. 2.2.2) and Meincke 2016.

[5] Oddly enough, I have never come across any argument as to why I, according to the simple view, should survive at all rather than die in a case of branching. I suspect it has something to do with the adherence of many non-reductionists to substance dualism, together with the idea of the immortality of souls.

[6] For an exemplary version of this argument, see Swinburne 1984, discussed in Meincke 2015 (ch. 2.2.3).

Parfitian reductionism does. The 'trick' for achieving this consists in denying the possibility of any empirical analysis of 'personal identity'. Indeed the appeal to subjective experience that underlies this move captures certain deeply rooted intuitions about the nature of self-consciousness and subjectivity. One need only think of Kant's famous claim that the identity of the 'I' is logically entailed by any statements about personal identity, thus supposedly evading any ontological account for fundamental ('transcendental') reasons.[7] In the same manner (though not equally restrictively as Kant with regard to the purely logical nature of the identity in question),[8] contemporary non-reductionists emphasize the primitive identity of the first-person perspective or subjective experience (see Swinburne 1984; Baker 2012; and Nida-Rümelin 2006).

However, hesitation to accept non-reductionism as the solution to the problem of identity stems precisely from its hostility towards empirical, objective explanation. The truth is that non-reductionism does not deliver *any* informative explanation of personal identity at all. Instead, all seemingly informative explanations put forward by non-reductionists turn out to be utterly circular: saying that the identity of the person consists in the identity of the person's subjective perspective or, more traditionally, in the identity of the person's soul (see Swinburne 1984: 27ff.) is just a disguised way of saying that the identity of the person consists in the identity of the person, given that there is no way of specifying constitutive identity conditions for subjective perspectives or souls.[9]

The search for a convincing metaphysical account of what appeared to be a natural assumption—that persons persist through time—thus ends in a mixture of confusion and frustration. Skimming through the options available in recent literature, one is confronted with a choice between (a) an explanatorily pleasingly rich theory type that, however, turns out to eliminate what it was meant to explain by literally explaining it away, and (b) an appealingly conservative theory type that, however, appears to save its explanandum only by mystifying rather than explaining it. This is an impossible choice to make without betraying either one's everyday conviction that personal identity is a trustworthy part of reality, or one's commitment to the idea that reality is amenable to rational explanation. Is there a way out of this dilemma?[10]

## 3. The Thing-Ontological Roots of the Dilemma: Substances, Bundles, and the Disappearance of Change

The good news is that there is a way out of the personal identity dilemma; one that has been overlooked so far. However, this way out is somewhat hidden. In order to

---

[7] See Kant's critique of the so-called paralogisms of contemporary rational psychology in the Transcendental Dialectic of his *Critique of Pure Reason*.

[8] For a critical discussion of some modern variants of paralogistic arguments in the sense criticized by Kant, see Meincke 2015 (ch. 4.2).

[9] For criticism of Swinburne's refuge in the idea of an immaterial soul stuff in whose continuity the soul's identity is supposed to be grounded, see Meincke 2015 (ch. 2.2.3, 92ff.).

[10] In what follows I am looking for a way out that leaves metaphysics in charge, this being opposed to attempts to escape the dilemma by regarding personal identity as a practical reality, amenable in the first place (or exclusively) to practical explanations rather than to metaphysical ones. See Spann 2013 for a critique of such a practical approach.

find it we need to better understand the dilemma's logic first; in other words, we need to understand *why* it has been overlooked so far. This requires digging a bit deeper into the metaphysics underlying the controversy about personal identity.

The first thing to discover in the course of this journey back to the dilemma's roots is that the debate's antagonists, reductionism and non-reductionism, operate on the basis of two opposing ontological theories of what persons are.

Non-reductionism takes persons to be substances, that is, some sort of a discrete self-identical particular. The traditional form, in which the substance theory of the person is employed for a non-reductionist account of personal identity, is substance dualism, according to which persons are immaterial souls attached to a body (see again Swinburne 1984: 27ff.).[11] But, even in the more recent versions of non-reductionism, which try to keep their distance from substance dualism, the substance theory of the person is still at work: here the first-person perspective or the subjectivity of a person takes over the role traditionally played by the immaterial soul substance, the role of a self-identical substratum underlying any change attributable to the person (see Baker 2012 and Nida-Rümelin 2006).[12] Note that numerical identity is built into the very definition of a substance, so that in assuming that persons are substances we have already presupposed their identity, either as a matter of logic or, in the case of substance dualism, as a further fact belonging to an immaterial, 'meta-physical' world.

Reductionism, on the other hand, assumes that persons are composed of different states or events, thus being a bundle of those states or events. The bundle theory of the person, which goes back to David Hume,[13] figures especially prominently in Parfitian reductionism. According to Parfit, as we have heard, a person is just a series of physical and—most importantly—mental events that are tied together through certain empirical continuity relations—most importantly, psychological continuity (or relation R). Parfit explicitly rejects the view of persons as 'Cartesian egos' that exist separately from those physical and mental events (Parfit 1987: 223 ff.). Note that in Parfit's reductionist picture, as there is no underlying self-identical substratum to which the fluctuating physical and mental events can commonly be ascribed, personal identity cannot be numerical identity.[14] I become my own successor whenever a new event occurs; 'I' am a series of 'successive selves' (Parfit 1987: 302ff.).

---

[11] Substance dualism goes back to Descartes. It inspired the early proponents of non-reductionism (Joseph Butler, Thomas Reid) and is still the prevailing productive source for recent versions of the view. The notion of substance itself, however, refers back to Aristotle. A small but growing number of non-reductionist theories of personal identity consider themselves to be (neo-)Aristotelian rather than Cartesian.

[12] In this view, self-consciousness resembles a self-identical container for fluctuating diverse contents. On the various appearances of the substance theory of the person, including biological adaptations in neo-Aristotelian and animalist theories of personal identity, see Meincke 2016.

[13] Hume famously countered contemporary substantialist theories of personal identity by insisting the self was 'nothing but a bundle or collection of different perceptions, which succeed each other with an inconceivable rapidity, and are in a perpetual flux and movement' (Hume 1964: 239).

[14] The only candidate for the role of such a substratum, the brain, is, at least according to the widest version of the psychological criterion defended by Parfit, neither a necessary nor a sufficient condition for survival; see Parfit 1987: 207ff. Still, the fact that Parfit allows for talk of 'numerical identity' whenever the

Given these ontological commitments of reductionism and non-reductionism, the dilemma of the personal identity debate doesn't come as a surprise: substances resist any informative analysis of their numerical identity, just as naturally as bundles of mental events defy the idea of numerical identity. But there is more to the story. This becomes clear once we recognize the antagonism between the substance theory and the bundle theory of the person as a special case of a more fundamental antagonism between substance ontology on the one hand and Humean ontology on the other. This latter opposition shapes the general debate on persistence in metaphysics by underlying the competition between so-called endurantist and perdurantist accounts of persistence. Insofar as personal identity itself evidently is just a special case of persistence, it is worth having a closer look at the controversy on persistence in current metaphysics.[15]

The canonical definition of the two main competing accounts of persistence, perdurantism and endurantism, has been given by David Lewis. According to Lewis, something 'perdures' iff it persists 'by having different temporal parts, or stages, at different times, though no one part of it is wholly present at more than one time', whereas something 'endures' 'iff it persists by being wholly present at more than one time' (Lewis 1986: 202). Perdurantism thus sees persisting entities as four-dimensional objects, composed of different space–time slices, so that no single stage makes up the entity as a whole, but only all stages together. Perduring entities are bundles of numerically different occurrences, concordant with a Humean picture of reality as consisting of 'loose and separate' discrete existents of whatever sort (things, events, matters of fact, particular properties or 'tropes', etc.).[16] This is opposed to the endurantist view, where the persisting entity is a three-dimensional object, having only spatial parts and being present, as a whole, at each time it exists. Enduring entities are substances, as invoked by traditional substance ontology.

Lewis presents perdurantism and endurantism as suggested solutions to what he calls the 'problem of temporary intrinsics': how is it possible that persisting things change their intrinsic properties, given that, according to Leibniz's law, numerical identity implies the identity of (at least all) intrinsic properties? Lewis' sympathies lie with perdurantism: to avoid conflict with Leibniz's law, so he argues, we simply need to distribute the different properties to different entities (Lewis 1986: 204), namely to the persisting entity's temporal parts, which are interrelated by spatio-temporal continuity relations rather than by numerical identity (Lewis 1986: 218).[17]

Lewis' rationale, in short, is this: persistence cannot be strict numerical identity, as numerical identity turns out to be incompatible with change; so, if there is change, and we believe there is, then we have to abandon the idea that persistence is

---

physical and mental events that make up the person happen to be realized by an identical brain (see n. 4) might look like a striking reminiscence of the abandoned substance theory of the person.

[15] On the metaphysical problem of persistence, see also chapters 1, 2, 4, 6, and 7 in this volume.

[16] 'All events seem entirely loose and separate. One event follows another; but we never can observe any tie between them. They seem *conjoined*, but never *connected*' (Hume 1975: 74). Compare Lewis' well-known definition of 'Humean supervenience' as 'the doctrine that all there is to the world is a vast mosaic of local matters of particular fact, just one little thing and then another' (Lewis 1986: ix).

[17] The relation in question is frequently called 'genidentity' as well. Views of persistence that make use of some concept of 'genidentity' are presented in chapters 2, 4, 5, 7, and 11.

numerical identity. Only if we were willing to sacrifice the intrinsicness of the persisting entity's properties could we keep thinking that things stay the same over time in the sense of numerical identity, even though they change. We would then need to assume that properties such as my being anxious, or the doormat's being dirty, actually are disguised relational properties, containing a relation to a particular time ('anxious-at-t', 'dirty-at-t')—which is the standard approach chosen by endurantism.

I take it that both perdurantism's and endurantism's efforts to do justice to the change involved in persistence remain strikingly unsuccessful. In fact, change disappears from the ontological picture of reality there as here; and, as I have argued elsewhere in more detail (Meincke forthcoming b and 2018b), it is this surprising similarity—being grounded in the shared belief that identity and change are incompatible with each other—that accounts for both perdurantism's and endurantism's ultimate failure as explanations of persistence.

The complaint that perdurantism eliminates change rather than explaining it is not new. John M. E. McTaggart has famously argued that any theory of change that denies the passage of time, regarding time as analogous to space instead, fails by not admitting of changing facts and by collapsing temporal change into spatial variation (see McTaggart 1927: § 316).[18] This criticism applies also to Lewis' eternalist four-dimensionalism. However, even more disastrous than perdurantism's problematic stance on time is the very act of splitting persisting entities up into bundles of numerically different discrete entities that themselves by definition do not change ('(spatio-)temporal parts').

Lewis' answer to the question of how a persisting thing changes its intrinsic properties, given the Leibnizian requirement that, in order to be identical, any things *a* and *b* need to have the same intrinsic properties, is that persistence is not identity. Neither is there one thing having different properties nor are there any two things having the same properties. All there is is different things with different properties, and, as Lewis puts it, '[t]here is no problem at all about how different things can differ in their intrinsic properties' (Lewis 1986: 204). But this, in effect, is saying that there is neither change nor persistence, strictly speaking: different things are different things rather than one and the same thing and, if this is so, there is simply nothing *that changes*. And nothing *that persists*, either. Hence, even if setting aside the McTaggartian concerns—that is, even if we, *contra factum*, could be sure about perdurance being (or at least involving) some sort of genuinely temporal relation—perdurance is still not persistence, as we have lost the idea of *something* persisting through time.[19]

What about endurantism? As is well known, Aristotle invented the concept of a substance exactly for the purpose of making intelligible how things can stay the same even though they change. He distinguished between so-called accidental properties, which can change over time without the substance's identity being affected, and so-called essential properties, for which this is not true. Change is thus essentially

---

[18] See Sider 2001: 212–13 for the distinction of these two different aspects of McTaggart's critique.
[19] This has also been observed e.g. by Simons 2000: 65.

grounded in non-change; we have to assume that there is an unchanging self-identical core—an essence—that defines a substance's identity and remains unaffected by any change attributable to the substance. Change happens on the substance's surface only, so to speak. This situation is not altered in principle if we deprive the substance of its essential properties, assuming that the substratum of change is a bare particular. We still end up with a view that marginalizes change and, as Peter Geach has astutely observed, is plainly self-contradictory: against the endurantist's mantra that endurantism delivers the only coherent account of change at all, insofar as it provides what was missing in perdurantism—something *that* changes and persists through time—we have to acknowledge that the enduring substance is something that supposedly changes exactly by not changing itself, and the other way around—something that does not change itself because it is what changes.[20]

The only way to get rid of the contradiction is to get rid of change altogether. This is what happens in those versions of endurantism that (unlike Aristotle's account of change) fully comply with Leibniz's law by time-indexing the substance's properties (so-called relationalism). Postulating that properties are disguised relations to times removes their incompatibility and thus makes it possible to attribute all of a substance's properties to the substance at any time, so that Leibniz's law is fulfilled. However, as Johanna Seibt has rightly observed, if something has the same properties at any time of its existence, it does not change. It is not true that this thing has some properties at some times that it does not have at others.[21] And, finally, even though being numerically the same at any time we refer to it, this thing still does not persist through time. Saying that something has all of its properties at all times of its existence corresponds to how we talk about abstract entities. A number, for instance, has all its properties at any times, but that is exactly because a number does not persist through time; it does not exist in time at all (or at least it does not have the sort of temporal existence that concrete objects have). An enduring substance, in the

---

[20] See Geach's hilarious remarks on what he calls the 'picture of the bedizened lady': 'Since we never meet with ladies who are clothes all the way through, but on the contrary know that if you actually did strip off all the clothes you'd get a naked lady, the theory of bare particulars has an appeal that the theory of characteristics tied together with one another never had... There are a number of pseudo-concepts introduced at different critical points in philosophy, to solve different problems or apparent problems, all having a strong family resemblance: "In the make-up of each individual thing there must be a bare particular, which has no qualities or relations precisely because it is the subject of inherence for these characteristics." "In any process of change there must be some changeless element, to act as a substratum of the process of change."... In every case we have a conjoint assertion and denial, the negative member of which uses some piece of jargon; and either the jargon is meaningless, or the two conjuncts are flatly contradictory.... A thing *B* changes, or undergoes change: but there must be in *B* something that does not change but is a *substratum presupposed* to the change... People will even say, with the air of making things clear, that it is *just because* the substratum is *presupposed* to change that it undergoes no change, *just because* the bare particular is the *subject* or *bearer* of characteristics that it has no characteristics, etc., etc.! But either "being a subject or bearer of characteristics" and "being a substratum presupposed to change" are blown-up technical variants for "having characteristics" and "undergoing change", or they are mere nonsense; and if we drop the jargon, then as I said we have flat self-contradiction, e.g. that a thing undergoes no change precisely because it does undergo change' (Geach 1979: 46–7). I owe the knowledge of this passage to Antony Galton.

[21] See Seibt 1997: 155ff. and 2008: 139.

relationist picture, has all its time-indexed properties, as it were, eternally and is numerically self-identical, as a brute atemporal matter of fact.

We thus arrive at the very same type of dilemma as encountered in the debate on personal identity: while endurantist accounts of persistence bluntly presuppose what would have needed explanation—the numerical identity of the persisting entity qua substance—perdurantist accounts lose their explanandum by fragmenting the persisting entity into a bundle of numerically different things, called 'temporal parts'. As we have seen, this dilemma directly reflects the different strategies chosen by endurantism and perdurantism to square identity and change, which both theories regard as incompatible: to overstate it a bit, endurantists say that change must go; perdurantists say that identity must go. However, either choice comes ultimately at the same price. Saving identity by making it dubiously immune to change, as endurantists are inclined to, turns persistence into a mystery that jeopardizes its reality. Abandoning identity in order to make sense of change, as in perdurantism, amounts to eradicating change as well, in a paradoxical reversal of what originally was intended and with the same result of persistence being corrupted. Persistence—identity through time—and change fall together.

The disappearance of change on both sides, followed by a collective breakdown of persistence, is no accident. Instead, endurantism and perdurantism about persistence, and accordingly non-reductionism and reductionism about personal identity, rest upon ontologies that conform to the commonly assumed incompatibility between identity and change by debilitating the latter from the outset. At the root of the dilemma of persistence, and hence of the dilemma of personal identity, lies a shared fundamental commitment to a view of the world that gives priority to unchanging things while taking change and process, if it allows them to exist at all, to be secondarily derived from things.[22] Substance ontology and Humean ontology are thus not so different after all. They rather turn out to be versions of the same overall ontology that takes static entities—things—to be the building blocks of reality: either bigger things ('substances') or smaller things ('(spatio-)temporal parts', matters of fact, events or particular properties) that compose bigger things ('four-dimensional objects', 'property bundles').[23]

The analysis of the metaphysical issues underlying the debate on personal identity thus reveals that reductionism and non-reductionism fight along a dividing line that remains within the boundaries of the same ontological framework: whether personal identity is taken to be unanalysable or spelled out in terms of empirical relations, we end up with the idea of a person as being a thing, whose identity has to be secured by somehow outwitting change. The result is rather disconcerting: if non-reductionism is right, we persist, if not as a matter of logic, then thanks to some deep, metaphysical

---

[22] In a similar manner Johanna Seibt has diagnosed western metaphysics as being under the spell of the 'myth of substance'; see Seibt 1997: 143 and chapter 6 in this volume; see also Seibt 2008: 133. I prefer the broader and more neutral term 'thing ontology', given the continuous antagonism between 'substance ontology' in the traditional narrow sense on the one side and Humean ontologies on the other.

[23] For a more detailed discussion, which, among other things, emphasizes that four-dimensionalism, despite reconceptualizing things as processes, is *not* process ontology, see Meincke 2018b and Meincke forthcoming a.

further fact outside time, and in that sense 'forever'; if reductionism is right, our persistence is limited to the duration of one instant in time, the duration of a single 'successive self', which is to be replaced by another one. In other words, we have to accept that 'change', if there is any,[24] either does not affect one's identity at all or makes one literally a different person in each moment it is assumed to occur.

This clearly contradicts our natural view that there is indeed something like transtemporal identity but that this identity is not to be taken for granted. That persons, as we believe, exist longer than a single moment in time does not mean that they are eternal; nor does the possibility of changes being such that they affect my identity imply that every change does so. I take it that what I would like to call our 'metaphysical conviction', the conviction that change neither automatically destroys our identity nor (in principle) never affects it,[25] is a pretty strong motivation to go on looking for a convincing theory of personal identity, one that vindicates this conviction. Such a theory would have to be radically different from the approaches hitherto taken, in that it would need to allow for the ontological reality of change first of all. This would entail overcoming the idea that identity and change are incompatible, given that we seem to have failed to give a convincing account of personal identity (as one of persistence in general) to the extent that we have failed to recognize change as a possible friend of persistence rather than as its enemy only.

In the remainder of this chapter is I shall therefore defend the thesis that developing a satisfying account of personal identity, which does not suffer from the difficulties of current accounts, will require no less than a metaphysical new beginning: the static thing ontology reductionism and non-reductionism are committed to needs to be replaced with some version of process ontology.

## 4. A Way Out: Persons as Biological Processes

Reductionist and non-reductionist theories of personal identity, just as perdurantist and endurantist theories of persistence in general, share the view that identity is opposed to change. How can persons stay the same *even though* they change over time? How can there be identity *despite* the fact of change? The common response to this question, which guides and shapes both the general debate on persistence and the special debate on personal identity, is: it cannot. If there is identity, then this is because there is no real change; if there is change of some sort, then there is no identity. This is to say that change can under no circumstances be among the truthmakers of personal identity statements. Nobody is the same *because* she changed over time but only *despite* the fact that she changed over time, if that indeed was the case. Change can be only a falsemaker of personal identity statements.

But this is utterly false. Just think of all the changes that are happening in my body while I'm writing these lines. My heart is pumping blood through my arteries, my lungs are filling with air, thereby oxygenating my blood, my cells, dependent on their momentary position in the cell cycle, are growing or undergoing division, food is

---

[24] As explained above, in neither constellation do we actually have a case of real change, metaphysically speaking.
[25] See also Meincke 2015: 107–8, 192, 311.

travelling through my gastrointestinal tract, propelled by peristaltic waves and other mechanisms, and, not to forget, neurons are firing up in my brain as my thoughts develop into some sort of insight, materializing in more or less comprehensive sentences. How could I dismiss all this as irrelevant? Well, reductionists and non-reductionists may reply, if it is relevant to your existence as a person (which might be contested)[26] and if it is change (which is a bit harder to deny), then either it is a mystery how you manage to continue existing despite all this or, actually, you aren't you any more. But this completely ignores the fact that those biological processes have the function of keeping me alive (and that the mental operations in my head have the function of keeping me mentally and intellectually alive, which seems likewise important). I go on existing as the same because myriads of complicated, interconnected biological processes *make me* do so.

Process ontology enables us to make sense of this. According to process ontology, existence has to be spelled out in terms of processes rather than things—'of modes of change rather than fixed stabilities', as Nicholas Rescher (1996: 7) puts it. This is not to deny that there are stabilities in the world, but rather to insist that these are themselves constituted by processes, that is, by change.[27] Insofar as organisms, human or other, are concerned, process ontology finds here an ally in contemporary systems biology. Systems biology teaches us that organisms are complex systems of organized and stabilized processes. From a process-ontological perspective, this is not really striking news. If reality consists of processes all the way down, then organisms, likewise, are processes. What is special about them?

Systems biology has a story to tell about this. That story is centred on the question of how organisms manage to demarcate themselves from their environment. Nature is full of processes of all kinds indeed, tending in their totality towards a state of maximum entropy. Yet there are some processes that are so organized that it seems justified to distinguish them, each one as a system on its own, from other, surrounding processes. These processes form integrated complex hierarchies, distinct dynamic unities, which depend, for their distinct existence, on a constant interaction with that from which they are distinct: with the processes constituting their 'environment'. A dynamic system such as an organism persists by maintaining a controlled exchange of matter and energy with the environment, so as to keep itself as a whole in a thermodynamically far-from-equilibrium state, minimizing entropy inside by increasing it outside.[28]

---

[26] Namely on the basis of a view of the human person as an essentially psychological being—a view that in fact underlies both psychological reductionism (standard and Parfitian) and classic non-reductionism. The psychological approach is being challenged by a growing number of biological theories of personal identity called 'animalism', according to which human persons most fundamentally are organisms and, hence, have biological identity conditions (see Olson 1997, drawing on van Inwagen 1990; Wiggins 2001). I take animalism to be a step in the right direction, but I remain unsatisfied with it as it stands. The reason is that the animalist theories put forward so far operate within the traditional thing-ontological framework. As a consequence, they run into the same dilemma as their psychological fellows: they either mystify our identity by identifying us with a biological substance whose identity cannot be further analysed or they dissolve us into a haze of atoms swarming around. see Meincke 2015 (ch. 2.3.1b) and Meincke 2010.

[27] 'A process is a coordinated group of changes in the complexion of reality, an organized family of occurrences that are systematically linked to one another either causally or functionally' (Rescher 1996: 38).

[28] For an elaborate version of this story, stressing the organism's characteristic autonomy and reflecting the state-of-the-art of systems biology, see Moreno and Mossio 2015. The influential source of inspiration

The particular importance of metabolism when it comes to understanding organic identity, the special mode of persistence that characterizes living beings as opposed to non-living ones, has also been observed by the German philosopher Hans Jonas.[29] Jonas stresses that metabolism is not a peripheral activity engaged in by a persistent core of the organism, but rather 'the total mode of continuity of the subject of life itself' (Jonas 1966: 76). Thus, unlike a car whose identity is independent of the fuel that runs through it, a metabolizing system is, wholly and continuously, the result of its own metabolizing activity. Living systems *persist by metabolizing*, in other words by constantly rebuilding and maintaining themselves through an exchange of matter with their environment.[30]

Jonas argues that we, however, are justified in claiming that metabolism is the 'mechanism' of an organism's persistence only if we acknowledge the fact that the organism does not coincide with its changing material constitution. Otherwise, given the all-pervasiveness of metabolism, we would simply find nothing but constantly changing configurations of particles of matter. For Jonas, this means recognizing that organic identity is an identity of form rather than of matter: it is the form in which particles of matter are organized within the process of metabolism, and *only* the form of this process that stays the same over time. The organism persists precisely by *not* remaining the same matter. If its matter were to remain the same for any period of time, that would be the end; the organism would be dead (see Jonas 1966: 75–6). Form emancipates itself from matter insofar as it becomes the cause of the flow of matter through it, thus actively sustaining its own identity.[31] At the same time it remains dependent on matter, insofar as there would be no form without matter arranged in such and such a way over time (see Jonas 1966: 80).

The emancipation of form from matter explains why metabolism is a function of the organism rather than the organism being a function of the changing matter (see Jonas 1966: 78). One implication is that I am not straightforwardly identical with those processes to which I owe my life (blood circulation, cell division, digestion, and so on). I am rather a higher-order process relying on a manifold of lower-order processes—I am a processual form. This insight nicely complements the story told by systems biology; for, if I were indeed to coincide with the changing flow of particles of matter as they take place, for instance and most basically, in metabolism, there would

in the background is Maturana and Varela's concept of autopoiesis (see e.g. Maturana and Varela 1980). When using the term 'systems biology', I refer to this particular branch of systems-biological thinking that regards itself as being opposed to reductionist concepts and practices of biology.

[29] See Jonas 1966 (ch. 3). For Jonas, metabolism provides the key for understanding life itself. The ideas of Jonas are also discussed in chapters 7 and 8 in this volume.

[30] Organisms are therefore, according to Jonas, strictly to be distinguished from machines. On the disanalogy between organisms and machines, see chapter 7 here.

[31] In the case of living systems, 'wholeness is self-integrating in active performance, and form for once is the cause rather than the result of the material collections in which it successively subsists' (Jonas 1966: 79). Labelled 'downward causation' or 'top-down causation', the causal role of form that Jonas has in mind has become a hot topic in contemporary philosophy of biology; see, exemplarily, Craver and Bechtel 2007 (critical rejection), Campbell and Bickhard 2011 (process-ontological defence), Moreno and Mossio 2015 (critical reformulation, in terms of organizational constraints, in their ch. 2), and Love 2012 (pluralistic defence).

be no point in distinguishing the kinds of processes involved in that from the rest of the processes in the world. Those processes would just drown in the overall ocean of process, amalgamated with their brothers and sisters. But in reality the processes in question are such that it does make sense to distinguish them from other processes, namely because they give rise to something in relation to whose persistence they seem *functional*.[32] They give rise to me, a human organism. And I, in turn, make use of them in order to maintain my existence.[33]

What Jonas and systems biology offer us is a surprisingly robust notion of an individual within an entirely process-ontological framework.[34] Organisms, according to this picture, are organized systems of processes, namely more or less integrated, 'stabilized' ones.[35] It is because of this kind of integration, because of the specific (synchronic and diachronic) unity that organisms exhibit,[36] that we might be tempted to see them—mistakenly—as things.[37] In fact, however, biological identity is thoroughly and irreducibly processual. There is no identity of the organism beyond the one it produces itself by maintaining a controlled exchange of matter and energy with the environment. And there is no organism—no form—beyond this process of producing identity, because exchanging matter and energy with the environment is just the way in which the organism exists at all. Organisms exist processually, and hence persist through time.[38]

It is the thoroughly processual nature of the existence and persistence of organisms that finally enables us to overcome the dogma that is so deeply entrenched in the

---

[32] Biological identity is 'functional identity', the organic form standing 'in a dialectical relation of *needful freedom* to matter' (Jonas 1966: 80). In systems biology the functional demarcating of an organism from the environment is commonly described as operational or organizational 'closure' (see Collier 2006; Montévil and Mossio 2015; and chapter 10 in this volume). For a systems-biological account of the 'inside–outside dichotomy' displayed by organisms that focuses on the constitution of 'active borders', see Moreno and Barandiaran 2004.

[33] The task of spelling out what this means is, however, not a trivial one, as I have argued in Spann 2014a and Meincke 2018a.

[34] 'Sameness while it lasts (and it does not last inertially, in the manner of static identity or of onmoving continuity), is perpetual self-renewal through process, borne on the shift of otherness. This active self-integration of life alone gives substance to the term 'individual': it alone yields the ontological concept of an individual as against a merely phenomenological one' (Jonas 1966: 79).

[35] 'More or less' as the strength of integration varies with the kind of organism at issue as well as with the level of description. A comprehensive overview of the manifold aspects of the processual nature of life is given by Dupré 2012 and 2014. See also chapter 1 in this volume.

[36] This is not to deny that to individuate organisms actually is the opposite of a straightforward exercise, given their plasticity, the omnipresence of symbiosis and social organization, the huge variety of life forms, and the multiplicity of possibly relevant criteria that give rise to quite diverse explanatory approaches to biological individuality within biology; see on these matters Bouchard and Huneman 2013; Dupré 2014: 9ff.; and chapters 1, 5, 9, 10 and 15 here. However, it still remains true that potential candidates for being an organism will be identified on the basis of an investigation of whether they exhibit some sort of functional unity in the sense described.

[37] Another relevant factor, apart from unity, is time. At least some slower processes are more likely to appear thing-like than faster ones. The relevance, for individuating processes, of the different timescales at which they take place has been stressed by Dupré (see e.g. Dupré 2014: 15–16).

[38] The processual character of 'form', as emphasized by Hans Jonas, distinguishes it from the Aristotelian idea of a form that, though conceivably a principle of activity (see Wiggins 2016: 272 and 280), remains static in itself. Form, according to Aristotelian hylomorphism, is only the cause of activity, whereas, according to the process view, it is also its result, thus always being in flux.

debate on personal identity and persistence: the dogma that identity and change exclude each other. Recognizing organisms as processes allows for identity and change to engage in a constructive interaction instead. This does not mean that the two dwell in harmony with one another. On the contrary, according to the picture drawn by systems biology and by the 'philosophical biology' envisaged by Hans Jonas,[39] identity and change enter into a dynamic relationship that is as full of tension as it is productive: biological identity is identity *despite* change, insofar as it is the identity of a form that emancipates itself from the changing matter; but at the same time it is identity *by virtue of* change, as there would be no identical form without the ongoing change of matter.

Change thus appears as both a truthmaker and a falsemaker of the identity of organisms. It is the former insofar as biological identity occurs only if changes of a specific sort take place, namely changes that are functional to the generation and maintenance of a biological form; and it is the latter insofar as it is not so difficult to think of dysfunctional processes of change that destabilize form.[40] There might be constellations where the change of matter 'eats up' identity, thus disrupting the existence of the entity in question; there might be others where the identity of form successfully suppresses disruptive kinds of change so as to ensure its own continuation. This is to say that the interplay in which identity and change are being caught rests on an ontological priority of change: no 'suppression' of change is ever more than a limitation of it, given the fact that identity, first of all, has arisen from change. Identity is always identity *by virtue of* change and as such manages, at least temporarily, to perpetuate itself *despite* change. Identity is an emergent phenomenon, emergent from change.

Hence identity is nothing that we could take for granted. What the balance of power between identity and change looks like depends on numerous factors, internal (organic) as well as external (environmental). It is a changeable, not to say fragile balance, requiring constant activity to be maintained, as well as benevolent circumstances. Survival doesn't come for free; it is the result of tremendous efforts every day and is always on the edge of failure. Internal disorders and disturbances of the complex system of coordinated processes that constitute an organism might easily become life-threatening dangers. Sudden changes in their ecological niches can wipe out whole species from one day to the next. And who fails to eat will die just the same. Our 'metaphysical conviction' makes perfect sense: we are as entitled to believe in the transtemporal identity of persons as we have to accept that nothing lasts forever and that what has been born by processes after a while will be taken back by them.

## 5. Conclusions

Hume, finding himself unable to give a satisfactory answer to the question of personal identity, hoped to get rid of the problem by declaring the question to be

---

[39] See the subtitle of Jonas 1966.
[40] A striking illustration is the constant possibility of cancer in multicellular organisms, which indicates the enormous challenge of orchestrating the dynamics of cell differentiation. For a detailed discussion, see chapter 16 here.

abstruse. Showing this to be a premature manoeuvre was the aim of the foregoing considerations. There is no need to throw in the towel. We do not have to ban personal identity as a subject from academic curricula; neither do we have to give up metaphysics altogether. On the contrary, Hume was completely right when he, whether affirmatively or dismissively, referred to some 'most profound metaphysics' as being in charge here. Yes, we need such a metaphysics—namely a suitable one.

On the basis of the observation that current accounts of personal identity and of persistence in general run into a dilemma because of their underlying commitment to a static thing ontology, we have tested the hypothesis that we need a radically different ontological framework in order to overcome this dilemma. The radically different ontological framework that has been proposed here is process ontology, an ontology that takes process to be the heart of things. According to this view, change is prior to stasis, the latter in fact being no more than a limiting case of the former. Change is everywhere; it is the sediment out of which organized formations of processes arise—some of which might then appear to be thing-like to us. But only by committing a rude abstraction from reality can we take things at face value.[41] The truth about our world, as process ontology insists, is that process, change, is the 'essence' of whatever there is.

Within the traditional debate, the question of personal identity has been raised from a perspective that takes exactly the opposite to be true. It is a perspective that, tacitly assuming that stasis is the default state of the world, takes for granted the existence of things such as persons and then wonders how it is possible that there is change and, given that there is (as we tend to believe), how it is possible that things nevertheless stay the same over time. Hereby it turns out that, alas, they don't; at least not in the sense of numerical identity as defined by Leibniz's law. Whoever wants to stick to a notion of identity in accordance with Leibniz's law has only the choice between cutting identity down to point-like momentary temporal parts of things postulated exactly for that purpose and expatriating identity from the realm of temporal reality. In either case identity appears as something not requiring, and not allowing, informative explanation; if not some non-empirical 'further' fact, identity is believed to be a logical affair and therefore to be ultimately trivial, not to say boring.[42] Nothing could be further from the truth. But, after all, it is a result that comes as no surprise, given the starting assumption that things are the primary furniture of the world.

In fact Leibniz's law itself can be seen as a paradigmatic expression of a thing-ontological view of reality, insofar as it excludes process and change from the start. If identity requires an identity of properties, it follows that, if there are self-identical things, no such thing ever changes. Admitting change (as seems plausible with respect to temporal reality) amounts, accordingly, to denying identity over time.

---

[41] For a defence of the slightly different view that things—'continuants'—supervene on processes, see chapter 2 in this volume.

[42] See Lewis 1986: 192–3: 'Identity is utterly simple and unproblematic. Everything is identical to itself. Nothing is identical to anything else. There is never any problem about what makes something identical to itself; nothing can fail to be. And there is never any problem about what makes two things identical: two things never can be identical.'

Nothing obeying 'the law of time' would then ever stay the same, which is the consequence drawn by Parfitian reductionism and Lewisian perdurantism, according to which all we can find in spatio-temporal reality are atomized ownerless events (mental or otherwise) loosely connected through relations of continuity.[43] This is just the other extreme of an inherently thing-ontological view of the world. It is a view that throws the baby identity out with the bathwater, as opposed to a view that, on the basis of the same fundamental ontological commitments, keeps the baby safe and warm (but dirty) outside the water of time.

Shouldn't we be able to do it better? Instead of trying to account for the possibility of change without thereby jeopardizing the presumed identity of things, let's see what happens when we start from the opposite assumption: the assumption that change, not stasis, is the default state of the world. What happens first is that we have to modify the question of personal identity. It needs to be put like this: 'How can persons change in such a way that they persist, at least for a certain time?' Or: 'What kind of changes are needed in order for there to be the same person over a certain time?' Reformulating the question in this way enables us to account for identity *in terms of* change, that is, to make a constructive use of change in an explanation of transtemporal identity rather than keeping change outside, as something that, if it were allowed in, would simply and straightforwardly annihilate identity.

If 'things'—persons included—are actually organized hierarchies of stabilized processes (as it turns out), then there is no difficulty in making sense of the fact that transtemporal identity can be lost, just as it needs (first of all) to be gained: processes, and clusters of processes, stabilize and destabilize. This is what it means to be a dynamic system: to rely on the dynamics of constitutive processes that themselves rely on environmental processes they are entangled with. However, the fact that identity needs to be gained and can be lost does not undermine its being identity. There is really something that stays the same over time in some robust sense. This is the specific arrangement of processes—the processual 'form'—and it stays the same over time for exactly as long as it displays some kind of successful activity to maintain itself, something that can be described by biology as part of a scientific investigation into the mechanisms of organic stabilization and destabilization.[44] The process view thus allows for an informative explanation of our identity over time that evades the unfortunate alternative between either mystification or elimination and thereby meets our metaphysical conviction.

And it has further advantages, which I would like to briefly mention here.

One is that there simply is no 'problem of temporary intrinsics'. While one might find it puzzling how one and the same *thing* can have different intrinsic properties at

---

[43] Strikingly enough, even Whiteheadian process ontology is in danger of relapsing into thing ontology by construing process from fundamental discrete events ('actual entities'; see Whitehead 1978, esp. 61ff., and the critique by Rescher 1996: 89–90; see also chapter 2 in this volume). It is here that the Leibnizian heritage of Whitehead's process ontology turns out to be unfortunate rather than helpful (on this issue, see also chapters 1 and 6 in this volume).

[44] Depending on the level of abstraction and on explanatory purposes, the details of such descriptions might look different, potentially resulting in diverging diachronic identity statements about organisms (see n. 37). Yet such divergent results are never arbitrary but rather suggested by biological reality (see Dupré 2014 for a defence of this moderate form of pluralism under the title 'promiscuous realism').

different times, there is in principle no mystery to one and the same *process* having different intrinsic properties at different times, as this is just part of what it is to be a process. This is another way of saying that the identity of a process cannot be an identity that conforms to Leibniz's law. According to the process view, the fact that Leibniz's law excludes process and change from the start counts against its applicability to temporal reality and the persisting entities therein or, more accurately and weightily, against a view of the world according to which it would be so applicable. The conclusion to be drawn from this is that we should stay away from Leibniz's law in our account of persistence, while still working with a robust notion of identity along the lines explained.[45]

Another advantage of the bioprocess view is that it allows us to relax about vagueness. That natural processes tend to have vague boundaries goes without saying. The crucial point is that there's nothing wrong with that; like the change of intrinsic properties, vagueness just lies in the nature of processes. And because there is nothing wrong with it, there is no need to give up on identity altogether (as suggested by Parfit) when it comes to making sense of the widely discussed puzzle cases of personal identity. Admitting that the beginning as well as the end of the existence of a person might be (or unavoidably are) fuzzy does no harm, if we conceive of persons as higher-order processes emergent from lower-level processes. This fuzziness does not undermine the person's identity at other times, nor does it involve an ontological mystery. Compare this to the situation of a thing. At the same time, to exist and not to exist (as one is forced to say of a persisting entity) in vague cases is quite an odd state to be in, for a thing—but not necessarily for a process. Also, there are still informative things to say in vague cases from a bioprocessual perspective. Rather than being caught in a logical perplexity with no resources for solving the matter, as in the notorious scenarios of the personal identity debate, we now find questions of vagueness transforming themselves naturally into questions of processual constitution, which to some considerable extent is just an empirical issue.[46] Focusing, within a bioprocessual framework, on organization and functional unity rather than on naked continuity is especially promising with regard to branching cases.[47]

'Well', someone might say now, 'everything that you told us may be true, and indeed it sounds quite convincing; but you still haven't presented a solution to the personal identity dilemma, as you haven't been talking about personal identity at all.

---

[45] See also Meincke 2018b. Spelling out the basic metaphysics of process identity is a core task for a prospective process account of persistence. Roughly, this means understanding certain processes as continuants and analysing transtemporal identity as a metaphysical and inherently temporal relation realized by continuant processes by virtue of their specific organization, as opposed to both the atemporal primitive identity of a substance and mere continuity between series of atomic entities of some kind. For a recent discussion of whether processes can be continuants, see Steward 2013 and 2015; and Stout 2016. Pioneering work towards a comprehensive ontology of processes has been done by Antony Galton (see e.g. Galton 2012, 2006; Galton and Mizoguchi 2009) and by Johanna Seibt (see e.g. Seibt 1997, 2008 and chapter 6 in this volume).

[46] This is not meant to imply that an empirical analysis dissolves vagueness in any case, but only that we can at least provide empirical or empirically grounded explanations and descriptions of vague cases.

[47] Branching happens all the time in living nature; but there are many different sorts of branchings with different sorts of functions, including branchings that are dysfunctional for a living system.

All you have proposed is a sophisticated process account of biological identity, but biological identity isn't personal identity.' Here is my reply: personal identity is about persons, and I find it hard to deny that *human* persons at least are some sort of organism, that is entities described and investigated by biologists. This is not to say that mental capacities aren't crucial when it comes to distinguishing persons from non-persons, nor do I want to deny the 'transcendental' logic inherent to subjectivity, as stressed by Kant and his followers. However, first, mental capacities are not reserved only to humans, and so, if 'personal identity' is meant to refer to the identity of human persons exclusively, the analysis would have to focus on something different as being the relevant distinctive feature—the *differentia specifica*—anyway. And, second, I agree with Kant that the transcendental logic of subjectivity does not provide sufficient resources for the *ontological* understanding of personal identity.[48] I take it that these resources, at least as the basic constitution of (human) personal identity is concerned, are provided by biology.

This chapter's ambitions thus have been located at a more fundamental level, following broadly this line of argument: in order for there to be a human person, there must be a human organism first of all, whatever the exact relationship between the two may be; so, if persons are organisms, let's see what follows for the concept of personal identity from a processual account of what an organism is. What we have found pursuing this path has been instructive enough and invites to be complemented by further investigations into the more specifically 'personal' aspects of personal identity. In this regard, my suggestion is to focus in particular on the phenomenon of personality in human persons.[49] Here we will actually encounter the same processual constitution that is characteristic of the underlying biological level, including the distinctive dialectic between identity and change: personal identity in the sense of keeping (and 'being') the same personality over time is the process of maintaining identity *despite and by virtue of* change (see Spann 2014b and Meincke 2016), a process that critically involves interactions with other persons, that is, with the social environment in which human persons are situated. Persons are processual from head to toe. I leave it to future endeavours to combine the bioprocessual approach as sketched here with an elaborated non-Cartesian holistic view of the human person so as to complete the 'most profound metaphysics' required, as Hume rightly surmised, for a way out of the personal identity dilemma.

# Acknowledgements

This chapter was funded by the European Research Council, grant agreement number 324186 ('A Process Ontology for Contemporary Biology'). I am indebted to John Dupré, Antony

---

[48] However, unlike Kant, I don't think that from this it follows that there cannot be any ontological account of personal identity at all. For a portrait of some of the endeavours, within the post-Kantian idealistic and phenomenological tradition of philosophy, to overcome Kant's negative verdict on the possibility of such an ontological account (and of metaphysics in general), see Meincke 2015 (ch. 3.3 and 4.3).

[49] That this crucial aspect of personal identity has increasingly and no less unjustly become neglected in recent metaphysical debates is the result of a growing popularity of the substance theory of the person: see Spann (2014b) and Meincke (2016).

Galton, and David Wiggins for inspiring discussions and for helpful comments on an earlier draft. I also would like to thank two anonymous referees for their helpful comments.

# References

Baker, L. R. (2012). Personal Identity: A Not-So-Simple Simple View. In G. Gasser and M. Stefan (eds), *Personal Identity: Complex or Simple?* (pp. 179–91). Cambridge: Cambridge University Press.

Bouchard, F. and Huneman, P. (2013). *From Groups to Individuals: Evolution and Emerging Individuality*. Cambridge, MA: MIT Press.

Campbell, R. J. and Bickhard, M. H. (2011). Physicalism, Emergence and Downward Causation. *Axiomathes* 21: 33–56.

Collier, J. (2006). Autonomy and Process Closure as the Basis for Functionality. *Annals of the New York Academy of Science* 901: 280–90.

Craver, C. F. and Bechtel, W. (2007). Top-Down Causation without Top-Down Causes. *Biology & Philosophy* 22, 547–63.

Dupré, J. A. (1993). *The Disorder of Things: Metaphysical Foundations of the Disunity of Science*. Cambridge, MA: Harvard University Press.

Dupré, J. A. (2012). *Processes of Life: Essays in the Philosophy of Biology*. Oxford: Oxford University Press.

Dupré, J. A. (2014). Animalism and the Persistence of Human Organisms. *The Southern Journal of Philosophy* 52: 6–23.

Galton, A. (2006). On What Goes On: The Ontology of Processes and Events. In B. Bennett and C. Fellbaum (eds), *Formal Ontology in Information Systems* (pp. 4–11). Amsterdam: IOS Press.

Galton, A. (2012). The Ontology of States, Processes, and Events. In M. Okada and B. Smith (eds), *Proceedings of the Fifth Interdisciplinary Ontology Meeting* (pp. 35–45). Tokyo: Keio University Press.

Galton, A. and Mizoguchi, R. (2009). The Water Falls but the Waterfall Does Not Fall: New Perspectives on Objects, Processes and Events. *Applied Ontology* 4: 71–107.

Geach, P. (1979). *Truth, Love and Immortality: An Introduction to McTaggart's Philosophy*. London: University of California Press.

Hume, D. (1964). *A Treatise of Human Nature*, vol. 1, ed. by A. D. Lindsay. London: Dent and Dutton.

Hume, D. (1966). *A Treatise of Human Nature*, vol. 2, ed. by A. D. Lindsay. London: Dent and Dutton.

Hume, D. (1975). *Enquiries Concerning Human Understanding and Concerning the Principles of Morals*, ed. by P. H. Nidditch. Oxford: Clarendon.

Jonas, H. (1966). *The Phenomenon of Life: Toward a Philosophical Biology*. Evanston: Northwestern University Press.

Lewis, D. K. (1986). *On the Plurality of Worlds*. Oxford: Blackwell.

Love, A. (2012). Hierarchy, Causation and Explanation: Ubiquity, Locality and Pluralism. *Interface Focus* 2: 115–25.

Meincke, A. S. (2010). Körper oder Organismus? Eric T. Olsons Cartesianismusvorwurf gegen das Körperkriterium transtemporaler personaler Identität. *Philosophisches Jahrbuch* 117: 88–120.

Meincke, A. S. (2015). *Auf dem Kampfplatz der Metaphysik: Kritische Studien zur transtemporalen Identität von Personen*. Münster: Mentis.

Meincke, A. S. (2016). Personale Identität ohne Persönlichkeit? Anmerkungen zu einem vernachlässigten Zusammenhang. *Philosophisches Jahrbuch* 123: 114–45.

Meincke, A. S. (forthcoming a): Dispositionalism and the Problem of Persistence. In A. S. Meincke: *Dispositionalism. Perspectives from Metaphysics and the Philosophy of Science*. Dordrecht: Springer.

Meincke, A. S. (forthcoming b): How to Stay the Same While Changing: Personal Identity as a Test Case for Reconciling 'Analytic' and 'Continental' Philosophy through Process Ontology. In R. Booth and O. Downing (ed.), *Analytic-Bridge-Continental + (ABC+) Process Philosophy*. Berlin: de Gruyter.

Meincke, A. S. (2018a). Bio-Powers and Free Will. Unpublished manuscript.

Meincke, A. S. (2018b). The Disappearance of Change. Unpublished manuscript.

Maturana, H. R. and Varela, F. J. (1980). *Autopoiesis and Cognition: The Realization of the Living*. Dordrecht: Reidel.

McTaggart, J. M. E. (1927). *The Nature of Existence*, vol. 2. Cambridge: Cambridge University Press.

Montévil, M. and Mossio, M. (2015). Biological Organisation as Closure of Constraints. *Journal of Theoretical Biology* 372: 179–91.

Moreno, A. and Barandiaran, X. (2004). A Naturalized Account of the Inside-Outside Dichotomy. *Philosophica* 73: 11–26.

Moreno, A. and Mossio, M. (2015). *Biological Autonomy: A Philosophical and Theoretical Enquiry*. Dordrecht: Springer.

Nida-Rümelin, M. (2006). *Der Blick von innen: Zur transtemporalen Identität bewusstseinsfähiger Wesen*. Frankfurt: Suhrkamp.

Nozick, R. (1981). *Philosophical Explanations*. Cambridge, MA: Harvard University Press.

Olson, E. T. (1997). *The Human Animal: Personal Identity without Psychology*. New York: Oxford University Press.

Parfit, D. (1987). *Reasons and Persons*, 3rd edn. Oxford: Oxford University Press.

Rescher, N. (1996). *Process Metaphysics: An Introduction to Process Philosophy*. Albany, NY: SUNY Press.

Seibt, J. (1997). Existence in Time: From Substance to Process. In J. Faye, et al. (eds), *Perspectives on Time* (pp. 143–82). Dordrecht: Kluwer.

Seibt, J. (2008). Beyond Endurance and Perdurance: Recurrent Dynamics. In C. Kanzian (ed.), *Persistence* (pp. 133–64). Frankfurt: Ontos.

Sider, T. (2001). *Four-Dimensionalism: An Ontology of Persistence and Time*. Oxford: Oxford University Press.

Simons, P. (2000). Continuants and Occurrents. *Aristotelian Society* 74: 59–75.

Spann, A. S. (née Meincke). (2013). Ohne Metaphysik, bitte?! Transtemporale personale Identität als praktische Wirklichkeit. In G. Gasser and M. Schmidhuber (eds), *Personale Identität, Narrativität und praktische Rationalität* (pp. 241–65). Münster: Mentis.

Spann, A. S. (née Meincke). (2014a). Bio-Agency: Können Organismen handeln? In A. S. Spann and D. Wehinger (eds), *Vermögen und Handlung: Der dispositionale Realismus und unser Selbstverständnis als Handelnde* (pp. 191–224). Münster: Mentis.

Spann, A. S. (née Meincke). (2014b). Persönlichkeit und personale Identität: Zur Fragwürdigkeit eines substanztheoretischen Vorurteils. In O. Friedrich and M. Zichy (eds), *Persönlichkeit: Neurowissenschaftliche und neurophilosophische Fragestellungen* (pp. 163–87). Münster: Mentis.

Steward, H. (2013). Processes, Continuants, and Individuals. *Mind* 122: 781–812.

Steward, H. (2015). What Is a Continuant? *Proceedings of the Aristotelian Society* 89: 109–23.

Stout, R. (2016). The Category of Occurrent Continuants. *Mind* 125: 41–62.

Swinburne, R. (1984). Personal Identity: The Dualist Theory. In S. Shoemaker and R. Swinburne (eds), *Personal Identity* (pp. 1–66). Oxford: Basil Blackwell.
Van Inwagen, P. (1990). *Material Beings*. Ithaca, NY: Cornell University Press.
Whitehead, A. N. (1978). *Process and Reality: An Essay in Cosmology*. New York: Free Press.
Wiggins, D. (2001). *Sameness and Substance Renewed*. Cambridge: Cambridge University Press.
Wiggins, D. (2016). Activity, Process, Continuant, Substance, Organism. *Philosophy* 91: 269–80.

# Index

Note: page numbers followed by 'n' refer to notes; those followed by 'f' refer to figures.

abstraction  4, 31, 32n, 33, 38, 56–8, 71, 201, 249, 251, 264, 265, 296, 372, 373n44
adaptation  180, 187–91, 194, 196, 208n16, 209, 296, 362n12
additivity  121n, 123
affordances  174, 176, 177, 180, 181, 182, 350
Agar, W. E.  11n8, 228, 229, 230, 239
agency  154, 167–82, 309, 350
  agent theories  167, 175–7, 181
  agential dynamics  172–5
  and environment  174
  and evolution  177–81
  goals  172–4
  immanence  176–7, 181
  means  173
  ontological surprise  168–70
  phases  170–2
Alberts, B.  141, 314n
Alizon, S.  37
allorecognition  108
animalism  368n26
aphids  21–2, 103–4, 213–14
Aquinas, Thomas  65, 66
Aristotle
  being  131
  and causation  67
  concept of substance  6, 194, 362n11, 364
  four causes  72
  homeomereity  120
  noun/verb distinction  50
  and purpose  229n4
  and substantialism  5–6
Armstrong, D.  62, 71
atomism  5, 6
attractor states  332
autocatalytic cycles  205n10, 289, 291, 292, 293
automereity (self-partedness)  120–5, 121f
autonomy  20, 145n, 152, 169, 170, 174–5, 200n, 323, 368n28
autopoiesis  145n, 154, 368n28
Axel, R.  345

B-theory of time  89n
bacteria  21, 64, 146, 308
  *B. aphidicola*  21–2, 103–4, 213
  *E. coli*  155n20, 207–8
  *M. xanthus*  210–11
  *V. fischeri*  103, 189–91, 193, 196, 215
  *see also* microbes

Baer, K. E. von  147n14, 265–6
Balázs, B.  274
Bapteste, E.  52, 96, 108, 109, 187, 194, 196
Barandiaran, X.  154, 173n10, 204, 370n32
Bary, A. de  102
baryogenesis  171
Batterman, R.  172
Bayesian brain  347
Bechtel, W.  312
Bénard convection cells *see* dissipative structures
Bennett, J.  63, 72–3
Bergson, H.  xi, 268n4
Bernard, C.  145, 146n12, 168
Bertalanffy, L. von  10, 14n, 31, 76, 145, 150, 154, 162, 252n7, 266, 296
  stream of life conception (SLC)  148–9
Bich, L.  208
Bickhard, M. H.  39, 65, 76, 125, 127, 133, 144n9, 199, 200, 203–5, 207, 350n8, 369n31
biochemistry  54, 78, 86, 106–7, 149–51, 168, 314–18: *see also* macromolecular biology; molecular biology
biofilms  21, 99, 130, 202, 210–11, 215–16, 283
biological atomism  329
bioluminescence  189–91, 193, 195–6, 215
biosynthesis  290–1
Birch, C.  22, 228, 304–8
Black, M.  139n
Bohm, D.  xi, 15n
Bohr, N.  272
*Botryllus schlosseri*  107–8, 107f, 109
Boveri, T.  324
Bowler, P.  228
Brandon, R.  187, 188, 196
Broad, C. D.  115, 167
Burnet, F. M.  228

Campbell, R. J.  15n, 125, 128n, 144n9, 199, 201, 203, 204, 207, 369n31
cancer  240, 321–32, 371n40
  fields of cancerization  330
  latency period  325
  morphogenetic fields  329–31
  as process  323–6
  relational ontology of levels  327–9
  somatic mutation theory (SMT) of  324, 327–8
  tissue organization field theory (TOFT) of  327, 328

Cannon, W. 145
capacities 311–15
　component capacities 312–13, 314, 315
　enzymes and 314, 316
　integrated capacities 313–14, 315
　proteins and 314–15, 316, 317
carcinogenesis 322, 323, 324, 325–6
　evolution of interpretive models 327–8
　reversibility of 330–1
Carnap, R. 77, 134
Castro, J. B. 342
categorial inferences 77–8
causation 54, 187
　in biology 63–5, 67–8, 73
　causes, types of 72
　dispositionalism as theory of 61–73
　downward 27, 328, 369n31
chemotaxis 207–9, 210
chreods 229, 230, 251–2, 254
Christensen, W. D. 154, 199, 205
chronobiology 20
cinema: and biological research 274
Clarke, E. 130–1, 194–5
Clef, J. von 348–9
Cobb, J. B. 22, 304–8
cohesion 13, 14, 84–6, 203, 207
　cohesive whole 130, 191, 203, 204
collaboration 199, 200, 201, 202, 217, 311, 313, 315
　eukaryotic 211–12
　eumetazoan 212–13
　and multicellular systems 209–13
　and processes, organizations of 203n4
Collier, J. D. 14, 84
Comandon, J. 268
commensalism *see* symbiosis
complex adaptive systems (CAS) theory 65
connectedness 191–2: *see also* ecological interdependence
continuants 25, 49–50, 51–2, 88n4, 89
　and abstraction 57–8
　and processes 55–6, 374n45
　*see also* endurantism
continuism 98
Craver, C. 276
cryptic (hidden) variability 247–8, 256–9
Cuvier, G. 149
cybernetics 147, 253–4
　servomechanisms 147
cytoskeleton 270, 330

Darwin, C. 6, 78, 181
Davidson, D. 54, 83, 85
Dawkins, R. 33, 35n, 141, 147n14, 188, 226
death 63, 143, 144, 149, 153–4, 156
Democritus 5
Dennett, D. 347
design 28, 140, 147, 157, 159–61, 230, 233, 239, 285

determinism xi, 8, 28, 54, 81, 140, 161, 231, 249, 271, 328
developmental biology 81, 140, 160, 226, 264–78
　developmental contextualism 231
　developmental genetics 275–6
　developmental (ontogenetic) niches 233, 236, 237
　process vs substance views of 248–50
developmental systems theory (DST) 34, 65, 70, 225–41
　account of evolution 225
　developmental dynamics 234–5
　dynamic interactionism 231, 234
　epigenesis 233–4
　heredity 225
　life cycles 225, 227–8, 239
　ontology for 235–7
　parity thesis 232
　as process biology 227–30
　as process theory of organism 237–40
Dewey, J. 6, 231n
Di Paolo, E. 169
dialectical materialism 6
Dickison, M. 226
disease: physiological vs ontological conceptions 36–7
dispositionalism 61–73
dissipative structures 12, 25, 84, 125, 126, 144, 148–53, 156, 159, 160, 162, 169, 204–5: *see also* stream of life conception (SLC); thermodynamics
DNA xiii, 17n14, 19n, 32, 33, 125, 160, 179–80, 233, 249, 307–8, 317–18
Dobzhansky, T. 180n21, 255, 256
Doolittle, W. F. 287–8, 288f
Dowe, P. 73
*Drosophila melanogaster* 64, 252–3, 256, 269
Duboule, D. 293n8
Dunn, G. A. 268
Dupré, J. 132, 187
　biological individuality 188
　collaboration 199, 201, 217
　concept of organism 201–2, 214, 215
　life/non-life distinction 216–17
　promiscuous individualism 24, 200, 201, 202, 213–17
　promiscuous realism 23–4, 194, 202, 373n44
dynamic systems theory (DyST) 235, 250–3
　coupled dynamic systems 251
dynamics, *see* general process theory (GPT)

early universe: phase transitions 171
ecological interdependence 20–5, 27, 28, 39–40
ecological model of the world 303–18
ecology 20–1, 78, 181, 182
ecosystems 85, 304–5, 315–16

Eddington, A. 143
efference copy 347
Eigen, M. 290–1, 293
Eisenstein, S. 274
Ellena, J.-C. 341
embodiment theories 343n, 348n
embryogenesis 251–2, 269: *see also* morphogenesis; organogenesis
embryology *see* developmental biology
emergence/emergent properties 69, 119, 126, 133, 170–1, 174, 191, 192, 233–4, 293, 371, 374
endurantism 52, 88–9, 363–6
Engels, F. 6
enzymology 314–16
epigenesis 225, 231, 233–4
epigenetics 33, 233–4, 252f, 322
  and cancer 322, 323–4, 326, 327, 330
  epigenetic inheritance 34, 158–9, 236–7
  epigenetic landscape 250–5, 252f
Epstein, J. 274
essentialism 5, 23–6, 78–9, 89, 98, 249n
eternalism 89n, 129, 364
events: static/dynamic 70–3
evolution 34–6
  and agency 177–81
  DST and 225
  epigenetic/exogenetic inheritance 236–7
  of explanantia 293–5
  of processes 293–5
  replicator/interactor view of 177, 188–9, 226–7
  of translation 290–2, 293
  *see also* modern synthesis (evolutionary theory)
explanatory reciprocity 176–7, 181
extended evolutionary synthesis (EES) 246
extended phenotype 35n

Fair, D. 73
*Fasciola gigantica* flatworms 79–81, 80f, 89, 91, 92: *see also* symbiosis
Fechner, G. 145
feedback loops/mechanisms 70, 84, 119, 126–7, 133, 134, 147, 206, 251, 289, 339, 351
Feeney, K. A. 82
Firestein, S. 345
flames *see* dissipative structures
flexibility (plasticity) 253–5, 256
fluid dynamics 171–2
Ford, D. 225, 231, 234
four-dimensionalism, *see* perdurantism
Fraser, S. 273
Frege, G. 56, 83
fuel-food analogy 145–6
Fulda, F. 172n7
functions
  evolution of 292–3, 292f

functional interdependence 206, 209
  and structures 31–2

Gamble, E. 341
Geach, P. 365
general process theory (GPT) 113–34
  application in philosophy of biology 128–33
  categorial features of processes 115–16
  emergent constituting constrainers (ECCs) 125
  levelled mereology 117–19
  relationships among processes 117–19
  typology of processes 120–8
genes 32, 64, 160–1, 234
  as agents xi, xiii, 290f
  gene externalization 287–8, 288f
  gene regulatory networks (GRNs) 81, 276
  gene trees 285–6, 286f
genetics 32–4
  developmental 275–6
  genetic assimilation 256
  genetic replication 33–4
  genetic variability 255
  polygeny 65
  population 9n, 240n8
  *see also* genes; genomes
genetic program xiii, 28, 33, 140, 159, 160, 178, 180, 246, 249, 270, 271, 328
genidentity 91, 96–109, 100f, 158, 229–30
  definition of 97–8
  and definition of organisms 105–8
  DST and 238
  internal organization 99–102, 103, 105, 106
  of life cycles 238
  and life history theory 240
  and processes 55, 57–8
  and symbiosis 102–5
genomes 19n, 32–3, 160, 161, 179–80, 236
  genome regulation 179
  genomic instability 330
  viral 288–9
Ghiselin, M. 34–5, 78
Gibson, J. J. 174, 350
Gilbert, S. F. 20, 21, 64, 102, 103, 150, 239, 246, 250, 252, 256n8, 266, 271, 287, 310
Godfrey-Smith, P. 178, 192, 225, 238
Goldstein, K. 174
Golgi apparatus 16
Goodwin, B. xiii, 64, 76, 172
Gottlieb, G. 225, 231
Gray, R. D. 226–7, 232, 239
Great Red Spot, Jupiter 25
Griesemer, J. R. 33, 89, 158, 264, 267, 272, 274
Guay, A. 238

Haag, J. W. 152
Haeckel, E. 265

Haldane, J. B. S. 150
Haldane, J. S. 8, 31, 146n12, 150, 158n26
Hall, B. K. 233, 234, 271
Hardy-Weinberg law of equilibrium 259
Hawaiian Bobtail squid: and *V. fischeri see* bacteria
Hegel, G. W. F. 6
Helmholtz, H. von 142, 143
Henderson, L. 150
Henning, B. G. 141, 303, 310-11
Henning, H. 340-1
Heraclitus 5, 24, 148, 149n16, 150, 162, 187
Herz, R. S. 348-9
His, W. 276
Hobbes, T. 140n
holobiont 21, 287: *see also* microbes; symbiosis
homeomereity (like-partedness) 120-5
homeorhesis 253-5, 257
homeostasis 145, 147, 253-4
Hopwood, N. 274
Horsten, L. 83
host imprinting 226
Hull, D. 34-5, 78, 96, 188
  definition of species 194
  genidentity 99-102, 100f
  individuation 99-100
  interactors 130, 226
  replicator/interactor view of evolution 226
Hume, D. 63, 366
  bundle theory of the person 362
  on causation 61, 62, 67
  and identity 357, 362, 363n16, 371-2
hurricanes *see* dissipative structures
Huxley, T. H. 149
hylomorphism 370n38
hypercycles 290-2, 291f, 293

identity 98, 370, 371, 375
  continuity-based conception of 99-102
  criteria of 82-6
  cross-generational 158-9
  diachronic 97, 155-9, 192-3
  endurance as 88
  functional 370n32
  numerical 77, 79, 88-9, 359-60, 362-4
  personal 357-75
  qualitative 82, 88
  transtemporal 367, 373
  *see also* genidentity; persistence; personal identity
imaging
  automated image processing and data analysis 274
  fluorescent 269, 270
  4D 273
  in toto 269
  in vivo 267-70
  quantitation of 275
  static/dynamic 272
  3D 273
  time-lapse 266, 268, 269, 272, 273
immune system
  discontinuity theory of immunity 106n
  and organisms 106-8, 107f
individuality 186-8, 238-9, 310-11
  functional 206n12
  notion of 188, 190, 192
individuation 21, 23-4, 99-100, 131-2
  causal cohesion account of 84
  of processes 82-6
  of rivers 24, 25
inheritance
  epigenetic 34, 158-9, 236-7
  exogenetic 236-7
innate behaviour 225, 230
interactionism 231, 234-5
  interactors and replicators 130, 188-9
intersecting processes 283-96
  evolution of 293-5
  and evolutionary theory 289-93
  merging of 289, 293-4, 295
  and phylogenetic networks 287-9
intrinsically disordered proteins (IDPs) 316n, 317
invariance 56-7

Jaeger, J. 76, 251, 275
James, W. 6
Johnson, M. xi
Johnson, W. E. 49
Johnstone, J. 143
Jonas, H. 146, 156, 157n
  metabolism 369
  systems biology 370, 371
  thermodynamic predicament of organisms 169, 170

Kant, I. 4, 66, 361
Kapp, R. 147
Kauffman, S. 81, 145, 205n10
Keller, E. F. 32, 140, 145, 160, 180, 217
Keller, P. 273
Khairy, K. 273
Kim, J. 26, 70-1
King, A. 233, 236, 237
Kistler, M. 73
Kitcher, P. 232n
Kripke, S. A. 98
Ku, C. 285, 286

La Mettrie, J. O. de 147n14
Ladyman, J. xi
Lakoff, G. xi
Landecker, H. 269-70, 274
Laplace, P.-S. 142

lateral gene transfer (LGT) 286, 287–9
Lavoisier, A. 142
Lehrman, D. S. 225, 230–1, 236
Leibniz, G. W. 6, 98
  Leibniz's Law (LL) 88, 365, 372, 374
Lerner, R. 225, 231, 234
Leucippus 5
Levin, M. 323, 330
Lewin, K. 55, 91, 97, 158
Lewis, D. 50, 89
  on causation 62, 63
  on events 70–1, 73
  and identity 372n42
  perdurance/endurance 363, 364
  problem of temporary intrinsics 373–4
Lewontin, R. 6, 64, 141, 177–9, 246n, 249
life cycles 18–20, 28, 39–40
  of biofilms 21
  DST and 225, 227–8, 239
  of *Fasciola gigantica* flatworms 79–81, 80*f*, 89, 91, 92
  of frogs 18–19, 18*f*
  genidentity of 238
life history theory 81, 239–40
life/non-life distinction 202, 216–17
Lillie, R. S. 11n8, 150
Lindquist, S. 257–9
lineages 24, 34–6, 78, 104, 130, 161, 191–2, 201, 202, 214, 225–8, 238, 240, 249, 255, 283, 285–9, 294–5, 322
Lippincott-Schwartz, J. 276–7
Lloyd, E. 189
Locke, J. 6, 98, 156n23
Lombard, R. 71
Love, A. 264, 265, 274, 369n31
Lowe, J. 52, 83
Luengo-Oroz, M. A. 275
Lycan, W. G. 341, 342

McClintock, B. 258
machine conception of the organism (MCO) 9, 28, 33, 140–8, 150–63, 304–8, 311–13, 315: *see also* mechanism
Mackie, J. L. 54
macromolecular biology 303–18: *see also* biochemistry; molecular biology
McShea, D. W. 191
McTaggart, J. M. E. 364
Marmodoro, A. 69
Martin, C. B. 63, 65–6
Martin, W. 293
matter/antimatter 171
Maturana, H. 145n, 289, 368n28
Maxwell, J. C. 143
Mayer, R. von 142
Mayr, E. 84, 180n21

mechanicism 8, 10, 28–30, 139–41, 161, 304–8: *see also* machine conception of the organism (MCO); Newtonian paradigm
mechanisms 29–30, 32n, 126, 127n
microbiology 37–38, 196: *see also* bacteria; biofilms; microbes
medicine 36–8, 58, 332
Megason, S. 273, 275, 277
meiosis 54, 237, 293–4
metabolism 8, 15–18, 22, 24, 28, 33, 34, 35, 39, 54, 63, 76, 81–2, 85, 105, 116, 145–63, 169, 202, 205–9, 212, 216–17, 368–70
metaphors: in scientific understanding 139–63
Méthot, P.-O. 37
metrology 58n12
microbes 21, 22n, 24, 37–8, 99, 103, 104, 131, 132, 201–2, 214–16, 238: *see also* bacteria
microscopy 272–3
  in vitro 268
  time-lapse 266
mitosis 19, 54, 118, 233
mitotic spindle 16
mobile genetic elements (MGEs) 287
modal properties 58
modern synthesis (evolutionary theory) 9n, 168, 177, 180–1, 247, 249, 255, 256
molecular biology xiii, 26, 28, 140, 150–1, 155n20
  central dogma of 160, 255
  *see also* biochemistry; macromolecular biology
Molnar, G. 70
Monk, N. 275
monogenomic differentiated cell lineages (MDCLs) 201, 202, 213, 214
Moore, C. 234–5
Moreno, A. 64, 145n, 152, 170, 204, 206–7
morphogenesis 266, 276, 277, 331: *see also* embryogenesis; organogenesis
Morrison, M. 172
mosaicism 213
Mossio, M. 64, 145n, 152, 170, 206–7
Muller, S. J. 84
Munro, E. 276
mutual manifestation 65–6, 67–9, 70, 72
mutualism *see* symbiosis

Nanney, D. L. 233–4
natural selection 35–6, 78, 118, 178, 179, 187, 188, 191, 196, 217, 226, 230, 294
Needham, J. 141n4, 228, 329
Neisser, U. 347, 350
Newtonian paradigm 175–8: *see also* mechanism
niche construction theory 35–6, 155, 178n17, 237, 246n, 322

Noble, D. 322
normativity 154n, 162, 173n10-11, 174, 177, 208, 212
non-equilibrium thermodynamics
 see thermodynamics
nucleosynthesis 171

object theories 175-7, 180-1
objectcy 167, 175, 178
objects
 continuant/occurrent duality 49-50
 reification of xi-xii
 see also things
occurrents 49-54, 58: see also perdurantism
Okasha, S. 232n
olfaction 337-53
 cognitive neuroscience and 345-6
 and complexity of flavours 343
 and flavour perception 346
 forecasting and stimulus input 348-51
 input determination of 339-43
 neural basis of 343-8
 odour objects 340
 olfactory illusions 348-9
 olfactory objecthood 340-1
 olfactory pathway 344-5, 344f
 and perceptual biases 346, 349, 352
O'Malley, M. A. 20, 104n, 170, 186, 199, 201-2, 215, 216-17, 310n, 311, 323
O'Neill, J. S. 82
Oparin, A. 150
order 143, 144, 151-2, 159-61, 169, 174, 203n7, 325-6: see also organization; self-organization
organicism 7-11, 14n, 22, 31, 76, 140, 146n12, 148, 158n26, 168, 229, 250, 266, 270-3, 329
organisms
 agency of 154, 168, 246n, 309-10
 concept of 201-2, 214, 215, 248
 definition of 105-8
 functional integration 191-3, 214-16
 immune system and 106-8, 107f
 process theory of 8-11, 148-52, 169-70, 237-40
 vs machines 9, 16, 28, 141-163, 369n30
 thermodynamics of 142-5, 169, 170
 see also machine conception of the organism (MCO); stream of life conception (SLC)
organism-environment relationship 8, 20-1, 32, 33, 35, 84-5, 130, 154-5, 169-70, 174, 177-80, 192, 206-15, 230-7, 246n, 253-60, 305-8, 313-18, 322
organismality 207-9, 214, 215, 218
organization 154, 159, 199-218, 289-90:
 see also order; self-organization

organizational closure 64, 152, 169-70, 205-7, 213, 216, 370n32
organogenesis 271: see also embryogenesis; morphogenesis
Oyama, S. 35, 65, 225, 231-2, 235, 246, 249

parasitism see symbiosis
Parfit, D. 359-61, 362, 373
Parmenides 5
parthood 117-19
 mereological nihilism 91
 mereological universalism 91
pathogens/pathogenicity 37-8
perception
 anticipation in 349-51
 ecological theory of 174
 input-output model 339-43
 olfaction 337-53
 percepts, traditional notion of 340
 perceptual experiences 339-43
 perceptual representation 353
 perceptual stability 352
perdurantism 11-12, 49-52, 58, 64, 73, 89-92, 273, 363-4, 365, 366, 373: see also occurrents
persistence 55, 86-9, 128-9, 155-9, 321-2, 363-4, 365-7: see also identity
personal identity 357-75
 as biological process 367-71
 dilemma of 358-61
 non-reductionism and 360-1, 362, 366-7
 reductionism and 359-60, 366
 solution to dilemma of 361-7
phase-contrast microscope 273
phase transitions 171-2
phenotypic variability, see plasticity
photosynthesis 116, 118f, 119, 119f
phylogenetics 283, 285-96
 gene trees 285-6, 286f
 phylogenetic frameworks 290f
 phylogenetic networks 287-9
physiology 8, 10, 14, 17n15, 27, 155
 history of 142, 145, 150-1
 processes 54-5, 228, 238, 253-4
 structures and functions 31-2
Pickering, A. 309
Plankar, M. 326
plasticity (phenotypic) 246-7, 253-5, 256, 259
Plato 5, 50
pleiotropy 70
pluralism 23-4, 35-6, 200-2, 296, 327-8, 373n44
Pradeu, T. 238
predictive coding theory 347
preformationism xi, xiii, 160, 231, 249-50
pregnancy 67-70
presentism 11, 129
Prigogine, I. 144, 161

process philosophy: history of 5–7
process structuralism xiii, 76
processes, life as hierarchy of xi, 3, 11, 16, 26–7, 64, 76, 118, 131, 151, 159, 199, 201, 248, 270–1, 321–3, 327, 368, 373
processes, subjectless 12, 114–17
proteins 17, 28, 32, 70, 81, 125, 151, 160, 233, 257–60, 270, 290, 292–3, 303–18

quantum physics 14–15, 53, 129, 140, 303, 307, 318
Quine, W. V. O. 77, 82, 83, 129n

Ramsey, F. P. 54
recapitulation theory 265
reductionism xiii, 8, 26–7, 29, 73, 140, 161, 194, 271, 323, 328, 357, 358, 366, 367
  Parfitian 359–61, 362, 373
  psychological 359–60
Reichenbach, H. 97
relativity theory 15, 53, 89n
representation 264–78
  automated image processing and data analysis 274
  descriptive models 274–7
  *see also* imaging
  digital revolution 274–5
reproduction 12, 25, 33, 35, 36, 101, 105, 127, 158–9, 191, 322
Rescher, N. 7n4, 14, 52, 65, 149n16, 153, 303, 368, 373n43
respiration: analogy with combustion 142
retrodiction 285–6, 288–9, 294
Richardson, R. C. 312
Ries, J. 268
robustness 81, 84, 105, 147, 206, 247, 253–5, 257, 293, 327, 332
Rosen, R. 155n20, 350
Rosenberg, A. 63
Ross, D. xi
Rubner, M. 142
Russell, B. 67
Russell, E. S. 8–9, 10n, 150, 168
Rutherford, S. L. 257–9

Salmon, W. 62, 73, 187
Schrödinger, E. 143, 145, 149, 159–60, 161n, 168, 249
Schuster, P. 290–1, 293
scientific domains: ontological explanation for 76–8
Seeley, W. P. 342
Seibt, J. 7n4, 11, 12, 52, 65, 77, 86, 87, 88, 90, 96n, 365, 366n22
self-maintenance 203–9, 212, 216–17
  minimal 204–7

recursive 207–9, 212, 216–17
self-maintaining dynamics 126–7
self-organization 144, 151, 159, 160, 161, 215–16, 270, 276: *see also* order; organization
Sellars, W. 65, 114, 115
sequence–structure–function (SSF) paradigm 314–15, 316, 317
Shapiro, J. 180
Shaw, G. B. 17n15
Sherrington, C. 150
Sinnott, E. 150
Smith, K. 226
Smolin, L. 175, 176
Sober, E. 23, 177, 186, 188, 190, 192
souls 17n15, 360n5, 361, 362
Specchia, V. 258
species 34–5, 78–9, 99–106, 124, 189, 191–6, 210, 258, 285, 290*f*
Spencer, H. 145
stability 151, 206–7
  dynamic vs static 144
static interactionism 231, 234
Stegmann, U. 232
Stein, R. 315–17
Sterelny, K. 226
Stoffregen, T. 174
Stramer, B. M. 268
Strawson, P. F. 25, 26n, 52
stream of life conception (SLC) 148–63; *see also* dissipative structures; thermodynamics
structures: and functions 31–2
substance dualism 360n, 362: *see also* souls
substance 5–6, 49–50, 168–9
  and atomism 6
  Aristotelian 5–6, 194, 362n11, 364
  concept of 362n11, 364
  endurance of 87–9
  and essence 23, 24, 88
  and individuality 86
  and mechanism 28–30, 141, 153, 313
  myth of 11, 113, 115, 128, 129, 366n22
  ontology 5–6, 11–12, 14, 39, 76–8, 98, 306–7, 366
  view of development 249
superorganisms 20, 21, 186, 190, 239, 310
Suppes, P. 62
symbiosis 102–5, 189–96, 202, 308–15
  biological individuality and 190–1
  bioluminescence 189–91, 193, 195–6, 215
  and capacities 311–15, 316, 317
  commensalism 20, 21, 102–3, 238
  genidentity and 102–5
  and individuality 310–11
  mutualism 20, 21, 102–3, 309
  obligate endosymbiosis 213–14
  omnipresence of 39, 64

symbiosis (*cont.*)
  parasitism 20, 37, 79-81, 89, 91, 92, 102-3, 104-5, 226
  symbiotic interactions 103-5
  symbiotic merging 102-5
  symbiotic partnerships 20-1, 23-4, 25-6
  symbiotic splitting 104
  termites 189, 309-11
  *see also* bacteria; microbes
synchronization 326n
systems
  identity of 203
  multicellular systems and collaboration 209-13
  open vs closed 144-5
  self-maintaining systems 203-4
systems biology 289-93, 295-6, 332, 368-70

Tabery, J. G. 235
temporal parts 9, 11-12, 19, 49-51, 52, 57, 71, 73, 79, 82, 86-91, 120-2, 128, 132, 265, 363, 364, 366, 372
Theoretical Biology Club 8n
thermodynamics 15-16, 25, 39, 126, 141n4, 142-5, 148, 149, 152, 153-4, 156, 159, 160, 161-3, 169, 203-6, 208, 321, 368
things
  and processes 11-14
  as space-time worms 11-12
  *see also* objects
Thom, R. 254
Thompson, E. 170
Thomson, W. (Lord Kelvin) 143
Thorpe, W. H. 228
thought experiments
  living/robotic dogs 16-17
  Theseus' ship xii, 156-7
three-dimensionalism (endurantism) 52, 88-9, 363-6
translation: evolution of 290-2, 293
Turner, J. S. 168, 309-10
two-component signal transduction (TCST) 207-8

Ullah, M. 271
uniformitarianism 294

vagueness 22, 25, 35, 359, 374
Varela, F. J. 145n, 169, 289, 343n, 368n28
Venit, E. P. 191

Virchow, R. 324
viruses 19, 21, 104, 216-17, 288-9
viscosity 171
vitalism 8, 143, 167-8, 170, 182, 270
von Dassow, G. 276

Waddington, C. H. 9-10&n, 36, 38, 150, 266
  canalization of development 257-60
  on cancer 329
  competence 229, 251-2
  cryptic (hidden) variability 247, 256-7, 259
  and DST 227-8
  epigenetics 233, 234, 250-3, 252f
  and epigenotype 225, 230, 233, 256
  homeorhesis 253-5, 257
  plasticity 246-7, 253-4
  robustness 253-4
Weiss, P. A. 10
Werndl, C. 215
West-Eberhard, M. J. 6n, 180, 259
West, M. J. 226, 233, 236, 237
Whewell, W. 149
Whitehead, A. N. 6, 22n, 28, 39, 52, 53, 55, 65, 113n2, 161, 231, 305n3, 310
  critique of 7, 13, 38, 114, 122, 129, 133, 134, 373n43
  fallacy of misplaced concreteness xi, 30
  influence on organicism 7-8, 10-11, 76, 140, 228-9, 250, 266n3
whirlpools *see* dissipative structures
Wiggins, D. 25, 52, 88, 90, 97, 98, 370n38
Wilkins, A. S. 293n8
Williams, G. C. 227n1
Wilson, D. S. 188, 190, 192
Wilson, J. 214
Winther, R. G. 132
Woese, C. 151
Wolkenhauer, O. 271
Wong, S. 269
Wood Wide Web 22n
Woodger, J. H. 9, 10n, 76, 139, 157, 228, 265

Xiong, F. 275, 277

Zemach, E. M. 121n